Cell Survival and Cell Death

SECOND EDITION

A subject collection from *Cold Spring Harbor Perspectives in Biology*

OTHER SUBJECT COLLECTIONS FROM *COLD SPRING HARBOR PERSPECTIVES IN BIOLOGY*

Calcium Signaling, Second Edition

Engineering Plants for Agriculture

Protein Homeostasis, Second Edition

Translation Mechanisms and Control

Cytokines: From Basic Mechanisms of Cellular Control to New Therapeutics

Circadian Rhythms

Immune Memory and Vaccines: Great Debates

Cell–Cell Junctions, Second Edition

Prion Biology

The Biology of the TGF-β Family

Synthetic Biology: Tools for Engineering Biological Systems

Cell Polarity

Cilia

Microbial Evolution

Learning and Memory

DNA Recombination

Neurogenesis

Size Control in Biology: From Organelles to Organisms

SUBJECT COLLECTIONS FROM *COLD SPRING HARBOR PERSPECTIVES IN MEDICINE*

The PTEN Family

Metastasis: Mechanism to Therapy

Genetic Counseling: Clinical Practice and Ethical Considerations

Bioelectronic Medicine

Function and Dysfunction of the Cochlea: From Mechanisms to Potential Therapies

Next-Generation Sequencing in Medicine

Prostate Cancer

RAS and Cancer in the 21st Century

Enteric Hepatitis Viruses

Bone: A Regulator of Physiology

Multiple Sclerosis

Cancer Evolution

The Biology of Exercise

Prion Diseases

Tissue Engineering and Regenerative Medicine

Chromatin Deregulation in Cancer

Malaria: Biology in the Era of Eradication

Antibiotics and Antibiotic Resistance

Cell Survival and Cell Death

SECOND EDITION

A subject collection from *Cold Spring Harbor Perspectives in Biology*

EDITED BY

Kim Newton
Genentech

James M. Murphy
Walter and Eliza Hall Institute of Medical Research

Edward A. Miao
Duke University

CSH PRESS

COLD SPRING HARBOR LABORATORY PRESS
Cold Spring Harbor, New York • www.cshlpress.org

Cell Survival and Cell Death, Second Edition

A subject collection from *Cold Spring Harbor Perspectives in Biology*
Articles online at cshperspectives.org

Executive Editor	Richard Sever
Managing Editor	Maria Smit
Senior Project Manager	Barbara Acosta
Permissions Administrator	Carol Brown
Production Editor	Diane Schubach
Production Manager/Cover Designer	Denise Weiss
Publisher	John Inglis

Front cover artwork: Our intimate connection with life and death is symbolized by the winged hourglass, reminding each of us how time flies swiftly from light into darkness. An instrument so revealing in its duality, the scythe both ends and renews life's circle. Face masks worn by plague doctors during epidemics connect cell death to pathogen defense with living and apoptotic cells at the heart of each practitioner. (Illustration created by Giovanni Luchetti, Genentech.)

Library of Congress Cataloging-in-Publication Data

Names: Newton, Kim, editor.
Title: Cell survival and cell death / edited by Kim Newton, Genentech, James M. Murphy, Walter and Eliza Hall Institute of Medical Research, and Edward A. Miao, University of North Carolina School of Medicine.
Description: Second edition. | Cold Spring Harbor, New York : Cold Spring Harbor Laboratory Press, [2019] | Series: Cold Spring Harbor perspectives in biology | A subject collection from Cold Spring Harbor Perspectives in Biology. | Includes bibliographical references and index. | Summary: "Cell death plays a critical role in development, normal physiology, and many diseases, including cancer. This new edition on the subject provides a state-of-the-art summary of our understanding of the signaling mechanisms that regulate cell survival and cell death, along with the molecular mechanisms underlying the various different death processes seen in cells" -- Provided by publisher.
Identifiers: LCCN 2019038852 (print) | LCCN 2019038853 (ebook) | ISBN 9781621823551 (hardcover) | ISBN 9781621823568 (epub) | ISBN 9781621823575 (kindle edition)
Subjects: LCSH: Cell death. | Cell physiology.
Classification: LCC QH671 .C453 2019 (print) | LCC QH671 (ebook) | DDC 571.9/36--dc23
LC record available at https://lccn.loc.gov/2019038852
LC ebook record available at https://lccn.loc.gov/2019038853

10 9 8 7 6 5 4 3 2 1

All World Wide Web addresses are accurate to the best of our knowledge at the time of printing.

Authorization to photocopy items for internal or personal use, or the internal or personal use of specific clients, is granted by Cold Spring Harbor Laboratory Press, provided that the appropriate fee is paid directly to the Copyright Clearance Center (CCC). Write or call CCC at 222 Rosewood Drive, Danvers, MA 01923 (978-750-8400) for information about fees and regulations. Prior to photocopying items for educational classroom use, contact CCC at the above address. Additional information on CCC can be obtained at CCC Online at www.copyright.com.

For a complete catalog of all Cold Spring Harbor Laboratory Press publications, visit our website at www.cshlpress.org.

Contents

Contents

Preface

MULTICELLULAR ORGANISMS ELIMINATE infected, damaged, or obsolete cells by activating genetically encoded cell death programs. These self-destruct programs are crucial to normal homeostasis and preservation of the overall health of the organism, but perturbations that enhance or suppress cell death may lead to disease. For example, excessive cell death is associated with neurodegeneration and other chronic inflammatory diseases, whereas too little cell death can promote cancer or susceptibility to infection. Much of this volume is devoted to the cellular signaling that underlies the three most studied cell death programs of apoptosis, pyroptosis, and necroptosis. This knowledge has led to the development of drugs designed to trigger the death of cancer cells and potential therapies for suppressing cell death that would otherwise exacerbate inflammation.

Apoptosis is a death program executed by cysteine proteases called caspases. It can be activated when internal cues alter the balance that exists between proapoptotic and antiapoptotic members of the BCL-2 family of cell death regulators. This volume begins by examining how proapoptotic BCL-2 family members permeabilize mitochondria to release factors that activate caspases and how drugs mimicking proapoptotic BCL-2 proteins have expanded the armamentarium against certain blood cancers. Apoptotic cells are dismantled into membrane-enveloped fragments that are rapidly cleared by phagocytes, so subsequent chapters explore the process of corpse engulfment and how dying cells influence living cells.

Other chapters describe how extracellular death ligands activate caspases to induce apoptosis in a cell-extrinsic manner. Necroptosis and pyroptosis, two death programs that, unlike apoptosis, are lytic in nature are also considered. For example, if caspases are inhibited, death ligands activate the kinase RIPK3 and its pseudokinase substrate MLKL to induce proinflammatory necroptosis. Pyroptosis is also proinflammatory and is driven by caspases that activate pore-forming proteins in the gasdermin family. The nature of these pores is described, as are the most studied triggers of gasdermin pores, the inflammasomes. These intracellular complexes assemble in response to different cellular insults and activate caspases that cleave the inhibitory domain from gasdermin D. Accumulating evidence that these cell death programs contribute to inflammation and various pathologies has sparked much interest in the therapeutic potential of inhibiting key pathway components.

Finally, there are chapters exploring the evolution of the different mammalian cell death programs and the pathogens that seek to subvert them. Cell death signaling mechanisms in plants and lower organisms are also reviewed. An interesting topic is whether these cell death programs arose from a common ancient pathway or by convergent evolution.

We thank all the contributing authors for the time they have devoted to these chapters, Barbara Acosta at Cold Spring Harbor Laboratory Press for her expert project management, Giovanni Luchetti for his creative cover artwork, and Richard Sever and Eric Baehrecke for suggesting that the time was right to document the exciting recent advances in our understanding of cell death and survival.

KIM NEWTON
JAMES M. MURPHY
EDWARD A. MIAO

The Evolutionary Origins of Programmed Cell Death Signaling

Kay Hofmann

Institute for Genetics, University of Cologne, Cologne D-50674, Germany

Correspondence: kay.hofmann@uni-koeln.de

Programmed cell death (PCD) pathways are found in many phyla, ranging from developmentally programmed apoptosis in animals to cell-autonomous programmed necrosis pathways that limit the spread of biotrophic pathogens in multicellular assemblies. Prominent examples for the latter include animal necroptosis and pyroptosis, plant hypersensitive response (HR), and fungal heterokaryon incompatibility (HI) pathways. PCD pathways in the different kingdoms show fundamental differences in execution mechanism, morphology of the dying cells, and in the biological sequelae. Nevertheless, recent studies have revealed remarkable evolutionary parallels, including a striking sequence relationship between the "HeLo" domains found in the pore-forming components of necroptosis and some types of plant HR and fungal HI pathways. Other PCD execution components show cross-kingdom conservation as well, or are derived from prokaryotic ancestors. The currently available data suggest a model, wherein the primordial eukaryotic PCD pathway used proteins similar to present-day plant R-proteins and caused necrotic cell death by direct action of Toll and IL-1 receptor (TIR) and HeLo-like domains.

Programmed cell death (PCD), either initiated cell autonomously in response to pathogens or stimulated from the outside through signaling molecules, is of crucial importance for the success of multicellular organisms. Most of current cell death research is focused on animals and plants, with mammals taking center stage, whereas substantial work is also devoted to model metazoans such as *Drosophila melanogaster* and *Caenorhabditis elegans*, or to *Arabidopsis thaliana* as the main model for plant cell death pathways. However, cell death pathways have also been described in other organisms such as filamentous fungi, and their molecular mechanisms are beginning to be unraveled. At first glance, PCD pathways in different kingdoms appear to work by fundamentally different rules, although a number of recent studies have revealed numerous mechanistic parallels and instances of clear evolutionary interkingdom relationships of cell death mediators. These findings suggest that at least a core pathway for PCD has existed in the common ancestor of metazoans, fungi, and plants. The following paragraphs will provide a synopsis of major cell death pathways in different kingdoms and will highlight the evolutionary processes leading to the diversification of PCD pathways observed today.

Cite this article as *Cold Spring Harb Perspect Biol* doi: 10.1101/cshperspect.a036442

PROGRAMMED CELL DEATH SYSTEMS IN DIFFERENT KINGDOMS

PCD modalities can be classified as "necrotic" (i.e., accompanied by membrane rupture and release of intracellular material), and "non-necrotic" without such leakage (Ashkenazi and Salvesen 2014). Because metazoan apoptosis is the major—if not the only—example for the latter type, non-necrotic cell death is usually referred to as "apoptotic." As compared with apoptosis, necrotic cell death is less intricate and far more widespread. Because cell death caused by major mechanical, chemical, or biological insults is typically associated with cell rupture, necrosis was initially considered to be a hallmark of non-PCD and it took a long time until "programmed necrosis" was accepted as a reality (Edinger and Thompson 2004). In animals, several different types of programmed necrosis exist, among which necroptosis and pyroptosis are the best understood and probably most important examples (Cookson and Brennan 2001; Bergsbaken et al. 2009; Vandenabeele et al. 2010). Cells dying by programmed necrosis release intracellular contents, several components of which are interpreted as "damage-associated molecular patterns" (DAMPs) by the immune system, resulting in inflammation (Schaefer 2014; Roh and Sohn 2018). Apoptosis, in contrast, is a more complicated process because it has to reliably kill the cell while at the same time preventing the leakage of intracellular material and DAMPs. This is no easy task, considering that cellular compartments such as the lysosome and the mitochondrion contain enzymes and oxidants with the potential to damage cell membranes. During apoptosis, the cell is broken into a number of smaller, membrane-enclosed vesicles called "apoptotic bodies," which are subsequently removed by phagocytic processes (Elmore 2007; Nagata 2018). Apoptosis is therefore ideally suited for developmentally scheduled cell death, a physiological process not supposed to alert the immune system (Fuchs and Steller 2011).

Nonmetazoan forms of PCD are not easily classified as "apoptosis" or "programmed necrosis," because the dying cells look morphologically different and the signaling cascades and death effectors appear—at least at first glance—unrelated to their metazoan counterparts. In the absence of circulating phagocytic cells, a permanent containment characteristic of metazoan apoptosis is unlikely; whether this containment is important in the absence of an inflammatory system is not clear. There is at least one class of PCD pathway in plants called the "hypersensitive response" (HR), which has been shown to be associated with DAMP release, therein resembling metazoan programmed necrosis pathways (Morel and Dangl 1997; Balint-Kurti 2019). The HR is a part of the so-called "effector triggered immunity" (ETI) system, which gets activated on the detection of pathogen-derived proteins within the host cell. The HR cell death is called "hypersensitive" because it exceeds the damage directly inflicted by the pathogen; its function is to limit the spread of biotrophic pathogens. The production and release of DAMPs—among them the small molecule salicylic acid (SA)—serves the purpose to alert other parts of the plants of the ongoing infection (Balint-Kurti 2019).

Outside of animals and plants, PCD pathways exist (Ameisen 2002), but only a few of them have been characterized in molecular detail. Filamentous fungi belonging to the *Ascomycetes* possess a number of functionally analogous, but molecularly diverse cell death pathways required for a process called "heterokaryon incompatibility" (HI) (Saupe 2000; Daskalov et al. 2017). Filamentous ascomycetes grow an extensive network of hyphae, which can both branch off and merge back—provided that the merging hyphae are genetically identical. Multiple HI systems prevent the successful fusion of hyphae emanating from genetically different individuals, thereby safeguarding against the spread of pathogens (Daskalov et al. 2017). For a successful hyphal fusion, each of the available HI systems has to be "disarmed" separately—usually by the two fusion partners being homozygous at a polymorphic sensor locus. Triggering only one of the HI systems is sufficient to cause localized cell death near the point of fusion. The formation of intrahyphal septa prevents the

Cite this article as *Cold Spring Harb Perspect Biol* doi: 10.1101/cshperspect.a036442

spreading of the cell death over the entire hyphal system. For several fungal species, in particular for the HI model organism *Podospora anserina*, multiple HI systems and their sensor proteins have been described (Saupe 2000; Daskalov et al. 2017; Gonçalves et al. 2017).

MOLECULAR FEATURES OF PROGRAMMED DEATH PATHWAYS

This section provides a synopsis of major PCD pathways, which are mechanistically understood to some degree. In particular, the molecular architecture of the key components is summarized, as this information is necessary to appreciate the evolutionary ancestry of PCD pathways.

Apoptosis

A simplified version of the apoptosis pathway is shown in Figure 1A, and detailed reviews can be found in Elmore (2007) and Nagata (2018). The crucial step in apoptotic cell death induction is the activation of caspase-3, a cysteine protease specifically cleaving a number of different substrates, which in combination bring about the apoptotic phenotype. Apoptosis has, quite appropriately, been called "death by a thousand cuts" (Martin and Green 1995). The activation of caspase-3 happens by proteolytic processing of an inactive precursor; the activating enzyme caspase-9 belongs to the same protease class as caspase-3 and several other proteases involved in cell death signaling (see below). Caspase-9, in turn, is activated by formation of a multiprotein complex called the "apoptosome." The central component of the apoptosome is APAF1, an ATPase of the STAND class (Danot et al. 2009), which is able to sense the presence of cytochrome *c* released from mitochondria. On binding to cytochrome *c*, APAF1 undergoes a major conformational change, leading to the exposure of its amino-terminal oligomerization domain. This so-called CARD (caspase activation and recruitment domain) will then hetero-oligomerize with another CARD domain found at the amino terminus of the inactive precursor of caspase-9. In this oligomeric state, two prox-

imal caspase-9 molecules can cleave and thereby activate each other. There are several pathways leading to apoptosome activation, either cell autonomously or responding to external stimuli via death receptors (Elmore 2007). These upstream pathways use other caspases (caspase-8, caspase-10) and other hetero-oligomerization domains, such as the "death domain" (DD) and the "death effector domain" (DED), which connect the apoptotic signaling components and can lead to the activation of caspases via induced oligomerization and cleavage.

Necroptosis

A simplified version of the core necroptosis pathway is shown in Figure 1B, and more detailed descriptions can be found in Newton and Manning (2016), Weinlich et al. (2017), and Petrie et al. (2019). The key step committing a cell to necroptosis is the oligomerization and activation of the protein kinase RIPK3. RIPK3 possesses a central RHIM (RIP homotypic interaction motif), which is important for recruitment of RIPK3 to other RHIM-containing proteins, in particular to the related protein kinase RIPK1. Recently, the RIPK1–RIPK3 "necrosome" complex was shown to form a RHIM-based amyloid fibril with alternating strands of RIPK1 and RIPK3 (Mompeán et al. 2018). Once part of the oligomeric necrosome, RIPK3 recruits and phosphorylates a protein called MLKL (mixed lineage kinase domain-like). MLKL is an inactive pseudo-kinase with an additional amino-terminal four-helix bundle (4HB) domain required for cell death execution. On phosphorylation of MLKL, the protein oligomerizes, associates with the cell membrane, causes ion influx, and eventually cells rupture. How exactly these events are timed and interconnected remains a matter of debate. It is nowadays assumed that membrane-associated MLKL oligomers form membrane pores, either on their own or with the help of other cellular factors (Petrie et al. 2019). Many aspects of necroptotic cell death can be mimicked by ectopic expression of the isolated MLKL amino-terminal domain, thereby obviating the need for upstream signaling and RIPK3 activity.

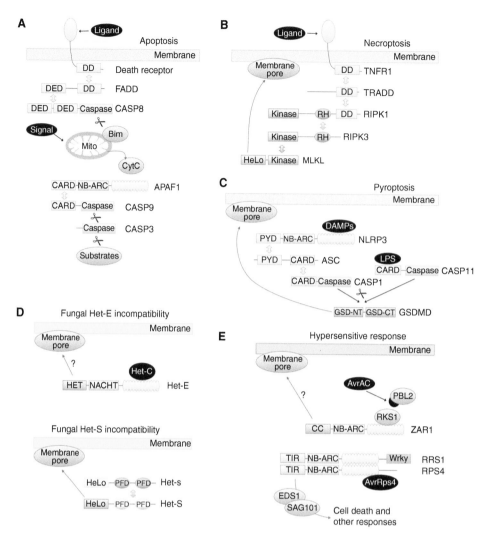

Figure 1. Signaling domains and their functions in programmed cell death (PCD) pathways. This figure shows simplified versions of major PCD pathways, focusing on the interaction properties of the domains. The arrangement of the proteins in homo-oligomeric complexes is not shown. Coloring: six-helix death-fold domains (death domain [DD], death effector domain [DED], caspase activation and recruitment domain [CARD], pyrin domain [PYD]) are shown in cyan, RIP homotypic interaction motifs (RHIMs) in orange, and Toll and IL-1 receptor (TIR) domains in purple. The central STAND ATPase domains (both NB-ARC and NACHT type) are blue and repetitive sensor domains (leucine-rich repeat [LRR], WD40) are green. Caspase domains are yellow and supposedly pore-forming domains (gasdermin amino-terminal domain, HET domain, and HeLo/coiled-coil [CC]) are shown in red. All other domain types are shown in gray. Homotypic oligomerization is indicated by double arrows colored by domain type. Proteolytic cleavage is indicated by a scissors symbol, whereas translocation events are shown as red arrows. Cell death stimuli are shown on a black background. (A) Apoptosis. Both the extrinsic pathway, triggered by ligand binding to a death receptor, and the intrinsic pathway initiated by a mitochondrial signal are shown. (B) Necroptosis. Only the main components of the canonical pathway, triggered by TNF-receptor type 1 (TNFR1) ligation, is shown here. (C) Pyroptosis. Both major pathways are shown: caspase-11 (human: caspase-4/5) triggering by intracellular lipopolysaccharide (LPS), and caspase-1 activation by signalosome signaling. (D) Fungal heterokaryon incompatibility (HI). Two HI systems are shown: Het-E (from one fusion partner) being triggered by Het-C (contributed by the other partner), and Het-S (from one fusion partner) being recruited to an amyloid formed by Het-s (from the other fusion partner). (E) Plant hypersensitive response. One example for each class of R-protein is shown: The CC-NB-LRR (CNL)-based ZAR1 resistosome is triggered by the ZAR1/RKS1 complex recognizing PBL2, which has been previously modified by the pathogen effector AvrAC. The dimer of the two TIR-NB-LRR (TNL)-based STAND proteins RPS4 and RRS1 is triggered by binding to the pathogen effector AvrRps4 and signals cell death via the EDS1/SAG101 complex.

Cite this article as *Cold Spring Harb Perspect Biol* doi: 10.1101/cshperspect.a036442

Pyroptosis

The two main pyroptosis pathways are shown in Figure 1C, more detailed descriptions can be found in Kovacs and Miao (2017), Man et al. (2017), and Shi et al. (2017). The key event in executing pyroptotic cell death is the proteolytic cleavage of gasdermin D (GSDMD) at a central position, separating the cell-killing amino-terminal domain from the inhibitory carboxy-terminal domain. Several (nonapoptotic) caspases are able to cleave GSDMD, depending on the initial trigger and the upstream signaling pathway. One subpathway responds to intracellular lipopolysaccharide (LPS) and cleaves GSDMD by caspase-11 in the mouse and caspase-4/5 in humans. Another pathway depends on inflammasome activation and uses the major proinflammatory caspase-1. Once activated, these caspases will not only cleave GSDMD but also other proteins with accessible cleavage sites, most importantly the proform of interleukin (IL)-1β, thereby forming the active mature form of this proinflammatory cytokine. The liberated amino-terminal domain of GSDMD is thought to undergo a major conformational change, leading to its oligomerization and formation of a large membrane pore, which allows the release of the processed IL-1β. Besides GSDMD, other members of the gasdermin family have cleavable amino-terminal domains that support pore formation (Feng et al. 2018). Recently, the structure of the gasdermin A3 pore has been solved by cryo-electron microscopy (EM) and was shown to form a 108-stranded β-barrel, consisting of 27 gasdermin units (Ruan et al. 2018). Other proteases have also been reported to cause gasdermin-dependent pyroptotic cell death (Xia et al. 2019). In cells infected by the bacterial pathogen *Yersinia pestis*, the apoptotic caspase-8 can be activated through a multiprotein complex called "RIPoptosome," which in turn leads to GSDMD and GSDME processing by the activated caspase, resulting in pyroptosis (Orning et al. 2018; Sarhan et al. 2018). The neutrophil-specific elastase ELANE, a serine-protease unrelated to caspases, has been shown to cleave GSDMD at an alternative site, which also results in the generation of a cytotoxic amino-terminal fragment (Kambara et al. 2018).

Heterokaryon Incompatibility

A simplified depiction of two fungal HI systems is shown in Figure 1D, and more comprehensive descriptions can be found in Saupe (2000), Daskalov et al. (2017), and Gonçalves et al. (2017). A number of different HI pathways have been described in model fungi. Some of these pathways are "allelic," meaning that they are triggered if the fused hyphae are heterozygous for a particular polymorphic sensor gene. Other systems are triggered by the interaction of two different gene products, each of them contributed by one of the fused cells. The best understood pathway is probably the allelic Het-S system in *P. anserina* (Seuring et al. 2012; Riek and Saupe 2016). Het-s and Het-S are two alleles of a gene encoding a potentially toxic two-domain protein. The carboxy-terminal domain can initiate formation of an amyloid structure, which is not toxic by itself, but can cluster multiple copies of the amino-terminal domain. On clustering, the amino-terminal domain can oligomerize, insert into the plasma membrane, and cause the loss of membrane integrity. The protein version encoded by the Het-s allele is able to initiate an amyloid structure by its carboxy-terminal "prion-forming domains" (PFDs); however, the amino-terminal "HeLo" domain of the Het-s protein is not able to permeabilize the membrane. In contrast, the protein encoded by the Het-S allele has a functional amino-terminal domain, but is not able to initiate amyloid formation owing to a mutation in the PFD region. When both Het-S and Het-s encoded proteins encounter each other during fusion of incompatible cells, the Het-s protein will initiate amyloid formation, whereas the Het-S protein can extend these amyloids, thereby triggering the membrane pore formed by the Het-S amino-terminal domain. A fundamentally different incompatibility mechanism, which is also relevant for the discussion of cell death evolution, is found in the non-allelic Het-E/Het-C system and its relatives. Here, the central component is the Het-E protein, a STAND-type ATPase with a similar architecture as the apoptosome component APAF1 and the NLR components of the inflammasome. The carboxy-terminal WD40-repeat region of Het-E can sense the presence of particular alleles of the (unrelated) Het-C gene product. On binding to Het-C, the STAND ATPase undergoes a conformational change leading to the exposure of the Het-E amino-terminal region, usually referred to as the "HET domain." Unlike the situation in apoptosis and inflammasome activation, these HET domains are not thought to be recruitment domains, but rather to directly disintegrate the membrane, leading to a necrotic type of cell death (Paoletti and Clavé 2007). Many other fungal HI systems exist, but only a few of them have been studied for their cell-killing mechanism (Fig. 2).

Hypersensitive Response

Two examples of plant defense signaling leading to cell death are shown in Figure 1E. More detailed descriptions can be found in Coll et al.

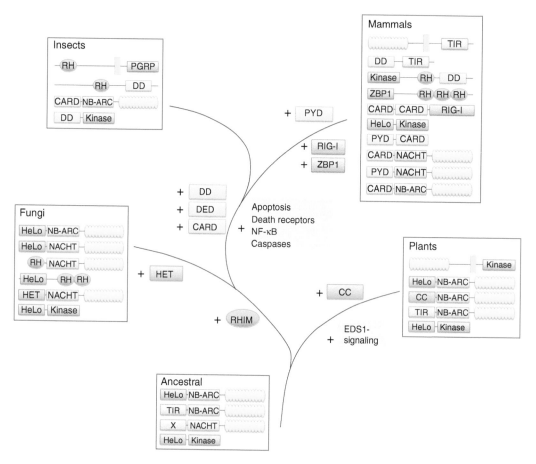

Figure 2. Evolutionary origins of cell death protein architectures. This figure shows generic protein architectures found in metazoans, fungi, and plants—and what is predicted to have existed in an ancestral (early eukaryote) organism. Domain coloring identical to Figure 1. The ancestral system is predicted to mainly have used STAND ATPases, most likely of the NB-ARC type, with amino-terminal Toll and IL-1 receptor (TIR) and HeLo-like domains. However, NACHT-ATPases and a mixed-lineage kinase domain-like (MLKL) HeLo-domain kinase might also have been present. In the plant lineage, components of EDS1 and probably many other relevant proteins have been acquired. The coiled-coil [CC]-domain found in present plant CC-NB-LRR (CNL) proteins was either acquired or, more likely, evolved from the HeLo domain. RIP homotypic interaction motifs (RHIMs) are first seen in the opisthokont lineage, whereas the six-helix death-fold domains are specific for the metazoan lineage. A more detailed description of early cell death evolution is given in the main text.

Cite this article as *Cold Spring Harb Perspect Biol* doi: 10.1101/cshperspect.a036442

(2011), Jones et al. (2016), and Balint-Kurti (2019). Although plant defense mechanisms against pathogens have been studied intensively, and protective cell death by the HR is one important branch of this defense, the exact cell death mechanism and the death-specific signaling components are much less clear. The central orchestrators of intracellular defense pathways are the "R-proteins" (resistance proteins). Most R-proteins belong to the class of STAND-type ATPases with a leucine-rich repeat (LRR) domain at the carboxyl terminus and an effector/signaling domain at the amino terminus. Like all other STAND-type ATPases, R-proteins are thought to undergo a structural rearrangement on binding to its cognate ligand, which may be bacterial effectors secreted into the host cell or any other molecule indicative of biotic stress. Depending on the nature of the amino-terminal domain, different downstream signaling pathways can be engaged, eventually leading to the induction of defense genes, the generation and release of the long-range signaling molecule SA, and/or to the induction of hypersensitive cell death. The factors required for cell death execution are not known and depend on the R-protein, which might belong to one of several subtypes. The TNL proteins carry an amino-terminal Toll and IL-1 receptor (TIR) domain and signal cell death via the EDS1-SAG101-NRG1 pathway (Lapin et al. 2019) although the details of that mechanism are not known. A second class of R-proteins are the "CC-NB-LRR" (CNL) proteins, which carry an amino-terminal domain that was initially considered to form a coiled-coil (CC) structure, although this is not necessarily true. At least some members of the CNL-type R-proteins can induce cell death dependent on the amino-terminal domain (Collier et al. 2011), although the exact mechanism remains unresolved.

RECURRING MOTIFS AND MECHANISMS IN CELL DEATH SIGNALING

When comparing the mechanisms of the cell death pathways described above, in particular the domain architecture of the key proteins involved, a number of recurring features suggesting a common evolutionary history becomes apparent.

STAND ATPases in Cell Death and Immunity

One of the most obvious recurring features in cell death signaling is the use of STAND-type ATPases as a signaling hub (Leipe et al. 2004; Danot et al. 2009). The name STAND, originally an acronym for "signal transducing ATPases with numerous domains," encompasses two major subfamilies of large ATPases of similar core architecture, which are both important for cell death signaling. One class has a central "NB-ARC" ATPase domain (named after the proteins APAF1/R-Proteins/Ced-4), the other class has a central "NACHT" ATPase domain (named after NAIP/CIIA/Het-E/TEP1). Both classes contain carboxy-terminal sensor regions, typically consisting of repeat-forming domains, such as LRRs, WD40 repeats, or tetratricopeptide repeats (TPRs). At the amino terminus, different effector domains can be found. Both NB-ARC and NACHT ATPases work similarly. In the resting state, the proteins are monomeric and show a "closed" conformation, which shields the amino-terminal effector domain. On binding of the carboxy-terminal sensor domain to its cognate stimulus, the central ATPase domain undergoes a conformational change, causing an oligomerization of the ATPase and leading to the exposure of the amino-terminal effector domains. This process is best understood for the NB-ARC protein APAF1, the central component of the apoptosome (Cheng et al. 2016; Dorstyn et al. 2018). Here, the active conformation has the shape of a heptameric "wheel" formed by seven ATPase domains; the amino-terminal CARD domains of the seven APAF1 units are found clustered above the plane of the ATPase wheel. Ced-4, the APAF1 homolog from the nematode *C. elegans*, forms an octameric ring with a similar domain arrangement (Qi et al. 2010; Huang et al. 2013).

The vast majority of eukaryotic NB-ARC and NACHT proteins are known or suspected to be involved in innate immunity and cell death signaling. As mentioned before, the (sole) human NB-ARC protein APAF1 and its nematode homolog Ced-4 use their amino-terminal CARD domain for apoptosome formation. Mammals possess several NACHT proteins with amino-terminal CARD or pyrin domains (PYDs), which are the key components of "inflammasomes," protein complexes similar to the apoptosome but activating caspase-1 rather than caspase-9 and thereby leading to pyroptosis (Broz and Dixit 2016).

The relationship between the mammalian STAND ATPases and the plant R-proteins is readily visible and has stimulated a number of analyses comparing animal and plant cell-autonomous immunity (Maekawa et al. 2011; Jones et al. 2016; Urbach and Ausubel 2017). Recently, it was shown by cryo-EM that plant R-proteins can also form wheel-like oligomers. In the example of ZAR1, an R-protein of the CNL class, the wheel has a pentameric structure called the "ZAR1 resistosome" to emphasize the analogy to the metazoan apoptosome and inflammasome complexes (Wang et al. 2019). In the activated ZAR1 resistosome, the five individual "CC-type" effector domains undergo a conformational rearrangement and form a pore-like α-barrel with an additional funnel-like structure formed by the first α-helices of each CC domain. This funnel appears to be crucial for cell death induction by activated ZAR1 (Wang et al. 2019). Although there is no formal proof that the CC domain of ZAR1 forms a membrane pore sufficient for ion influx or even cell rupture, it appears that the plant R-proteins use a more direct way to cell death than their metazoan counterparts, which rely on further downstream signaling.

Several of the fungal HI systems also make use of STAND ATPases. The Het-E protein mentioned above, but also Het-D and Het-R all contain a central NACHT ATPase domain, followed by a carboxy-terminal sensor domain consisting of WD40-repeats. The amino-terminal effector domains (HET-domains) are only found in filamentous fungi and are thought to directly form membrane pores on oligomerization (Paoletti and Clavé 2007). This mode of action would be analogous to what has been suggested for ZAR1 and possibly other CC-type R-proteins. When comparing the domain architectures of APAF1 (CARD, NB-ARC, WD40) with that of ZAR1 (CC, NB-ARC, LRR) or HET-E (HET, NACHT, WD40), it becomes obvious that these proteins have the same overall architecture but did not evolve by divergent evolution from a common ancestor. It can be assumed that early eukaryotes already contained STAND-ATPases because most extant bacteria encode several such ATPases. However, the classical bacterial STAND-ATPases do not belong to the NB-ARC or NACHT subtype and have no connection to cell death signaling (Leipe et al. 2004). A careful phylogenetic analysis of the STAND family concluded that plant and animal STAND proteins evolved independently from bacterial precursors, in at least two separate events (Urbach and Ausubel 2017). However, STAND ATPases and other innate immunity proteins are clearly subject to evolution by "domain swapping," and when analyzing eukaryotic proteomes available nowadays, virtually all combinations of effector domains, ATPase subtype, and carboxy-terminal sensor domains can be observed. Therefore, alternative evolutionary hypotheses should not be discounted, for example, a pervasive "mixing and matching" of domains in early eukaryotic evolution, with a subsequent "fixing" of select domain architectures that proved most beneficial for the particular eukaryotic lineage.

Evolution of Effector and Oligomerization Domains

The formation of transient multiprotein complexes, leading to the recruitment of adaptor and effector proteins to activated receptors, is a recurring feature of apoptosis, necroptosis, and pyroptosis signaling. In metazoan systems, the prevalent oligomerization domain types are the DD, the DED, the CARD, and the PYD. All four domain types are distantly related to each other (Hofmann 1999; Park et al. 2007a; Kersse et al. 2011) and share a common structural fold consisting of six α-helices, often referred to as the "death fold" or "six-helix bundle fold." The principal function of all four domain classes is to recruit another domain of the same class.

Despite the relationship between DD, DED, CARD, and PYD, interactions across the class boundaries are rare. It is interesting to note that RHIM motifs and TIR domains fulfil analogous roles of "recruitment by oligomerization," but are not related to the death-fold superfamily and do not share the six-helix bundle structure (Nanson et al. 2019).

Despite exhaustive bioinformatical searches (Hofmann 1999; unpubl. data), no members of the death-fold superfamily could be detected outside of metazoans, with the exception of viruses and other pathogens, which have probably acquired these domains from their metazoan hosts (Thome et al. 1997; Lamkanfi and Dixit 2010). It thus appears that the six-helix death fold arose during metazoan evolution. It is also remarkable that all characterized members of the death-fold superfamily reside in proteins involved in cell death and innate immunity signaling; other pathways requiring oligomerization appear to use other domains for this purpose. A possible reason for this pathway specificity might lie in the availability of multiple interaction surfaces of the fold, which can be used simultaneously and thus support the formation of higher-order oligomeric structures called "filaments" (Park et al. 2007b; Kersse et al. 2011; Hauenstein et al. 2015). Such filaments can be nucleated by a di- or trimerization and then grow by using the remaining interaction surfaces—analogous to an amyloid but without the β-stack structure typical of real amyloids. These cooperatively formed higher-order structures are instrumental for inflammasome formation (Hauenstein et al. 2015) and support the "all-or-nothing" characteristics required for life/death decisions.

Evolution of TIR Domains

TIR domains have a particularly interesting evolutionary history. In mammals, TIR domains are found at the cytoplasmic end of Toll-like receptors (TLRs) and receptors for IL-1 cytokines—all of them receptors that alert a cell to a danger situation from the outside. Their function is the recruitment of other TIR domains found in "adaptor proteins" such as MYD88, TIRAP,

TRAM, and TRIF. Typical downstream events of mammalian TIR signaling include the up-regulation of defense genes via transcription factors of the NF-κB (nuclear factor κ light chain enhancer) or IRF (interferon regulatory factor) families, but also necroptosis induced via the RHIM motif of TRIF. Despite a completely unrelated structure, the role of TIR domains is remarkably similar to members of the six-helix death-fold family. This analogy is further emphasized by the ability of TIR domains to form filaments akin to those of the death-fold domains (Ve et al. 2017; Nanson et al. 2019). Because there clearly is no evolutionary relationship between these domain classes, oligomerization through TIR and death-fold domains is another example of convergent evolution.

In contrast to animals, plants do not use TIR domains in their surface receptors, but rather as the amino-terminal effector domains of one class of intracellular R-proteins (TIR-NB-LRR or TNL type). Surprisingly, plants lack TIR-containing adaptor proteins and the TNL proteins signal via a complex of EDS1 (enhanced disease susceptibility 1) and SAG101 (senescence-associated gene 101) through an as-yet uncharacterized mechanism (Lapin et al. 2019). However, plant TIR domains share with their metazoan counterparts the capacity to mediate dimerization between different R-proteins (Bernoux et al. 2011; Williams et al. 2014; Zhang et al. 2017). An interesting twist to the enigma of TIR signaling was introduced recently by detecting an NAD-cleaving activity for the TIR domain of the human SARM1 (sterile α and TIR motif) protein (Essuman et al. 2017), as well as in several bacterial TIR proteins (Essuman et al. 2018). SARM1 has a role in mediating neurodegeneration and the NAD depletion by its TIR domain was suggested to be a main factor in this process (Essuman et al. 2017). It is unlikely that other human TIR domains share this enzymatic activity, because they either lack the active site glutamate residue or have a structure that places this residue outside the catalytic cleft. However, it remains possible that in the presence of suitable binding partners or posttranslational modifications, the structure may be converted into an active form. Plant TIR domains, in contrast, tend to show conservation in the active site region, and the few available structures appear to support catalysis. It is therefore an intriguing possibility that either NAD depletion, or the generation of (cyclic) ADP-ribose as the NAD degradation product, play a role in cell death signaling by R-proteins of the TIR-NB-LRR architecture. Given that TIR proteins are abundant in extant bacteria and appear to be generally catalytically active (Essuman et al. 2018), it is highly likely that both animal and plant TIR domains evolved from a prokaryotic ancestor and later gained the capacity for oligomerization, although (mostly) losing their catalytic activity in the process.

Evolutionary History of Necroptosis Execution

The execution phase of necroptosis is characterized by MLKL phosphorylation and oligomerization, initiated by formation of a mixed amyloid structure formed by the RHIM motifs of RIPK1 and RIPK3 (Mompeán et al. 2018; Petrie et al. 2019), or possibly by similar structures formed by RIPK3 with other RHIM proteins such as TRIF or ZBP1/DAI. Despite its fundamentally different structure, the role of RHIMs in necroptosis signaling is analogous to those of six-helix death-fold domains or TIR domains. All of these domains form homotypic oligomers and have the tendency to create higher-order superstructures. In the case of RHIM, this superstructure is a "real" amyloid (Li et al. 2012; Mompeán et al. 2018). In evolution, RHIM-based oligomerization is far more widespread than the few examples known in mammals. A bioinformatical analysis showed a large array of RHIM-containing proteins in nonmammalian metazoans. In many of these cases, the RHIM appears to substitute for TIR or six-helix death-fold domains, which are found in the mammalian version of these proteins (Kajava et al. 2014). This finding suggests that RHIM motifs work similar to TIRs and death-fold domains and actually can replace them.

Interestingly, the same bioinformatical analyses found an evolutionary relationship between RHIM motifs and the PFDs used in the Het-S system of HI (Kajava et al. 2014). Like RHIMs,

the fungal PFDs form an amyloid of similar structure (Wasmer et al. 2008; Riek and Saupe 2016), supporting a common evolutionary origin of these two oligomerization systems. This evolutionary parallel is further underscored by the finding that the "HeLo domain," the cell-killing moiety of the Het-S system, is related to the functionally analogous amino-terminal domain of MLKL (Daskalov et al. 2016). It is thus very likely that the metazoan necroptosis system evolved from a simpler precursor similar to Het-S, with amyloid-forming domain and killing domain within the same polypeptide. During vertebrate evolution, the two functionalities were probably split into separate proteins. It is difficult to decide from the available data whether the common precursor was using a one-component amyloid (as in Het-S) or a two-component version (as in RIPK1/RIPK3). Because even the one-component amyloid uses two alternating strands—PFD1 and PFD2 of Het-S—it appears more likely that the extant Het-S system is a degenerate version of a former two-component systems, now perfectly adapted to the task of detecting heterozygosity.

Evolutionary History of Pyroptosis and Caspases

The decisive step in pyroptosis execution is the cleavage of GSDMD by caspase-1 or other caspases with similar cleavage specificities. Caspase-1 itself is activated by inflammasomes, which can be induced by many proinflammatory stimuli. The fact that caspase-1 also activates IL-1β makes pyroptosis the major modality for releasing processed IL-1β and thus causing the inflammatory phenotype (Man et al. 2017; Green 2019). The other GSDMD-cleaving caspases 4, 5, and 11 are directly activated by intracellular LPS. The details of this activation are not fully understood. All of the GSDMD-cleaving caspases are thought to oligomerize via their amino-terminal CARD domains, but the involvement of filaments of higher-order structures has not been reported.

Neither gasdermins nor "proper" caspases are found outside of metazoans; the IL-1 cytokine family is even restricted to vertebrates.

Cite this article as *Cold Spring Harb Perspect Biol* doi: 10.1101/cshperspect.a036442

Thus, pyroptosis, like apoptosis, appears to be a relatively recent addition to the arsenal of PCD pathways. On the other hand, the metazoan caspases are related to two other classes of cysteine proteases, the metacaspases and paracaspases, which have a much wider evolutionary distribution and are even found in bacteria (Koonin and Aravind 2002). Initially, the finding of metacaspases in yeasts, together with the finding that overexpression of some mammalian apoptosis proteins killed yeast cells, gave rise to speculations that caspase-dependent cell death or even apoptosis might be conserved in lower eukaryotes (Váchová and Palková 2007). By now, it is known that metacaspases, despite their evolutionary relationship to caspases, have a totally different cleavage specificity (cleavage after arginine rather than aspartate) and cannot replace the function of proper metazoan caspases (Tsiatsiani et al. 2011). Nevertheless, metacaspases might have a more general role in DAMP processing during necrotic cell death. A recent study showed that plant metacaspases process a cytoplasmic immunomodulatory plant protein, the active part of which PEP1 (plant elicitor peptide 1) is secreted from damaged cells and acts as a defense signal (Hander et al. 2019). This situation is comparable to the IL-1β release during pyroptosis but without any sequence or structural similarity between PEP1 and the metazoan interleukin.

Another question of evolutionary relevance concerns the origins of the gasdermin family. Of the six gasdermins known in the human genome, four are relatively closely related to each other (GSDMA, GSDMB, GSDMC, GSDMD). This subfamily is fast evolving, with some additional members in rodents and other mammals. Common to all these proteins is the architecture with an amino-terminal toxic domain, a carboxy-terminal inhibitory domain, and a protease cleavage site in the middle. Two more gasdermins (DFNA5/GSDME and Pejvakin/PJVK) form a second subfamily, which is somewhat more distantly related, in particular in the quite divergent inhibitory region. GSDME, which unlike gasdermin A-D has homologs in fish, behaves like a classical gasdermin; it is cleaved by caspases and its amino-terminal domain is able

to cause pyroptosis (Wang et al. 2017; Jiang et al. 2019). In contrast, the sixth gasdermin, Pejvakin/PJVK, does not appear to be toxic (Feng et al. 2018) but has homologs in invertebrates. The exact role of PJVK is not known; the human gene is implicated in nonsyndromic hearing loss (Harris et al. 2017) and has been proposed to regulate pexophagy, the autophagic removal of peroxisomes (Defourny et al. 2019). Based on these considerations, it appears likely that the metazoan gasdermin family evolved from a Pejvakin-like precursor and possibly acquired the cell-killing activity of the amino-terminal later on. However, it remains possible that the proto-gasdermin gene encoded a cell-killing protein and that Pejvakin lost this activity. As a third possibility, Pejvakin might still be able to kill cells, but requires a specific activation mode awaiting to be discovered. A much older evolutionary history—or possibly a horizontal acquisition—is suggested by the recently published structure of the mouse gasdermin A3 pore (Ruan et al. 2018), which is probably a good model for other gasdermin pores as well. Both the pore structure and the conformational change undergone by the GSDMA3 amino-terminal domain on pore formation strongly resemble the pores of bacterial cytolysins such as Pneumolysin and Perfringolysin O, but also the pores of mammalian perforin and the membrane attack complex (MAC) of the complement system (Ruan et al. 2018). Despite the structural similarities of these pores, there is no overt sequence similarity, which makes it difficult to judge whether these proteins are truly related or just further examples of convergent evolution.

HOW IT ALL BEGAN

An inventory of present-day cell death components, their domain architecture, and their interrelationship should, in principle, allow the reconstruction of key evolutionary events shaping cell death signaling. However, the fast evolution of some components, the pervasive domain shuffling, and the incomplete knowledge of death signaling outside the classical model organisms make this task difficult if not impossible. The sequence of evolutionary events proposed here is in accordance with the available data, but there is no certainty that it reflects the history correctly.

Because STAND-type ATPases are abundant in both pro- and eukaryotes, and fulfill similar purposes in multiple kingdoms, it is safe to assume that STAND-based signaling was available in early eukaryotes including the last common ancestor to animals, fungi, and plants. At the carboxyl terminus of extant STAND ATPases, many different sensor domains are found, typically repeat domains. In plants, most of the sensing appears to be performed by LRR domains, which are also common in mammalian STAND ATPases. However, when including invertebrates and fungi (which are more closely related to animals than plants), a much greater sensor diversity is observed, including WD40, TPR, and ankyrin repeats. One published study concluded that the NB-ARC subtype (as used in plant R-proteins) and the NACHT subtype (as used in animal inflammasomes and fungal Het-E) evolved independently from a bacterial ancestor (Urbach and Ausubel 2017). This study assumed bacteria-specific subfamilies of the STAND superfamily as the precursor, from which eukaryotic NACHT and NB-ARC subtypes evolved. Whereas this is undoubtedly possible, there are plenty of bacterial NB-ARC and NACHT members in present-day genomes. This observation means that either the two independent events leading to NACHT and NB-ARC happened before the split of eukaryotes, or that the NACHT and NB-ARC proteins seen in extant bacteria are late horizontal acquisitions from a eukaryotic source. Most likely, early eukaryotes coded for NACHT and NB-ARC-type ATPases with a wide range of possible sensor motifs.

A most relevant question of cell death evolution is what kind of effector domains were used by these early signaling hubs. One prime candidate are the TIR domains, they clearly predate the advent of eukaryotes, and they were most likely enzymatically active like the extant bacterial versions (Essuman et al. 2018). Actually, the NAD-degrading function of TIR domains in the context of a STAND ATPase might have formed the basis for a very early eukaryotic cell death system, triggered by some pathogen- or damage-derived molecule and leading to cell death by NAD depletion. Another early cell death system could have used a HeLo-like domain as the effector. Although HeLo domains have not been detected in bacteria, several copies of this domain exist in plants, fungi, animals, and other eukaryotes. Because the HeLo domain appears to cause membrane pores autonomously, at least in fungi, this domain is an equally good candidate for an early PCD system. Interestingly, the two most likely architectures resemble the two classes of plant R-proteins: TIR-NB-LRR and CC-NB-LRR. However, it is well possible that the ancient versions of this system used another subtype of STAND ATPase, or another type of sensor repeat.

Cell death systems relying on caspase activity, such as apoptosis and pyroptosis, most likely did not develop before the advent of metazoans, although it cannot be excluded that there have been pyroptosis-like systems using another type of protease. However, there are no indications for such a system in the presently available data. Apoptosis signaling, at least in its present mammalian form, relies not only on caspases but also on numerous six-helix death-fold domains, all of which appear only in metazoans, suggesting that apoptosis is probably the youngest of the PCD systems discussed here. In contrast, necroptosis appears to be ancient, at least the downstream events. Necroptotic cell death is caused by the HeLo domain of MLKL, which is clearly ancient, whereas RHIM motifs are only found in animals and the related PFD of Het-s has not been found outside of fungi. Nevertheless, we recently identified an MLKL pseudokinase in plants, whose amino-terminal domain forms a 4HB similar to that of mammalian MLKL and is also able to cause cell death (Mahdi et al. 2019). Although plants appear to lack RIPK3 homologs and other RHIM-based signaling proteins, this finding suggests that at least the execution step of necroptosis predates the split of animals and plants.

When considering the later evolutionary steps of cell death pathways, there are several examples of intermediate steps being added, probably to allow more regulatory layers, or to

Cite this article as *Cold Spring Harb Perspect Biol* doi: 10.1101/cshperspect.a036442

further amplify the reaction—making PCD induction an "all-or-nothing" decision. One example is the (proposed) conversion of TIR domains from its original catalytic form, thought to exert a direct killing effect, into a signaling domain that recruits further TIR-containing proteins in large numbers. A similar change might have occurred to the HeLo domain, thought to be a direct cell-killing domain in fungi, into a possible signaling domain in plants. On a smaller scale, the addition of regulatory layers is also visible in apoptosis evolution among metazoans. Whereas in nematodes, the Ced-4 apoptosome directly recruits the effector caspase Ced-3, the mammalian APAF1 apoptosome recruits and activates an intermediate caspase (caspase-9), each molecule of which can then activate multiple copies of the mammalian effector caspase-3. It might even be possible that the RHIM motifs and six-helix death-fold domains, in their original form, did have a direct cell-killing effect and only later evolved into signaling and recruitment domains. However, this remains speculative because there are no data to support that idea. Taken together, it becomes clear that PCD pathways across kingdoms use a similar signaling logic and evolved from a common ancestral pathway, probably much simpler than the intricate multilayer systems observed in multicellular organisms.

ACKNOWLEDGMENTS

We thank Shuhua Chen for valuable discussions. This work was funded by a grant from the Deutsche Forschungsgemeinschaft (SFB 670).

REFERENCES

*Reference is also in this collection.

Ameisen JC. 2002. On the origin, evolution, and nature of programmed cell death: a timeline of four billion years. *Cell Death Differ* 9: 367–393. doi:10.1038/sj.cdd.4400950

Ashkenazi A, Salvesen G. 2014. Regulated cell death: signaling and mechanisms. *Annu Rev Cell Dev Biol* 30: 337–356. doi:10.1146/annurev-cellbio-100913-013226

Balint-Kurti P. 2019. The plant hypersensitive response: concepts, control and consequences. *Mol Plant Pathol* 20: 1163–1178.

Bergsbaken T, Fink SL, Cookson BT. 2009. Pyroptosis: host cell death and inflammation. *Nat Rev Microbiol* 7: 99–109. doi:10.1038/nrmicro2070

Bernoux M, Ve T, Williams S, Warren C, Hatters D, Valkov E, Zhang X, Ellis JG, Kobe B, Dodds PN. 2011. Structural and functional analysis of a plant resistance protein TIR domain reveals interfaces for self-association, signaling, and autoregulation. *Cell Host Microbe* 9: 200–211. doi:10.1016/j.chom.2011.02.009

Broz P, Dixit VM. 2016. Inflammasomes: mechanism of assembly, regulation and signalling. *Nat Rev Immunol* 16: 407–420. doi:10.1038/nri.2016.58

Cheng TC, Hong C, Akey IV, Yuan S, Akey CW. 2016. A near atomic structure of the active human apoptosome. *eLife* 5: e17755.

Coll NS, Epple P, Dangl JL. 2011. Programmed cell death in the plant immune system. *Cell Death Differ* 18: 1247–1256. doi:10.1038/cdd.2011.37

Collier SM, Hamel LP, Moffett P. 2011. Cell death mediated by the N-terminal domains of a unique and highly conserved class of NB-LRR protein. *Mol Plant Microbe Interact* 24: 918–931. doi:10.1094/MPMI-03-11-0050

Cookson BT, Brennan MA. 2001. Pro-inflammatory programmed cell death. *Trends Microbiol* 9: 113–114. doi:10.1016/S0966-842X(00)01936-3

Danot O, Marquenet E, Vidal-Ingigliardi D, Richet E. 2009. Wheel of life, wheel of death: a mechanistic insight into signaling by STAND proteins. *Structure* 17: 172–182. doi:10.1016/j.str.2009.01.001

Daskalov A, Habenstein B, Sabaté R, Berbon M, Martinez D, Chaignepain S, Coulary-Salin B, Hofmann K, Loquet A, Saupe SJ. 2016. Identification of a novel cell death-inducing domain reveals that fungal amyloid-controlled programmed cell death is related to necroptosis. *Proc Natl Acad Sci* 113: 2720–2725. doi:10.1073/pnas.1522361113

Daskalov A, Heller J, Herzog S, Fleißner A, Glass NL. 2017. Molecular mechanisms regulating cell fusion and heterokaryon formation in filamentous fungi. *Microbiol Spectr* 5. doi:10.1128/microbiolspec.FUNK-0015-2016

Defourny J, Aghaie A, Perfettini I, Avan P, Delmaghani S, Petit C. 2019. Pejvakin-mediated pexophagy protects auditory hair cells against noise-induced damage. *Proc Natl Acad Sci* 116: 8010–8017. doi:10.1073/pnas.1821844116

Dorstyn L, Akey CW, Kumar S. 2018. New insights into apoptosome structure and function. *Cell Death Differ* 25: 1194–1208. doi:10.1038/s41418-017-0025-z

Edinger AL, Thompson CB. 2004. Death by design: apoptosis, necrosis and autophagy. *Curr Opin Cell Biol* 16: 663–669. doi:10.1016/j.ceb.2004.09.011

Elmore S. 2007. Apoptosis: a review of programmed cell death. *Toxicol Pathol* 35: 495–516. doi:10.1080/01926230701320337

Essuman K, Summers DW, Sasaki Y, Mao X, DiAntonio A, Milbrandt J. 2017. The SARM1 Toll/interleukin-1 receptor domain possesses intrinsic NAD$^+$ cleavage activity that promotes pathological axonal degeneration. *Neuron* 93: 1334–1343.e5. doi:10.1016/j.neuron.2017.02.022

Essuman K, Summers DW, Sasaki Y, Mao X, Yim AKY, DiAntonio A, Milbrandt J. 2018. TIR domain proteins are an ancient family of NAD$^+$-consuming enzymes. *Curr Biol* 28: 421–430.e4. doi:10.1016/j.cub.2017.12.024

Feng S, Fox D, Man SM. 2018. Mechanisms of gasdermin family members in inflammasome signaling and cell death. *J Mol Biol* **430:** 3068–3080. doi:10.1016/j.jmb.2018.07.002

Fuchs Y, Steller H. 2011. Programmed cell death in animal development and disease. *Cell* **147:** 742–758. doi:10.1016/j.cell.2011.10.033

Gonçalves AP, Heller J, Daskalov A, Videira A, Glass NL. 2017. Regulated forms of cell death in fungi. *Front Microbiol* **8:** 1837. doi:10.3389/fmicb.2017.01837

Green DR. 2019. The coming decade of cell death research: five riddles. *Cell* **177:** 1094–1107. doi:10.1016/j.cell.2019.04.024

Hander T, Fernández-Fernández AD, Kumpf RP, Willems P, Schatowitz H, Rombaut D, Staes A, Nolf J, Pottie R, Yao P, et al. 2019. Damage on plants activates Ca^{2+}-dependent metacaspases for release of immunomodulatory peptides. *Science* **363:** eaar7486. doi:10.1126/science.aar7486

Harris SL, Kazmierczak M, Pangršič T, Shah P, Chuchvara N, Barrantes-Freer A, Moser T, Schwander M. 2017. Conditional deletion of pejvakin in adult outer hair cells causes progressive hearing loss in mice. *Neuroscience* **344:** 380–393. doi:10.1016/j.neuroscience.2016.12.055

Hauenstein AV, Zhang L, Wu H. 2015. The hierarchical structural architecture of inflammasomes, supramolecular inflammatory machines. *Curr Opin Struct Biol* **31:** 75–83. doi:10.1016/j.sbi.2015.03.014

Hofmann K. 1999. The modular nature of apoptotic signaling proteins. *Cell Mol Life Sci* **55:** 1113–1128. doi:10.1007/s000180050361

Huang W, Jiang T, Choi W, Qi S, Pang Y, Hu Q, Xu Y, Gong X, Jeffrey PD, Wang J, et al. 2013. Mechanistic insights into CED-4-mediated activation of CED-3. *Genes Dev* **27:** 2039–2048. doi:10.1101/gad.224428.113

Jiang S, Gu H, Zhao Y, Sun L. 2019. Teleost gasdermin E is cleaved by caspase 1, 3, and 7 and induces pyroptosis. *J Immunol* **203:** 1369–1382. doi:10.4049/jimmunol.1900383

Jones JD, Vance RE, Dangl JL. 2016. Intracellular innate immune surveillance devices in plants and animals. *Science* **354:** aaf6395. doi:10.1126/science.354.6316.1174-b

Kajava AV, Klopffleisch K, Chen S, Hofmann K. 2014. Evolutionary link between metazoan RHIM motif and prion-forming domain of fungal heterokaryon incompatibility factor HET-s/HET-s. *Sci Rep* **4:** 7436. doi:10.1038/srep07436

Kambara H, Liu F, Zhang X, Liu P, Bajrami B, Teng Y, Zhao L, Zhou S, Yu H, Zhou W, et al. 2018. Gasdermin D exerts anti-inflammatory effects by promoting neutrophil death. *Cell Rep* **22:** 2924–2936. doi:10.1016/j.celrep.2018.02.067

Kersse K, Verspurten J, Vanden Berghe T, Vandenabeele P. 2011. The death-fold superfamily of homotypic interaction motifs. *Trends Biochem Sci* **36:** 541–552. doi:10.1016/j.tibs.2011.06.006

Koonin EV, Aravind L. 2002. Origin and evolution of eukaryotic apoptosis: the bacterial connection. *Cell Death Differ* **9:** 394–404. doi:10.1038/sj.cdd.4400991

Kovacs SB, Miao EA. 2017. Gasdermins: effectors of pyroptosis. *Trends Cell Biol* **27:** 673–684. doi:10.1016/j.tcb.2017.05.005

Lamkanfi M, Dixit VM. 2010. Manipulation of host cell death pathways during microbial infections. *Cell Host Microbe* **8:** 44–54. doi:10.1016/j.chom.2010.06.007

Lapin D, Kovacova V, Sun X, Dongus JA, Bhandari DD, von Born P, Bautor J, Guarneri N, Rzemieniewski J, Stuttmann J, et al. 2019. A coevolved EDS1-SAG101-NRG1 module mediates cell death signaling by TIR-domain immune receptors. *Plant Cell*. doi:10.1105/tpc.19.00118

Leipe DD, Koonin EV, Aravind L. 2004. STAND, a class of P-loop NTPases including animal and plant regulators of programmed cell death: multiple, complex domain architectures, unusual phyletic patterns, and evolution by horizontal gene transfer. *J Mol Biol* **343:** 1–28. doi:10.1016/j.jmb.2004.08.023

Li J, McQuade T, Siemer AB, Napetschnig J, Moriwaki K, Hsiao YS, Damko E, Moquin D, Walz T, McDermott A, et al. 2012. The RIP1/RIP3 necrosome forms a functional amyloid signaling complex required for programmed necrosis. *Cell* **150:** 339–350. doi:10.1016/j.cell.2012.06.019

Maekawa T, Kufer TA, Schulze-Lefert P. 2011. NLR functions in plant and animal immune systems: so far and yet so close. *Nat Immunol* **12:** 817–826. doi:10.1038/ni.2083

Mahdi L, Huang M, Zhang X, Nakano RT, Kopp LB, Saur IML, Jacob F, Kovacova V, Lapin D, Parker JE, et al. 2019. Plant mixed lineage kinase domain-like proteins limit biotrophic pathogen growth. bioRxiv 681015. doi:10.1101/681015

Man SM, Karki R, Kanneganti TD. 2017. Molecular mechanisms and functions of pyroptosis, inflammatory caspases and inflammasomes in infectious diseases. *Immunol Rev* **277:** 61–75. doi:10.1111/imr.12534

Martin SJ, Green DR. 1995. Protease activation during apoptosis: death by a thousand cuts? *Cell* **82:** 349–352. doi:10.1016/0092-8674(95)90422-0

Mompeán M, Li W, Li J, Laage S, Siemer AB, Bozkurt G, Wu H, McDermott AE. 2018. The structure of the necrosome RIPK1-RIPK3 core, a human hetero-amyloid signaling complex. *Cell* **173:** 1244–1253.e10. doi:10.1016/j.cell.2018.03.032

Morel JB, Dangl JL. 1997. The hypersensitive response and the induction of cell death in plants. *Cell Death Differ* **4:** 671–683. doi:10.1038/sj.cdd.4400309

Nagata S. 2018. Apoptosis and clearance of apoptotic cells. *Annu Rev Immunol* **36:** 489–517. doi:10.1146/annurev-immunol-042617-053010

Nanson JD, Kobe B, Ve T. 2019. Death, TIR, and RHIM: self-assembling domains involved in innate immunity and cell-death signaling. *J Leukoc Biol* **105:** 363–375. doi:10.1002/JLB.MR0318-123R

Newton K, Manning G. 2016. Necroptosis and inflammation. *Annu Rev Biochem* **85:** 743–763. doi:10.1146/annurev-biochem-060815-014830

Orning P, Weng D, Starheim K, Ratner D, Best Z, Lee B, Brooks A, Xia S, Wu H, Kelliher MA, et al. 2018. Pathogen blockade of TAK1 triggers caspase-8-dependent cleavage of gasdermin D and cell death. *Science* **362:** 1064–1069. doi:10.1126/science.aau2818

Paoletti M, Clavé C. 2007. The fungus-specific HET domain mediates programmed cell death in *Podospora anserina*. *Eukaryot Cell* **6:** 2001–2008. doi:10.1128/EC.00129-07

Park HH, Lo YC, Lin SC, Wang L, Yang JK, Wu H. 2007a. The death domain superfamily in intracellular signaling of apoptosis and inflammation. *Annu Rev Immunol* **25**: 561–586. doi:10.1146/annurev.immunol.25.022106.141656

Park HH, Logette E, Raunser S, Cuenin S, Walz T, Tschopp J, Wu H. 2007b. Death domain assembly mechanism revealed by crystal structure of the oligomeric PIDDosome core complex. *Cell* **128**: 533–546. doi:10.1016/j.cell.2007.01.019

Petrie EJ, Czabotar PE, Murphy JM. 2019. The structural basis of necroptotic cell death signaling. *Trends Biochem Sci* **44**: 53–63. doi:10.1016/j.tibs.2018.11.002

Qi S, Pang Y, Hu Q, Liu Q, Li H, Zhou Y, He T, Liang Q, Liu Y, Yuan X, et al. 2010. Crystal structure of the *Caenorhabditis elegans* apoptosome reveals an octameric assembly of CED-4. *Cell* **141**: 446–457. doi:10.1016/j.cell.2010.03.017

Riek R, Saupe SJ. 2016. The HET-S/s prion motif in the control of programmed cell death. *Cold Spring Harb Perspect Biol* **8**: a023515. doi:10.1101/cshperspect.a023515

Roh JS, Sohn DH. 2018. Damage-associated molecular patterns in inflammatory diseases. *Immune Netw* **18**: e27. doi:10.4110/in.2018.18.e27

Ruan J, Xia S, Liu X, Lieberman J, Wu H. 2018. Cryo-EM structure of the gasdermin A3 membrane pore. *Nature* **557**: 62–67. doi:10.1038/s41586-018-0058-6

Sarhan J, Liu BC, Muendlein HI, Li P, Nilson R, Tang AY, Rongvaux A, Bunnell SC, Shao F, Green DR, et al. 2018. Caspase-8 induces cleavage of gasdermin D to elicit pyroptosis during *Yersinia* infection. *Proc Natl Acad Sci* **115**: E10888–E10897. doi:10.1073/pnas.1809542115

Saupe SJ. 2000. Molecular genetics of heterokaryon incompatibility in filamentous ascomycetes. *Microbiol Mol Biol Rev* **64**: 489–502. doi:10.1128/MMBR.64.3.489-502.2000

Schaefer L. 2014. Complexity of danger: the diverse nature of damage-associated molecular patterns. *J Biol Chem* **289**: 35237–35245. doi:10.1074/jbc.R114.619304

Seuring C, Greenwald J, Wasmer C, Wepf R, Saupe SJ, Meier BH, Riek R. 2012. The mechanism of toxicity in HET-S/HET-s prion incompatibility. *PLoS Biol* **10**: e1001451. doi:10.1371/journal.pbio.1001451

Shi J, Gao W, Shao F. 2017. Pyroptosis: gasdermin-mediated programmed necrotic cell death. *Trends Biochem Sci* **42**: 245–254. doi:10.1016/j.tibs.2016.10.004

Thome M, Schneider P, Hofmann K, Fickenscher H, Meinl E, Neipel F, Mattmann C, Burns K, Bodmer JL, Schröter M, et al. 1997. Viral FLICE-inhibitory proteins (FLIPs) prevent apoptosis induced by death receptors. *Nature* **386**: 517–521. doi:10.1038/386517a0

Tsiatsiani L, Van Breusegem F, Gallois P, Zavialov A, Lam E, Bozhkov PV. 2011. Metacaspases. *Cell Death Differ* **18**: 1279–1288. doi:10.1038/cdd.2011.66

Urbach JM, Ausubel FM. 2017. The NBS-LRR architectures of plant R-proteins and metazoan NLRs evolved in independent events. *Proc Natl Acad Sci* **114**: 1063–1068. doi:10.1073/pnas.1619730114

Váchová L, Palková Z. 2007. Caspases in yeast apoptosis-like death: facts and artefacts. *FEMS Yeast Res* **7**: 12–21. doi:10.1111/j.1567-1364.2006.00137.x

Vandenabeele P, Galluzzi L, Vanden Berghe T, Kroemer G. 2010. Molecular mechanisms of necroptosis: an ordered cellular explosion. *Nat Rev Mol Cell Biol* **11**: 700–714. doi:10.1038/nrm2970

Ve T, Vajjhala PR, Hedger A, Croll T, DiMaio F, Horsefield S, Yu X, Lavrencic P, Hassan Z, Morgan GP, et al. 2017. Structural basis of TIR-domain-assembly formation in MAL- and MyD88-dependent TLR4 signaling. *Nat Struct Mol Biol* **24**: 743–751. doi:10.1038/nsmb.3444

Wang Y, Gao W, Shi X, Ding J, Liu W, He H, Wang K, Shao F. 2017. Chemotherapy drugs induce pyroptosis through caspase-3 cleavage of a gasdermin. *Nature* **547**: 99–103. doi:10.1038/nature22393

Wang J, Hu M, Wang J, Qi J, Han Z, Wang G, Qi Y, Wang HW, Zhou JM, Chai J. 2019. Reconstitution and structure of a plant NLR resistosome conferring immunity. *Science* **364**: eaav5870.

Wasmer C, Lange A, Van Melckebeke H, Siemer AB, Riek R, Meier BH. 2008. Amyloid fibrils of the HET-s (218–289) prion form a β solenoid with a triangular hydrophobic core. *Science* **319**: 1523–1526. doi:10.1126/science.1151839

Weinlich R, Oberst A, Beere HM, Green DR. 2017. Necroptosis in development, inflammation and disease. *Nat Rev Mol Cell Biol* **18**: 127–136. doi:10.1038/nrm.2016.149

Williams SJ, Sohn KH, Wan L, Bernoux M, Sarris PF, Segonzac C, Ve T, Ma Y, Saucet SB, Ericsson DJ, et al. 2014. Structural basis for assembly and function of a heterodimeric plant immune receptor. *Science* **344**: 299–303. doi:10.1126/science.1247357

* Xia S, Hollingsworth LR IV, Wu H. 2019. Mechanism and regulation of gasdermin-mediated cell death. *Cold Spring Harb Perspect Biol* doi:10.1101/cshperspect.a036400

Zhang X, Bernoux M, Bentham AR, Newman TE, Ve T, Casey LW, Raaymakers TM, Hu J, Croll TI, Schreiber KJ, et al. 2017. Multiple functional self-association interfaces in plant TIR domains. *Proc Natl Acad Sci* **114**: E2046–E2052. doi:10.1073/pnas.1621248114

BAX, BAK, and BOK: A Coming of Age for the BCL-2 Family Effector Proteins

Tudor Moldoveanu[1,2] and Peter E. Czabotar[3,4]

[1]Department of Structural Biology, St. Jude Children's Research Hospital, Memphis, Tennessee 38105, USA

[2]Department of Chemical Biology and Therapeutics, St. Jude Children's Research Hospital, Memphis Tennessee 38105, USA

[3]Walter and Eliza Hall Institute of Medical Research, University of Melbourne, Parkville, Victoria 3052, Australia

[4]Department of Medical Biology, University of Melbourne, Parkville, Victoria 3010, Australia

Correspondence: tudor.moldoveanu@stjude.org; czabotar@wehi.edu.au

The BCL-2 family of proteins control a key checkpoint in apoptosis, that of mitochondrial outer membrane permeabilization or, simply, mitochondrial poration. The family consists of three subgroups: BH3-only initiators that respond to apoptotic stimuli; antiapoptotic guardians that protect against cell death; and the membrane permeabilizing effectors BAX, BAK, and BOK. On activation, effector proteins are converted from inert monomers into membrane permeabilizing oligomers. For many years, this process has been poorly understood at the molecular level, but a number of recent advances have provided important insights. We review the regulation of these effectors, their activation, subsequent conformational changes, and the ensuing oligomerization events that enable mitochondrial poration, which initiates apoptosis through release of key signaling factors such as cytochrome *c*. We highlight the mysteries that remain in understanding these important proteins in an endeavor to provide a comprehensive picture of where the field currently sits and where it is moving toward.

AWAKENING THE BCL-2 EFFECTORS IN APOPTOSIS

Commitment to mitochondrial apoptosis involves activation of effectors, which are uniquely able to directly mediate mitochondrial poration (Fig. 1; Czabotar et al. 2014; Moldoveanu et al. 2014). In most cells, the canonical effectors BAX and BAK are typically expressed in stable dormant conformations in the absence of overt cellular stress (e.g., chemotherapeutic insults, nutrient deprivation, growth factor with-drawal). BAK and BAX have distinct mechanisms of regulation, reflected primarily in their respective steady-state localization at the mitochondrial outer membrane (MOM) and in the cytosol in nonapoptotic live cells. Cellular stress triggers apoptotic stimuli to awaken or activate the effectors, which relocalize and change shape to execute mitochondrial poration thereby initiating apoptosis (Fig. 1). The underlying mechanism of mitochondrial poration by effectors is incompletely defined and debatable as discussed below.

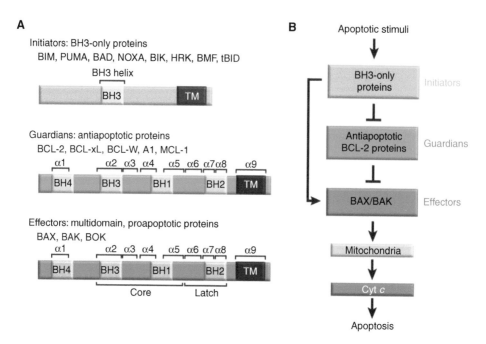

Figure 1. The BCL-2 family and the intrinsic pathway to apoptosis. (*A*) Family members of the BCL-2 protein family. The family is made up of three subgroups of proteins related to each other by regions of sequence homology, the so-called Bcl-2 homology (BH) domains. Regions of secondary structure and domains discussed in the text are labeled. (*B*) BH3-only proteins, which generally only possess the BH3 domain, are up-regulated on apoptotic stimuli to initiate signaling of the pathway. BH3-only proteins interact with both the effectors BAX and BAK, and the antiapoptotic guardians. Guardians can protect against apoptosis by sequestering both the BH3-only proteins, thus inhibiting effector activation, and by neutralizing activated effector proteins directly. If freed, activated effectors oligomerize at the mitochondrial outer membrane leading to permeabilization of this barrier. This enables the release of apoptogenic factors into the cytosol, primarily cytochrome *c* (cyt *c*), leading to caspase activation and ensuing apoptosis. Emerging evidence indicates that BOK is a third member of the effector subgroup with alternative mechanisms of regulation (discussed later).

Steady-State BAX on a Leash in the Cytosol

Association of BCL-2 family proteins with mitochondria is largely governed by their carboxy-terminal tails. Tail deletion abolishes mitochondrial localization and severely impacts apoptotic function (Ferrer et al. 2012; Llambi et al. 2016). Although not formally shown, the tails most likely form transmembrane (TM) anchoring helices or may involve association via hydrophobic and hydrophilic interactions with the outer leaflet of the MOM without traversing the bilayer (e.g., as proposed for the BIM carboxy-terminal tail) (Liu et al. 2019). BAX is the only folded BCL-2 family protein that preferentially accommodates its carboxy-terminal TM tail within the canonical hydrophobic groove,

protecting the hydrophobic character of this helix and thus enabling the protein to remain soluble in the cytoplasm (Table 1; Fig. 2A; Suzuki et al. 2000; Garner et al. 2016; Robin et al. 2018).

It has been reported that cytosolic BAX exists in an equilibrium between monomer and dimer with an estimated K_D of ~30 μM determined by estimating dissociation of BAX dimers into monomers over time from size-exclusion chromatography (Garner et al. 2016). Based on probing of BAX in solution by hydrogen deuterium exchange mass spectrometry and crystal structures, it has been proposed that BAX dimers form by burying the putative regulatory interfaces (canonical groove and rear site) thought to be involved in activation (Fig. 2B; Garner et al. 2016, vide infra). This configuration was

Cite this article as *Cold Spring Harb Perspect Biol* doi: 10.1101/cshperspect.a036319

Table 1. Summary of effector structures

Protein complex	PDB	References
Full-length hBAX	1F16	Suzuki et al. 2000
ΔNhBAKΔTM	2IMS	Moldoveanu et al. 2006
BIM SAHB: Full-length hBAX	2K7W	Gavathiotis et al. 2008
vMIA: Full-length hBAX	2LR1	Ma et al. 2012
BID BH3:hBAX	4BD2	Czabotar et al. 2013
BAX BH3:hBAX	4BD6	Czabotar et al. 2013
Octylmaltoside-induced hBAX dimer (Apo)	4BD7	Czabotar et al. 2013
BIM BH3-induced hBAX dimer (Apo)	4BD8	Czabotar et al. 2013
hBAX BH3-in-groove dimer (GFP)	4BDU	Czabotar et al. 2013
BID SAHB:ΔNhBAKΔTM	2M5B	Moldoveanu et al. 2013
hBAK core/latch dimer	4U2U	Brouwer et al. 2014
GFP-hBAK(α2–α5) (core domain dimer)	4U2V	Brouwer et al. 2014
BIM BH3:hBAXΔTM	4ZIE	Robin et al. 2015
BIM BH3mini:hBAXΔTM	4ZIF	Robin et al. 2015
BID BH3mini:hBAXΔTM	4ZIG	Robin et al. 2015
BIM BH3mini:hBAXΔTM	4ZIH	Robin et al. 2015
BID BH3:hBAXI66AΔTM	4ZII	Robin et al. 2015
Full-length hBAX P168G	4S0O	Garner et al. 2016
Full-length hBAX G67R	4S0P	Garner et al. 2016
BIM-RT:hBAKΔTM core/latch dimer	5VWV	Brouwer et al. 2017
BIM-RT:hBAKΔTM core/latch dimer	5VWW	Brouwer et al. 2017
BIM-h3Pc:hBAK monomer	5VWZ	Brouwer et al. 2017
BIM-h3Pc-RT:hBAK core/latch dimer	5VWY	Brouwer et al. 2017
chBOK	5WDD	Ke et al. 2018
hBOK	6CKV	Zheng et al. 2018
Full-length hBAX P168G	5W60	Robin et al. 2018
Full-length hBAX P168G	5W61	Robin et al. 2018
Full-length mBAX	5W62	Robin et al. 2018
3C10Ab: full-length hBAX P168G	5W5X	Robin et al. 2018

interpreted as an autoinhibitory mechanism whereby the dimer is resistant to activation by BH3-only proteins. BAX mutagenesis to break down the autoinhibitory dimer augmented membrane poration and apoptosis (Garner et al. 2016). In contrast, a recent study observed BAX monomers, not dimers in cells, and interpreted the BAX configuration seen in the crystal structure as a crystal-packing artifact (Dengler et al. 2019). Another study also found Bax to be monomeric in cells when solubilized in 1% CHAPS detergent (Dai et al. 2015). Some estimates of BAX cellular levels are lower than the reported dimerization K_D (e.g., in overexpression systems at ∼3 μM) (Dussmann et al. 2010), and in several tumor cell lines at ∼20,000–70,000 BAX molecules per cell (∼17–60 nM for average cell volume of ∼2000 μm³) (Dai et al. 2018). How do we recon-

cile these conflicting reports? One possible explanation is that cytosolic BAX dimers could be stabilized by additional mechanisms, perhaps through posttranslational modifications. Further studies may shed light on this phenomenon.

Dynamics of BAX Association with the Mitochondria

BAX exists in a dynamic equilibrium shuttling between the mitochondria and cytosol, a process termed retrotranslocation (Edlich et al. 2011). In nonapoptotic cells, this equilibrium is weighted such that BAX is primarily cytosolic. Retrotranslocation is thought to occur through interactions between BAX and antiapoptotic BCL-2 proteins such as BCL-xL, BCL-2, and MCL-1, with the TM and hydrophobic grooves of the

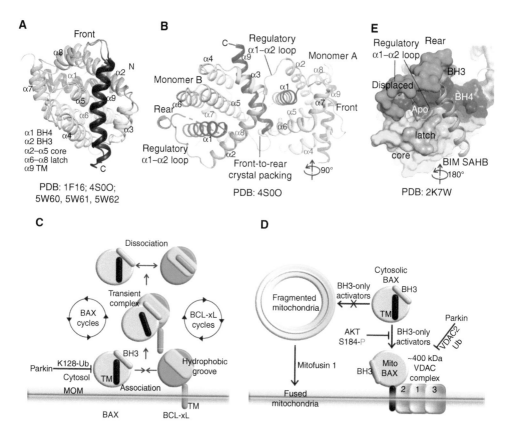

Figure 2. Regulated recruitment of cytosolic BAX to mitochondria. (*A*) The nuclear magnetic resonance and crystal structures of wild-type (WT) and P168G mutant full-length hBAX (FL BAX) revealed the dormant conformation of cytosolic BAX, with the carboxy-terminal transmembrane (TM) helix α9 bound to and occluding the BAX canonical hydrophobic groove (*front* view). The same colors have been used in structural figures throughout for different secondary structure elements and regions. PDB identifiers are included (see Table 1 for a summary of effector PDBs). (*B*) The crystal packing dimer of BAX P168G mutant. The proposed regulatory α1–α2 loop is highly dynamic and found at the rear site. (*C*) Schematic of retrotranslocation and possible inhibition of mitochondrial BAX by parkin ubiquitylation (Ub). (*D*) Mitochondrial recruitment is also thought to be regulated by mitochondrial dynamics and by VDAC complex via VDAC2 interactions. Parkin ubiquitylation of VDAC2 blocks BAX recruitment. BAX phosphorylation by AKT at position S184 blocked its mitochondrial association. (*E*) Chemically stapled BIM, BH3-stabilized α helix of BCL-2 protein (SAHB) has been proposed to displace the regulatory α1–α2 loop at the rear to induce allosteric changes that ultimately destabilized the TM interaction at the front activation groove to drive targeting of BAX to the mitochondria via TM (see *D*).

latter interacting with the BH3 region and TM of the former (Edlich et al. 2011; Todt et al. 2013). This process has not been elucidated structurally, but it appears that antiapoptotic BCL-2 proteins may act as chaperones of BAX, constantly sampling BAX conformations that promote mitochondrial association such as exposure of amino-terminal activation epitopes (6A7 is the antibody that recognizes one such epitope), carboxy-terminal TM association with MOM, and

BH3 exposure (Edlich et al. 2011; Todt et al. 2013). Reversal of mitochondrial association by retrotranslocation is thought to occur through transient formation of BAX:antiapoptotic BCL-2 protein complexes that shuttle back into the cytosol before dissociating (Fig. 2C). Remarkably, BAX retrotranslocation appears to occur even if BAX is membrane integral, presumably with the carboxy-terminal TM tail traversing the bilayer (Lauterwasser et al. 2016). It has also been proposed that BAK undergoes a similar shuttling mechanism, although in this case with the equilibrium pushed heavily toward membrane association (Todt et al. 2015). Mitochondrial association also seems to be dependent on mitochondrial morphology, as hyperfragmented mitochondria fail to support BAX association and mitochondrial poration (Renault et al. 2015). This failure has been attributed to the inability of BAX TM tails to interact with fragmented membranes and GTPase mitofusin 1 has been reported to counteract this by mediating mitochondrial fusion (Fig. 2D; Renault et al. 2015).

Several studies have also revealed a role for VDAC2 in the recruitment of BAX to mitochondria (Ma et al. 2014; Lauterwasser et al. 2016; Cakir et al. 2017; Chin et al. 2018; Bernardini et al. 2019). The size of digitonin-solubilized complexes of BAX and VDACs on native PAGE are estimated to be >400 kDa. In biochemical and genetic studies, VDAC2 deletion inhibited the mitochondrial localization and apoptotic function of BAX. Surprisingly, although BAK is also found in a similar high-molecular weight complex that includes VDAC2 (Lazarou et al. 2010; Chin et al. 2018), BAK-mediated apoptosis is not significantly impaired by VDAC2 deletion, but is actually potentiated (Cheng et al. 2003). Structures for these effector:VDAC complexes have not yet been solved, but once these details are understood it is likely that they will provide new strategies for modulating mitochondrial poration and apoptosis.

The Rear of BAX and Mitochondrial Association

The nuclear magnetic resonance (NMR) structure of human full-length BAX predicted that BH3 activators would be unable to displace the TM (helix α9) from the groove in the absence of an energy-driven triggering process to disengage the α9 (Suzuki et al. 2000). In the absence of additional energy, we now know that addition of BH3 activator peptides to full-length BAX induces large aggregates in solution over time (Sung et al. 2015; Garner et al. 2016). These aggregates are readily able to porate membranes (Sung et al. 2015; Garner et al. 2016). We do not know the structure of BAX in this aggregate, but it likely involves a dynamic heterogeneous interaction web mediated by intermolecular associations of displaced hydrophobic TM regions. Similar aggregates are not seen with BAX-ΔTM and BH3 activator peptides, which form 1:1 complexes amenable to structural characterization (Table 1, vide infra). These observations indirectly suggest that BH3 activators are able to displace the TM helix from the groove in solution.

A major point of contention in BAX activation over the past decade has centered on revealing how the TM is displaced from the groove. Allostery has been implicated in this process through ensembles of Bax conformers within the cytosol (Robin et al. 2018; Dengler et al. 2019). To discover allosteric mechanisms for TM displacement from the groove by BH3 domains, chemically stapled BIM BH3 has been deployed in structure–function studies (Gavathiotis et al. 2008, 2010), and chemically stapled BH3 peptides from BID and PUMA were designed to investigate BAX and BAK activation (Edwards et al. 2013; Leshchiner et al. 2013). These low-resolution studies by NMR and photoaffinity labeling mass spectrometry, suggested engagement by BH3 activators to a rear allosteric activation site on BAX (but not BAK) (Fig. 2E). Although the binding of unstapled BH3 activators to BAX is undetectable, stabilization through chemical stapling increased the affinity for the rear allosteric site (K_D in µM) (Walensky et al. 2006). Unfortunately, the low-resolution model of the rear activation interface has been refractory to surface probing by site-directed mutagenesis, as none of the substitutions tested have been able to block BAX activation in multiple laboratories (Okamoto et al. 2013; Peng et al. 2013; Garner et al. 2016; Dengler et al. 2019). For instance, mutagenesis to block autoinhibitory BAX dimers, which should impact BAX activation at the rear site directly, revealed a counterintuitive enhancement in BAX apoptotic activity compared with wild-type (WT) (Garner et al. 2016). It is possible that the models for BH3 activation at the rear site are incomplete, that this interface does not matter significantly for BAX activation, or that the staple itself is facilitating some of these events. From our experience, stapled peptides are most useful when they mimic all of the functional features observed with the unstapled counterparts while still offering enhanced helicity and affinity for the target (Moldoveanu et al. 2013). TM-deleted BAX and full-length BAK show significant interactions with activator BH3 peptides at the canonical groove, which is their main binding site common to both effectors (vide infra) (Edwards et al. 2013; Leshchiner et al. 2013). Future mechanistic high-resolution studies exploring complexes of BAX and BAK with full-length BH3-only proteins, which may show additional regulatory interaction interfaces as recently proposed for full-length BIM and BAX (Chi et al. 2019) and tBID and BAK (Dengler et al. 2019), may resolve these issues.

Getting in (and out of) the Groove—"Hit-and-Run" Effector Unleashing

A better characterized mechanism of BAK and BAX activation involves canonical engagement of their groove by BH3 activators, such as BID, BIM, and PUMA. Because the affinity of BH3 activator peptides for BAK and BAX is relatively low ($K_D \sim \mu M$), structural analysis by NMR and X-ray crystallography has lagged behind that of the much tighter complexes occurring between BH3 peptides and antiapoptotic BCL-2 proteins ($K_D \sim nM$) (Suzuki et al. 2000; Moldoveanu et al. 2006, 2013; Czabotar et al. 2013). The effector activation "hit-and-run" model is based on the weak affinities of these interactions, whereby direct activators bind transiently and are thus not found in stable complexes with effectors, especially in cell-based pull-down or gel-filtration chromatography assays in which protein levels are below the interaction K_D (Wei et al. 2000). In retrospect, it is not surprising that effectors were not found bound to direct activators in these early studies.

To this end, three breakthroughs toward stabilization of effector:activator complexes have been made to achieve their high-resolution structures: (1) domain-swapped dimers of BAX-ΔTM and BAK-ΔTM, which are more stable than monomers and crystallize more readily

in complex with activators (Czabotar et al. 2013; Robin et al. 2015; Brouwer et al. 2017); (2) chemically stabilized activator peptides with enhanced helicity and affinity for effectors (Moldoveanu et al. 2013); and (3) mutant activator peptides with enhanced solubility and affinity (Robin et al. 2015; Brouwer et al. 2017). These efforts have generated 11 high-resolution structures of effector:activator complexes (Table 1). These structures have captured BAK and BAX in several groove-destabilized conformations, whose intramolecular contacts throughout the groove are rearranged or disrupted relative to apo structures (Fig. 3A). Similar to their complexes with antiapoptotic BCL-2 proteins, these structures have revealed a six-turn activator helix bound via six to eight hydrophobic residues to hydrophobic pockets in effector grooves, stabilized by a conserved arginine-aspartate salt bridge (Fig. 3B). Extensive activator mutagenesis corroborated these structures and has been adopted to turn BH3-only sensitizers such as BAD into direct effector activators (Czabotar et al. 2013; Moldoveanu et al. 2013). Biochemical studies have suggested that most BH3 peptides are able to activate BAX and BAK (Kuwana et al. 2002, 2005; Kim et al. 2009; Du et al. 2011; Chen et al. 2015; Hockings et al. 2015), and sequence alignments in light of the structural studies corroborate that most BH3 peptides have some conserved features allowing them to partly engage canonical grooves (Fig. 3C). How full-length BH3-only proteins contribute to effector activation remains a pressing question. A recent biochemical study suggests that there are contributions from the carboxy-terminal mitochondrial targeting sequence of BIM in BAX activation (Chi et al. 2019).

Gluing Inhibitors at the Groove of BAK

Two studies revealed how the canonical activation groove can also act as a site of inhibition for BAK. The original study showed biochemically that when the BID BH3 carboxyl terminus was cross-linked to the carboxyl terminus of BAK-ΔTM via engineered cysteines, the corresponding complex was intrinsically inactive, yet it was activatable through competition with excess-free

Cite this article as *Cold Spring Harb Perspect Biol* doi: 10.1101/cshperspect.a036319

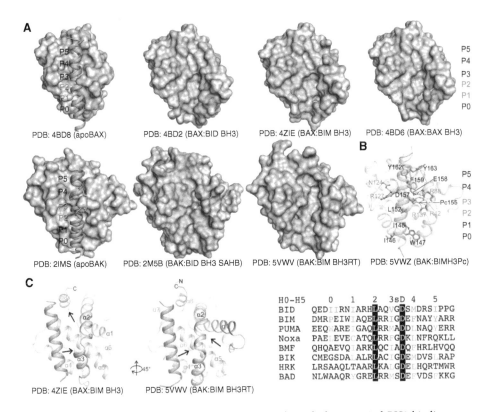

Figure 3. Structural basis of BAX and BAK direct activation through the canonical BH3 binding groove and selective human BAK inhibition. (*A*) BH3 peptide complexes with BAX (*top*) and BAK (*bottom*), aligned on apoBAK, reveal the overall similar arrangement of BH3 helices at the activation grooves. Six to eight hydrophobic residues of the BH3 peptides engage six hydrophobic pockets (numbered P0–P5). BAX and BAK grooves are occluded in apo forms at P1–P2, and P2–P3, respectively. Chemical stapling induced a slight shift of the P0–P3 portion of the BID BH3 helix at the activation groove of BAK. BH3 binding involves opening of the occluded grooves in BAX and BAK and formation of deep pockets not engaged by peptide residues (van der Walls contacts 4 Å from the nearest peptide atom) (orange). (*B*) Structure-guided design of BIM BH3-based molecular glue that locks BAK in an inactive conformation similar to apoBAK by introducing stabilizing salt bridges that block helix α1 release. (*C*) BAX and BAK engage BH3 peptides similarly at the canonical groove yet they show different peptide-induced changes, which manifest by formation of engorged cavities at the regions indicated by the arrows. Destabilized regions in BAX are at the groove region near helix α3 and α8, and in BAK are on either side of the carboxyl terminus of helix α2 and the amino terminus of helix α3. Rotation is relative to the *front* view. BH3 peptide alignments indicate the position of the hydrophobic residues (H0–H5) that make contact with hydrophobic pockets (P0–P5) in BAX and BAK. H0–H5 residues in BID and BIM (red) were deduced from structures of respective peptides in complex with BAX and BAK. Purple residues are hydrophobic residues found at the putative positions in other BH3 peptides expected to more weakly activate effectors compared with BH3 peptides from BID and BIM. A conserved BH3 aspartic acid (black highlight) forms a stabilizing salt bridge with a conserved arginine in the groove of BAX and BAK (*B*). A small residue (glycine or alanine, labeled s), allows tight contact between the BH3 helix and the groove next to the conserved salt bridge.

BID BH3 peptide (Moldoveanu et al. 2013). More recently, structure-guided engineering of BIM-like BH3 peptides selective for human BAK provided a strategy for BAK inhibition. This study rationally introduced nonnatural acidic amino acid residues at the BIM H3 position to interact with two buried basic residues deep in the P3 pocket of BAK (Brouwer et al. 2017). Although the K_D of these inhibitory complexes range from μM to nM, the peptides show molecular glue characteristics through stabilizing contacts that keep BAK inactive by preventing release of helix α1

(Fig. 3B). This study also shows that replacement of H0 amino acids from BIM-like to BID-like residues confers a significant boost in affinity (K_D ~1 µM to ~20 nM) indicating major contributions to stabilizing interactions at this position (Fig. 3B,C). Based on the original study, modifications in BH3 peptides to enhance affinity for effectors may be applicable more broadly to effector inhibition (Moldoveanu et al. 2013), and similar design strategies may be adopted in groove-targeted effector inhibitory small-molecule chemical probes (Moldoveanu et al. 2013; Brouwer et al. 2017).

HOW ACTIVE EFFECTORS EXECUTE MITOCHONDRIAL PORATION

From Monomers to Dimers, the Particle on Which the Oligomer Builds

Once activated and at the MOM, Bax and Bak go through a series of conformational changes that involve the release of their α1 helix (Hsu and Youle 1998; Llambi et al. 2011), partial exposure

of their BH3 domain (Dewson et al. 2008, 2012; Llambi et al. 2011) and disengagement of core (α2–α5) from latch (α6–α8) regions (Fig. 4A,B; Czabotar et al. 2013; Brouwer et al. 2014). The order of these events remains unclear, and some or all may occur simultaneously. Antiapoptotic guardians, if not already complexed with BH3-only relatives, can neutralize the activated forms of BAX and BAK at this step by binding to their exposed BH3 domains to prevent homodimerization (Oltvai et al. 1993; Wang et al. 1998; Fletcher et al. 2008; Llambi et al. 2011). Increasing concentrations of competing BH3-only proteins, decreasing concentrations of prosurvival proteins (e.g., by degradation) (Willis et al. 2005; Czabotar et al. 2007), or increasing concentrations of activated BAX and/or BAK lead to the formation of homodimers, or possibly in some settings BAX/BAK heterodimers (Dewson et al. 2012).

Homodimer formation is a key step in the process to oligomerization and pore formation. Early cross-linking studies showed that these

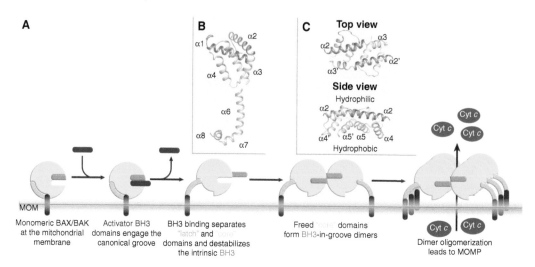

Figure 4. Structural transitions during BAX and BAK activation. (*A*) Activation of monomeric BAX and BAK leads to a series of structural transitions eventuating in dimerization, oligomerization, and ultimately pore formation. Of this sequence, high-resolution structural details are available for monomeric forms of the proteins (Fig. 2A), BH3 activation at the canonical groove (Fig. 3), core from latch detachment (*B*), and core domain dimers as in *C*. (*B*) One of the structurally characterized conformation changes is disengagement of the core domain from the latch region. (*C*) Once released, core domains dimerize through a symmetric BH3-in-groove protein–protein interface. The dimer produced has a hydrophilic surface dominated by α2–α3 and a hydrophobic surface lined by α4–α5. The hydrophobic surface is thought to engage the outer mitochondrial membrane as a step toward permeabilization. Structures shown throughout are for BAX, but similar transitions have been shown for BAK as indicated in the text.

formed through insertion of the BH3 of one monomer into the groove of a neighboring molecule (Dewson et al. 2008, 2012; Bleicken et al. 2010; Zhang et al. 2010). Subsequent crystallographic studies revealed the structure of these symmetric BH3-in-groove dimers (Czabotar et al. 2013; Brouwer et al. 2014), later also confirmed through biophysical proximity measurements (Bleicken et al. 2014; Mandal et al. 2016). A consequence of this transition from monomer to dimer is the conversion from a globular entity with a largely hydrophilic surface to a platform-like structure, which is amphipathic in nature (Fig. 4C). It has been proposed that the hydrophobic surface of this platform is responsible for interacting with lipids of the outer mitochondrial membrane surface as a step toward permeabilization (Czabotar et al. 2013; Brouwer et al. 2014) and/or to line the pore of the lumen subsequently formed (Bleicken et al. 2014; Mandal et al. 2016).

Building the Larger Oligomer and Forming a Pore

The precise nature by which BAX and BAK dimers form into larger oligomers and then permeabilize the outer mitochondrial membrane remains one of the principal questions within the field. From some studies, a daisy chain model of Bax and Bak oligomerization has been proposed (Bogner et al. 2010; Pang et al. 2012), with the front of one Bax monomer interacting with the rear of a neighbor. However, this model is inconsistent with the now well-characterized symmetric BH3-in-groove dimers, as a daisy chain interface is asymmetric, and thus is unlikely to represent the reality of the larger oligomer. Other studies showed that residues in the α6 of Bax and Bak could be cross-linked in a fashion consistent with these helices running parallel to each other in a larger complex (Dewson et al. 2009, 2012). Further cross-linking studies showed that linkage between dimers was not restricted to only this region, and that in fact a large number of residues across a range of different regions of the proteins could indeed be cross-linked (Zhang et al. 2010, 2016; Aluvila et al. 2014; Iyer et al. 2015; Uren et al. 2017).

In all cases, these interdimer cross-links do not saturate, unlike the previously observed intradimer cross-links, suggesting that these contacts do not necessarily represent defined protein–protein interfaces, as observed and structurally characterized for the BH3-in-groove dimer.

The mechanism by which larger BAX and BAK oligomers build, how they form pores, and the nature of the pores themselves has also been a matter of much conjecture. Atomic force microscopy, superresolution microscopy, and electron tomography show that BAX oligomers can form a variety of sizes and shapes including clusters, rings, lines, and arcs (Grosse et al. 2016; Salvador-Gallego et al. 2016; Ader et al. 2019). Recently, it was discovered that pores of extremely large apertures can be induced by BAX (McArthur et al. 2018; Riley et al. 2018). These "macropores" enable the release of mtDNA into the cytoplasm resulting in activation of the cGAS/STING pathway and leading to type I interferon production (Rongvaux et al. 2014; White et al. 2014). It is still unclear how BAX and BAK initiate these heterogeneic membrane ruptures, but once this event has occurred it is likely that oligomers line the pore in some manner to stabilize the lumen. Protein-lined membrane pores can be either proteinaceous, in which the lumen is entirely lined by protein, or toroidal, where the lumen is lined by both protein and lipid headgroups; mounting evidence supports the latter for BAX and BAK (Terrones et al. 2004; Qian et al. 2008; Aluvila et al. 2014; Bleicken et al. 2014; Salvador-Gallego et al. 2016; Li et al. 2017). Spectroscopic experiments have been used to understand the alignment of dimers on the pore lumen and within the oligomer, leading to two distinct models. One involves dimers straddling the membrane bilayer such that the α9 TM helices run antiparallel to each other, the so-called clamp model (Bleicken et al. 2014). A second model involves dimers concentrated on the upper edge of the lumen leaving α9 TM helices to enter the membrane from the cytosolic side of the membrane and thus run parallel to each other (Mandal et al. 2016). It is currently unclear which of the two models represents the situation in cells, with further work required to understand this important aspect of membrane permeabilization. For further details and discussion of these opposing models, we refer the reader to a recent comprehensive review by Bleicken et al. (2018).

TOWARD CHEMICAL PROBING CANONICAL EFFECTORS

BAX and BAK are potential therapeutic targets that may be activated or inhibited to initiate or block apoptosis in disease. Understanding their molecular mechanisms of action importantly informs the development of small-molecule chemical probes to selectively target their activity. Accordingly, the effector field is now following in the footsteps of groups that developed antagonist of antiapoptotic BCL-2 proteins as exquisite chemical probes and therapeutics (Oltersdorf et al. 2005; Tse et al. 2008; Lessene et al. 2013; Souers et al. 2013; Huang et al. 2019).

Several small molecules have been discovered in academic laboratories through screening efforts against BAX and BAK. It is early for many of these compounds, particularly compared with those targeting antiapoptotic proteins, and future applicability requires these to be carefully tested by multiple investigators for potency, selectivity, and a proven mechanism of action (Arrowsmith et al. 2015). Nonetheless, these published compounds potentially inform on the dynamics, accessibility, and drugability of effector binding sites. Compound binding at these sites may induce, block, or cooperate with putative mechanisms of effector activation, inhibition, functional homodimerization, and higher order oligomerization. We briefly summarize features of small-molecule effector modulators below (Fig. 5).

BAX Activators Aimed at Its Front and Rear Sites

Compound 106 (ZINC 14750348) was identified in virtual screens for binders of the front, canonical BAX activation site and induced BAX-dependent but not BAK-dependent cytotoxicity in $bak^{-/-}bax^{-/-}$ mouse embryonic fibroblasts (MEFs) expressing human BAX or human BAK (Fig. 5A; Zhao et al. 2014). Compound 106 destabilized BAX in thermal shift assays and induced BAX-dependent cyt c release in vitro and human tumor apoptosis in tissue culture and in vivo. Others have discovered additional BAX-selective activator compounds with in vitro and in vivo activity in virtual screens as supposed modulators of the front site interaction with the TM tail (Xin et al. 2014), although their direct interaction with, and functional activation of, BAX was not rigorously shown. Two of these compounds contain alkene groups that could form Michael addition adducts with many intracellular nucleophiles and therefore are not likely selective for BAX.

Compounds designed as rear site activators of BAX, including BAM7 (Gavathiotis et al. 2012) and its second iteration BTSA1 (Reyna et al. 2017), were identified through in silico and subsequent analog searches (Fig. 5A). These compounds displace BIM SAHB:BAX complexes with IC_{50} of 3 μM and 250 nM in fluorescence polarization assays, respectively. Evidence that they behave as bona fide BAX activators includes their induction of 6A7 antibody epitope exposure, oligomerization in solution over time similar to that induced by BH3 activators, membrane translocation, membrane permeabilization, BAX-selectivity over BAK in cells, and cell-based and in vivo activity against acute myeloid leukemia tumors (Reyna et al. 2017). On the other hand, their binding to BAX by 2D NMR is less compelling with very small chemical shift perturbations (<0.015 ppm) for most of the resonances deemed significant and overlapping the rear activation site (Gavathiotis et al. 2012; Reyna et al. 2017). Might these compounds engage the canonical activation site in BAX? We do not currently know, but this question should be included in experimental design whenever BAX modulators are pursued, as we expect that canonical groove engagement and destabilization to be essential in BAX activation. Interestingly, BTSA1 contains an imine group known to be prone to hydrolysis that may release the hydrazino-thiazole moiety (Gaulton et al. 2017). A caveat of effector activation in cells is that many chemicals may trigger effectors indirectly by up-regulating and activating the upstream BH3-only proteins.

Cite this article as *Cold Spring Harb Perspect Biol* doi: 10.1101/cshperspect.a036319

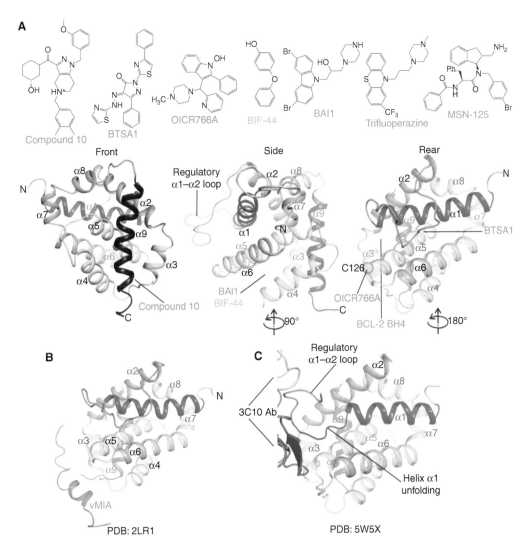

Figure 5. Effector activation and inhibition with molecules other than BH3 peptides. (*A*) Proposed small-molecule activators (red), inhibitors (green), and allosteric modulators (orange) of BAX and BAK membrane permeabilization, and their possible interaction sites. The postulated inhibitory site of interaction for BCL-2 BH4 SAHB is also indicated. (*B*) The human cytomegalovirus protein viral mitochondrial-localized inhibitor of apoptosis (vMIA) inhibits BAX through a short helical peptide. (*C*) Antibody-mediated activation of BAX involves binding of an epitope of the regulatory α1–α2 loop, which unfolds the carboxyl terminus of helix α1. The heavy and light chains of the 3C10 Ab are colored red and pink, respectively.

Such possibilities could be explored through profiling of BCL-2 proteins ± drug to ensure direct rather than indirect effector activation.

BAX Inhibitors Glue the Bottom of the Groove

BAX inhibitory compounds (BAIs) were discovered early (Bombrun et al. 2003), and were re- cently reported to inhibit activated truncated BID (tBID)-induced BAX-mediated liposome permeabilization with IC_{50} ~4 μM (Garner et al. 2019). BAIs bind to a pocket at the bottom of the canonical groove between helices α3, α4, α5, and α6 with K_D ~15 μM, established by microscale thermophoresis, although they induce small chemical shift perturbations (<0.015

ppm) in BAX detectable by 2D NMR (Fig. 5A; Garner et al. 2019). These inhibitors are thought to block the conformational changes associated with BAX activation, thus preventing its mitochondrial translocation and oligomerization, and selectively inhibit BAX- but not BAK-dependent apoptosis with an IC_{50} of ~2 μM (Garner et al. 2019). Interestingly, trifluoroperazine (TFP) is a distant structural analog of BAIs, discovered as an inhibitor of BAK-mediated liposome permeabilization with an IC_{50} ~10 μM (Song et al. 2014). TFP binds BAK-ΔTM with K_D ~70 μM in isothermal titration calorimetry assays, and blocks BAK oligomerization in liposomes, but its mode of binding to BAK has not been established (Song et al. 2014).

Other Means to Effector Structure–Function Modulation

Several modulators of BAK and BAX were discovered based on screening using liposome permeabilization assays. OICR766A activated BAX with an EC_{50} of ~100 nM (liposomes) and ~900 nM (mitochondria), was shown to induce oligomerization and membrane localization of BAX, and showed a strict requirement for residue C126 for activity (Fig. 5A; Brahmbhatt et al. 2016). OICR766A reportedly induced apoptosis in a BAX- or BAK-dependent manner and was inactive in $bak^{-/-}bax^{-/-}$ baby mouse kidney (BMK) cells, suggesting that it directly or indirectly activated BAX/BAK. MSN-125 and its analogs partially inhibited BAX-mediated liposome permeabilization with an IC_{50} of ~4 μM, BAX- or BAK-mediated mitochondrial poration with an IC_{50} of ~10 μM and ~20 μM, respectively, and inhibited BAK- and BAX-dependent apoptosis (Niu et al. 2017). Interestingly, MSN-125 appeared to act by preventing higher order oligomerization beyond dimerization as examined in cross-linking studies (Fig. 5A).

Allosteric modulation of effector activity has been reported with: (1) a helical peptide from human cytomegalovirus viral mitochondria-localized inhibitor of apoptosis (vMIA), which binds a surface pocket at the Bax α3–α4 and α5–α6 hairpins, presumably acting by allosterically blocking conformational changes, or

activation at the canonical groove (Fig. 5B; Ma et al. 2012); (2) antibodies that bind epitopes in the α1–α2 loop, inducing protein unfolding to directly activate BAK and BAX (Fig. 5C; Iyer et al. 2016; Robin et al. 2018); (3) a stapled BCL-2 BH4 helix peptide that binds a site at the bottom of the canonical groove composed of the α1, α1–α2 loop, and α5–α6 hairpin presumably blocking BAX activation (Fig. 5A; Barclay et al. 2015); and (4) a BIF-44 fragment binding a surface pocket at the center of α5–α6 helices, which synergizes with peptide-based BH3 activators (Fig. 5A; Pritz et al. 2017). These studies have revealed multiple modalities for modulating BAK and BAX conformations, potentially impacting their apoptotic function either independently or through synergism with BH3 activators. Ultimately, chemical probe effector modulators need to evolve via structure-based design to simplify and guide a structure–activity relationship (SAR). Moreover, effector binding site mutagenesis must be thoroughly performed to test the role of surface pocket engagement in effector modulation. Some of the sites may be undruggable because of their small size and/or peripheral involvement in effector modulation, factors that would limit the development of potent ($K_D \sim$ nM) and selective (e.g., ~30-fold differences for BAX vs. BAK) chemical probes.

POSTTRANSLATIONAL EFFECTOR MODIFICATIONS DOWN-REGULATE APOPTOSIS

Posttranslational modifications have been implicated in BAK and BAX activity, although this area of the field has not been exhaustively probed. It has been reported that phosphorylation of BAX S184 in the carboxy-terminal TM tail prevents BAX targeting to mitochondria in tumor cell lines (Gardai et al. 2004; Xin and Deng 2005; Wang et al. 2010; Quast et al. 2013). Through exposure of the canonical groove, this phosphorylation event could render cytosolic BAX a better receptor for BH3-only activators. Such unproductive BH3-only protein:BAX interactions would down-regulate apoptosis induced by BH3 mimetics that disrupt or prevent

Cite this article as *Cold Spring Harb Perspect Biol* doi: 10.1101/cshperspect.a036319

formation of complexes between pro- and anti-apoptotic BCL-2 proteins (Fig. 2D; Kale 2018).

Ubiquitylation of BAK, BAX, and BOK (vide infra) has been reported to down-regulate effector-mediated apoptosis (Cakir et al. 2017; Bernardini et al. 2019). Monoubiquitylation of BAK's single lysine, K113, in the canonical groove by the E3 ligase parkin during mitophagy has been shown to down-regulate the ability of BAK to permeabilize membranes in vitro and in cells by interfering with activation, dimerization, and oligomerization (Bernardini et al. 2019). In contrast, VDAC2 ubiquitination by parkin inhibited BAX mitochondrial localization, possibly through steric hindrance of the interaction between BAX and ubiquitylated VDAC2 and/or through reduced BAX recruitment because of reduced VDAC2 levels (Bernardini et al. 2019), although this has not been formally tested. BAX regulation by ubiquitylation is more complex and context-dependent and may involve direct (Cakir et al. 2017) and indirect targeting of the BAX apoptotic axis (Fig. 2C,D; Bernardini et al. 2019). BAX K128 has been implicated in ubiquitylation and degradation by ectopic parkin expression, which may protect against BAX-mediated apoptosis. K128R BAX was resistant to parkin-mediated degradation suggesting that pharmacologic targeting of parkin may modulate BAX-dependent apoptosis (Cakir et al. 2017). In contrast, VDAC2 ubiquitination by parkin inhibited BAX mitochondrial localization by presumed steric hindrance of the interaction between BAX and ubiquitylated VDAC2 (Bernardini et al. 2019). Future high-resolution studies may shed light on the molecular mechanisms of inhibition of effectors by posttranslational modifications.

BOK—THE LONE WOLF EFFECTOR

BOK is a BCL-2 family member that most resembles BAX and BAK at the sequence level, but for many years its role in apoptosis was unclear largely because its upstream triggers were unknown (Ke et al. 2012; Carpio et al. 2015; Fernández-Marrero et al. 2016). Early studies indicated that it was predominately located at the endoplasmic reticulum (ER) in complex with IP$_3$ receptors rather than at the mitochondria (Echeverry et al. 2013; Schulman et al. 2013, 2016). Additionally, a role for BOK in mitochondrial fusion has also been suggested (Schulman et al. 2019). However, recently, an effector role for BOK has emerged from genetic, biochemical, and structural studies (Fig. 6). Genetically, BOK has been implicated as an effector in studies that combined deletion of BAK, BAX, and BOK in hematopoiesis and during embryonic development (Ke et al. 2015, 2018). Combined deletion of BAK and BOK or BAX and BOK only revealed a possible role for BOK in ovaries (Ke et al. 2013). In lethally irradiated mice transplanted with fetal liver cells from $bak^{-/-}bax^{-/-}$ double knockout (DKO) and $bak^{-/-}bax^{-/-}bok^{-/-}$ triple knockout (TKO) mice, TKO reconstituted mice had more peripheral blood lymphocytes, and lymphoid infiltration, supporting BOK's functional redundancy with BAK and BAX (Ke et al. 2015). The most compelling genetic demonstration for a role of BOK as an effector was the comparative investigation of DKO and TKO mice (Ke et al. 2018). These suggested that intrinsic apoptosis defects are greatly exacerbated in TKOs, which ultimately contributed to widespread embryonic and perinatal lethality. Surprisingly, a small fraction of TKO mice lived normally to adulthood, showing no apparent phenotypes, suggesting that intrinsic apoptosis is dispensable for normal organ development and homeostasis. Other forms of cell death may be redundant with intrinsic apoptosis, although neither extrinsic apoptosis, necroptosis, pyroptosis, or autophagy were up-regulated in the TKO mice during development.

Biochemical demonstration for the effector role of BOK involved discovery of BOK's degradation machinery, the ER-associated degradation (ERAD) gp78 E3 ligase proteasome system (Llambi et al. 2016). Accordingly, proteasome inhibition stabilized and up-regulated BOK protein (Llambi et al. 2016; Zheng et al. 2018), and this could also be achieved through a combination of certain ER stress drugs (tunicamycin + PERK inhibitors), which dismantle the gp78 E3 ligase complex, inhibition of VCP AAA-ATPase, which blocks substrate shuttling from

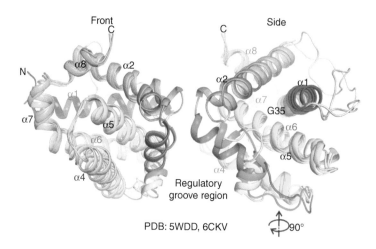

Figure 6. Dormant BOK is intrinsically unstable. Overlay of chicken and human BOK structures reveals three possible arrangements of the regulatory groove region between helices α3–α4. Additionally, human BOK has instability at the α1 helix, through G35, whose mutation to alanine significantly inhibits membrane association and permeabilization, and blocks BAK- and BAX-independent apoptosis.

the E3 ligase to the proteasome, or simply through knockdown of the gp78 subunits or VCP (Llambi et al. 2016). Unlike BAX and BAK, BOK interactions with BCL-2 family proteins appear to be weak and functionally fruitless. Mitochondrial poration and apoptosis induction by BOK does not appear to be enhanced by direct activators such as activated BID, or inhibited by antiapoptotic BCL-2 proteins (Llambi et al. 2016; Zheng et al. 2018). However, it should be noted that some studies report that activated BID or BID BH3 can promote BOK activity in liposome settings (Fernández-Marrero et al. 2017; Zheng et al. 2018). Also, it has been reported that antiapoptotic BCL-2 proteins may modulate BOK activity through their TM region (Stehle et al. 2018), more so than through their canonical grooves, which have weak affinity for BOK BH3 ($K_D \sim \mu M$) (Llambi et al. 2016).

Structural studies of apoBOK-ΔTM suggest that dormant BOK is poised for spontaneous conformational changes, captured by NMR spectroscopy and X-ray crystallography, providing a potential explanation for BOK poration of the mitochondrial membrane in the absence of direct activation (Fig. 6; Ke et al. 2018; Zheng et al. 2018). Like BAX and BAK, BOK can form large pores that are likely toroidal in nature

(Fernández-Marrero et al. 2017). Our understanding of BOK function and activity lags far behind that of BAX and BAK. There is much that remains to be discovered, for example, what are the contributions of BOK to apoptosis in cells in which BAX and/or BAK are present, and does BOK permeabilize membranes in a similar fashion to BAX and BAK. This is likely an area in which significant discoveries will be made in the field in coming years.

FUTURE DIRECTIONS

Mitochondrial poration is moving into a new structure–function phase with the pursuit of high-resolution structural details for effector conformations and complexes at membranes. Targets include structures for effectors alone and posttranslationally modified effector: activator complexes, effector:antiapoptotic complexes, and effector:VDAC complexes. Innovative multipronged high-resolution approaches and careful functional probing will be necessary to achieve these aims, and these efforts will fuel the discovery of new chemical probes for apoptosis research. We anticipate a rich decade ahead in effector research as we delve deeper into mechanisms underlying the key event of MOM permeabilization.

Cite this article as *Cold Spring Harb Perspect Biol* doi: 10.1101/cshperspect.a036319

REFERENCES

*Reference is also in this collection.

Ader NR, Hoffmann PC, Ganeva I, Borgeaud AC, Wang C, Youle RJ, Kukulski W. 2019. Molecular and topological reorganizations in mitochondrial architecture interplay during Bax-mediated steps of apoptosis. *eLife* **8:** e40712. doi:10.7554/eLife.40712

Aluvila S, Mandal T, Hustedt E, Fajer P, Choe JY, Oh KJ. 2014. Organization of the mitochondrial apoptotic BAK pore: oligomerization of the BAK homodimers. *J Biol Chem* **289:** 2537–2551. doi:10.1074/jbc.M113.526806

Arrowsmith CH, Audia JE, Austin C, Baell J, Bennett J, Blagg J, Bountra C, Brennan PE, Brown PJ, Bunnage ME, et al. 2015. The promise and peril of chemical probes. *Nat Chem Biol* **11:** 536–541. doi:10.1038/nchembio.1867

Barclay LA, Wales TE, Garner TP, Wachter F, Lee S, Guerra RM, Stewart ML, Braun CR, Bird GH, Gavathiotis E, et al. 2015. Inhibition of Pro-apoptotic BAX by a noncanonical interaction mechanism. *Mol Cell* **57:** 873–886. doi:10.1016/j.molcel.2015.01.014

Bernardini JP, Brouwer JM, Tan IK, Sandow JJ, Huang S, Stafford CA, Bankovacki A, Riffkin CD, Wardak AZ, Czabotar PE, et al. 2019. Parkin inhibits BAK and BAX apoptotic function by distinct mechanisms during mitophagy. *EMBO J* **38:** e99916. doi:10.15252/embj.201899916

Bleicken S, Classen M, Padmavathi PV, Ishikawa T, Zeth K, Steinhoff HJ, Bordignon E. 2010. Molecular details of Bax activation, oligomerization, and membrane insertion. *J Biol Chem* **285:** 6636–6647. doi:10.1074/jbc.M109.081539

Bleicken S, Jeschke G, Stegmueller C, Salvador-Gallego R, García-Sáez AJ, Bordignon E. 2014. Structural model of active Bax at the membrane. *Mol Cell* **56:** 496–505. doi:10.1016/j.molcel.2014.09.022

Bleicken S, Assafa TE, Stegmueller C, Wittig A, Garcia-Saez AJ, Bordignon E. 2018. Topology of active, membrane-embedded Bax in the context of a toroidal pore. *Cell Death Differ* **25:** 1717–1731. doi:10.1038/s41418-018-0184-6

Bogner C, Leber B, Andrews DW. 2010. Apoptosis: embedded in membranes. *Curr Opin Cell Biol* **22:** 845–851. doi:10.1016/j.ceb.2010.08.002

Bombrun A, Gerber P, Casi G, Terradillos O, Antonsson B, Halazy S. 2003. 3,6-Dibromocarbazole piperazine derivatives of 2-propanol as first inhibitors of cytochrome *c* release via Bax channel modulation. *J Med Chem* **46:** 4365–4368. doi:10.1021/jm034107j

Brahmbhatt H, Uehling D, Al-Awar R, Leber B, Andrews D. 2016. Small molecules reveal an alternative mechanism of Bax activation. *Biochem J* **473:** 1073–1083. doi:10.1042/BCJ20160118

Brouwer JM, Westphal D, Dewson G, Robin AY, Uren RT, Bartolo R, Thompson GV, Colman PM, Kluck RM, Czabotar PE. 2014. Bak core and latch domains separate during activation, and freed core domains form symmetric homodimers. *Mol Cell* **55:** 938–946. doi:10.1016/j.molcel.2014.07.016

Brouwer JM, Lan P, Cowan AD, Bernardini JP, Birkinshaw RW, van Delft MF, Sleebs BE, Robin AY, Wardak A, Tan IK, et al. 2017. Conversion of Bim-BH3 from activator to inhibitor of Bak through structure-based design. *Mol Cell* **68:** 659–672.e9. doi:10.1016/j.molcel.2017.11.001

Cakir Z, Funk K, Lauterwasser J, Todt F, Zerbes RM, Oelgeklaus A, Tanaka A, van der Laan M, Edlich F. 2017. Parkin promotes proteasomal degradation of misregulated BAX. *J Cell Sci* **130:** 2903–2913. doi:10.1242/jcs.200162

Carpio MA, Michaud M, Zhou W, Fisher JK, Walensky LD, Katz SG. 2015. BCL-2 family member BOK promotes apoptosis in response to endoplasmic reticulum stress. *Proc Natl Acad Sci* **112:** 7201–7206. doi:10.1073/pnas.1421063112

Chen HC, Kanai M, Inoue-Yamauchi A, Tu HC, Huang Y, Ren D, Kim H, Takeda S, Reyna DE, Chan PM, et al. 2015. An interconnected hierarchical model of cell death regulation by the BCL-2 family. *Nat Cell Biol* **17:** 1270–1281. doi:10.1038/ncb3236

Cheng EH, Sheiko TV, Fisher JK, Craigen WJ, Korsmeyer SJ. 2003. VDAC2 inhibits BAK activation and mitochondrial apoptosis. *Science* **301:** 513–517. doi:10.1126/science.1083995

Chi X, Pemberton J, Nguyen D, Osterlund EJ, Liu Q, Brahmbhatt H, Zhang Z, Lin J, Leber B, Andrews D. 2019. The carboxy-terminal sequence of Bim enables Bax activation and killing of unprimed cells. bioRxiv doi:10.1101/554907.

Chin HS, Li MX, Tan IKL, Ninnis RL, Reljic B, Scicluna K, Dagley LF, Sandow JJ, Kelly GL, Samson AL, et al. 2018. VDAC2 enables BAX to mediate apoptosis and limit tumor development. *Nat Commun* **9:** 4976. doi:10.1038/s41467-018-07309-4

Czabotar PE, Lee EF, van Delft MF, Day CL, Smith BJ, Huang DC, Fairlie WD, Hinds MG, Colman PM. 2007. Structural insights into the degradation of Mcl-1 induced by BH3 domains. *Proc Natl Acad Sci* **104:** 6217–6222. doi:10.1073/pnas.0701297104

Czabotar PE, Westphal D, Dewson G, Ma S, Hockings C, Fairlie WD, Lee EF, Yao S, Robin AY, Smith BJ, et al. 2013. Bax crystal structures reveal how BH3 domains activate Bax and nucleate its oligomerization to induce apoptosis. *Cell* **152:** 519–531. doi:10.1016/j.cell.2012.12.031

Czabotar PE, Lessene G, Strasser A, Adams JM. 2014. Control of apoptosis by the BCL-2 protein family: implications for physiology and therapy. *Nat Rev Mol Cell Biol* **15:** 49–63. doi:10.1038/nrm3722

Dai H, Ding H, Meng XW, Peterson KL, Schneider PA, Karp JE, Kaufmann SH. 2015. Constitutive BAK activation as a determinant of drug sensitivity in malignant lymphohematopoietic cells. *Genes Dev* **29:** 2140–2152. doi:10.1101/gad.267997.115

Dai H, Ding H, Peterson KL, Meng XW, Schneider PA, Knorr KLB, Kaufmann SH. 2018. Measurement of BH3-only protein tolerance. *Cell Death Differ* **25:** 282–293. doi:10.1038/cdd.2017.156

Dengler MA, Robin AY, Gibson L, Li MX, Sandow JJ, Iyer S, Webb AI, Westphal D, Dewson G, Adams JM. 2019. BAX activation: Mutations near its proposed non-canonical BH3 binding site reveal allosteric changes controlling mitochondrial association. *Cell Rep* **27:** 359–373.e6. doi:10.1016/j.celrep.2019.03.040

Dewson G, Kratina T, Sim HW, Puthalakath H, Adams JM, Colman PM, Kluck RM. 2008. To trigger apoptosis, Bak exposes its BH3 domain and homodimerizes via BH3:

groove interactions. *Mol Cell* **30**: 369–380. doi:10.1016/j.molcel.2008.04.005

Dewson G, Kratina T, Czabotar P, Day CL, Adams JM, Kluck RM. 2009. Bak activation for apoptosis involves oligomerization of dimers via their α6 helices. *Mol Cell* **36**: 696–703. doi:10.1016/j.molcel.2009.11.008

Dewson G, Ma S, Frederick P, Hockings C, Tan I, Kratina T, Kluck RM. 2012. Bax dimerizes via a symmetric BH3: groove interface during apoptosis. *Cell Death Differ* **19**: 661–670. doi:10.1038/cdd.2011.138

Du H, Wolf J, Schafer B, Moldoveanu T, Chipuk JE, Kuwana T. 2011. BH3 domains other than Bim and Bid can directly activate Bax/Bak. *J Biol Chem* **286**: 491–501. doi:10.1074/jbc.M110.167148

Dussmann H, Rehm M, Concannon CG, Anguissola S, Würstle M, Kacmar S, Völler P, Huber HJ, Prehn JH. 2010. Single-cell quantification of Bax activation and mathematical modelling suggest pore formation on minimal mitochondrial Bax accumulation. *Cell Death Differ* **17**: 278–290. doi:10.1038/cdd.2009.123

Echeverry N, Bachmann D, Ke F, Strasser A, Simon HU, Kaufmann T. 2013. Intracellular localization of the BCL-2 family member BOK and functional implications. *Cell Death Differ* **20**: 785–799. doi:10.1038/cdd.2013.10

Edlich F, Banerjee S, Suzuki M, Cleland MM, Arnoult D, Wang C, Neutzner A, Tjandra N, Youle RJ. 2011. Bcl-x_L retrotranslocates Bax from the mitochondria into the cytosol. *Cell* **145**: 104–116. doi:10.1016/j.cell.2011.02.034

Edwards AL, Gavathiotis E, LaBelle JL, Braun CR, Opoku-Nsiah KA, Bird GH, Walensky LD. 2013. Multimodal interaction with BCL-2 family proteins underlies the proapoptotic activity of PUMA BH3. *Chem Biol* **20**: 888–902. doi:10.1016/j.chembiol.2013.06.007

Fernández-Marrero Y, Ke F, Echeverry N, Bouillet P, Bachmann D, Strasser A, Kaufmann T. 2016. Is BOK required for apoptosis induced by endoplasmic reticulum stress? *Proc Natl Acad Sci* **113**: E492–E493. doi:10.1073/pnas.1516347113

Fernández-Marrero Y, Bleicken S, Das KK, Bachmann D, Kaufmann T, Garcia-Saez AJ. 2017. The membrane activity of BOK involves formation of large, stable toroidal pores and is promoted by cBID. *FEBS J* **284**: 711–724. doi:10.1111/febs.14008

Ferrer PE, Frederick P, Gulbis JM, Dewson G, Kluck RM. 2012. Translocation of a Bak C-terminus mutant from cytosol to mitochondria to mediate cytochrome C release: Implications for Bak and Bax apoptotic function. *PLoS ONE* **7**: e31510. doi:10.1371/journal.pone.0031510

Fletcher JI, Meusburger S, Hawkins CJ, Riglar DT, Lee EF, Fairlie WD, Huang DC, Adams JM. 2008. Apoptosis is triggered when prosurvival Bcl-2 proteins cannot restrain Bax. *Proc Natl Acad Sci* **105**: 18081–18087. doi:10.1073/pnas.0808691105

Gardai SJ, Hildeman DA, Frankel SK, Whitlock BB, Frasch SC, Borregaard N, Marrack P, Bratton DL, Henson PM. 2004. Phosphorylation of Bax Ser184 by Akt regulates its activity and apoptosis in neutrophils. *J Biol Chem* **279**: 21085–21095. doi:10.1074/jbc.M400063200

Garner TP, Reyna DE, Priyadarshi A, Chen HC, Li S, Wu Y, Ganesan YT, Malashkevich VN, Cheng EH, Gavathiotis E. 2016. An autoinhibited dimeric form of BAX regulates the BAX activation pathway. *Mol Cell* **63**: 485–497. doi:10.1016/j.molcel.2016.06.010

Garner TP, Amgalan D, Reyna DE, Li S, Kitsis RN, Gavathiotis E. 2019. Small-molecule allosteric inhibitors of BAX. *Nat Chem Biol* **15**: 322–330. doi:10.1038/s41589-018-0223-0

Gaulton A, Hersey A, Nowotka M, Bento AP, Chambers J, Mendez D, Mutowo P, Atkinson F, Bellis LJ, Cibrian-Uhalte E, et al. 2017. The ChEMBL database in 2017. *Nucleic Acids Res* **45**: D945–D954. doi:10.1093/nar/gkw1074

Gavathiotis E, Suzuki M, Davis ML, Pitter K, Bird GH, Katz SG, Tu HC, Kim H, Cheng EH, Tjandra N, et al. 2008. BAX activation is initiated at a novel interaction site. *Nature* **455**: 1076–1081. doi:10.1038/nature07396

Gavathiotis E, Reyna DE, Davis ML, Bird GH, Walensky LD. 2010. BH3-triggered structural reorganization drives the activation of proapoptotic BAX. *Mol Cell* **40**: 481–492. doi:10.1016/j.molcel.2010.10.019

Gavathiotis E, Reyna DE, Bellairs JA, Leshchiner ES, Walensky LD. 2012. Direct and selective small-molecule activation of proapoptotic BAX. *Nat Chem Biol* **8**: 639–645. doi:10.1038/nchembio.995

Grosse L, Wurm CA, Bruser C, Neumann D, Jans DC, Jakobs S. 2016. Bax assembles into large ring-like structures remodeling the mitochondrial outer membrane in apoptosis. *EMBO J* **35**: 402–413. doi:10.15252/embj.201592789

Hockings C, Anwari K, Ninnis RL, Brouwer J, O'Hely M, Evangelista M, Hinds MG, Czabotar PE, Lee EF, Fairlie WD, et al. 2015. Bid chimeras indicate that most BH3-only proteins can directly activate Bak and Bax, and show no preference for Bak versus Bax. *Cell Death Dis* **6**: e1735. doi:10.1038/cddis.2015.105

Hsu YT, Youle RJ. 1998. Bax in murine thymus is a soluble monomeric protein that displays differential detergent-induced conformations. *J Biol Chem* **273**: 10777–10783. doi:10.1074/jbc.273.17.10777

* Huang DCS, Gong J-N, Fairbrother WJ. 2019. BH3 mimetic drugs to treat cancers. *Cold Spring Harb Perpect Biol* doi:10.1101/cshperspect.a036327

Iyer S, Bell F, Westphal D, Anwari K, Gulbis J, Smith BJ, Dewson G, Kluck RM. 2015. Bak apoptotic pores involve a flexible C-terminal region and juxtaposition of the C-terminal transmembrane domains. *Cell Death Differ* **22**: 1665–1675. doi:10.1038/cdd.2015.15

Iyer S, Anwari K, Alsop AE, Yuen WS, Huang DC, Carroll J, Smith NA, Smith BJ, Dewson G, Kluck RM. 2016. Identification of an activation site in Bak and mitochondrial Bax triggered by antibodies. *Nat Commun* **7**: 11734. doi:10.1038/ncomms11734

Kale J, Kutuk O, Brito GC, Andrews TS, Leber B, Letai A, Andrews DW. 2018. Phosphorylation switches Bax from promoting to inhibiting apoptosis thereby increasing drug resistance. *EMBO Rep* **19**. doi:10.15252/embr.201745235

Ke F, Voss A, Kerr JB, O'Reilly LA, Tai L, Echeverry N, Bouillet P, Strasser A, Kaufmann T. 2012. BCL-2 family member BOK is widely expressed but its loss has only minimal impact in mice. *Cell Death Differ* **19**: 915–925. doi:10.1038/cdd.2011.210

Ke F, Bouillet P, Kaufmann T, Strasser A, Kerr J, Voss AK. 2013. Consequences of the combined loss of BOK and

BAK or BOK and BAX. *Cell Death Dis* **4**: e650. doi:10 .1038/cddis.2013.176

Ke F, Grabow S, Kelly GL, Lin A, O'Reilly LA, Strasser A. 2015. Impact of the combined loss of BOK, BAX and BAK on the hematopoietic system is slightly more severe than compound loss of BAX and BAK. *Cell Death Dis* **6**: e1938. doi:10.1038/cddis.2015.304

Ke FFS, Vanyai HK, Cowan AD, Delbridge ARD, Whitehead L, Grabow S, Czabotar PE, Voss AK, Strasser A. 2018. Embryogenesis and adult life in the absence of intrinsic apoptosis effectors BAX, BAK, and BOK. *Cell* **173**: 1217– 1230.17. doi:10.1016/j.cell.2018.04.036

Kim H, Tu HC, Ren D, Takeuchi O, Jeffers JR, Zambetti GP, Hsieh JJ, Cheng EH. 2009. Stepwise activation of BAX and BAK by tBID, BIM, and PUMA initiates mitochondrial apoptosis. *Mol Cell* **36**: 487–499. doi:10.1016/j.molcel .2009.09.030

Kuwana T, Mackey MR, Perkins G, Ellisman MH, Latterich M, Schneiter R, Green DR, Newmeyer DD. 2002. Bid, Bax, and lipids cooperate to form supramolecular openings in the outer mitochondrial membrane. *Cell* **111**: 331–342. doi:10.1016/S0092-8674(02)01036-X

Kuwana T, Bouchier-Hayes L, Chipuk JE, Bonzon C, Sullivan BA, Green DR, Newmeyer DD. 2005. BH3 do- mains of BH3-only proteins differentially regulate Bax- mediated mitochondrial membrane permeabilization both directly and indirectly. *Mol Cell* **17**: 525–535. doi:10.1016/j.molcel.2005.02.003

Lauterwasser J, Todt F, Zerbes RM, Nguyen TN, Craigen W, Lazarou M, van der Laan M, Edlich F. 2016. The porin VDAC2 is the mitochondrial platform for Bax retrotrans- location. *Sci Rep* **6**: 32994. doi:10.1038/srep32994

Lazarou M, Stojanovski D, Frazier AE, Kotevski A, Dewson G, Craigen WJ, Kluck RM, Vaux DL, Ryan MT. 2010. Inhibition of Bak activation by VDAC2 is dependent on the Bak transmembrane anchor. *J Biol Chem* **285**: 36876– 36883. doi:10.1074/jbc.M110.159301

Leshchiner ES, Braun CR, Bird GH, Walensky LD. 2013. Direct activation of full-length proapoptotic BAK. *Proc Natl Acad Sci* **110**: E986–E995. doi:10.1073/pnas .1214313110

Lessene G, Czabotar PE, Sleebs BE, Zobel K, Lowes KN, Adams JM, Baell JB, Colman PM, Deshayes K, Fair- brother WJ, et al. 2013. Structure-guided design of a se- lective BCL-X_L inhibitor. *Nat Chem Biol* **9**: 390–397. doi:10.1038/nchembio.1246

Li MX, Tan IKL, Ma SB, Hockings C, Kratina T, Dengler MA, Alsop AE, Kluck RM, Dewson G. 2017. BAK α6 permits activation by BH3-only proteins and homooli- gomerization via the canonical hydrophobic groove. *Proc Natl Acad Sci* **114**: 7629–7634. doi:10.1073/pnas .1702453114

Liu Q, Oesterlund EJ, Chi X, Pogmore J, Leber B, Andrews DW. 2019. Bim escapes displacement by BH3-mimetic anti-cancer drugs by double-bolt locking both Bcl-XL and Bcl-2. *eLife* **8**: e37689. doi:10.7554/eLife.37689

Llambi F, Moldoveanu T, Tait SW, Bouchier-Hayes L, Te- mirov J, McCormick LL, Dillon CP, Green DR. 2011. A unified model of mammalian BCL-2 protein family inter- actions at the mitochondria. *Mol Cell* **44**: 517–531. doi:10 .1016/j.molcel.2011.10.001

Llambi F, Wang YM, Victor B, Yang M, Schneider DM, Gingras S, Parsons MJ, Zheng JH, Brown SA, Pelletier S, et al. 2016. BOK is a non-canonical BCL-2 family effector of apoptosis regulated by ER-associated degradation. *Cell* **165**: 421–433. doi:10.1016/j.cell.2016.02.026

Ma J, Edlich F, Bermejo GA, Norris KL, Youle RJ, Tjandra N. 2012. Structural mechanism of Bax inhibition by cyto- megalovirus protein vMIA. *Proc Natl Acad Sci* **109**: 20901–20906. doi:10.1073/pnas.1217094110

Ma SB, Nguyen TN, Tan I, Ninnis R, Iyer S, Stroud DA, Menard M, Kluck RM, Ryan MT, Dewson G. 2014. Bax targets mitochondria by distinct mechanisms before or during apoptotic cell death: a requirement for VDAC2 or Bak for efficient Bax apoptotic function. *Cell Death Differ* **21**: 1925–1935. doi:10.1038/cdd.2014.119

Mandal T, Shin S, Aluvila S, Chen HC, Grieve C, Choe JY, Cheng EH, Hustedt EJ, Oh KJ. 2016. Assembly of Bak homodimers into higher order homooligomers in the mitochondrial apoptotic pore. *Sci Rep* **6**: 30763. doi:10 .1038/srep30763

McArthur K, Whitehead LW, Heddleston JM, Li L, Padman BS, Oorschot V, Geoghegan ND, Chappaz S, Davidson S, San Chin H, et al. 2018. BAK/BAX macropores facilitate mitochondrial herniation and mtDNA efflux during apoptosis. *Science* **359**: eaao6047. doi:10.1126/science .aao6047

Moldoveanu T, Liu Q, Tocilj A, Watson M, Shore G, Gehring K. 2006. The X-ray structure of a BAK homodimer reveals an inhibitory zinc binding site. *Mol Cell* **24**: 677–688. doi:10.1016/j.molcel.2006.10.014

Moldoveanu T, Grace CR, Llambi F, Nourse A, Fitzgerald P, Gehring K, Kriwacki RW, Green DR. 2013. BID-induced structural changes in BAK promote apoptosis. *Nat Struct Mol Biol* **20**: 589–597. doi:10.1038/nsmb.2563

Moldoveanu T, Follis AV, Kriwacki RW, Green DR. 2014. Many players in BCL-2 family affairs. *Trends Biochem Sci* **39**: 101–111. doi:10.1016/j.tibs.2013.12.006

Niu X, Brahmbhatt H, Mergenthaler P, Zhang Z, Sang J, Daude M, Ehlert FGR, Diederich WE, Wong E, Zhu W, et al. 2017. A small-molecule inhibitor of Bax and Bak oligomerization prevents genotoxic cell death and pro- motes neuroprotection. *Cell Chem Biol* **24**: 493–506.e5. doi:10.1016/j.chembiol.2017.03.011

Okamoto T, Zobel K, Fedorova A, Quan C, Yang H, Fair- brother WJ, Huang DC, Smith BJ, Deshayes K, Czabotar PE. 2013. Stabilizing the pro-apoptotic BimBH3 helix (BimSAHB) does not necessarily enhance affinity or bio- logical activity. *ACS Chem Biol* **8**: 297–302. doi:10.1021/ cb3005403

Oltersdorf T, Elmore SW, Shoemaker AR, Armstrong RC, Augeri DJ, Belli BA, Bruncko M, Deckwerth TL, Dinges J, Hajduk PJ, et al. 2005. An inhibitor of Bcl-2 family pro- teins induces regression of solid tumours. *Nature* **435**: 677–681. doi:10.1038/nature03579

Oltvai ZN, Milliman CL, Korsmeyer SJ. 1993. Bcl-2 hetero- dimerizes in vivo with a conserved homolog, Bax, that accelerates programed cell death. *Cell* **74**: 609–619. doi:10.1016/0092-8674(93)90509-O

Pang YP, Dai H, Smith A, Meng XW, Schneider PA, Kauf- mann SH. 2012. Bak conformational changes induced by ligand binding: Insight into BH3 domain binding and

Bak homo-oligomerization. *Sci Rep* **2**: 257. doi:10.1038/srep00257

Peng R, Tong JS, Li H, Yue B, Zou F, Yu J, Zhang L. 2013. Targeting Bax interaction sites reveals that only homo-oligomerization sites are essential for its activation. *Cell Death Differ* **20**: 744–754. doi:10.1038/cdd.2013.4

Pritz JR, Wachter F, Lee S, Luccarelli J, Wales TE, Cohen DT, Coote P, Heffron GJ, Engen JR, Massefski W, et al. 2017. Allosteric sensitization of proapoptotic BAX. *Nat Chem Biol* **13**: 961–967. doi:10.1038/nchembio.2433

Qian S, Wang W, Yang L, Huang HW. 2008. Structure of transmembrane pore induced by Bax-derived peptide: evidence for lipidic pores. *Proc Natl Acad Sci* **105**: 17379–17383. doi:10.1073/pnas.0807764105

Quast SA, Berger A, Eberle J. 2013. ROS-dependent phosphorylation of Bax by wortmannin sensitizes melanoma cells for TRAIL-induced apoptosis. *Cell Death Dis* **4**: e839. doi:10.1038/cddis.2013.344

Renault TT, Floros KV, Elkholi R, Corrigan KA, Kushnareva Y, Wieder SY, Lindtner C, Serasinghe MN, Asciolla JJ, Buettner C, et al. 2015. Mitochondrial shape governs BAX-induced membrane permeabilization and apoptosis. *Mol Cell* **57**: 69–82. doi:10.1016/j.molcel.2014.10.028

Reyna DE, Garner TP, Lopez A, Kopp F, Choudhary GS, Sridharan A, Narayanagari SR, Mitchell K, Dong B, Bartholdy BA, et al. 2017. Direct activation of BAX by BTSA1 overcomes apoptosis resistance in acute myeloid leukemia. *Cancer Cell* **32**: 490–505.e10. doi:10.1016/j.ccell.2017.09.001

Riley JS, Quarato G, Cloix C, Lopez J, O'Prey J, Pearson M, Chapman J, Sesaki H, Carlin LM, Passos JF, et al. 2018. Mitochondrial inner membrane permeabilisation enables mtDNA release during apoptosis. *EMBO J* **37**: e99238. doi:10.15252/embj.201899238

Robin AY, Krishna Kumar K, Westphal D, Wardak AZ, Thompson GV, Dewson G, Colman PM, Czabotar PE. 2015. Crystal structure of Bax bound to the BH3 peptide of Bim identifies important contacts for interaction. *Cell Death Dis* **6**: e1809. doi:10.1038/cddis.2015.141

Robin AY, Iyer S, Birkinshaw RW, Sandow J, Wardak A, Luo CS, Shi M, Webb AI, Czabotar PE, Kluck RM, et al. 2018. Ensemble properties of Bax determine its function. *Structure* **26**: 1346–1359.e5. doi:10.1016/j.str.2018.07.006

Rongvaux A, Jackson R, Harman CC, Li T, West AP, de Zoete MR, Wu Y, Yordy B, Lakhani SA, Kuan CY, et al. 2014. Apoptotic caspases prevent the induction of type I interferons by mitochondrial DNA. *Cell* **159**: 1563–1577. doi:10.1016/j.cell.2014.11.037

Salvador-Gallego R, Mund M, Cosentino K, Schneider J, Unsay J, Schraermeyer U, Engelhardt J, Ries J, Garcia-Saez AJ. 2016. Bax assembly into rings and arcs in apoptotic mitochondria is linked to membrane pores. *EMBO J* **35**: 389–401. doi:10.15252/embj.201593384

Schulman JJ, Wright FA, Kaufmann T, Wojcikiewicz RJ. 2013. The Bcl-2 protein family member Bok binds to the coupling domain of inositol 1,4,5-trisphosphate receptors and protects them from proteolytic cleavage. *J Biol Chem* **288**: 25340–25349. doi:10.1074/jbc.M113.496570

Schulman JJ, Wright FA, Han X, Zluhan EJ, Szczesniak LM, Wojcikiewicz RJ. 2016. The stability and expression level of Bok are governed by binding to inositol 1,4,5-trisphos-

phate receptors. *J Biol Chem* **291**: 11820–11828. doi:10.1074/jbc.M115.711242

Schulman JJ, Szczesniak LM, Bunker EN, Nelson HA, Roe MW, Wagner LE II, Yule DI, Wojcikiewicz RJH. 2019. Bok regulates mitochondrial fusion and morphology. *Cell Death Differ*. doi:10.1038/s41418-019-0327-4

Song SS, Lee WK, Aluvila S, Oh KJ, Yu YG. 2014. Identification of inhibitors against BAK pore formation using an improved in vitro assay system. *Bull Korean Chem Soc* **35**: 419–424. doi:10.5012/bkcs.2014.35.2.419

Souers AJ, Leverson JD, Boghaert ER, Ackler SL, Catron ND, Chen J, Dayton BD, Ding H, Enschede SH, Fairbrother WJ, et al. 2013. ABT-199, a potent and selective BCL-2 inhibitor, achieves antitumor activity while sparing platelets. *Nat Med* **19**: 202–208. doi:10.1038/nm.3048

Stehle D, Grimm M, Einsele-Scholz S, Ladwig F, Johänning J, Fischer G, Gillissen B, Schulze-Osthoff K, Essmann F. 2018. Contribution of BH3-domain and transmembrane-domain to the activity and interaction of the pore-forming Bcl-2 proteins Bok, Bak, and Bax. *Sci Rep* **8**: 12434. doi:10.1038/s41598-018-30603-6

Sung TC, Li CY, Lai YC, Hung CL, Shih O, Yeh YQ, Jeng US, Chiang YW. 2015. Solution structure of apoptotic BAX oligomer: oligomerization likely precedes membrane insertion. *Structure* **23**: 1878–1888. doi:10.1016/j.str.2015.07.013

Suzuki M, Youle RJ, Tjandra N. 2000. Structure of Bax: coregulation of dimer formation and intracellular localization. *Cell* **103**: 645–654. doi:10.1016/S0092-8674(00)00167-7

Terrones O, Antonsson B, Yamaguchi H, Wang HG, Liu J, Lee RM, Herrmann A, Basañez G. 2004. Lipidic pore formation by the concerted action of proapoptotic BAX and tBID. *J Biol Chem* **279**: 30081–30091. doi:10.1074/jbc.M313420200

Todt F, Cakir Z, Reichenbach F, Youle RJ, Edlich F. 2013. The C-terminal helix of Bcl-x$_L$ mediates Bax retrotranslocation from the mitochondria. *Cell Death Differ* **20**: 333–342. doi:10.1038/cdd.2012.131

Todt F, Cakir Z, Reichenbach F, Emschermann F, Lauterwasser J, Kaiser A, Ichim G, Tait SW, Frank S, Langer HF, et al. 2015. Differential retrotranslocation of mitochondrial Bax and Bak. *EMBO J* **34**: 67–80. doi:10.15252/embj.201488806

Tse C, Shoemaker AR, Adickes J, Anderson MG, Chen J, Jin S, Johnson EF, Marsh KC, Mitten MJ, Nimmer P, et al. 2008. ABT-263: a potent and orally bioavailable Bcl-2 family inhibitor. *Cancer Res* **68**: 3421–3428. doi:10.1158/0008-5472.CAN-07-5836

Uren RT, O'Hely M, Iyer S, Bartolo R, Shi MX, Brouwer JM, Alsop AE, Dewson G, Kluck RM. 2017. Disordered clusters of Bak dimers rupture mitochondria during apoptosis. *eLife* **6**: e19944. doi:10.7554/eLife.19944

Walensky LD, Pitter K, Morash J, Oh KJ, Barbuto S, Fisher J, Smith E, Verdine GL, Korsmeyer SJ. 2006. A stapled BID BH3 helix directly binds and activates BAX. *Mol Cell* **24**: 199–210. doi:10.1016/j.molcel.2006.08.020

Wang K, Gross A, Waksman G, Korsmeyer SJ. 1998. Mutagenesis of the BH3 domain of BAX identifies residues critical for dimerization and killing. *Mol Cell Biol* **18**: 6083–6089. doi:10.1128/MCB.18.10.6083

Cite this article as *Cold Spring Harb Perspect Biol* doi: 10.1101/cshperspect.a036319

Wang Q, Sun SY, Khuri F, Curran WJ, Deng X. 2010. Mono- or double-site phosphorylation distinctly regulates the proapoptotic function of Bax. *PLoS ONE* **5:** e13393. doi:10.1371/journal.pone.0013393

Wei MC, Lindsten T, Mootha VK, Weiler S, Gross A, Ashiya M, Thompson CB, Korsmeyer SJ. 2000. tBID, a membrane-targeted death ligand, oligomerizes BAK to release cytochrome c. *Genes Dev* **14:** 2060–2071.

White MJ, McArthur K, Metcalf D, Lane RM, Cambier JC, Herold MJ, van Delft MF, Bedoui S, Lessene G, Ritchie ME, et al. 2014. Apoptotic caspases suppress mtDNA-induced STING-mediated type I IFN production. *Cell* **159:** 1549–1562. doi:10.1016/j.cell.2014.11.036

Willis SN, Chen L, Dewson G, Wei A, Naik E, Fletcher JI, Adams JM, Huang DC. 2005. Proapoptotic Bak is sequestered by Mcl-1 and Bcl-xL, but not Bcl-2, until displaced by BH3-only proteins. *Genes Dev* **19:** 1294–1305. doi:10.1101/gad.1304105

Xin M, Deng X. 2005. Nicotine inactivation of the proapoptotic function of Bax through phosphorylation. *J Biol Chem* **280:** 10781–10789. doi:10.1074/jbc.M500084200

Xin M, Li R, Xie M, Park D, Owonikoko TK, Sica GL, Corsino PE, Zhou J, Ding C, White MA, et al. 2014. Small-molecule Bax agonists for cancer therapy. *Nat Commun* **5:** 4935. doi:10.1038/ncomms5935

Zhang Z, Zhu W, Lapolla SM, Miao Y, Shao Y, Falcone M, Boreham D, McFarlane N, Ding J, Johnson AE, et al. 2010. Bax forms an oligomer via separate, yet interdependent, surfaces. *J Biol Chem* **285:** 17614–17627. doi:10.1074/jbc.M110.113456

Zhang Z, Subramaniam S, Kale J, Liao C, Huang B, Brahmbhatt H, Condon SG, Lapolla SM, Hays FA, Ding J, et al. 2016. BH3-in-groove dimerization initiates and helix 9 dimerization expands Bax pore assembly in membranes. *EMBO J* **35:** 208–236. doi:10.15252/embj.201591552

Zhao G, Zhu Y, Eno CO, Liu Y, Deleeuw L, Burlison JA, Chaires JB, Trent JO, Li C. 2014. Activation of the proapoptotic Bcl-2 protein Bax by a small molecule induces tumor cell apoptosis. *Mol Cell Biol* **34:** 1198–1207. doi:10.1128/MCB.00996-13

Zheng JH, Grace CR, Guibao CD, McNamara DE, Llambi F, Wang YM, Chen T, Moldoveanu T. 2018. Intrinsic instability of BOK enables membrane permeabilization in apoptosis. *Cell Rep* **23:** 2083–2094.e6. doi:10.1016/j.celrep.2018.04.060

Cracking the Cell Death Code

Carla V. Rothlin[1,2,4] and Sourav Ghosh[2,3,4]

[1]Department of Immunobiology, School of Medicine, Yale University, New Haven, Connecticut 06520, USA
[2]Department of Pharmacology, School of Medicine, Yale University, New Haven, Connecticut 06520, USA
[3]Department of Neurology, School of Medicine, Yale University, New Haven, Connecticut 06520, USA

Correspondence: carla.rothlin@yale.edu; sourav.ghosh@yale.edu

Cell death is an invariant feature throughout our life span, starting with extensive scheduled cell death during morphogenesis and continuing with death under homeostasis in adult tissues. Additionally, cells become victims of accidental, unscheduled death following injury and infection. Cell death in each of these occasions triggers specific and specialized responses in the living cells that surround them or are attracted to the dying/dead cells. These responses sculpt tissues during morphogenesis, replenish lost cells in homeostasis to maintain tissue/system function, and repair damaged tissues after injury. Wherein lies the information that sets in motion the cascade of effector responses culminating in remodeling, renewal, or repair? Here, we attempt to provide a framework for thinking about cell death in terms of the specific effector responses that accompanies various modalities of cell death. We also propose an integrated threefold "cell death code" consisting of information intrinsic to the dying/dead cell, the surroundings of the dying cell, and the identity of the responder.

… Mrityor ma amritam gamaya
(Take me from death to beyond death)

> —Pavamana Mantra, Brihadaranyaka
> Upanishad

The inevitability of cell death makes a convincing argument for studying it, not for the sake of trying to prevent it, but rather in the context of understanding its significance to life. Although cell death itself is useful, the response of other cells to the dead cell is critical for the life of the organism. The myriad contexts of cell death and the array of specific responses to it remain a great mystery.

EVER PRESENT CELL DEATH

What should I do and where should I go? A thief has taken hold of my flesh!
For there in my bed-chamber Death dwells, and wherever I turn, there too is Death

> —Epic of Gilgamesh, Tablet XI

Cell death occurs during embryonic development, in adult tissues in homeostasis, as well as in all manners of damage such as sterile injury or infection. The response to cell death in development, homeostasis, and injury is distinct. Sometimes, dead cells are simply shed, but at other

[4]These authors contributed equally to this work.

Cite this article as *Cold Spring Harb Perspect Biol* doi: 10.1101/cshperspect.a036343

times, dead/dying cells are actively removed. Occasionally, the dead cell makes way to new types of cells. In other instances, a new cell of the same type takes the place of the dead cell. Death may occur in limited numbers and slowly. Replacement may involve counting the number of cells being lost and its coordination with stem cell proliferation and differentiation to replace the lost cells (renewal or compensation). Or, the loss of cells may occur unexpectedly and extensively wherein repair may require the recruitment and concerted function of a large number of cells of different types. The universality and constancy of cell death, along with the existence of a variety of effector responses to cell death that are highly specialized and specific, point to some kind of a biological code. Yet, our comprehension of this code remains extremely limited.

INFORMATION IS ENCODED IN CELL DEATH

Have you ever wondered if death is the same for all living beings, be they animals, human beings included, or plants, from the grass you walk on to the hundred-meter-tall Sequoiadendron giganteum, will the death that kills a man who knows he's going to die be the same as that of a horse who never will.

> —As Intermitencias da Morte (Death with Interruptions), Jose Saramago

Attempts to categorize cell death, so far, have primarily relied on the molecular mechanism of execution of cell death. Careful dissection of the many ways cells die has culminated in more than 15 types of distinct cell death modalities including, but not limited to, extrinsic apoptosis, intrinsic apoptosis, immunogenic cell death (ICD), autolysis or lysosome-dependent cell death, autophagic cell death, autosis, entosis, ferroptosis, methuosis, mitotic catastrophe, necroptosis, netosis, oncosis, paraptosis, parthanatos, pyroptosis, sarmoptosis, or mitochondrial permeability transition-driven cell death and phagoptosis (Galluzzi et al. 2018). This bewildering array presents a numerical challenge in trying to understand the information encoded in each type of cell death and how that information is extrapolated to impart specificity of effector responses. Even death by a single mechanism can result in vastly different responses by other cells in the tissue. For example, the response to apoptosis could be removal (in development or postnatal neuronal death), renewal (in adult tissue), inflammation and then repair (when there is apoptosis following sterile injury), and inflammation coupled to an adaptive immune response and repair (when for example there is apoptosis of a virally infected cell). Therefore, the execution mechanism, by itself, is likely not sufficient to code for the effector response. Instead, we speculate "Why cells die?" This essentially teleological question is not intended to assign purpose to cell death, but rather to provide a complementary conceptual framework to categorize the basic types of cell death and the consequences of cell death.

WHY CELLS DIE?

The sorrows of existence, Oh "Ghalib," what is its cure but death
The flame of the lamp burns in all its shades, till morning comes

> —Till Morning Comes, Mirza Asadullah Beg Khan "Ghalib"

The simplest context of thinking about cell death is perhaps accidental cell death during an injury. But, cell death also occurs in a programmed manner, not only in homeostasis but also during embryonic development. Below, we follow the chronology of an organism from birth to adulthood in terms of its encounters with cell death and posit why cells die.

Cell Death during Embryogenesis and Postnatal Development

Extensive cell death is a feature of developmental morphogenesis (Saunders 1966). Cells die to create a new form or function, or eliminate a function. Many cells, such as those of the larval tissue of insects or amphibians, die and are removed to reshape the tissue/organ during metamorphosis (Suzanne and Steller 2013). In humans, interdigital mesenchymal cells die during the seventh and eighth week of gestation (Jordan et al. 2012). Failure to die can manifest

Cite this article as *Cold Spring Harb Perspect Biol* doi: 10.1101/cshperspect.a036343

as developmental abnormalities. For example, syndactyly involves fusion of soft tissues and failure of differentiation between adjacent digits because of the absence of apoptosis in the interdigital mesenchyme (Hernández-Martínez and Covarrubias 2011; Jordan et al. 2012). Up to 50% of newborn neurons die during embryonic and perinatal brain development in vertebrates (Yeo and Gautier 2004). Children with autism spectrum disorder were reported to have a significantly increased number of neurons in the basal and central nuclei of the amygdala (Avino et al. 2018). The evolutionary benefit of neuronal death is not known, but it may be to insert randomness into what is essentially a programmed and stereotypical developmental pattern. An individual may be born with the same exact number and organization of neurons, yet become distinct as neurons die randomly (as it relates to comparison between individuals). A similar argument can be made for cell death during digit

development wherein an invariant, evolutionarily conserved pattern is tinkered with to create new form and function in some species, for example, webbed feet in ducks makes way for toes in chickens (Yokouchi et al. 1996; Zou and Niswander 1996). Cells also die to eliminate a function. A classic example is the elimination of autoreactive T cells in the thymus (Daley et al. 2017). Here, we propose the term "type I cell death," wherein a cell dies to create a new function or to elimination an existing function (Fig. 1).

Type I death occurs when the cells receive specific instruction to do so. This instruction may come as the disappearance of a signal such as growth factor depletion or loss of substrate attachment (anoikis), or through the appearance of a specific factor such as signal through the Fas receptor. The molecular mechanism of execution is apoptosis—both caspase-8-dependent cell-extrinsic apoptosis and cas-

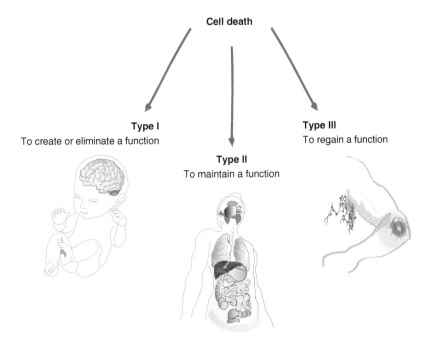

Figure 1. Classification of cell death based on associated outcomes. Cell death associated with the creation or elimination of a function is defined as type I. An example is represented by the cell death associated with embryonic and perinatal brain development. Type II cell death defines the demise of cells associated with homeostatic renewal of tissues in adulthood. This is exemplified by cell death associated with replenishment of adult organs such as the intestine. Unscheduled cell death is defined as type III cell death and is associated with inflammation, immunity (in case of infection), and tissue repair. This is represented in the context of an arm injury.

pase-9-dependent cell-intrinsic apoptosis. More importantly, type I cell death is associated with a response that is removal only. The process of removal of a dead cell involves its phagocytosis by a responding cell. This specialized phagocytosis is termed "efferocytosis" and the responding cell is termed "efferocyte." There is no need to replace the dying cell with one of its kind. A cell with a different function will take its place.

Cell Death during Tissue/Organ Homeostasis

Once development is complete, cell death continues to occur in the adult organism. In a healthy adult human being, billions of cells are scheduled to die every day (Arandjelovic and Ravichandran 2015). Cells die to prevent their loss-of-function from endangering or compromising tissue/system function, that is, self-sacrifice for a greater cause. This is perhaps most frequent at the mucosal barriers and other hazardous localizations where cells are more prone to being damaged and therefore at higher risk of progressive deterioration of function. However, this theory should be generally true for other locations. An exception can be logically expected when unique information is stored within the cells or their connections (such as adult neurons) or cells that are tightly coupled for their function (such as cardiomyocytes). Indeed, cell death is observed in most tissues except for the majority of central nervous system (CNS) neurons and cardiomyocytes. In the CNS, cells have been reported to die in specialized areas associated with neurogenesis (Whitman and Greer 2009; Kuhn 2015; Ryu et al. 2016). This type of cell death is an intrinsic part of tissue homeostasis. We propose the term "type II cell death" for the death of a cell during tissue/organ homeostasis, which presumably prevents the loss-of-function/maintains-the-function of the tissue/organ through rejuvenation (Fig. 1).

Type II cell death appears to be a matter of intrinsic choice for the cell. It may be the result of accumulated mutations in its genetic material that may adversely affect its function or duplication. An example would be death caused by DNA-damage or replication errors in cells. The execution mechanism is intrinsic, caspase-9-de-

pendent apoptosis. This is also expected to be the primary form of cell death characterized by telomere shortening and mitotic catastrophe such as when cells reach the Hayflick limit. Another form of type II cell death presages loss of metabolic function under conditions of energetic stress. Modalities of cell death in this category may be autophagic cell death, autosis, autolysis, entosis, parthanatos, paraptosis, ferroptosis, or methuosis.

Under homeostatic conditions, cell death requires replacement by another identical cell. Survival depends on the ability to not only remove, but also to replace the dead cells and maintain uninterrupted tissue/organ function. Therefore, type II cell death is associated with proliferation. It is possible that a counting mechanism is encoded in the efferocyte that then communicates this information to the local stem or cycling cell. Either the absence of renewal or ectopic renewal can be problematic—one leads to loss of tissue function, whereas the other may lead to tumorigenesis. We cannot formally rule out the possibility that nearest neighbors play a role in sensing, but not removal, of dead cells and this information accounts for induced proliferation. For example, loss of cell–cell contact with a dead cell might spur cell division and contact-inhibition will serve as the stop signal. Alternatively, the dying cell itself can be the source of a quantum of mitogenic information that directs a defined number of proliferative events (Ryoo et al. 2004). However, the dead cell still needs to be cleared to make space, and failure to clear a dying cell has been reported to prevent cell death (Reddien et al. 2001). Thus, Occam's razor would dictate that the efferocyte simultaneously elaborates signals to replenish. Induction of growth factors coupled to efferocytosis provides important evidence in support of this theory (Morimoto et al. 2001; Golpon et al. 2004).

Cell Death during Injury

Finally, death may be unscheduled, including both accidental cell death and unscheduled but regulated cell death. Here, cells die as a result of externally caused damage, infection, or because

Cite this article as *Cold Spring Harb Perspect Biol* doi: 10.1101/cshperspect.a036343

of the response to infection. We term this type of unscheduled death as type III cell death (Fig. 1). Examples include nonprogrammed cell death (necrosis) as well as programmed cell death (pyroptosis, necroptosis, and ICD). The response to unscheduled cell death is distinct from that to scheduled cell death. Unscheduled insults lead to a sequence of coordinated mechanisms that can include inflammation and immunity (the latter in case of infected, but not in sterile injury) before culminating, ideally, in tissue regeneration and restoration of lost functions (Medzhitov 2008). The information regarding the cause of cell death is made available to the efferocyte, which in turn, either directly or indirectly through a secondary effector, orchestrates the tissue's response to cell death. Inappropriate or unnecessary responses can manifest as ulcers or fibrosis.

MOLECULAR PATTERNS ASSOCIATED WITH THE DEAD/DYING CELL

The doctor said: Such-and-such indicates that there is such-and-such inside you; but if that is not confirmed by the analysis of this-and-that, then it must be assumed that you have such-and-such... a floating kidney, chronic catarrh, or appendicitis.

—*The Death of Ivan Ilyich, Leo Tolstoy*

Teleology aside, there must be a molecular code that distinguishes these three categories of cell death to enable efferocytes to drive a specific effector response. As described above, this code requires more information than simply the molecular execution pathway. Ruslan Medzhitov predicted the presence of an invariant signal (such as phosphatidylserine [PtdSer]) together with a unique signature in the dying cell as a cell death code (Medzhitov 2008). Although PtdSer is a hallmark of cell death, the nature of the variable signal associated with the different causes of cell death remains poorly defined.

PtdSer is present in all eukaryotic cells and within the cell surface plasma membrane is asymmetrically distributed so that its vast majority faces the cytoplasm. This is achieved through phospholipid translocases that specifically move phospholipids from the outer to the inner leaflet of the lipid bilayer or vice versa in

an ATP-dependent manner—flippases and floppases, respectively. Scramblases can negate this asymmetry by translocating phospholipids in both directions (Nagata et al. 2016). The exposure of PtdSer on the outer leaflet during apoptosis is the result of inhibition of flippases and activation of scramblases. Cleavage of flippases by apoptotic caspases render them inactive, reducing the ability to translocate PtdSer to the inner leaflet (Segawa et al. 2014). In contrast, cleavage of scramblases such as Xkr8 is essential for their activity (Suzuki et al. 2013). The net result is the increase in the amount of PtdSer on the surface of apoptotic cells (Fig. 2) and the activation of PtdSer-receptors (e.g., MERTK, TIM4, BAI1) on efferocytes leading to the clearance of the dead cell. A recent study has extended the function of exposed PtdSer as an "eat me" signal to necroptotic cells (Zargarian et al. 2017), yet whether this phospholipid functions as an universal "eat me" signal remains to be explored.

The variable molecular signatures of cell death represent the proxy of the elements that bring about death or, in other words, the reason for dying. As such, they should be associated with either type I, type II, or type III cell death. The most well-established variable signature is provided by damage-associated molecular patterns (DAMPs), which are classically associated with type III cell death. A series of DAMPs, such as alarmins and cytokines, are released during unscheduled cell death (Fig. 2). Although many DAMPs are commonly associated with type III cell death, some DAMPs can impart specificity to the effector response. For example, type III cell death can trigger a response characteristic of sterile injury wherein adaptive immune response is not engaged. When a specific pathogen is involved in causing cell death, the innate immune response is complemented with the induction of Th1, Th2, or Th17 adaptive immune responses. For example, intracellular lipopolysaccharide (LPS) directly activates caspase-11 in rodents or caspases-4 and -5 in humans, which subsequently cleave and activate the pore-forming protein gasdermin D (GSDMD). The rupture of the plasma membrane leads to pore-induced intracellular traps (PITs) wherein live bacteria are contained in pyroptotic cell

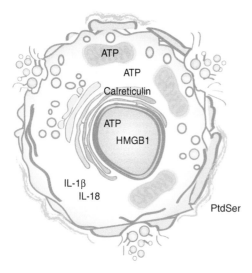

Figure 2. Invariant and unique molecular patterns associated with cell death. Phosphatidylserine ([PtdSer], indicated as green textured line) is a phospholipid found in all eukaryotic cells this is asymmetrically distributed in the plasma membrane of alive cells. Although PtdSer is enriched in the inner leaflet of live cells, it becomes invariantly exposed on the cell surface of apoptotic cells. Whether PtdSer is an universal molecular pattern of all cell death modalities remains unknown. Other molecular patterns of cell death are characterized by their distinct subcellular localization in live cells before their release when cells die. For example, calreticulin is found in the endoplasmic reticulum and HMGB1 in the nucleus. In contrast, ATP localization is less restricted, being present in mitochondria, the cytosol, and the nucleus. Molecular patterns associated with cell death can also be induced on specific cell death modalities. Activation of the inflammasome during pyroptosis leads to the production of interleukin (IL)-1β and IL-18.

corpses (Jorgensen et al. 2016a,b). The loss of K^+ through GSDMD pores also leads to the activation of NLRP3 and the release of interleukin (IL)-1β and IL-18 (Rathinam et al. 2019). IL-1β and IL-18, together with eicosanoids, promote the recruitment of neutrophils and clearing of PITs. These DAMPs are also critical for the ensuing Th1/Th17 responses (Rathinam et al. 2019). In contrast, release of TSLP, IL-33, and IL-25 on injury to epithelial barriers leads to the the induction of type 2 immunity including Th2 adaptive immune response (Han et al. 2017).

Alarmins can be found in different intracellular compartments (Fig. 2). The distinct subcellular localization of alarmins, for example, ATP in the cytosol, calreticulin in the endoplasmic reticulum (ER) and HMGB1 in the nucleus, may function as the basis for a temporal code in the progression of cell death. Apoptotic caspases activate the channel Pannexin I that allows the release of ATP while the plasma membrane is still intact, promoting the recruitment of phagocytes to the "death bed" (Elliott et al. 2009). This might be the first alarmin released in preparation of the impending cell death. Once released into the extracellular space, alarmins exert distinct functions from the ones associated with their intracellular localization. The intracellular metabolite ATP, not only induces the migration of phagocytes (Elliott et al. 2009) but at higher concentrations can also lead to activation of the inflammasome NLRP3 in the presence of pathogen-associated molecular patterns (PAMPs) (Ghiringhelli et al. 2009). In contrast, the translocation of ER-localized calreticulin or release of nuclear HMGB1 should require additional organelle damage that provides a time-delay mechanism for their release.

Not only the sequence of release of alarmins, but their integration with the timing of PtdSer exposure, may encode specific effector functions. Chemotherapeutic agents have been shown to promote the active translocation of calreticulin before PtdSer exposure, favoring the uptake of the corpse by dendritic cells (DCs) and thereby promoting an immunogenic response (Obeid et al. 2007; Panaretakis et al. 2009). However, when calreticulin is displayed as PtdSer is exposed, it potentiates the uptake of the dead cell by macrophages (Gardai et al. 2005) favoring an anti-inflammatory response. The signal integration between the receptors activated by the exposed/released DAMPs and the sensing of PtdSer remains to be explored.

Similarly, metabolites released during the process of cell death could reveal whether the cell is dying to preserve a function (type II) or dying as a result of injury (type III) and specify different effector functions. For example, the release of lactic acid from damaged skeletal muscle cells may instruct satellite cell proliferation and myogenesis or promote myoblast differentiation (Willkomm et al. 2014; Tsukamoto et al. 2018). In contrast, uric acid and cholesterol crystals, resulting from lytic cell death, induce inflammation through the NLRP3 inflammasome (Martinon et al. 2006; Duewell et al. 2010).

Can there be other forms of cell-intrinsic information indicating which cell is dying and why? It is conceivable that even the "identity" of the dead/dying cell may inform the efferocyte why the cell is dying. For example, the identity of a cell may predestine it to a favored mode of cell death. Terminally differentiated cells may prefer to die by means different from cycling cells. A terminally differentiated cell, such as a neuron or a cardiomyocyte, cannot be replaced easily unlike a fibroblast or an epithelial cell. At the same time, the neuron or cardiomyocyte is not at risk of transforming into a cancer cell unlike an epithelial or stromal cell. The identity of the dead cell may also be disclosed by the release of molecules unique to the cell type. For example, heart injury and cardiomyocyte death are associated with the release of the actin-associated cardiac troponins. Interestingly, the muscle actin-binding protein myosin II, was recently shown to potentiate the binding of F-actin released from necrotic cells to DNGR1 on DCs and promote cross-presentation of antigens (Schulz et al. 2018). Therefore, the cellular identity of a dying neuron or cardiomyocyte itself may specify type III cell death.

Apart from the death masks described above, additional information may also be encoded by *when* a cell dies. Although DNA damage leads to cell death, various inducers of DNA damage are known to arrest and subsequently kill cells at various phases of the cell cycle such as at the G_1, S, or G_2/M checkpoints. Mitotic poisons kill cells through mitotic catastrophe in M phase. Other chemicals such as retinoic acid, 5-FU, and methylxanthine-derived drugs can lead to G_1, S, or G_2/M arrest, respectively, and subsequent cell death (Shapiro and Harper 1999). G_1, S, or G_2/M arrest are mediated by molecules that can be shared, but also include molecules that work at specific checkpoints. For example, p53 functions at the G_1 checkpoint, but Nbs1 and Chk1 function at the S checkpoint

and BRCA1 at the G_2/M checkpoint (Sancar et al. 2004). Whether this information is available to the efferocyte and whether it can inform the efferocyte of the cause of cell death remains entirely unknown.

THE ENVIRONS OF CELL DEATH

The word death is not pronounced in New York, in Paris, in London, because it burns the lips. The Mexican, in contrast, is familiar with death, jokes about it, caresses it, sleeps with it, celebrates it; it is one of his favorite toys and his most steadfast love. True, there is perhaps as much fear in his attitude as in that of others, but at least death is not hidden away; he looks at it face to face with impatience, disdain or irony.

—El Laberinto de la Soledad y otras obras (The Labyrinth of Solitude and other works), Octavio Paz

In keeping with the sentiment alluded to above, we postulate that environmental signals, in addition to the type of cell death, influence the funeral rites (Fig. 3). Changes in environmental signals can specify type I cell death to instruct the elimination of a function. In seasonally breeding male song birds, the levels of testosterone increase during the breeding period and the song center significantly increases in size. After mating, the levels of testosterone decline resulting in a wave of apoptosis (Thompson 2011). Similarly, sex hormones drive the hyperplasia of the pituitary and mammary glands during pregnancy and their decline postpartum and postlactation is associated with pituitary and mammary involution (Scheithauer et al. 1990; McNally and Stein 2017). Efferocytes are responsible for removal of these dead cells. Typically in adult vertebrates, efferocytosis is associated with proliferation and renewal or repair. Removal *sans* renewal or repair is seen primarily in development. Therefore, the responses to these dying cells in adults resemble those during development. It is intriguing to hypothesize that efferocytes exposed to sex hormones are reprogrammed to respond by solely removing the dead cells rather than triggering their replacement.

Environmental signals present during type II or type III cell death may be physical cues. For

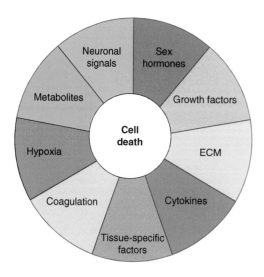

Figure 3. Environmental signals and cell death. A panoply of signals, spanning from hormones and cytokines to neurotransmitters, metabolites, and beyond can coexist with dead cells. The processing of cell death recognition together with distinct environmental signals contributes to the specification of the outcomes of cell death. ECM, Extracellular matrix.

example, loss of cell–cell contact owing to cell death or changes in the stiffness of the extracellular matrix may be an environmental signal sensed by the effectors. It has been described that cell–cell contacts and growth factors such as epidermal growth factor (EGF) establish a tunable system that guides cell-cycle decision making (Kim et al. 2009). A critical threshold of environmental EGF concentration defines the contact-uninhibited from contact-inhibited states of epithelial cells. Conversely, the strength of cell–cell contacts is proportional to the concentration of EGF required to drive cells into cell cycle. Matrix stiffening can reduce the threshold concentration of environmental EGF required to override contact inhibition (Kim and Asthagiri 2011). Perhaps local matrix stiffness or loss of cellular contacts are similarly integrated with the type of cell death to shape the transcriptional output of the efferocyte that responds to cell death. Even more common may be chemical signals. Clotting factors, oxygen concentration (hypoxia), or cytokines and metabolites produced by responding or bystander cells could

inform on the cause of cell death insofar as to interpret it as type II or III cell death. We have previously shown that sensing apoptotic cells together with sensing the cytokine IL-4 enables macrophages to promote healing in the context of type III cell death (Bosurgi et al. 2017). In response to helminth infections, tissue-resident macrophages recruit neutrophils to the damaged site. Following the successful control of the pathogen, neutrophils die by apoptosis. The coincident detection of apoptotic neutrophils and the hallmark cytokine produced during helminth infections—IL-4—induces a tissue repair response in macrophages (Bosurgi et al. 2017). Of note, this response is distinct from that of type II, apoptotic death of aged neutrophils under physiological conditions. In the latter case, the response is the increase production of granulocyte colony-stimulating factor (G-CSF) to replenish the neutrophil pool (Furze and Rankin 2008). The shift in the response from induced proliferation of the lost cells to induction of tissue repair constitutes an example of how environmental cues, such as cytokines, can diversify the response to cell death.

Phagocytosis may also sometimes be restricted to a part of, rather than the whole cell. Such is the case of pruning of presynaptic axonal buttons by microglia during brain development, wherein PtdSer exposure is restricted to the membrane destined to be eaten. Local, environmental cues have been shown to modulate synaptic pruning. IL-33 was found to be expressed by synapse-associated astrocytes in the spinal cord and thalamus and to increase the engulfment of synapses by microglia (Vainchtein et al. 2018). The necessity for invoking a "third-party"-derived molecule for effective pruning of synapses underscores the functional integration of cells in the nervous system rather than a "private" conversation between the phagocyte and the cargo for physiological development of neuronal circuits.

The nature of the environmental signal might be also tissue-specific. Minutti et al. showed that surfactant protein A (SP-A) potentiates IL-4-induced polarization of lung macrophages in response to damage elicited by helminth infection (Minutti et al. 2017). In con-

Cite this article as *Cold Spring Harb Perspect Biol* doi: 10.1101/cshperspect.a036343

trast, C1q was found to potentiate the repair response of macrophages induced by IL-4 in the liver after bacterial infection. This suggests that the integration of cell death sensing with the prevailing cytokine and tissue-specific environmental signals could specify tissue-specific responses. The distinct tissue repair responses adapted to the requirements of specific tissues is an important area of future investigations.

Type III cell death is associated with signature environmental signals. The clotting factor thrombin was recently shown to cleave and activate the cytokine IL-1α in macrophages (Burzynski et al. 2019), leading to the infiltration of immune cells and subsequent wound healing. Hypoxia is also closely associated with an inflammatory state. Although vascular flow is increased during the onset of inflammation, inflamed areas are hypoxic. This is caused by higher interstitial pressure adversely affecting oxygen diffusion into the tissue and also because of increased oxygen consumption in the damaged tissue. Neutrophils increase their oxygen consumption during phagocytosis—a process termed "respiratory burst." This, in turn, can induce a transcriptional program associated with hypoxia in intestinal epithelial cells (Campbell et al. 2014). Hypoxia can also affect transcription factors (i.e., HIF1α), leading to the production of growth factors and promoting endothelial proliferation and growth (Semenza 2000; Apte et al. 2019). Similarly, lactic acid produced in the wound bed leads to the stabilization of HIF1α and the expression of vascular endothelial growth factor (VEGF) by macrophages further contributing to angiogenesis (Constant et al. 2000). Of note, comparable responses underlie pathological angiogenesis that fuels the growth of tumors (Colegio et al. 2014). Thus, thrombin, hypoxia, and lactic acid operate in concert in the induction of tissue repair responses to unscheduled, type III cell death. Yet, how the response to clotting factors, hypoxia, and lactic acid is integrated with the sensing of dead cells remains unexplored.

Another interesting environmental correlate may be neuronal signals. On intestinal bacterial infection, the extrinsic sympathetic nervous system was shown to induce a tissue repair program in muscularis macrophages (Gabanyi et al. 2016). Pain has both been positively, as well as negatively, correlated with wound healing. Some studies claim that pain enables improved wound healing, whereas others indicate that pain is detrimental to wound healing. The latter includes the recent characterization of a microdeletion in an *FAAH* pseudogene in a patient who shows pain and anxiety insensitivity, elevated amounts of anandamide in circulation, and accelerated skin wound healing (Habib et al. 2019). Macrophages express the anandamide receptors CB_1 and CB_2 (Mai et al. 2015) and may be the direct substrates of neuronal signal-induced wound repair.

WHO CARES ABOUT THE DEAD CELL

… it's the others who would live my death
—The Mandarins, Simone de Beauvoir

The macrophage has been front and center of cell death literature as the prime sensors and disposers of dead cells. Even so, evidence suggests that there may be functional redundancies between mononuclear phagocytes and other efferocytes, especially during development. For example, the germline genetic ablation of *Csf1r*, which encodes for the receptor involved in macrophage differentiation, leads to a significant reduction in the number of mononuclear phagocytes in both mice and rats (Dai et al. 2002; Pridans et al. 2018). However, the effects of microglial loss on brain development were less severe in the rat than in the mouse. Despite the absence of microglia, no major abnormalities were observed in the rat brain with the exception of enlargement of lateral ventricles (Pridans et al. 2018). In contrast, both species displayed severe abnormalities in teeth eruption, bone, and reproductive functions (Dai et al. 2002; Pridans et al. 2018). These results indicate that there are redundancies in efferocytosis although such redundancies might be tissue-specific across species. Interestingly, other cell types such as fibroblasts, epithelial cells, and endothelial cells also express efferocytosis receptors (Arandjelovic and Ravichandran 2015). Studies including Monks et al. (2005), Juncadella et al.

(2013), and Mesa et al. (2015) show that epithelial cells can function as efferocytes.

Since many different cells can partake in the sensing and disposal of the dead cells, we propose that the effector function to cell death is, at least in part, determined by the identity of the sensor/efferocyte (Fig. 4). Furthermore, even macrophages resident in different tissues use distinct efferocytosis receptor repertoire and respond to efferocytosis through distinct transcriptional signatures (A-Gonzalez et al. 2017). Thus, the transcriptional program unleashed because of sensing/efferocytosis should depend on the cellular identity of the efferocyte and/or the identity of the efferocytosis receptors.

Dead/dying neurons are removed both during embryonic and postnatal development, as well as in adults. Theoretically, it is possible that the efferocytosis of dead/dying neurons is entrusted to different cell types in development and in adulthood. Alternatively, the efferocyte uses distinct efferocytosis receptors for disposal of the dead cells. Importantly, efferocytosis of embryonic and postnatal neurons during their programmed death culminates in a removal-only response and morphogenesis. It is likely that adult neurogenesis, in contrast, is associated with a counting mechanism mediated by efferocytes and a signal for replenishment as loss of the efferocytosis receptor MERTK in microglia leads to increased neural stem cell proliferation (Fourgeaud et al. 2016).

In keeping with the extraordinary diversity of cell death sensors, macrophages present themselves in several "avatars" (appear in different forms; the term "avatars" originates from the incarnations of Vishnu in different forms in different eras). Differentially polarized macrophages use distinct receptors for phagocytosis and the outcome of phagocytosis too can be quite different. FcγRI-, FcγRII-, and CD13-mediated phagocytosis were significantly increased in IL-10 polarized macrophages, compared with nonpolarized (M_0) macrophages (Mendoza-Coronel and Ortega 2017). Conversely, IL-4 polarization of macrophages enhanced the phagocytosis of *Escherichia coli* but not FcγR- or CD13-mediated phagocytosis (Mendoza-Coronel and Ortega 2017). IFN-γ-polarized macrophages and IL-4-polarized macrophages

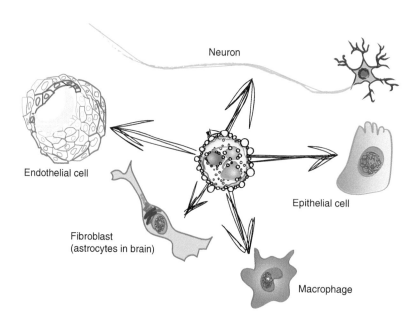

Figure 4. Who cares about dead cells? An array of different cells are endowed with the capacity to efferocytose or sense dead cells. The identity of the cellular sensors involved in the diverse tissues and contexts in which cells can die in part dictates the specific outcomes of the efferocytes that respond to dead cells.

were both more efficient in the phagocytosis of *Porphyromonas gingivalis*, but only the IFN-γ-polarized macrophages were better able to kill the bacteria (Lam et al. 2016). Could it be that there is some specificity for a particular polarization of macrophages for efferocytosis? Could efferocytosis in differentially polarized macrophages or through different efferocytosis receptors encode distinct functional outcomes? Dexamethasone was shown to up-regulate MERTK-mediated uptake of dead cells in macrophages, whereas polyI:C up-regulated AXL-mediated efferocytosis (Zagorska et al. 2014). How this is linked to function remains unclear.

Anticancer immunity may represent a function encoded by the specificity of phagocytes. Although DCs and macrophages are both avid phagocytes, DCs are better at antigen-presentation and in their ability to engage the adaptive immune system. Macrophages, although capable of presentation of foreign antigen on MHCII, are the professional phagocytes involved mostly in scavenging and disposal of type I cell death or for signaling proliferation during homeostatic removal of cells undergoing type II cell death. DCs, in contrast, are typically involved in the context of type III cell death. Therefore, the removal of dead cancer cells by DCs might be critical for generating an anticancer immune response, whereas macrophage-dependent clearance may abet immune escape. We speculate that both cell-intrinsic and environmental signals associated with type II cell death are involved in tumorigenesis and tumor progression. Tumors have been designated as "a wound that never heals." Elimination or blockade of these signals may be important in arresting the propagation of this "wound" that continuously signals for replenishment. Furthermore, triggering a response associated with type III cell death may be what is optimal for eliminating the tumor.

Finally, successful repair and "restitution ad integrum" may involve a coordinated sequence of cell death sensing by phagocytes. The sensing of endothelial cell death by platelets could trigger coagulation and hemostasis. This initial event is followed by inflammation and neutrophil recruitment. Neutrophils often die by a distinct cell death modality while fighting a pathogen—netosis. This changes as the pathogen is cleared and neutrophils continuing to arrive at the wound site die because of lack of growth factor by apoptosis. Macrophages clear the dead cells, including dead neutrophils. We have previously reported that recognition of apoptotic neutrophils by macrophages spurs macrophage polarization to a tissue-repair phenotype (Bosurgi et al. 2017). Finally, the fibroblasts themselves die to allow remodeling of the extracellular matrix and vasculature. As discussed before, epithelial and stromal cells also express cell death sensors. But how the sensing of cell death by these cells may lead to their differential functions during wound repair, such as in fibroblast proliferation and differentiation, epithelial proliferation, and re-epithelialization or angiogenesis, remains unclear. It is possible that aberrations in the sequential cell death recognition in an injured tissue are the basis of wound repair defects spanning from failure of wound closure or ulcers to abnormal growth/overgrowth of unwanted cell types or fibrosis.

CONCLUDING REMARKS

When you're standing on the crossroads that you cannot comprehend
Just remember that death is not the end

—Death Is Not the End, Bob Dylan

In summary, cell death is a conserved feature of developmental remodeling/morphogenesis, tissue turnover/renewal, as well as injury/repair. Inappropriate effector responses to cell death can manifest as developmental defects, tissue degeneration, tumorigenesis and tumor progression, ulcerative damages, infections, fibrosis, and a host of other pathologies. We propose a simple cell death code categorized in a manner consistent with outcomes of cell death, namely, removal, renewal, and repair/regeneration. We also posit that the integration of information contained in a dying/dead cell, environmental signals associated with cell death and the identity of the responding phagocyte constitutes the molecular and cellular basis of this code.

Understanding the fundamental principles guiding the response to cell death may allow modulation of the effector response to better engage physiological processes and to avoid/reduce pathological outcomes of cell death.

ACKNOWLEDGMENTS

We apologize to those whose work we were unable to cite because of space limitations. We thank Ruslan Medzhitov whose work, in part, has inspired this review. We thank Ed Miao for the Bob Dylan quote. This work was supported by grants from the National Institutes of Health (NIH-NIAID R01 AI089824 and NIH-NCI R01 CA212376) and the Kenneth Rainin Foundation. C.V.R is a Howard Hughes Medical Institute (HHMI) Faculty Scholar.

REFERENCES

A-Gonzalez N, Quintana JA, García-Silva S, Mazariegos M, González de la Aleja A, Nicolás-Ávila JA, Walter W, Adrover JM, Crainiciuc G, Kuchroo VK, et al. 2017. Phagocytosis imprints heterogeneity in tissue-resident macrophages. *J Exp Med* 214: 1281–1296. doi:10.1084/jem .20161375

Apte RS, Chen DS, Ferrara N. 2019. VEGF in signaling and disease: beyond discovery and development. *Cell* 176: 1248–1264. doi:10.1016/j.cell.2019.01.021

Arandjelovic S, Ravichandran KS. 2015. Phagocytosis of apoptotic cells in homeostasis. *Nat Immunol* 16: 907–917. doi:10.1038/ni.3253

Avino TA, Barger N, Vargas MV, Carlson EL, Amaral DG, Bauman MD, Schumann CM. 2018. Neuron numbers increase in the human amygdala from birth to adulthood, but not in autism. *Proc Natl Acad Sci* 115: 3710–3715. doi:10.1073/pnas.1801912115

Bosurgi L, Cao YG, Cabeza-Cabrerizo M, Tucci A, Hughes LD, Kong Y, Weinstein JS, Licona-Limon P, Schmid ET, Pelorosso F, et al. 2017. Macrophage function in tissue repair and remodeling requires IL-4 or IL-13 with apoptotic cells. *Science* 356: 1072–1076. doi:10.1126/science .aai8132

Burzynski LC, Humphry M, Pyrillou K, Wiggins KA, Chan JNE, Figg N, Kitt LL, Summers C, Tatham KC, Martin PB, et al. 2019. The coagulation and immune systems are directly linked through the activation of interleukin-1α by thrombin. *Immunity* 50: 1033–1042.e6. doi:10.1016/j .immuni.2019.03.003

Campbell EL, Bruyninckx WJ, Kelly CJ, Glover LE, McNamee EN, Bowers BE, Bayless AJ, Scully M, Saeedi BJ, Golden-Mason L, et al. 2014. Transmigrating neutrophils shape the mucosal microenvironment through localized oxygen depletion to influence resolution of inflammation. *Immunity* 40: 66–77. doi:10.1016/j.immuni.2013.11.020

Colegio OR, Chu NQ, Szabo AL, Chu T, Rhebergen AM, Jairam V, Cyrus N, Brokowski CE, Eisenbarth SC, Phillips GM, et al. 2014. Functional polarization of tumour-associated macrophages by tumour-derived lactic acid. *Nature* 513: 559–563. doi:10.1038/nature13490

Constant JS, Feng JJ, Zabel DD, Yuan H, Suh DY, Scheuenstuhl H, Hunt TK, Hussain MZ. 2000. Lactate elicits vascular endothelial growth factor from macrophages: a possible alternative to hypoxia. *Wound Repair Regen* 8: 353–360. doi:10.1111/j.1524-475X.2000.00353.x

Dai XM, Ryan GR, Hapel AJ, Dominguez MG, Russell RG, Kapp S, Sylvestre V, Stanley ER. 2002. Targeted disruption of the mouse colony-stimulating factor 1 receptor gene results in osteopetrosis, mononuclear phagocyte deficiency, increased primitive progenitor cell frequencies, and reproductive defects. *Blood* 99: 111–120. doi:10 .1182/blood.V99.1.111

Daley SR, Teh C, Hu DY, Strasser A, Gray DHD. 2017. Cell death and thymic tolerance. *Immunol Rev* 277: 9–20. doi:10.1111/imr.12532

Duewell P, Kono H, Rayner KJ, Sirois CM, Vladimer G, Bauernfeind FG, Abela GS, Franchi L, Nunez G, Schnurr M, et al. 2010. NLRP3 inflammasomes are required for atherogenesis and activated by cholesterol crystals. *Nature* 464: 1357–1361. doi:10.1038/nature08938

Elliott MR, Chekeni FB, Trampont PC, Lazarowski ER, Kadl A, Walk SF, Park D, Woodson RI, Ostankovich M, Sharma P, et al. 2009. Nucleotides released by apoptotic cells act as a find-me signal to promote phagocytic clearance. *Nature* 461: 282–286. doi:10.1038/nature08296

Fourgeaud L, Través PG, Tufail Y, Leal-Bailey H, Lew ED, Burrola PG, Callaway P, Zagórska A, Rothlin CV, Nimmerjahn A, et al. 2016. TAM receptors regulate multiple features of microglial physiology. *Nature* 532: 240–244. doi:10.1038/nature17630

Furze RC, Rankin SM. 2008. The role of the bone marrow in neutrophil clearance under homeostatic conditions in the mouse. *FASEB J* 22: 3111–3119. doi:10.1096/fj.08-109876

Gabanyi I, Muller PA, Feighery L, Oliveira TY, Costa-Pinto FA, Mucida D. 2016. Neuro-immune interactions drive tissue programming in intestinal macrophages. *Cell* 164: 378–391. doi:10.1016/j.cell.2015.12.023

Galluzzi L, Vitale I, Aaronson SA, Abrams JM, Adam D, Agostinis P, Alnemri ES, Altucci L, Amelio I, Andrews DW, et al. 2018. Molecular mechanisms of cell death: recommendations of the nomenclature committee on cell death 2018. *Cell Death Differ* 25: 486–541. doi:10 .1038/s41418-017-0012-4

Gardai SJ, McPhillips KA, Frasch SC, Janssen WJ, Starefeldt A, Murphy-Ullrich JE, Bratton DL, Oldenborg PA, Michalak M, Henson PM. 2005. Cell-surface calreticulin initiates clearance of viable or apoptotic cells through trans-activation of LRP on the phagocyte. *Cell* 123: 321–334. doi:10.1016/j.cell.2005.08.032

Ghiringhelli F, Apetoh L, Tesniere A, Aymeric L, Ma Y, Ortiz C, Vermaelen K, Panaretakis T, Mignot G, Ullrich E, et al. 2009. Activation of the NLRP3 inflammasome in dendritic cells induces IL-1β-dependent adaptive immunity against tumors. *Nat Med* 15: 1170–1178. doi:10.1038/nm .2028

Golpon HA, Fadok VA, Taraseviciene-Stewart L, Scerbavicius R, Sauer C, Welte T, Henson PM, Voelkel NF. 2004.

Life after corpse engulfment: phagocytosis of apoptotic cells leads to VEGF secretion and cell growth. *FASEB J* 18: 1716–1718. doi:10.1096/fj.04-1853fje

Habib AM, Okorokov AL, Hill MN, Bras JT, Lee MC, Li S, Gossage SJ, van Drimmelen M, Morena M, Houlden H, et al. 2019. Microdeletion in a *FAAH* pseudogene identified in a patient with high anandamide concentrations and pain insensitivity. *Br J Anaesth* 123: e249–e253. doi:10.1016/j.bja.2019.02.019

Han H, Roan F, Ziegler SF. 2017. The atopic march: current insights into skin barrier dysfunction and epithelial cell-derived cytokines. *Immunol Rev* 278: 116–130. doi:10.1111/imr.12546

Hernández-Martínez R, Covarrubias L. 2011. Interdigital cell death function and regulation: new insights on an old programmed cell death model. *Dev Growth Differ* 53: 245–258. doi:10.1111/j.1440-169X.2010.01246.x

Jordan D, Hindocha S, Dhital M, Saleh M, Khan W. 2012. The epidemiology, genetics and future management of syndactyly. *Open Orthop J* 6: 14–27. doi:10.2174/1874325001206010014

Jorgensen I, Lopez JP, Laufer SA, Miao EA. 2016a. IL-1β, IL-18, and eicosanoids promote neutrophil recruitment to pore-induced intracellular traps following pyroptosis. *Eur J Immunol* 46: 2761–2766. doi:10.1002/eji.201646647

Jorgensen I, Zhang Y, Krantz BA, Miao EA. 2016b. Pyroptosis triggers pore-induced intracellular traps (PITs) that capture bacteria and lead to their clearance by efferocytosis. *J Exp Med* 213: 2113–2128. doi:10.1084/jem.20151613

Juncadella IJ, Kadl A, Sharma AK, Shim YM, Hochreiter-Hufford A, Borish L, Ravichandran KS. 2013. Apoptotic cell clearance by bronchial epithelial cells critically influences airway inflammation. *Nature* 493: 547–551.

Kim JH, Asthagiri AR. 2011. Matrix stiffening sensitizes epithelial cells to EGF and enables the loss of contact inhibition of proliferation. *J Cell Sci* 124: 1280–1287. doi:10.1242/jcs.078394

Kim JH, Kushiro K, Graham NA, Asthagiri AR. 2009. Tunable interplay between epidermal growth factor and cell–cell contact governs the spatial dynamics of epithelial growth. *Proc Natl Acad Sci* 106: 11149–11153. doi:10.1073/pnas.0812651106

Kuhn HG. 2015. Control of cell survival in adult mammalian neurogenesis. *Cold Spring Harb Perspect Biol* 7: a018895. doi:10.1101/cshperspect.a018895

Lam RS, O'Brien-Simpson NM, Holden JA, Lenzo JC, Fong SB, Reynolds EC. 2016. Unprimed, M1 and M2 macrophages differentially interact with *Porphyromonas gingivalis*. *PLoS ONE* 11: e0158629. doi:10.1371/journal.pone.0158629

Mai P, Tian L, Yang L, Wang L, Yang L, Li L. 2015. Cannabinoid receptor 1 but not 2 mediates macrophage phagocytosis by G$_{(\alpha)i/o}$/RhoA/ROCK signaling pathway. *J Cell Physiol* 230: 1640–1650. doi:10.1002/jcp.24911

Martinon F, Pétrilli V, Mayor A, Tardivel A, Tschopp J. 2006. Gout-associated uric acid crystals activate the NALP3 inflammasome. *Nature* 440: 237–241. doi:10.1038/nature04516

McNally S, Stein T. 2017. Overview of mammary gland development: a comparison of mouse and human. *Methods Mol Biol* 1501: 1–17. doi:10.1007/978-1-4939-6475-8_1

Medzhitov R. 2008. Origin and physiological roles of inflammation. *Nature* 454: 428–435. doi:10.1038/nature07201

Mendoza-Coronel E, Ortega E. 2017. Macrophage polarization modulates FcγR- and CD13-mediated phagocytosis and reactive oxygen species production, independently of receptor membrane expression. *Front Immunol* 8: 303. doi:10.3389/fimmu.2017.00303

Mesa KR, Rompolas P, Zito G, Myung P, Sun TY, Brown S, Gonzalez DG, Blagoev KB, Haberman AM, Greco V. 2015. Niche-induced cell death and epithelial phagocytosis regulate hair follicle stem cell pool. *Nature* 522: 94–97.

Minutti CM, Jackson-Jones LH, García-Fojeda B, Knipper JA, Sutherland TE, Logan N, Ringqvist E, Guillamat-Prats R, Ferenbach DA, Artigas A, et al. 2017. Local amplifiers of IL-4Rα-mediated macrophage activation promote repair in lung and liver. *Science* 356: 1076–1080. doi:10.1126/science.aaj2067

Monks J, Rosner D, Geske FJ, Lehman L, Hanson L, Neville MC, Fadok VA. 2005. Epithelial cells as phagocytes: apoptotic epithelial cells are engulfed by mammary alveolar epithelial cells and repress inflammatory mediator release. *Cell Death Differ* 12: 107–114.

Morimoto K, Amano H, Sonoda F, Baba M, Senba M, Yoshimine H, Yamamoto H, Ii T, Oishi K, Nagatake T. 2001. Alveolar macrophages that phagocytose apoptotic neutrophils produce hepatocyte growth factor during bacterial pneumonia in mice. *Am J Respir Cell Mol Biol* 24: 608–615. doi:10.1165/ajrcmb.24.5.4292

Nagata S, Suzuki J, Segawa K, Fujii T. 2016. Exposure of phosphatidylserine on the cell surface. *Cell Death Differ* 23: 952–961. doi:10.1038/cdd.2016.7

Obeid M, Tesniere A, Ghiringhelli F, Fimia GM, Apetoh L, Perfettini JL, Castedo M, Mignot G, Panaretakis T, Casares N, et al. 2007. Calreticulin exposure dictates the immunogenicity of cancer cell death. *Nat Med* 13: 54–61. doi:10.1038/nm1523

Panaretakis T, Kepp O, Brockmeier U, Tesniere A, Bjorklund AC, Chapman DC, Durchschlag M, Joza N, Pierron G, van Endert P, et al. 2009. Mechanisms of pre-apoptotic calreticulin exposure in immunogenic cell death. *EMBO J* 28: 578–590. doi:10.1038/emboj.2009.1

Pridans C, Raper A, Davis GM, Alves J, Sauter KA, Lefevre L, Regan T, Meek S, Sutherland L, Thomson AJ, et al. 2018. Pleiotropic impacts of macrophage and microglial deficiency on development in rats with targeted mutation of the *Csf1r* locus. *J Immunol* 201: 2683–2699. doi:10.4049/jimmunol.1701783

Rathinam VAK, Zhao Y, Shao F. 2019. Innate immunity to intracellular LPS. *Nat Immunol* 20: 527–533. doi:10.1038/s41590-019-0368-3

Reddien PW, Cameron S, Horvitz HR. 2001. Phagocytosis promotes programmed cell death in *C. elegans*. *Nature* 412: 198–202. doi:10.1038/35084096

Ryoo HD, Gorenc T, Steller H. 2004. Apoptotic cells can induce compensatory cell proliferation through the JNK and the Wingless signaling pathways. *Dev Cell* 7: 491–501. doi:10.1016/j.devcel.2004.08.019

Ryu JR, Hong CJ, Kim JY, Kim EK, Sun W, Yu SW. 2016. Control of adult neurogenesis by programmed cell death in the mammalian brain. *Mol Brain* 9: 43. doi:10.1186/s13041-016-0224-4

Sancar A, Lindsey-Boltz LA, Ünsal-Kaçmaz K, Linn S. 2004. Molecular mechanisms of mammalian DNA repair and the DNA damage checkpoints. *Annu Rev Biochem* **73:** 39–85. doi:10.1146/annurev.biochem.73.011303.073723

Saunders JW Jr. 1966. Death in embryonic systems. *Science* **154:** 604–612. doi:10.1126/science.154.3749.604

Scheithauer BW, Sano T, Kovacs KT, Young WF Jr, Ryan N, Randall RV. 1990. The pituitary gland in pregnancy: a clinicopathologic and immunohistochemical study of 69 cases. *Mayo Clin Proc* **65:** 461–474. doi:10.1016/S0025-6196(12)60946-X

Schulz O, Hanč P, Böttcher JP, Hoogeboom R, Diebold SS, Tolar P, Reis ESC. 2018. Myosin II synergizes with F-actin to promote DNGR-1-dependent cross-presentation of dead cell-associated antigens. *Cell Rep* **24:** 419–428. doi:10.1016/j.celrep.2018.06.038

Segawa K, Kurata S, Yanagihashi Y, Brummelkamp TR, Matsuda F, Nagata S. 2014. Caspase-mediated cleavage of phospholipid flippase for apoptotic phosphatidylserine exposure. *Science* **344:** 1164–1168. doi:10.1126/science.1252809

Semenza GL. 2000. HIF-1: mediator of physiological and pathophysiological responses to hypoxia. *J Appl Physiol (1985)* **88:** 1474–1480. doi:10.1152/jappl.2000.88.4.1474

Shapiro GI, Harper JW. 1999. Anticancer drug targets: cell cycle and checkpoint control. *J Clin Invest* **104:** 1645–1653. doi:10.1172/JCI9054

Suzanne M, Steller H. 2013. Shaping organisms with apoptosis. *Cell Death Differ* **20:** 669–675. doi:10.1038/cdd.2013.11

Suzuki J, Denning DP, Imanishi E, Horvitz HR, Nagata S. 2013. Xk-related protein 8 and CED-8 promote phosphatidylserine exposure in apoptotic cells. *Science* **341:** 403–406. doi:10.1126/science.1236758

Thompson CK. 2011. Cell death and the song control system: a model for how sex steroid hormones regulate naturally-occurring neurodegeneration. *Dev Growth Differ* **53:** 213–224. doi:10.1111/j.1440-169X.2011.01257.x

Tsukamoto S, Shibasaki A, Naka A, Saito H, Iida K. 2018. Lactate promotes myoblast differentiation and myotube hypertrophy via a pathway involving MyoD in vitro and enhances muscle regeneration in vivo. *Int J Mol Sci* **19:** 3649. doi:10.3390/ijms19113649

Vainchtein ID, Chin G, Cho FS, Kelley KW, Miller JG, Chien EC, Liddelow SA, Nguyen PT, Nakao-Inoue H, Dorman LC, et al. 2018. Astrocyte-derived interleukin-33 promotes microglial synapse engulfment and neural circuit development. *Science* **359:** 1269–1273. doi:10.1126/science.aal3589

Whitman MC, Greer CA. 2009. Adult neurogenesis and the olfactory system. *Prog Neurobiol* **89:** 162–175. doi:10.1016/j.pneurobio.2009.07.003

Willkomm L, Schubert S, Jung R, Elsen M, Borde J, Gehlert S, Suhr F, Bloch W. 2014. Lactate regulates myogenesis in C2C12 myoblasts in vitro. *Stem Cell Res* **12:** 742–753. doi:10.1016/j.scr.2014.03.004

Yeo W, Gautier J. 2004. Early neural cell death: dying to become neurons. *Dev Biol* **274:** 233–244. doi:10.1016/j.ydbio.2004.07.026

Yokouchi Y, Sakiyama J, Kameda T, Iba H, Suzuki A, Ueno N, Kuroiwa A. 1996. BMP-2/-4 mediate programmed cell death in chicken limb buds. *Development* **122:** 3725–3734.

Zagorska A, Traves PG, Lew ED, Dransfield I, Lemke G. 2014. Diversification of TAM receptor tyrosine kinase function. *Nat Immunol* **15:** 920–928. doi:10.1038/ni.2986

Zargarian S, Shlomovitz I, Erlich Z, Hourizadeh A, Ofir-Birin Y, Croker BA, Regev-Rudzki N, Edry-Botzer L, Gerlic M. 2017. Phosphatidylserine externalization, "necroptotic bodies" release, and phagocytosis during necroptosis. *PLoS Biol* **15:** e2002711. doi:10.1371/journal.pbio.2002711

Zou H, Niswander L. 1996. Requirement for BMP signaling in interdigital apoptosis and scale formation. *Science* **272:** 738–741. doi:10.1126/science.272.5262.738

Cite this article as *Cold Spring Harb Perspect Biol* doi: 10.1101/cshperspect.a036343

Phagocyte Responses to Cell Death in Flies

Andrew J. Davidson and Will Wood

Centre for Inflammation Research, University of Edinburgh, Queen's Medical Research Institute, Edinburgh EH16 4TJ, United Kingdom

Correspondence: andrew.davidson@ed.ac.uk

Multicellular organisms are not created through cell proliferation alone. It is through cell death that an indefinite cellular mass is pared back to reveal its true form. Cells are also lost throughout life as part of homeostasis and through injury. This detritus represents a significant burden to the living organism and must be cleared, most notably through the use of specialized phagocytic cells. Our understanding of these phagocytes and how they engulf cell corpses has been greatly aided by studying the fruit fly, *Drosophila melanogaster*. Here we review the contribution of *Drosophila* research to our understanding of how phagocytes respond to cell death. We focus on the best studied phagocytes in the fly: the glia of the central nervous system, the ovarian follicle cells, and the macrophage-like hemocytes. Each is explored in the context of the tissue they maintain as well as how they function during development and in response to injury.

LIFE, DEATH, AND THE FRUIT FLY

Cell death is an unavoidable part of life. During embryogenesis and development, animals generate an excess of cells, which are then selectively culled through programmed cell death to help shape the emerging organism. Even after birth, deliberate killing of defective or even otherwise viable cells continues throughout adult life with the purpose of maintaining tissue homeostasis and good health. Cell death also occurs unpredictably in response to injury and infection. Ultimately, a significant fraction of our cells have to die for us to live. However, clearing all this cell debris presents a further challenge to every living organism. This is achieved through the engulfment of the dead cells by the living through phagocytosis (also known as "efferocytosis"). Many cells within a tissue have some inherent ability to take up cellular debris in this way in what is termed "nonprofessional phagocytosis." However, as animals have become larger and more complex, their burden of dead cells has increased exponentially. This has necessitated the enlistment of professional phagocytes, which are specifically tasked with removing cell corpses.

Cell death and the clearance thereof has been studied with great success in the nematode *Caenorhabditis elegans* and many of the seminal findings discussed in this review have their origin in this organism. For example, the highly conserved cell death abnormality (CED) pathways were first discovered in the worm through screens to identify mutants in which cell corpses

Cite this article as *Cold Spring Harb Perspect Biol* doi: 10.1101/cshperspect.a036350

were not cleared (Hedgecock et al. 1983; Ellis et al. 1991). This led to the characterization of two parallel, partially redundant pathways acting in the phagocyte to drive engulfment (Ellis et al. 1991). Both these CED pathways converge on the Rho-GTPase, CED-10 (Rac1), the activity of which stimulates the actin cytoskeleton to promote phagocytic cup formation (Kinchen et al. 2005). CED-2/CED-5/CED-12 form one arm of the pathway and act downstream of a number of partially redundant corpse-recognition receptors, including the phosphatidylserine receptor and integrins (Ellis et al. 1991; Gumienny et al. 2001; Wu et al. 2001; Wang et al. 2003; Hsu and Wu 2010). CED-2, CED-5, and CED-12 were subsequently identified in mammals as crkII, Dock180, and ELMO, respectively, and together form an unconventional Rho-GEF that activates Rac1 (Wu and Horvitz 1998a; Gumienny et al. 2001; Wu et al. 2001; Brugnera et al. 2002).

Alternatively, CED-1/CED-6/CED-7 can independently promote CED-10 activation and engulfment (Kinchen et al. 2005). CED-1 (Draper in flies, MEGF10 in mammals) is a receptor involved in corpse recognition, which acts upstream of the adaptor protein CED-6 (Zhou et al. 2001; Callebaut et al. 2003; Freeman et al. 2003). CED-6 (GULP in mammals) binds the NPxY motif within the cytoplasmic tail of CED-1 via its PTB domain (Su et al. 2002). CED-7 is an ABC transporter (ABC1 in mammals) and is interestingly required in both the apoptotic corpse and the phagocyte for effective clearance (Wu and Horvitz 1998b). In the dying cell, it acts to help expose the "eat-me" signal, phosphatidylserine, on the outer leaflet of the plasma membrane (Hamon et al. 2000). This, in turn, is detected by CED-1 in the phagocyte during engulfment (Venegas and Zhou 2007). However, the role of CED-7 plays in the phagocyte is less clear. Furthermore, how CED-1/CED-6 connect to Rac activation to aid corpse uptake is also not known.

The nematode has revealed much about the basics underlying the clearance of cell corpses. However, it lacks the professional class of phagocytes required to engulf and clear the sheer volume of cellular debris found in more complex organisms such as ourselves. The invariant nature of programmed cell death in the nematode is also at odds with the more plastic apoptosis found in most animals. Alternatively, the use of the fruit fly, *Drosophila melanogaster*, offers many advantages for studying the clearance of cell corpses. Like the worm, the fly is highly genetically tractable, aiding the identification and dissection of molecular pathways. Moreover, compared with the nematode, there is an order of magnitude more cell death in the developing fly, which is addressed in part through the possession of professional phagocytes (Abrams et al. 1993). Furthermore, although the patterns of programmed cell death are broadly similar between individuals, the exact cells removed are not precisely delineated as they are in the worm (Abrams et al. 1993; Rogulja-Ortmann et al. 2007). This requires a far greater degree of plasticity from not only the tissues from which cells will be removed, but also from the responding phagocytes themselves. This additional complexity is much closer to the scenario found in mammals. Pioneering studies in *Drosophila* revealed the transcriptional induction of programmed cell death and the inhibitory control of apoptosis (White et al. 1994; Hay et al. 1995). However, this review focuses on the contribution of *Drosophila* research to our understanding of how both professional and nonprofessional phagocytes respond to and clear cell death. The phagocytic glia of the central nervous system (CNS), the follicular cells of the ovary, and the macrophage-like hemocytes are the three best studied fly phagocytes, and their mode of engulfment will be compared and contrasted in this review. *Drosophila* possess conserved homologs of all the CED pathway members and a particular focus will be placed on the findings that have emerged from their study in these three phagocytic cell types. We will also highlight the central role of the fly CED-1 homolog, Draper, as it operates in all three of these phagocytes and in each step of corpse engulfment. Finally, an emphasis will also be placed on the tissue context in which these phagocytes operate as this has profound consequences for how cellular debris is ultimately cleared.

PHAGOCYTIC GLIA OF THE FLY CENTRAL NERVOUS SYSTEM

The developing CNS of the fly embryo contain specialized phagocytic glial cells that act to clear apoptotic neurons generated during development and metamorphosis (Sonnenfeld and Jacobs 1995; Cantera and Technau 1996). Glia are nonneuronal cells that play many different supportive roles within the brain. Unlike vertebrates, flies do not have CNS resident macrophages such as microglia, making glia the sole phagocytes within the fly brain.

The developing CNS is shaped by waves of neuronal programmed cell death, which begin during mid–late embryogenesis (Abrams et al. 1993; Rogulja-Ortmann et al. 2007). Once ensheathed, the developing CNS of the embryo (and the apoptotic corpses therein) is isolated from the patroling macrophages (hemocytes) that populate the interstitial space above the CNS (Sonnenfeld and Jacobs 1995). It then falls to the astrocyte-like glia embedded within the embryonic CNS to clear the apoptotic neurons (Kurant et al. 2008). Although classed as "nonprofessional phagocytes," the phagocytic activity of these glia is comparable to that of the macrophages. Despite their exclusion from the CNS, macrophages do engulf some neuronal debris, presumably occurring before ensheathment is completed at the interface between these immune cells and the CNS (Sonnenfeld and Jacobs 1995).

Clearance of embryonic neuronal debris by glia requires the *Drosophila* CED-1 homolog, Draper (Freeman et al. 2003). However, Draper is not required for the uptake of apoptotic debris by the embryonic glia and is instead required for processing of the engulfed corpses (Kurant et al. 2008). Therefore, the failure of *draper* mutant glia to effectively clear embryonic neuronal debris is presumably because of an inability to turn over engulfed corpses, preventing further phagocytosis. A related receptor, SIMU, is required for glial phagocytosis in the embryo and acts upstream of Draper during engulfment (Kurant et al. 2008). Although the extracellular domains of SIMU and Draper share similarities, the former lacks the intracellular cytoplasmic

signaling domains possessed by Draper. Both are highly expressed in glia as part of their developmental differentiation and are sufficient to promote the phagocytic function of these cells (Shklyar et al. 2014). It is proposed that SIMU acts as a tethering receptor that binds directly to the phosphatidylserine exposed on dying cells (Shklyar et al. 2013). Although Draper also binds phosphatidylserine, this is seemingly not required for the interaction between glia and apoptotic debris and instead potentially triggers downstream signaling required for subsequent corpse processing. Surprisingly, SIMU is not expressed in postembryonic glia and is not necessary for clearing apoptotic debris during the wave of neuronal programmed cell death occurring during metamorphosis (Hilu-Dadia et al. 2018). Instead, it is Draper that is required for glial engulfment of apoptotic neurons (Hilu-Dadia et al. 2018). Glia also play an active role in pruning, whereby superfluous neuronal projections are removed (Fig. 1; Awasaki and Ito 2004). Although pruning is not associated with neuronal death, it uses the same engulfment machinery required by glia to clear apoptotic debris (Awasaki et al. 2006; Williams et al. 2006). Furthermore, localized caspase activity (which usually drives apoptosis) within the degenerating dendrites is necessary for their clearance during pruning (Williams et al. 2006).

Beyond clearing degenerating neurons as part of development, glia also engulf dying neurons arising from damage to the adult CNS (MacDonald et al. 2006). One powerful model of this response is the glia-mediated clearance of severed olfactory receptor neurons undergoing Wallerian degeneration in the adult fly antennal lobe (MacDonald et al. 2006). In contrast to the embryo, in the adult CNS, it is ensheathing glia and not astrocyte-like glia that clear degenerating axons (Doherty et al. 2009). Whether this difference is because of a dramatic switch in the role of astrocyte-like glia, a further subdivision of this cell type between the embryo and the adult or simply a change in phagocytic glia morphology is not clear. In response to axonal injury, these glia up-regulate their phagocytic machinery, extend infiltrating membrane processes into the antennal lobe and specifically target the

Central nervous   Synaptic pruning

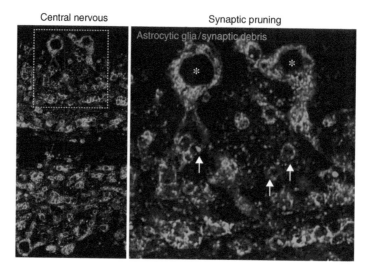

Figure 1. Phagocytic glia clear synaptic debris within the fly central nervous system. (*Left* panel) Low-magnification image of pupal ventral nerve cord undergoing synaptic pruning (6 h APF). (*Right* panel) Magnified image of region outlined by dashed box. Astrocyte-like glia (mCD8-GFP driven by alrm-GAL4, green) and synaptic debris (anti-Brp staining, red) are visualized. Asterisks highlight two astrocyte-like glia in which examples of phagocytic events are seen. Note the extensions of these two cells, highlighted by arrows, in which the anti-Brp red staining is enclosed in the green glial phagocytic cups. (Images provided by Yunsik Kang and Marc Freeman.)

cleaved axons for engulfment (MacDonald et al. 2006). Draper is one such up-regulated receptor and in its absence glia fail to infiltrate toward the severed axons, which then persist within the lobe long after injury (MacDonald et al. 2006). Draper possesses a carboxy-terminal immunoreceptor tyrosine-based activation motif (ITAM), a signaling motif commonly found in the cytoplasmic tail of mammalian immunoreceptors including Fc, T-cell, and B-cell receptors (Ziegenfuss et al. 2008). The ITAM of Draper is phosphorylated by the src family kinase, Src42a, leading to association with the Syk family kinase, Shark (Ziegenfuss et al. 2008). Loss of either Src42a or Shark phenocopies loss of Draper, whereby glia fail to clear axonal damage within the antennal lobe (Ziegenfuss et al. 2008). The proposed model emerging from this work is that Draper ligand binding promotes receptor clustering and subsequent ITAM phosphorylation by Src42a. The activated Draper receptor then recruits Shark, which in turn triggers downstream signaling and ultimately engulfment. Interestingly, a second, inhibitory splice variant of Draper (Draper-II) is expressed in glia following glial recruitment to

severed axons (Logan et al. 2012). Draper-II lacks an ITAM and instead possesses an immunoreceptor tyrosine-based inhibitory motif (ITIM). Draper-II acts in a negative feedback loop to suppress Draper activity via recruitment of the SHIP phosphatase, corkscrew, to dephosphorylate Shark (Logan et al. 2012). It is proposed that the delayed expression of Draper-II is timed to coincide with the clearance of axonal damage and so prevent excessive, potentially damaging, glial activity.

Beyond Draper, the clearance of severed axons by glia also requires other CED pathway components, including crk (CED-2/crkII), mbc (CED-5/Dock180), and *Drosophila* CED-12 (ELMO) (Ziegenfuss et al. 2012). Together, these three form an unconventional Rho-GEF, which activates the RhoGTPase, Rac1 (CED-10), which in turn mobilizes the actin cytoskeleton and drives phagocytic cup formation. Whereas Draper and Rac1 are required for the initial infiltration of glial processes toward the damaged axons, crk/mbc/dCED-12 are only required for the subsequent uptake of cellular debris (Ziegenfuss et al. 2012). Because the output of the crk/mbc/dCED-12 complex is

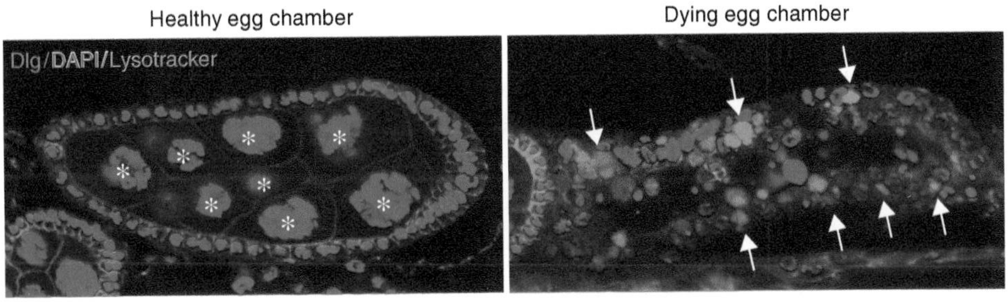

Figure 2. The follicle cells of the ovary phagocytose dying nurse cells. Egg chambers stained with DAPI (nuclei, blue), anti-Dgl (cell boundaries, green), and Lysotracker (acidified phagosomes, red). (*Left* panel) Healthy egg chamber containing nurse cells (asterisks) surrounded by follicle cells. (*Right* panel) Dying egg chamber in which the nurse cells have been engulfed by the follicle cells. The numerous acidified mature phagosomes within the follicle cells (arrows) are indicative of engulfment. (Images provided by Sandy Serizier and Kim McCall.)

activated Rac1, the requirement for Rac activity for glial recruitment implies the presence of a distinct, earlier activator. The sevenless receptor tyrosine kinase (sev RTK) pathway members downstream of receptor kinase (DRK), daughter of sevenless (DOS), and son of sevenless (SOS) are also required for clearance of axonal damage by glia (Lu et al. 2014). Although the sev RTK pathway conventionally activates the Rho GTPase, Ras, glial expression of dominant-negative ras failed to block engulfment. Instead within the glia, DRK/DOS/SOS appear to be working in parallel with crk/mbc/dCED-12 to activate Rac1. Loss of either pathway blocks glial engulfment rather than glial recruitment to axonal injury. However, the combined disruption of both results in a complete failure of the glia to respond to the damaged axons. Confusingly, this implies that these two pathways are functioning redundantly during the initial infiltration of glia toward axonal damage, but nonredundantly during engulfment. No physical interaction between Draper and DRK has been detected and so remains unknown whether Draper is directly responsible for sev RTK pathway activation. However, what is clear is that Draper plays a central role in orchestrating glial engulfment of axonal debris. Importantly, the research discussed in this section has expanded our understanding of the intracellular signaling occurring downstream of Draper and, in all probability, other CED-1 homologs.

The study of phagocytic glia of the fly have taught us much about how the CNS is kept clear of cell corpses arising from development or injury. These glia function within a highly insulated, compact, and complicated tissue where collateral damage from careless or excessive phagocytosis can have severe consequences. One particular advantage offered by *Drosophila* is the opportunity to study glia within their in vivo environment and so fully capture their complex role within the brain

PHAGOCYTIC FOLLICLE CELLS OF THE *DROSOPHILA* OVARY

Within the *Drosophila* ovary, hundreds of egg chambers progress through the well-defined stages of oogenesis. Within each egg chamber, the oocyte is nourished by 15 nurse cells and surrounding these germ line cells are the somatic epithelial follicle cells (Fig. 2). Post "dumping" of their cytoplasmic contents into the oocyte, nurse cells undergo programmed cell death and are cleared by a phagocytic subset of the follicle cells (Timmons et al. 2016). Experimentally, nutrient deprivation can induce premature nurse cell death, which again leads to their engulfment by the follicle cells (Fig. 2). Because the egg chambers of the fly ovary are sealed off from professional phagocytes such as the macrophages, these follicle cells are solely responsible for the clearance of dying nurse cells. Ultimately, they serve as an important model of nonprofessional phagocytosis including Sertoli cells found in the testes.

As in the glia, efficient nurse cell clearance by follicle cells depends on the fly CED homologs, including Draper (CED-1), dCED-12, Rac1 (CED-10), and a recently identified fly CED-7 homolog, Eato, which act within their canonical parallel pathways (Etchegaray et al. 2012; Timmons et al. 2016; Santoso et al. 2018). Draper, in particular, is enriched at the interface between the engulfing phagocyte and the dying nurse cells (Etchegaray et al. 2012). Integrins (specifically αPS3/βPS heterodimers) are also involved in nurse cell clearance and are actively trafficked to the apical surface of the follicle cells where they accumulate with Draper (Meehan et al. 2015, 2016). The combined loss of Draper and integrins yields a stronger defect in clearance than loss of either one alone, implying that they work independently from one another (Meehan et al. 2016). Nevertheless, together, these receptors likely trigger phagocytosis by allowing follicle cells to physically interact with the nurse cells. Whereas integrins are only involved in the initial uptake of cellular debris, Draper is additionally required for subsequent phagosome maturation/acidification (Meehan et al. 2016). Interestingly, follicle cells lacking both Draper and integrins still retain some residual phagocytic ability, implying there are players still yet to be identified (Meehan et al. 2016).

Overexpression of Draper alone in the phagocytic follicle cells is sufficient to promote the inappropriate engulfment of otherwise healthy nurse cells (Etchegaray et al. 2012). Strikingly, it has emerged that neither apoptosis nor autophagy are necessary for the death of the nurse cells (Peterson and McCall 2013). It appears that the follicle cells induce the death of the nurse cells nonautonomously via their engulfment machinery in what is termed "phagoptosis" (Timmons et al. 2016). The existence of phagoptosis turns what we understand about the interaction between dying cell and phagocyte on its head. Instead of simply responding to and clearing cell corpses, phagocytes can play an active role in inducing cell death raising fascinating questions about selectivity and implementation. Exactly how follicle cells trigger the death of the nurse cells is yet to be addressed, but it appears to require the lysosomal pathway (Peterson and McCall 2013; Timmons et al. 2016). As a self-contained tissue isolated from any other phagocytes and where cell death can be easily induced experimentally, the *Drosophila* ovary represents an ideal setting to dissect the process of phagoptosis.

PHAGOCYTIC MACROPHAGES OF THE FLY IMMUNE SYSTEM

Macrophage-like blood cells (hemocytes) are the professional phagocytes of the fly and represent the cellular component of the fly's immune system. They are an extremely multifunctional cell type, involved in the inflammatory response to wounds, the engulfment of pathogens, and the deposition of extracellular matrix (Stramer et al. 2005; Vlisidou et al. 2009; Matsubayashi et al. 2017). Like their mammalian counterparts, *Drosophila* macrophages also have a vital role in clearing cellular debris generated during embryogenesis and metamorphosis, as well as that arising from tissue damage. In contrast to the other two types of phagocytes discussed in this review, macrophages are highly motile during embryogenesis and metamorphosis. Although they show low motility during larval and adult stages, this is because they are instead pumped around the larval/adult body cavity by the fly's open circulatory system. Importantly, this allows macrophages to move toward and clear cellular debris that would be otherwise out of reach to immobile cells.

The embryonic macrophages extend large, dynamic, actin-rich protrusions known as "lamellipods" to move toward and engulf pathogenic intruders and cellular debris (Davidson et al. 2019). Macrophages originate from the head mesoderm and migrate along predetermined routes to populate the entire embryo (Tepass et al. 1994; Wood et al. 2006). During this developmental dispersal, these highly phagocytic cells clear apoptotic corpses that have been eliminated through programmed cell death as part of embryogenesis (Tepass et al. 1994). The engulfment of these first apoptotic corpses triggers the up-regulation of phagocytic receptors

Cite this article as *Cold Spring Harb Perspect Biol* doi: 10.1101/cshperspect.a036350

such as the CED-1 homolog, Draper, and the CD36 homolog, croquemort (Franc et al. 1999; Weavers et al. 2016a). This in turn undoubtedly heightens their phagocytic activity and reinforces their macrophage identity. Remarkably, in the absence of the key blood cell differentiation factor, serpent, expression of either crq, Draper or SIMU is sufficient to partially restore the dispersal and phagocytic ability of macrophages within the embryo (Shlyakhover et al. 2018).

Draper is a promiscuous receptor and has several reported ligands. One such ligand is Pretaporter, which during apoptosis is released from the endoplasmic reticulum, at which point it relocates to the cell surface and is subsequently recognized by phagocytes via Draper (Kuraishi et al. 2009). However, this initial finding has yet to be advanced. Macroglobulin complement-related (mcr), is another proposed Draper ligand (Lin et al. 2017). However, as of yet no physical interaction been Draper and mcr has been detected. By far the best characterized Draper ligand thus far is phosphatidylserine, which is one of the best known "eat me" signals exposed on the outer leaflet of cells undergoing apoptosis (Tung et al. 2013). Furthermore, this receptor/ligand interaction is consistent with that found with CED-1 in the worm (Venegas and Zhou 2007). The overexpression of Draper in hemocyte-derived, *Drosophila* "S2" cell lines is sufficient to increase apoptotic corpse uptake in vitro (Williamson and Vale 2018). During S2 cell engulfment of apoptotic corpses or phosphatidylserine-coated beads in vitro, Draper is strongly enriched at the phagocytic cup (Williamson and Vale 2018). This promotes phosphorylation of Draper's cytoplasmic tail, including on tyrosine residues within the ITAM motif. As in phagocytic glia, the phosphorylation of Draper's ITAM motif triggers intracellular signaling, including recruitment of Shark, which becomes enriched at S2 cell phagocytic cups during corpse engulfment (Williamson and Vale 2018). Initially it was reported that the knockdown of Draper severely impairs macrophage phagocytosis of cell corpses both in vitro and within the embryo (Manaka et al. 2004). However, surprisingly, Draper is not required for

embryonic macrophage uptake of apoptotic cells in vivo, as *draper* mutant macrophages are full of engulfed corpses (Kurant et al. 2008; Evans et al. 2015). In fact, *draper* mutant macrophages are excessively vacuolated, containing a higher apoptotic load than their wild-type counterparts, implying a downstream corpse processing defect (Evans et al. 2015). The related receptor, SIMU, has instead been implicated in corpse uptake, acting upstream of Draper (Kurant et al. 2008). Furthermore, the loss of SIMU results in a failure to clear apoptotic debris from the embryo (Roddie et al. 2019). However, even the combined loss of both SIMU and Draper fails to block all engulfment.

Another phagocytic receptor, the fly homolog of the vertebrate scavenger receptor CD36 named croquemort (crq), is first detected in macrophages during their embryonic dispersal and is required for efficient clearance of apoptotic corpses (Franc et al. 1996, 1999). However, the downstream effectors of crq and how they feed into the established CED pathways is poorly understood. A screen for genes required for efficient clearance of corpses by the embryonic macrophages identified Pallbearer, which acts through an E3 ubiquitin ligase complex to target proteins for proteasomal degradation (Silva et al. 2007). Exactly which proteins Pallbearer earmarks for destruction and how this contributes to corpse uptake is not known and this finding has not been further advanced. In the same screen, the junctophilin Undertaker/Retinophilin was also implicated in clearance of apoptotic corpses by embryonic macrophages (Cuttell et al. 2008). Undertaker/Retinophilin was found to work downstream of Draper and dCED-6 and regulate intracellular calcium release from the endoplasmic reticulum. Interestingly, transient increases in macrophage intracellular calcium are observed with every phagocytic event and are required for the up-regulation of Draper via JNK activity (Weavers et al. 2016a). This suggests a feedback loop whereby Draper increases its own expression via Undertaker and intracellular calcium release.

Although clearing apoptotic debris arising from programmed cell death is a major role of the embryonic macrophages, they are also highly responsive to tissue damage induced by wounding. When laser ablation is used to wound the overlying epithelium, macrophages chemotax to the site of injury where they engulf cellular debris (Fig. 3). Given the instantaneous nature of laser-induced damage and the lack of cleaved-caspase staining, it is assumed that the cellular debris within the wound is necrotic (Weavers et al. 2016a). As the presence of macrophages is not required for wound closure itself, the clearance of these necrotic cells is likely to be a primary function of the recruited immune cells (alongside phagocytosis of any invading pathogens) (Stramer et al. 2005). Interestingly, during their developmental dispersal throughout the embryo, macrophages will prioritize the engulfment of apoptotic cells over recruitment to wounds (Moreira et al. 2010). However, this is likely attributable to their immaturity because until these macrophages have engulfed an apoptotic corpse, they are unresponsive to wounds (Weavers et al. 2016a). These early phagocytic events are vital for functionalizing (or "priming") these macrophages, without which they fail to mount the appropriate inflammatory response (Weavers et al. 2016a). In ΔH99 mutant embryos, which are deficient in programmed cell death, macrophages disperse throughout the embryo without ever encountering an apoptotic corpse. Although their basal migration is normal, these cells are only weakly recruited to wounds (Weavers et al. 2016a). Reintroduction of apoptotic corpses into ΔH99 mutant embryos via UV irradiation, rescues the inflammatory response. The engulfment of apoptotic corpses activates JNK signaling (see above), which upregulates the expression of Draper (Weavers et al. 2016a). Elevated Draper is required for robust macrophage recruitment to wounds and exogenously expressed Draper alone is sufficient to rescue inflammation in the absence of engulfment (Evans et al. 2015; Weavers et al. 2016a). Interestingly, too much cell death also impairs the ability of macrophages to mount a subsequent inflammatory response (Roddie et al. 2019). Therefore, it appears that the amount of corpse uptake by macrophages plays a critical role in defining their future behavior.

When the embryo is wounded, one of the earliest damage signals released is hydrogen peroxide (Moreira et al. 2010; Razzell et al. 2013). This is detected within these immune cells by a redox-sensitive cysteine within the src family kinase, Src42a (Evans et al. 2015). Src42a is known to phosphorylate the ITAM motif within the cytoplasmic tail of Draper (Ziegenfuss et al. 2008; Evans et al. 2015). This, in turn, leads to the recruitment of the downstream effector kinase Shark and recruitment to the wound. Although mathematical modeling of macrophage

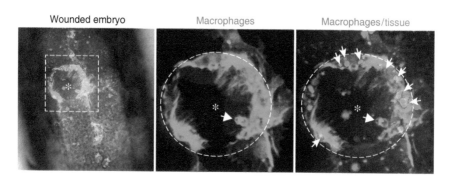

Figure 3. Inflammatory macrophages engulf debris at a wound. (*Left* panel) Low-magnification image of the ventral surface of a stage 15 embryo. An epithelial wound (asterisk) has been generated by laser ablation resulting in macrophage recruitment (LifeAct-GFP driven by sn-GAL4, green). (*Middle* panel) Magnified image of region outlined by the dashed box. Macrophages encircle the wound edge (dashed circle) and phagocytose debris (arrow). (*Right* panel) Same image with inclusion of whole embryo tissue label (sqh-mCherry-Moesin, red). Internalized labeled tissue is present in macrophages (barbed arrows) originating from wound and earlier apoptotic events.

Cite this article as *Cold Spring Harb Perspect Biol* doi: 10.1101/cshperspect.a036350

behavior in response to wounding suggests that hydrogen peroxide is not the de facto chemoattractant that is guiding macrophages to the wound, it is important in triggering the inflammatory response (Razzell et al. 2013; Weavers et al. 2016b). Draper's responsiveness to hydrogen peroxide combined with its requirement for the inflammatory recruitment of macrophages to the wound implies it is a chemotactic receptor (Evans et al. 2015). Thus, Draper is necessary for chemotaxis to wounds, is involved in phagocytosis (albeit not strictly required), and needed for efficient corpse processing.

It is increasingly clear from studies in *Drosophila* that the inflammatory function of macrophages is intricately intertwined with their role in engulfing cell corpses. Combined with its powerful genetics and excellent in vivo imaging, the great advantage offered by *Drosophila* for macrophage research is the ability to challenge these immune cells with a full array of different stimuli within the context of a living organism. Such an approach will be necessary if we are to fully appreciate the consequences of cell corpse uptake on the inflammatory response and vice versa. Furthermore, the increasing use of the pupae for live imaging is offering the opportunity to visualize macrophage behavior in a whole variety of contexts not possible in the

embryo (Thuma et al. 2018). As such, there remains much to learn from the fly about these truly multifunctional phagocytes.

Draper: The Many Headed Hydra

In each of the three *Drosophila* phagocytes discussed within this review, the CED-1 homolog Draper is a recurring protagonist (Fig. 4). Furthermore, in response to cell death, Draper is acting as a multifunctional "Hydra" playing a central role in all aspects of corpse clearance, including chemotaxis, phagocytosis, and phagosome maturation/acidification (Fig. 5). As discussed throughout this review, Draper has been implicated in corpse recognition, uptake, and processing in each of the phagocytes in the fly, highlighting its key role in shepherding cellular debris through the entire engulfment process (Freeman et al. 2003; Kurant et al. 2008; Evans et al. 2015; Meehan et al. 2016). Quite how this one receptor is capable of contributing to all these functions is intriguing and perplexing. One might think that Draper's promiscuity offers one possible solution, whereby the recognition of different ligands triggers different molecular and, therefore, cellular, responses. Furthermore, hydrogen peroxide release on wounding activates Draper via Src42, bypassing

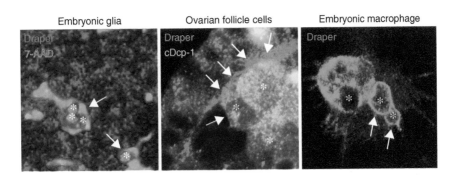

Figure 4. Draper localization during corpse clearance in multiple fly phagocytes. (*Left* panel) Draper (anti-Draper, green) localizes to membranes of embryonic glia (arrows) engulfing apoptotic corpses (7-AAD staining, purple) in developing central nervous system. Asterisks highlight engulfed corpses. (*Middle* panel) Draper (anti-Draper, pink) is enriched (arrows) at the interface between the engulfing follicle cells and the dying nurse cells (anti-cDcp-1, asterisks and orange). (*Right* panel) Draper (Draper-GFP driven by sn-GAL4) localizes to phagocytic cups (arrows) of embryonic macrophages engulfing apoptotic corpses (negative fluorescence and asterisks). (The glia image was provided by Mary Logan and the ovarian follicle image was provided by Sandy Serizier and Kim McCall.)

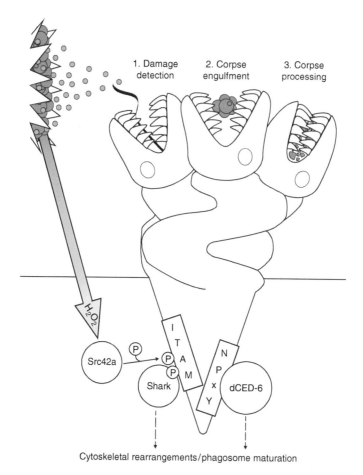

Figure 5. Draper is a multifunctional "Hydra," participating in all stages of corpse clearance. (1) Draper is required to detect tissue damage (red circles) perhaps via detection of an as-of-yet unidentified ligand released from the wound (blue circles). Hydrogen peroxide emanating from the wound (orange/yellow arrow) bypasses the requirement for Draper ligand binding by directly activating Src42a. Activated Src42a phosphorylates the ITAM domain, which leads to the recruitment of Shark and downstream signaling. This draws the phagocyte toward the debris via cytoskeletal rearrangements. (2) Draper localizes to phagocytic cups during engulfment of cell corpses (red circles) triggering ITAM phosphorylation via Src42a. Shark recruitment and subsequent downstream signaling promotes the cytoskeletal rearrangements required for engulfment. (3) Draper is also required for phagosome maturation and efficient corpse processing via its NPxY motif, which associates with dCED-6/GULP.

the need for receptor engagement (Evans et al. 2015). Nevertheless, how different stimuli can evoke different responses from the same receptor is difficult to explain. Remarkably, much of Draper's extracellular sequence is dispensable for clearance of neuronal damage by glia in the CNS (Logan et al. 2012). This result awaits further exploration in all of the established roles of Draper and the different phagocytes. It is also possible that tethering receptors such as SIMU

relieve Draper of the necessity of interacting with cellular debris directly. However, in the absence of definitive answers, such findings further complicate our understanding of Draper.

Intracellularly, Draper possesses two well-defined signaling motifs: an NPxY motif and an ITAM domain. Each of these motifs interact with different downstream effectors and therefore might drive the different functions of Draper. For instance, although the ITAM motif

Cite this article as *Cold Spring Harb Perspect Biol* doi: 10.1101/cshperspect.a036350

is necessary for the inflammatory recruitment to wounds, it is not required for the processing of engulfed corpses (Evans et al. 2015). However, whether this holds true for the other phagocytes remains to be confirmed.

Beyond clearance of cellular debris, Draper has also been implicated in mediating cell–cell competition and in the caspase-independent, autophagic cell death of the salivary glands (Li and Baker 2007; McPhee et al. 2010). Cell competition is the process that drives the elimination of cells of one genotype by the cells of a different genotype within the same tissue. The requirement for Draper in this process might prove to be via phagoptosis, whereby the "winner" cells kill the "loser" cells through Draper-mediated engulfment (Li and Baker 2007). How the cells of the salivary glands use Draper to promote their own caspase-independent programmed cell death is less clear. However, to date, no known phagocytes have been implicated in the clearance of these cells, which might otherwise be driving their own removal via autophagy (McPhee et al. 2010). Within this setting, it is possible that Draper is required for the maturation of autolysosomes similar to its requirement for corpse processing during engulfment. However, even if the salivary glands are able to devour themselves, there remains a need for a yet-to-be-identified phagocyte to engulf whatever remains and Draper will undoubtedly have a role in this uptake too.

CONCLUDING REMARKS

In many ways, *Drosophila* is the ideal model to study the clearance of cellular debris. The fly combines a sophisticated, yet well-defined development and anatomy (and the inevitable associated cell death) with incredible genetic tractability. Furthermore, the fly possesses a variety of phagocytes operating within a range of different tissues, with each presenting unique demands and constraints. The best studied of these phagocytes include the phagocytic glia of the fly CNS, the follicle cells of the *Drosophila* ovary, and the macrophage-like hemocytes of the fly innate immune system. These have all been intensively studied over the past few de-

cades and so each offer abundant molecular tools as well as a wealth of preexisting knowledge. A general theme emerging from the in vivo study of engulfment within the fly is that not only are phagocytes carefully matched to their quarry, but the mode of cell death is tailored to suit their designated phagocyte. For example, the motility of the macrophages allows them to clear developmental debris from a wide array of different tissues alongside the ability to rapidly concentrate phagocytosis in response to wounding. Tellingly, the voracious macrophages are excluded from particularly delicate tissues such as the CNS and the ovary, where specialized phagocytes operate. Despite the extent to which these various phagocytes differ in their morphology, mode of engulfment, tissue residency and the target cells they ultimately clear, they share clear commonalities in their underlying molecular pathways. More specifically, the CED pathways and, in particular, the CED-1 homolog Draper, are driving clearance of cell corpses in each of these phagocytes. How such phagocytic plasticity is derived from the same set of molecular players is an intriguing open question, but one which *Drosophila* and its phagocytes are well placed to address.

ACKNOWLEDGMENTS

We would like to thank Yunsik Kang, Marc Freeman, Mary Logan, Sandy Serizier, and Kim McCall for contributing images for the figures presented in this review.

REFERENCES

Abrams JM, White K, Fessler LI, Steller H. 1993. Programmed cell death during *Drosophila* embryogenesis. *Development* 117: 29–43.

Awasaki T, Ito K. 2004. Engulfing action of glial cells is required for programmed axon pruning during *Drosophila* metamorphosis. *Curr Biol* 14: 668–677. doi:10 .1016/j.cub.2004.04.001

Awasaki T, Tatsumi R, Takahashi K, Arai K, Nakanishi Y, Ueda R, Ito K. 2006. Essential role of the apoptotic cell engulfment genes *draper* and *ced-6* in programmed axon pruning during *Drosophila* metamorphosis. *Neuron* 50: 855–867. doi:10.1016/j.neuron.2006.04.027

Brugnera E, Haney L, Grimsley C, Lu M, Walk SF, Tosello-Trampont AC, Macara IG, Madhani H, Fink GR, Ravichandran KS. 2002. Unconventional Rac-GEF activity is mediated through the Dock180-ELMO complex. *Nat Cell Biol* **4:** 574–582. doi:10.1038/ncb824

Callebaut I, Mignotte V, Souchet M, Mornon J-P. 2003. EMI domains are widespread and reveal the probable orthologs of the *Caenorhabditis elegans* CED-1 protein. *Biochem Biophys Res Commun* **300:** 619–623. doi:10.1016/S0006-291X(02)02904-2

Cantera R, Technau GM. 1996. Glial cells phagocytose neuronal debris during the metamorphosis of the central nervous system in *Drosophila melanogaster*. *Dev Genes Evol* **206:** 277–280. doi:10.1007/s004270050052

Cuttell L, Vaughan A, Silva E, Escaron CJ, Lavine M, Van Goethem E, Eid JP, Quirin M, Franc NC. 2008. Undertaker, a *Drosophila* Junctophilin, links Draper-mediated phagocytosis and calcium homeostasis. *Cell* **135:** 524–534. doi:10.1016/j.cell.2008.08.033

Davidson AJ, Millard TH, Evans IR, Wood W. 2019. Ena orchestrates remodelling within the actin cytoskeleton to drive robust *Drosophila* macrophage chemotaxis. *J Cell Sci* **132:** jcs224618. doi:10.1242/jcs.224618

Doherty J, Logan MA, Tasdemir OE, Freeman MR. 2009. Ensheathing glia function as phagocytes in the adult *Drosophila* brain. *J Neurosci* **29:** 4768–4781. doi:10.1523/jneurosci.5951-08.2009

Ellis RE, Jacobson DM, Horvitz HR. 1991. Genes required for the engulfment of cell corpses during programmed cell death in *Caenorhabditis elegans*. *Genetics* **129:** 79–94.

Etchegaray JI, Timmons AK, Klein AP, Pritchett TL, Welch E, Meehan TL, Li C, McCall K. 2012. Draper acts through the JNK pathway to control synchronous engulfment of dying germline cells by follicular epithelial cells. *Development* **139:** 4029–4039. doi:10.1242/dev.082776

Evans IR, Rodrigues FS, Armitage EL, Wood W. 2015. Draper/CED-1 mediates an ancient damage response to control inflammatory blood cell migration in vivo. *Curr Biol* **25:** 1606–1612. doi:10.1016/j.cub.2015.04.037

Franc NC, Dimarcq JL, Lagueux M, Hoffmann J, Ezekowitz RA. 1996. Croquemort, a novel *Drosophila* hemocyte/macrophage receptor that recognizes apoptotic cells. *Immunity* **4:** 431–443. doi:10.1016/S1074-7613(00)80410-0

Franc NC, Heitzler P, Ezekowitz RA, White K. 1999. Requirement for croquemort in phagocytosis of apoptotic cells in *Drosophila*. *Science* **284:** 1991–1994. doi:10.1126/science.284.5422.1991

Freeman MR, Delrow J, Kim J, Johnson E, Doe CQ. 2003. Unwrapping glial biology: Gcm target genes regulating glial development, diversification, and function. *Neuron* **38:** 567–580. doi:10.1016/S0896-6273(03)00289-7

Gumienny TL, Brugnera E, Tosello-Trampont AC, Kinchen JM, Haney LB, Nishiwaki K, Walk SF, Nemergut ME, Macara IG, Francis R, et al. 2001. CED-12/ELMO, a novel member of the crkII/dock180/rac pathway, is required for phagocytosis and cell migration. *Cell* **107:** 27–41. doi:10.1016/S0092-8674(01)00520-7

Hamon Y, Broccardo C, Chambenoit O, Luciani MF, Toti F, Chaslin S, Freyssinet JM, Devaux PF, McNeish J, Marguet D, et al. 2000. ABC1 promotes engulfment of apoptotic cells and transbilayer redistribution of phosphatidylserine. *Nat Cell Biol* **2:** 399–406. doi:10.1038/35017029

Hay BA, Wassarman DA, Rubin GM. 1995. *Drosophila* homologs of baculovirus inhibitor of apoptosis proteins

function to block cell death. *Cell* **83:** 1253–1262. doi:10.1016/0092-8674(95)90150-7

Hedgecock EM, Sulston JE, Thomson JN. 1983. Mutations affecting programmed cell deaths in the nematode *Caenorhabditis elegans*. *Science* **220:** 1277–1279. doi:10.1126/science.6857247

Hilu-Dadia R, Hakim-Mishnaevski K, Levy-Adam F, Kurant E. 2018. Draper-mediated JNK signaling is required for glial phagocytosis of apoptotic neurons during *Drosophila* metamorphosis. *Glia* **66:** 1520–1532. doi:10.1002/glia.23322

Hsu TY, Wu YC. 2010. Engulfment of apoptotic cells in *C. elegans* is mediated by integrin α/SRC signaling. *Curr Biol* **20:** 477–486. doi:10.1016/j.cub.2010.01.062

Kinchen JM, Cabello J, Klingele D, Wong K, Feichtinger R, Schnabel H, Schnabel R, Hengartner MO. 2005. Two pathways converge at CED-10 to mediate actin rearrangement and corpse removal in *C. elegans*. *Nature* **434:** 93–99. doi:10.1038/nature03263

Kuraishi T, Nakagawa Y, Nagaosa K, Hashimoto Y, Ishimoto T, Moki T, Fujita Y, Nakayama H, Dohmae N, Shiratsuchi A, et al. 2009. Pretaporter, a *Drosophila* protein serving as a ligand for Draper in the phagocytosis of apoptotic cells. *EMBO J* **28:** 3868–3878. doi:10.1038/emboj.2009.343

Kurant E, Axelrod S, Leaman D, Gaul U. 2008. Six-microns-under acts upstream of Draper in the glial phagocytosis of apoptotic neurons. *Cell* **133:** 498–509. doi:10.1016/j.cell.2008.02.052

Li W, Baker NE. 2007. Engulfment is required for cell competition. *Cell* **129:** 1215–1225. doi:10.1016/j.cell.2007.03.054

Lin L, Rodrigues F, Kary C, Contet A, Logan M, Baxter RHG, Wood W, Baehrecke EH. 2017. Complement-related regulates autophagy in neighboring cells. *Cell* **170:** 158–171.e8. doi:10.1016/j.cell.2017.06.018

Logan MA, Hackett R, Doherty J, Sheehan A, Speese SD, Freeman MR. 2012. Negative regulation of glial engulfment activity by Draper terminates glial responses to axon injury. *Nat Neurosci* **15:** 722–730. doi:10.1038/nn.3066

Lu TY, Doherty J, Freeman MR. 2014. DRK/DOS/SOS converge with Crk/Mbc/dCed-12 to activate Rac1 during glial engulfment of axonal debris. *Proc Natl Acad Sci* **111:** 12544–12549. doi:10.1073/pnas.1403450111

MacDonald JM, Beach MG, Porpiglia E, Sheehan AE, Watts RJ, Freeman MR. 2006. The *Drosophila* cell corpse engulfment receptor Draper mediates glial clearance of severed axons. *Neuron* **50:** 869–881. doi:10.1016/j.neuron.2006.04.028

Manaka J, Kuraishi T, Shiratsuchi A, Nakai Y, Higashida H, Henson P, Nakanishi Y. 2004. Draper-mediated and phosphatidylserine-independent phagocytosis of apoptotic cells by *Drosophila* hemocytes/macrophages. *J Biol Chem* **279:** 48466–48476. doi:10.1074/jbc.M408597200

Matsubayashi Y, Louani A, Dragu A, Sanchez-Sanchez BJ, Serna-Morales E, Yolland L, Gyoergy A, Vizcay G, Fleck RA, Heddleston JM, et al. 2017. A moving source of matrix components is essential for de novo basement membrane formation. *Curr Biol* **27:** 3526–3534.e4. doi:10.1016/j.cub.2017.10.001

McPhee CK, Logan MA, Freeman MR, Baehrecke EH. 2010. Activation of autophagy during cell death requires the engulfment receptor Draper. *Nature* **465**: 1093–1096. doi:10.1038/nature09127

Meehan TL, Kleinsorge SE, Timmons AK, Taylor JD, McCall K. 2015. Polarization of the epithelial layer and apical localization of integrins are required for engulfment of apoptotic cells in the *Drosophila* ovary. *Dis Model Mech* **8**: 1603–1614. doi:10.1242/dmm.021998

Meehan TL, Joudi TF, Timmons AK, Taylor JD, Habib CS, Peterson JS, Emmanuel S, Franc NC, McCall K. 2016. Components of the engulfment machinery have distinct roles in corpse processing. *PLoS ONE* **11**: e0158217. doi:10.1371/journal.pone.0158217

Moreira S, Stramer B, Evans I, Wood W, Martin P. 2010. Prioritization of competing damage and developmental signals by migrating macrophages in the *Drosophila* embryo. *Curr Biol* **20**: 464–470. doi:10.1016/j.cub.2010.01.047

Peterson JS, McCall K. 2013. Combined inhibition of autophagy and caspases fails to prevent developmental nurse cell death in the *Drosophila melanogaster* ovary. *PLoS ONE* **8**: e76046. doi:10.1371/journal.pone.0076046

Razzell W, Evans IR, Martin P, Wood W. 2013. Calcium flashes orchestrate the wound inflammatory response through DUOX activation and hydrogen peroxide release. *Curr Biol* **23**: 424–429. doi:10.1016/j.cub.2013.01.058

Roddie HG, Armitage EL, Coates JA, Johnston SA, Evans IR. 2019. Simu-dependent clearance of dying cells regulates macrophage function and inflammation resolution. *PLoS Biol* **17**: e2006741. doi:10.1371/journal.pbio.2006741

Rogulja-Ortmann A, Luer K, Seibert J, Rickert C, Technau GM. 2007. Programmed cell death in the embryonic central nervous system of *Drosophila melanogaster*. *Development* **134**: 105–116. doi:10.1242/dev.02707

Santoso CS, Meehan TL, Peterson JS, Cedano TM, Turlo CV, McCall K. 2018. The ABC transporter *Eato* promotes cell clearance in the *Drosophila melanogaster* ovary. *G3 (Bethesda)* **8**: 833–843. doi:10.1534/g3.117.300427

Shklyar B, Levy-Adam F, Mishnaevski K, Kurant E. 2013. Caspase activity is required for engulfment of apoptotic cells. *Mol Cell Biol* **33**: 3191–3201. doi:10.1128/MCB.00233-13

Shklyar B, Sellman Y, Shklover J, Mishnaevski K, Levy-Adam F, Kurant E. 2014. Developmental regulation of glial cell phagocytic function during *Drosophila* embryogenesis. *Dev Biol* **393**: 255–269. doi:10.1016/j.ydbio.2014.07.005

Shlyakhover E, Shklyar B, Hakim-Mishnaevski K, Levy-Adam F, Kurant E. 2018. *Drosophila* GATA factor serpent establishes phagocytic ability of embryonic macrophages. *Front Immunol* **9**: 266. doi:10.3389/fimmu.2018.00266

Silva E, Au-Yeung HW, Van Goethem E, Burden J, Franc NC. 2007. Requirement for a *Drosophila* E3-ubiquitin ligase in phagocytosis of apoptotic cells. *Immunity* **27**: 585–596. doi:10.1016/j.immuni.2007.08.016

Sonnenfeld MJ, Jacobs JR. 1995. Macrophages and glia participate in the removal of apoptotic neurons from the *Drosophila* embryonic nervous system. *J Comp Neurol* **359**: 644–652. doi:10.1002/cne.903590410

Stramer B, Wood W, Galko MJ, Redd MJ, Jacinto A, Parkhurst SM, Martin P. 2005. Live imaging of wound inflammation in *Drosophila* embryos reveals key roles for small GTPases during in vivo cell migration. *J Cell Biol* **168**: 567–573. doi:10.1083/jcb.200405120

Su HP, Nakada-Tsukui K, Tosello-Trampont AC, Li Y, Bu G, Henson PM, Ravichandran KS. 2002. Interaction of CED-6/GULP, an adapter protein involved in engulfment of apoptotic cells with CED-1 and CD91/low density lipoprotein receptor-related protein (LRP). *J Biol Chem* **277**: 11772–11779. doi:10.1074/jbc.M109336200

Tepass U, Fessler LI, Aziz A, Hartenstein V. 1994. Embryonic origin of hemocytes and their relationship to cell-death in *Drosophila*. *Development* **120**: 1829–1837.

Thuma L, Carter D, Weavers H, Martin P. 2018. *Drosophila* immune cells extravasate from vessels to wounds using Tre1 GPCR and Rho signaling. *J Cell Biol* **217**: 3045–3056. doi:10.1083/jcb.201801013

Timmons AK, Mondragon AA, Schenkel CE, Yalonetskaya A, Taylor JD, Moynihan KE, Etchegaray JI, Meehan TL, McCall K. 2016. Phagocytosis genes nonautonomously promote developmental cell death in the *Drosophila* ovary. *Proc Natl Acad Sci* **113**: E1246–E1255. doi:10.1073/pnas.1522830113

Tung TT, Nagaosa K, Fujita Y, Kita A, Mori H, Okada R, Nonaka S, Nakanishi Y. 2013. Phosphatidylserine recognition and induction of apoptotic cell clearance by *Drosophila* engulfment receptor Draper. *J Biochem* **153**: 483–491. doi:10.1093/jb/mvt014

Venegas V, Zhou Z. 2007. Two alternative mechanisms that regulate the presentation of apoptotic cell engulfment signal in *Caenorhabditis elegans*. *Mol Biol Cell* **18**: 3180–3192. doi:10.1091/mbc.e07-02-0138

Vlisidou I, Dowling AJ, Evans IR, Waterfield N, ffrench-Constant RH, Wood W. 2009. *Drosophila* embryos as model systems for monitoring bacterial infection in real time. *PLoS Pathog* **5**: e1000518. doi:10.1371/journal.ppat.1000518

Wang X, Wu YC, Fadok VA, Lee MC, Gengyo-Ando K, Cheng LC, Ledwich D, Hsu PK, Chen JY, Chou BK, et al. 2003. Cell corpse engulfment mediated by *C. elegans* phosphatidylserine receptor through CED-5 and CED-12. *Science* **302**: 1563–1566. doi:10.1126/science.1087641

Weavers H, Evans IR, Martin P, Wood W. 2016a. Corpse engulfment generates a molecular memory that primes the macrophage inflammatory response. *Cell* **165**: 1658–1671. doi:10.1016/j.cell.2016.04.049

Weavers H, Liepe J, Sim A, Wood W, Martin P, Stumpf MPH. 2016b. Systems analysis of the dynamic inflammatory response to tissue damage reveals spatiotemporal properties of the wound attractant gradient. *Curr Biol* **26**: 1975–1989. doi:10.1016/j.cub.2016.06.012

White K, Grether ME, Abrams JM, Young L, Farrell K, Steller H. 1994. Genetic control of programmed cell death in *Drosophila*. *Science* **264**: 677–683. doi:10.1126/science.8171319

Williams DW, Kondo S, Krzyzanowska A, Hiromi Y, Truman JW. 2006. Local caspase activity directs engulfment of dendrites during pruning. *Nat Neurosci* **9**: 1234–1236. doi:10.1038/nn1774

Williamson AP, Vale RD. 2018. Spatial control of Draper receptor signaling initiates apoptotic cell engulfment. *J Cell Biol* **217**: 3977–3992. doi:10.1083/jcb.201711175

Wood W, Faria C, Jacinto A. 2006. Distinct mechanisms regulate hemocyte chemotaxis during development and wound healing in *Drosophila melanogaster*. *J Cell Biol* 173: 405–416. doi:10.1083/jcb.200508161

Wu YC, Horvitz HR. 1998a. *C. elegans* phagocytosis and cell-migration protein CED-5 is similar to human DOCK180. *Nature* 392: 501–504. doi:10.1038/33163

Wu YC, Horvitz HR. 1998b. The *C. elegans* cell corpse engulfment gene ced-7 encodes a protein similar to ABC transporters. *Cell* 93: 951–960. doi:10.1016/S0092-8674(00)81201-5

Wu YC, Tsai MC, Cheng LC, Chou CJ, Weng NY. 2001. *C. elegans* CED-12 acts in the conserved crkII/DOCK180/Rac pathway to control cell migration and cell corpse engulfment. *Dev Cell* 1: 491–502. doi:10.1016/S1534-5807(01)00056-9

Zhou Z, Hartwieg E, Horvitz HR. 2001. CED-1 is a transmembrane receptor that mediates cell corpse engulfment in *C. elegans*. *Cell* 104: 43–56. doi:10.1016/S0092-8674(01)00190-8

Ziegenfuss JS, Biswas R, Avery MA, Hong K, Sheehan AE, Yeung YG, Stanley ER, Freeman MR. 2008. Draper-dependent glial phagocytic activity is mediated by Src and Syk family kinase signalling. *Nature* 453: 935–939. doi:10.1038/nature06901

Ziegenfuss JS, Doherty J, Freeman MR. 2012. Distinct molecular pathways mediate glial activation and engulfment of axonal debris after axotomy. *Nat Neurosci* 15: 979–987. doi:10.1038/nn.3135

Multitasking Kinase RIPK1 Regulates Cell Death and Inflammation

Kim Newton

Department of Physiological Chemistry, Genentech, South San Francisco, California 94080, USA

Correspondence: knewton@gene.com

Receptor-interacting serine threonine kinase 1 (RIPK1) is a widely expressed kinase that is essential for limiting inflammation in both mice and humans. Mice lacking RIPK1 die at birth from multiorgan inflammation and aberrant cell death, whereas humans lacking RIPK1 are immunodeficient and develop very early-onset inflammatory bowel disease. In contrast to complete loss of RIPK1, inhibiting the kinase activity of RIPK1 genetically or pharmacologically prevents cell death and inflammation in several mouse disease models. Indeed, small molecule inhibitors of RIPK1 are in phase I clinical trials for amyotrophic lateral sclerosis, and phase II clinical trials for psoriasis, rheumatoid arthritis, and ulcerative colitis. This review focuses on which signaling pathways use RIPK1, how activation of RIPK1 is regulated, and when activation of RIPK1 appears to be an important driver of inflammation.

Receptor-interacting serine threonine kinase 1 (RIPK1) was discovered more than two decades ago because of its ability to interact with the apoptosis-inducing death receptor Fas (Stanger et al. 1995). The carboxy-terminal death domain (DD) in RIPK1 (Fig. 1) binds to the intracellular DD of Fas or tumor necrosis factor receptor 1 (TNFR1) and to the DD in the adaptor proteins TRADD and FADD (Stanger et al. 1995; Hsu et al. 1996a; Chen et al. 2008; Ermolaeva et al. 2008; Pobezinskaya et al. 2008; Park et al. 2013). Another protein interaction motif in RIPK1, termed the RIP homotypic interaction motif (RHIM), mediates interactions with the RHIM-containing proteins RIPK3 (Sun et al. 2002), TRIF (also called TICAM-1) (Meylan et al. 2004; Kaiser and Offer-

mann 2005), and ZBP1 (also called DAI) (Kaiser et al. 2008). Analyses of RIPK1-deficient cells indicate that RIPK1 is dispensable for Fas-induced apoptosis, but is needed in some cell types for optimal activation of nuclear factor (NF)-κB-dependent gene transcription by TNFR1 and TLR3 (Toll-like receptor 3) (Ting et al. 1996; Kelliher et al. 1998; Meylan et al. 2004; Cusson-Hermance et al. 2005; Vanlangenakker et al. 2011; Dannappel et al. 2014; Takahashi et al. 2014; Newton et al. 2016b; Van et al. 2017; Cuchet-Lourenço et al. 2018). The kinase activity of RIPK1, however, is dispensable for NF-κB signaling (Ting et al. 1996; Lee et al. 2004; Berger et al. 2014; Newton et al. 2014; Polykratis et al. 2014), instead being required for RIPK1 to engage the cell death machinery downstream from TNFR1.

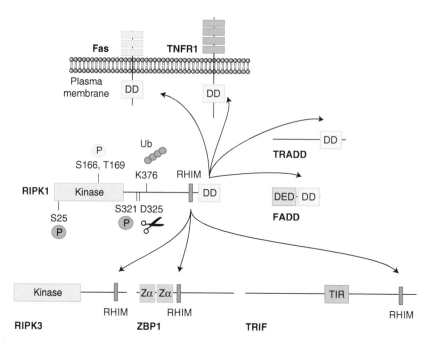

Figure 1. Domain organization and protein interactions of receptor-interacting serine threonine kinase 1 (RIPK1). RIPK1 has an amino-terminal kinase domain and carboxy-terminal RIP homotypic interaction motif (RHIM) and death domain (DD) motif. The RHIM and DD mediate interactions with other RHIM- or DD-containing proteins. Dimerization of RIPK1 via its DD promotes autophosphorylation of the kinase domain on Ser166. Thr169 in mouse RIPK1 is autophosphorylated as well, but this residue is not conserved in human RIPK1. Posttranslational modifications of RIPK1 that appear to limit activation of the kinase include ubiquitination (Ub) on mouse Lys376 (human Lys377), cleavage after mouse Asp325 (human Asp324) by caspase-8, phosphorylation on mouse Ser321 (human Ser320) by MK2, and phosphorylation on Ser25 by IκB kinase (IKK). DED, death effector domain; TIR, Toll/interleukin-1 receptor domain; Zα, Z-DNA-binding domain.

SIGNALING PATHWAYS THAT USE RIPK1

TNFR1 Signaling

RIPK1 is a component of what is termed TNFR1 complex I (Fig. 2; Micheau and Tschopp 2003). Ligated TNFR1 recruits TRADD and RIPK1, with TRADD serving as an adaptor for TRAF2 (Hsu et al. 1996b), which in turn binds to the E3 ubiquitin ligases cIAP1 and cIAP2 (Shu et al. 1996; Vince et al. 2009). The cellular inhibitors of apoptosis proteins (cIAPs) ubiquitinate RIPK1, themselves, and possibly other components of complex I (Bertrand et al. 2008; Mahoney et al. 2008; Varfolomeev et al. 2008), forming polyubiquitin chains in which the carboxy-terminal glycine of one ubiquitin is conjugated predominantly to Lys11 or Lys63 of a second ubiquitin (Dynek et al. 2010). This polyubiquitin

recruits several ubiquitin-binding proteins, including TAB2 and TAB3, adaptors for the kinase TAK1 (Cheung et al. 2004; Kanayama et al. 2004); NEMO, the regulatory subunit of the canonical IκB kinase (IKK) complex (Ea et al. 2006; Wu et al. 2006); and HOIL-1, HOIP, and Sharpin, which comprise the E3 ubiquitin ligase linear ubiquitin assembly complex (LUBAC) (Haas et al. 2009). LUBAC then modifies TNFR1, TRADD, RIPK1, and NEMO with Met1-linked polyubiquitin (Gerlach et al. 2011; Tokunaga et al. 2011; Draber et al. 2015), which further stabilizes complex I through interactions with NEMO (Lo et al. 2009; Rahighi et al. 2009). Hybrid polyubiquitin chains on RIPK1 containing both Lys63 and Met1 linkages are proposed to position TAK1 next to the IKK complex (Emmerich et al. 2016), so that TAK1 can phosphor-

Cite this article as *Cold Spring Harb Perspect Biol* doi: 10.1101/cshperspect.a036368

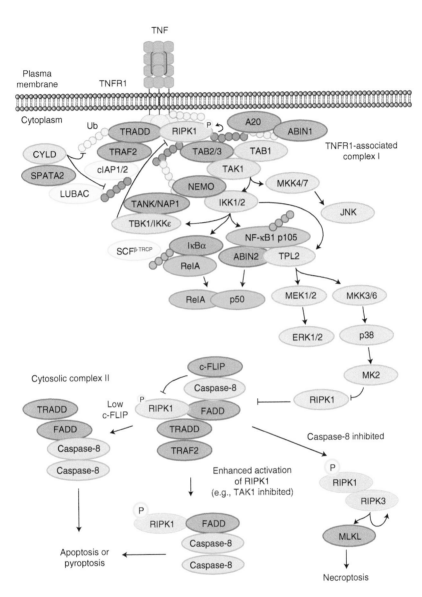

Figure 2. Receptor-interacting serine threonine kinase 1 (RIPK1) is a component of tumor necrosis factor receptor 1 (TNFR1) complex I and complex II. Binding of tumor necrosis factor (TNF) to TNFR1 triggers assembly of complex I, which activates nuclear factor (NF)-κB-dependent gene transcription and the mitogen-activated protein kinases ERK, JNK, and p38. Autophosphorylation of RIPK1 is detected in complex I in mouse cells, but it is unclear whether this is also true in human cells. Activation of the kinases TAK1, IKK, and TBK1/IKKε in complex I limits activation of RIPK1 and suppresses formation of cytosolic complex II. The translational inhibitor cycloheximide is thought to promote TNF-induced apoptosis by depleting c-FLIP from complex II, which facilitates the emergence of caspase-8 homodimers that drive cell death. Inhibitors of RIPK1 do not block this death, but they do prevent caspase-8-dependent cell death that is driven by perturbations of complex I that enhance activation of RIPK1, such as cIAP1/2 deficiency or inhibition of the IKK complex. The kinase activity of RIPK1 is also needed for RIPK3/MLKL-dependent necroptosis when the proteolytic activity of caspase-8/c-FLIP heterodimers is inhibited. Inhibitory phosphorylation of cytosolic RIPK1 by MK2 is thought to limit recruitment of additional RIPK1 into complex II. Kinases are colored blue, ubiquitin ligases are yellow, proteases are orange, adaptor proteins are purple, and transcription factors are pink. Met1-linked polyubiquitin (Ub) is colored gray, K63- or K11-linked chains are green, and K48-linked chains are brown.

ylate and activate the IKK2 catalytic subunit (Zhang et al. 2014). The IKK complex phosphorylates the NF-κB inhibitory proteins NF-κB1/p105 and IκBα, leading to their ubiquitination and proteasomal processing (Zhang et al. 2017). Degradation of IκBα and processing of NF-κB1/p105 to NF-κB1/p50 liberates NF-κB dimers composed largely of RelA and NF-κB1/p50. These dimers enter the nucleus and promote expression of their target genes. Processing of NF-κB1/p105 also liberates the kinase TPL2 to stimulate the MEK–ERK and MKK3/6–p38 kinase cascades (Eliopoulos et al. 2003; Pattison et al. 2016). Besides activating the IKK complex, TAK1 also stimulates the MKK7–JNK kinase cascade (Tournier et al. 2001; Shim et al. 2005).

NEMO and LUBAC in complex I are also needed to recruit TANK and NAP1, the ubiquitin-binding adaptors for the kinases TBK1 and IKKε (Lafont et al. 2018; Xu et al. 2018b). Phosphorylation of TBK1 and IKKε by the canonical IKK complex stimulates their activation (Clark et al. 2011). The major function of TBK1 and IKKε is to suppress the activation of RIPK1 and prevent TNF-induced cell death. Accordingly, lethal apoptosis in TBK1-deficient mouse embryos is prevented by TNFR1 deficiency (Bonnard et al. 2000), TNF deficiency (Matsui et al. 2006), or catalytically inactive RIPK1 (Xu et al. 2018b). TBK1 is proposed to phosphorylate RIPK1 directly on multiple residues to suppress RIPK1 autophosphorylation (Lafont et al. 2018; Xu et al. 2018b). TAK1 also suppresses RIPK1 activation after TNF treatment and this is independent of its role in activating NF-κB (Vanlangenakker et al. 2011; Dondelinger et al. 2013; Lamothe et al. 2013). Whether TAK1 phosphorylates RIPK1 directly is controversial (Dondelinger et al. 2015; Geng et al. 2017), but suppression of RIPK1 activation by TAK1-dependent IKK and MK2 activation is well documented (Dondelinger et al. 2015; Jaco et al. 2017; Menon et al. 2017).

IKK appears to phosphorylate RIPK1 in complex I directly on Ser25 (Dondelinger et al. 2015, 2019), but it could also influence phosphorylation of RIPK1 indirectly via the activation of TBK1 (Clark et al. 2011; Lafont et al. 2018; Xu et al. 2018b). Consistent with multiple IKK-dependent phosphorylation events restraining activation of RIPK1, mutation of RIPK1 Ser25 to Ala does not mimic inhibition of IKK (Dondelinger et al. 2019). MK2 is activated by the TAK1–MKK3/6–p38 kinase cascade and it phosphorylates cytosolic RIPK1 on murine Ser321/human Ser320 and possibly other residues (Jaco et al. 2017; Menon et al. 2017). Consistent with IKK and MK2 inhibiting RIPK1 by distinct mechanisms, combined IKK and MK2 inhibition sensitizes cells to TNF more than inhibiting just one of the kinases (Menon et al. 2017).

Genetic studies in mice support the idea that NEMO is critical for inhibiting the activation of RIPK1 and does so independent of its role in promoting NF-κB-dependent gene transcription. For example, *Nemo* deletion in intestinal epithelial cells causes colitis that is prevented by catalytically inactive RIPK1, whereas colitis is not induced by the combined loss of RelA, RelB, and c-Rel, the three NF-κB subunits with transactivation domains (Vlantis et al. 2016). Similarly, *Nemo* deletion in hepatocytes causes hepatocellular carcinoma that is prevented by catalytically inactive RIPK1, whereas hepatocellular carcinoma is not observed on deletion of *RelA*, *RelB*, and *c-Rel* (Kondylis et al. 2015). Although these experiments highlight an NF-κB-independent mechanism for RIPK1 inactivation by NEMO, perhaps via the combined actions of IKKα/β, TBK1, and IKKε, there is also evidence for NF-κB-dependent inactivation of RIPK1. For example, mice lacking either NEMO or RelA die during embryogenesis, but survive to birth if they express catalytically inactive RIPK1 (Vlantis et al. 2016; Xu et al. 2018a). Several prosurvival proteins that could impact RIPK1 activation are encoded by NF-κB target genes, including the cIAPs (Wang et al. 1998), X-linked inhibitor of apoptosis protein (XIAP) (Stehlik et al. 1998), A20 (Krikos et al. 1992), and cellular FLICE inhibitory protein (c-FLIP) (Kreuz et al. 2001).

The cIAPs contribute to the assembly of TNFR1 complex I as described above, but may also limit activation of RIPK1 by modifying RIPK1 with Lys48-linked polyubiquitin that targets it for proteasomal degradation (Varfolo-

Cite this article as *Cold Spring Harb Perspect Biol* doi: 10.1101/cshperspect.a036368

meev et al. 2008; Dynek et al. 2010; Annibaldi et al. 2018). Genetic studies in mice implicate XIAP and A20 in the suppression of RIPK1 activation (Wong et al. 2014; Onizawa et al. 2015; Newton et al. 2016a; Kattah et al. 2018; Rijal et al. 2018), but the underlying biochemical mechanism in each case is still being elucidated. The binding of A20 to Met1-linked polyubiquitin in TNFR1 complex I appears to be critical to limiting TNF-induced cell death (Draber et al. 2015; Yamaguchi and Yamaguchi 2015; Polykratis et al. 2019). A20 also appears to act in concert with the ubiquitin-binding protein ABIN-1 (Oshima et al. 2009; Dziedzic et al. 2018; Kattah et al. 2018). c-FLIP is the catalyti-

cally inactive paralog of the cysteine protease caspase-8 (previously called FLICE) (Irmler et al. 1997) and is discussed in more detail in the section on activation of RIPK1.

TLR3 and TLR4 Signaling

RIPK1 recruitment to TLR3, which detects double-stranded RNA (dsRNA) in the endosomal compartment, or to TLR4, which detects bacterial lipopolysaccharide (LPS) at the cell surface and is then endocytosed, is mediated by the adaptor TRIF (Fig. 3; Meylan et al. 2004). TRIF binding to RIPK1 and to the E3 ubiquitin ligase TRAF6 promotes TAK1- and LUBAC-depen-

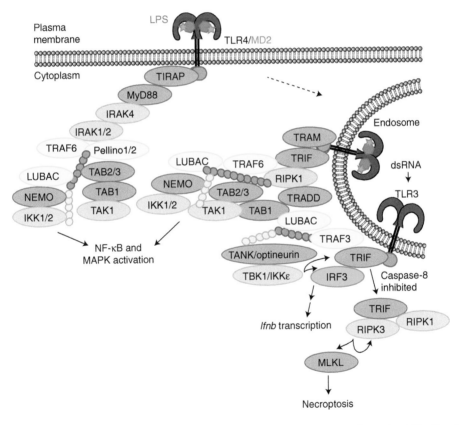

Figure 3. Receptor-interacting serine threonine kinase 1 (RIPK1) mediates TRIF-dependent NF-κB and mitogen-activated protein kinase (MAPK) activation by TLR3 and TLR4. The adaptor protein TRIF has a RHIM that recruits RIPK1 to TLR3 and TLR4 signaling complexes on endosomes. RIPK1 contributes to the activation of TAK1 and the IKK complex, but appears dispensable for TRIF-dependent activation of IRF3 and the expression of type I interferons, as well as TRIF-dependent activation of RIPK3/MLKL-dependent necroptosis. Inhibited RIPK1 can, however, block TRIF-dependent necroptosis. Both TLR3 and TLR4 trigger these different TRIF-dependent signals. RIPK1 is dispensable for TLR4 signaling via the adaptor MyD88.

dent IKK activation that leads to the expression of proinflammatory genes (Jiang et al. 2003, 2004; Sato et al. 2003; Meylan et al. 2004; Cusson-Hermance et al. 2005; Zinngrebe et al. 2016; Bakshi et al. 2017). Presumably, TRAF6 builds Lys63-linked polyubiquitin chains on RIPK1 that recruit TAK1, IKK, and LUBAC, but this has not been confirmed experimentally. Another E3 ubiquitin ligase, Pellino1, was implicated in TLR3-induced RIPK1 ubiquitination and IKK activation (Chang et al. 2009), but results obtained with cells expressing ligase inactive Pellino1 indicate that the E3 activity of Pellino1 is not required for either signaling event (Enesa et al. 2012). The adaptor TRADD, perhaps through binding to the DD of RIPK1, contributes to TLR3-induced NF-κB and mitogen-activated protein kinase (MAPK) activation in fibroblasts, but appears less critical in myeloid cells (Chen et al. 2008; Ermolaeva et al. 2008; Pobezinskaya et al. 2008). TLR3 and TLR4 also use TRIF to engage TANK- or optineurin-containing TBK1 complexes that promote activation of the transcription factor IRF3 and expression of *Interferon beta* (*Ifnb*) (Fitzgerald et al. 2003; Hemmi et al. 2004; McWhirter et al. 2004; Perry et al. 2004; Liu et al. 2015; Bakshi et al. 2017). However, neither TRAF6 nor RIPK1 is essential for TRIF-induced IFN-β expression (Sato et al. 2003; Jiang et al. 2004; Meylan et al. 2004; Cusson-Hermance et al. 2005).

Other Signaling Pathways Involved in Pathogen Defense

Recent genetic studies in mice have illuminated a role for RIPK1 in promoting proinflammatory gene expression during development, although the nature of the upstream activating signal(s) is unclear. For example, RIPK1 causes modest proinflammatory gene expression in mouse embryos lacking caspase-8 and the kinase RIPK3 (Kang et al. 2018). This aberrant gene expression is greatly exacerbated if LUBAC is compromised (Heger et al. 2018; Peltzer et al. 2018). Indeed, mice lacking caspase-8 and RIPK3 are viable (Kaiser et al. 2011; Oberst et al. 2011), whereas mice lacking caspase-8, RIPK3, and either HOIL-1 or Sharpin die in the perinatal pe-

riod (Rickard et al. 2014a; Peltzer et al. 2018). Mice lacking caspase-8, RIPK3, and OTULIN activity also die in the perinatal period, OTULIN being a deubiquitinating enzyme that preserves LUBAC activity (Heger et al. 2018). Depending on the mouse model, cytokine and chemokine expression is normalized by either complete loss of *Ripk1* (Peltzer et al. 2018) or by loss of a single allele of *Ripk1* (Heger et al. 2018), and perinatal lethality is either delayed (Peltzer et al. 2018) or prevented (Heger et al. 2018). A small molecule inhibitor of RIPK1 was unable to rescue mice lacking caspase-8, RIPK3, and HOIL-1 (Peltzer et al. 2018), so the kinase activity of RIPK1 may be dispensable for this proinflammatory gene expression.

IFN-β and IFN-inducible chemokines such as CXCL10 were expressed in a RIPK1-dependent manner in both of the aforementioned studies (Heger et al. 2018; Peltzer et al. 2018). Although RIPK1 is dispensable for TLR3- or TLR4-induced IFN-β production (Meylan et al. 2004; Cusson-Hermance et al. 2005), RIPK1, TRADD, and FADD are implicated in IFN-β production triggered by cytosolic RLRs (RIG-I-like receptors) (Balachandran et al. 2004; Michallet et al. 2008; Rajput et al. 2011). RLRs RIG-I and MDA5 recognize dsRNA and engage the mitochondrial antiviral signaling (MAVS) protein, which in turn recruits TRAF2, TRAF5, and TRAF6 to promote NEMO-dependent activation of IKK and TBK1 (Liu et al. 2013). Precisely how RIPK1 fits into this signaling pathway and whether its kinase activity is needed is unclear. It is tempting to speculate that aberrant RIPK1-dependent IFN-β signaling drives the lethal phenotype of mice lacking LUBAC activity, RIPK3 and caspase-8, but exploring this further will require genetic crosses that eliminate the IFN-α/β receptor or MAVS.

ACTIVATION OF RIPK1 PROMOTES CELL DEATH

The enzymatic activity of RIPK1 was first shown to promote cell death in Jurkat T cells treated with FasL (Fas ligand) and the pan-caspase inhibitor Z-VAD-FMK (Holler et al. 2000). Fas-induced apoptosis mediated by the adaptor

Cite this article as *Cold Spring Harb Perspect Biol* doi: 10.1101/cshperspect.a036368

FADD and caspase-8 was blocked, but now the cells died a necrotic death requiring the kinase activity of RIPK1. Jurkat T cells treated with TNF, CHX (cycloheximide), and Z-VAD-FMK also died in a manner requiring the kinase activity of RIPK1 (Holler et al. 2000). Although Jurkat T cells (and many other cell types) are not killed by TNF alone, addition of the translational inhibitor CHX is thought to promote TNFR1-dependent apoptosis by eliminating the labile protein c-FLIP (Micheau and Tschopp 2003). Caspase-8 then forms homodimers rather than caspase-8/c-FLIP heterodimers in a secondary, cytosolic complex II that also contains FADD, TRADD, TRAF2, and RIPK1 (Fig. 2). Activation of caspase-8 within complex II initiates the apoptotic caspase cascade that dismantles the cell (Micheau and Tschopp 2003). Note that the kinase activity of RIPK1 is dispensable for apoptosis induced by TNF and cycloheximide (Wang et al. 2008). It is only required when Z-VAD-FMK is incorporated into the death stimulus (Holler et al. 2000).

Z-VAD-FMK and other pan-caspase inhibitors block caspase-8-dependent apoptosis, but cell death is not prevented if cells express the kinase RIPK3 and the pseudokinase MLKL (mixed lineage kinase domain-like) (Cho et al. 2009; He et al. 2009; Zhang et al. 2009; Sun et al. 2012; Zhao et al. 2012; Murphy et al. 2013; Wu et al. 2013). Instead of dying by apoptosis, cells unleash a caspase-independent death program termed necroptosis (Degterev et al. 2005). The enzymatic activity of RIPK1, which is essential for TNFR1-induced necroptosis (Holler et al. 2000), is required for interactions between RIPK1 and RIPK3 in complex II (Cho et al. 2009; He et al. 2009). The only known substrate of RIPK1 is itself, with Ser166 and other residues within the kinase domain, being autophosphorylated in *trans* (Ting et al. 1996; Degterev et al. 2008). In primary mouse embryo fibroblasts (MEFs), autophosphorylated RIPK1 is detected in TNFR1 complex I within minutes of TNF stimulation (Newton et al. 2016b). Dimerization of RIPK1 via its DD promotes autophosphorylation (Meng et al. 2018), whereas the RHIM in RIPK1 is dispensable (Newton et al. 2016b). Autophosphorylation may induce

conformational changes that expose the RHIM of RIPK1 for interactions with the RHIM of RIPK3. RHIM–RHIM interactions between RIPK1 and RIPK3 in complex II are required for oligomerization and autophosphorylation of RIPK3 (Cho et al. 2009; He et al. 2009; Sun et al. 2012; Chen et al. 2013; Wu et al. 2014; Mompeán et al. 2018). Autophosphorylated RIPK3 then phosphorylates the carboxy-terminal pseudokinase domain of MLKL (Sun et al. 2012; Murphy et al. 2013; Xie et al. 2013; Rodriguez et al. 2016) and conformational changes in MLKL trigger its translocation to membranes and cell lysis (Fig. 2; Cai et al. 2014; Hildebrand et al. 2014; Su et al. 2014; Wang et al. 2014; Petrie et al. 2018).

Necroptosis is considered a host-defense mechanism for eliminating cells infected with viruses that encode inhibitors of caspase-8-dependent apoptosis (Chan et al. 2003; Cho et al. 2009; Guo et al. 2015). Death ligands such as FasL, TRAIL, and TNF are not the only triggers of necroptosis. Provided caspase-8 is inhibited, RIPK3-dependent necroptosis can also be triggered by TLRs, RIG-I, the cytosolic DNA sensor cGAS, the RHIM-containing protein ZBP1, or the T-cell receptor (Ch'en et al. 2008, 2011; He et al. 2011; Upton et al. 2012; Kaiser et al. 2013; Schock et al. 2017; Brault et al. 2018). It is unclear how the T-cell receptor engages the necroptosis machinery, but TLRs that signal using the adaptor MyD88, RIG-I, and cGAS all appear to induce necroptosis via autocrine TNF production (Kaiser et al. 2013; Brault et al. 2018). In contrast, TLR3 and TLR4 can engage RIPK3 directly via TRIF (Fig. 3; He et al. 2011; Kaiser et al. 2013; Buchrieser et al. 2018). ZBP1, which appears to sense certain virus infections by binding to viral RNA transcripts (Thapa et al. 2016; Maelfait et al. 2017; Guo et al. 2018), can also engage RIPK3 (Upton et al. 2012; Lin et al. 2016; Newton et al. 2016b; Nogusa et al. 2016). ZBP1 and TRIF can both activate RIPK3 in the absence of RIPK1 (Lin et al. 2016; Newton et al. 2016b), so it is interesting that inhibited RIPK1 blocks TLR3-induced necroptosis in macrophages (He et al. 2011; Kaiser et al. 2013; McComb et al. 2014). The ability of RIPK1 to suppress RIPK3 activation by ZBP1 and TRIF is discussed further in the section on the prosurvival function of RIPK1.

Many studies now indicate that activation of RIPK1 is not solely associated with necroptosis. For example, TNF induces excessive activation of RIPK1 in cells lacking cIAP1 and cIAP2 (Wang et al. 2008; Moulin et al. 2012; Dondelinger et al. 2013; Polykratis et al. 2014), TAK1 (Dondelinger et al. 2013; Lamothe et al. 2013), LUBAC (Berger et al. 2014; Kumari et al. 2014; Peltzer et al. 2014, 2018; Rickard et al. 2014a; Heger et al. 2018), IKK (Dondelinger et al. 2015), or ABIN-1 (Dziedzic et al. 2018; Kattah et al. 2018), and this results in caspase-8-dependent cell death. Active RIPK1 promotes the assembly of complex II, using its DD to interact with FADD. Activation of caspase-8 via FADD typically engages the apoptotic machinery, but a recent study showed that macrophages exposed to TNF and TAK1 inhibitor display a lytic form of cell death termed pyroptosis (Orning et al. 2018). In this setting, caspase-8 cleaves the protein gasdermin D (GSDMD) to produce an amino-terminal fragment that forms pores in the plasma membrane. Consequently, cell death that is prevented by RIPK1 inhibitors, such as Nec-1s (Degterev et al. 2008), could be caused by apoptosis, pyroptosis, or necroptosis. The type of cell death will depend on the activity of caspase-8 and the substrates that are available to it. Pyroptosis mediated by gasdermin pores is more rapid than apoptosis (Wang et al. 2017b), so it might be the dominant form of cell death in cells expressing both caspase-8 and GSDMD.

CASPASE-8 SUPPRESSES CELL DEATH BY CLEAVING RIPK1

Pan-caspase inhibitors such as Z-VAD-FMK are used to induce necroptosis in cultured cells, but exactly how inhibition of caspase-8 promotes activation of RIPK3 is not well understood. In apoptotic cells, caspase-8 cleaves RIPK1 after murine Asp325/human Asp324 (Lin et al. 1999) and this might limit interactions between RIPK1 and RIPK3 that would lead to necroptosis. It should be noted, however, that caspase-8 can suppress necroptosis even in the absence of apoptosis. For example, inhibition of caspase-8

with Z-VAD-FMK in mouse bone marrow–derived macrophages is sufficient to convert TNF from a nonlethal stimulus into a lethal necroptosis stimulus (Newton et al. 2016b). Indeed, necroptosis suppression appears to be mediated by the proteolytic activity of caspase-8/c-FLIP heterodimers rather than autoprocessed caspase-8 homodimers (Kang et al. 2008; Oberst et al. 2011). Consistent with c-FLIP suppressing both caspase-8-dependent apoptosis and RIPK3-dependent necroptosis, mice lacking c-FLIP die around embryonic day 11 (E11) (Yeh et al. 2000), but are viable if they also lack caspase-8 and RIPK3 (Dillon et al. 2012). Mice lacking caspase-8 or FADD also die around E11 (Varfolomeev et al. 1998; Yeh et al. 1998), but only need to lose RIPK3 or MLKL to be viable (Kaiser et al. 2011; Oberst et al. 2011; Alvarez-Diaz et al. 2016; Zhang et al. 2016). TNFR1- and RIPK1-dependent necroptosis, particularly in endothelial cells, appears to drive the lethal phenotype of FADD- or caspase-8-deficient embryos at E11 (Kang et al. 2004; Zhang et al. 2011; Dillon et al. 2014; Kaiser et al. 2014; Rickard et al. 2014b), although embryos lacking caspase-8 and TNFR1, or FADD and RIPK1, or caspase-8 and RIPK1 still succumb later in development (Zhang et al. 2011; Dillon et al. 2014; Kaiser et al. 2014; Rickard et al. 2014b) because ZBP1 and TRIF then drive RIPK3-dependent necroptosis (Newton et al. 2016b).

What do caspase-8/c-FLIP heterodimers cleave to prevent necroptosis? In mice, mutation of RIPK1 Asp325 to Ala to prevent RIPK1 cleavage by caspase-8 is lethal around E11 (Zhang et al. 2019a), which is what one would expect if RIPK1 has to be cleaved to prevent necroptosis. However, lethality is prevented by the combined loss of FADD and RIPK3, but not by the loss of RIPK3 alone (Zhang et al. 2019a). Therefore, RIPK1 D325A appears to trigger apoptosis mediated by FADD and caspase-8, rather than necroptosis. Perhaps transient assembly of TNFR1 complex II is sustained in the absence of RIPK1 cleavage and, as a consequence, the capacity of c-FLIP to curb caspase-8-dependent apoptosis is overwhelmed.

Caspase-8 can also cleave the deubiquitinating enzyme CYLD after Asp215 (O'Donnell et al. 2011). CYLD is recruited to TNFR1 complex I by its adaptor SPATA2, which binds to HOIP (Elliott et al. 2016; Kupka et al. 2016; Schlicher et al. 2016; Wagner et al. 2016). Cleavage of Lys63- and Met1-linked polyubiquitin by CYLD then promotes complex II assembly and cell death (Hitomi et al. 2008; Wang et al. 2008; Moquin et al. 2013; Draber et al. 2015; Callow et al. 2018). Genetic studies in mice are consistent with CYLD facilitating necroptosis. For example, catalytically inactive CYLD delays RIPK3-dependent skin lesions when FADD is deleted from keratinocytes (Bonnet et al. 2011), and prevents some, but not all, RIPK3-dependent lesions when FADD is deleted from intestinal epithelial cells (Welz et al. 2011). Nonetheless, cleavage of CYLD after Asp215 does not appear to be a critical function of caspase-8 because mutation of CYLD Asp215 to Ala is not sufficient for TNF-induced necroptosis in macrophages (Legarda et al. 2016).

CELL DEATH–INDEPENDENT OUTCOMES OF RIPK1 ACTIVATION

Activation of RIPK1 may also promote inflammation independent of its role in triggering cell death, although the underlying mechanistic details are less clear. For example, mice expressing catalytically inactive RIPK1 show increased paresis after infection with Zika virus (Daniels et al. 2019) and increased mortality after infection with West Nile virus (Daniels et al. 2017). These phenotypes appear cell death–independent because they are not observed in mice lacking both caspase-8 and MLKL. Instead, activation of RIPK1 in neurons appears to be important for inducing antiviral transcriptional programs (Daniels et al. 2019). In a similar vein, MLKL-deficient mouse macrophages treated with LPS plus Z-VAD-FMK up-regulate proinflammatory genes such as *Tnf* and *Ifnb* without dying by necroptosis, and this transcriptional response is blocked by inhibitors of RIPK1 (Najjar et al. 2016; Saleh et al. 2017). In another study, RIPK1 inhibitor Nec-1 blocked transcription of *Tnf* in RIPK3-deficient macrophages treated with compound A, a pan-IAP antagonist (Wong et al. 2014). Collectively,

these data suggest that inhibitors of RIPK1 could block several proinflammatory mechanisms, rather than just proinflammatory cell death.

THE PROSURVIVAL FUNCTION OF RIPK1

Despite being required for TNF-induced necroptosis, and in some contexts, TNF-induced apoptosis or pyroptosis, there is compelling genetic evidence that RIPK1 can also suppress caspase-8-dependent apoptosis and MLKL-dependent necroptosis (Fig. 4; Dannappel et al. 2014; Dillon et al. 2014; Kaiser et al. 2014; Orozco et al. 2014; Rickard et al. 2014b; Roderick et al. 2014; Takahashi et al. 2014; Raju et al. 2018). For example, mice lacking RIPK1 die around birth (Kelliher et al. 1998), unless both the apoptotic and necroptotic death programs are disabled by the loss of caspase-8 and RIPK3 (Dillon et al. 2014; Kaiser et al. 2014; Rickard et al. 2014b). RIPK1 may suppress caspase-8-dependent cell death by preventing interactions between TRADD and FADD (Anderton et al. 2019), but it is unclear whether this is because the DD of RIPK1 binds to and sequesters TRADD or FADD. Mice expressing RIPK1 with a K584R DD mutation are viable (Meng et al. 2018), but the role of the DD in the prosurvival function of RIPK1 remains ambiguous because this mutation is unlikely to block all DD interactions. For example, although human RIPK1 with the equivalent K599R mutation is unable to form homodimers, it can still interact with TNFR1, FADD, and TRADD (Meng et al. 2018). RIPK1 that completely lacks the carboxy-terminal DD is expressed poorly in mice (K Newton and VM Dixit, unpubl.), so lethality in the homozygous state cannot be attributed to a prosurvival function of the DD specifically.

Mutating the RHIM of RIPK1 causes ZBP1/RIPK3/MLKL-dependent perinatal lethality in mice without compromising the expression of RIPK1 (Lin et al. 2016; Newton et al. 2016b). Therefore, the RIPK1 RHIM is a critical brake on ZBP1-induced necroptosis. In macrophages, the RIPK1 RHIM also prevents TRIF from activating RIPK3 (Newton et al. 2016b). Biochemical evidence for RIPK1 simply sequestering the

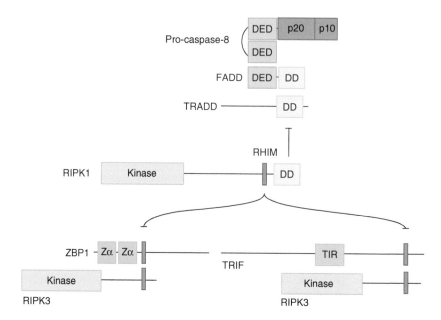

Figure 4. Model for the prosurvival function of receptor-interacting serine threonine kinase 1 (RIPK1). Mouse RIPK1 uses its RIP homotypic interaction motif (RHIM) to suppress activation of RIPK3-dependent necroptosis by TRIF and ZBP1. RIPK1 also suppresses FADD/caspase-8-dependent cell death, but whether this is dependent on the death domain (DD) of RIPK1 is unclear. The nature of the RHIM-dependent interactions that allow RIPK1 to limit activation of RIPK3 are also unknown.

other RHIM-containing proteins has proven elusive, prompting speculation that transient RHIM–RHIM interactions might hold RIPK3 in check. TNF-induced activation of MAPKs and NF-κB is normal in cells expressing RHIM mutant RIPK1 (Lin et al. 2016; Newton et al. 2016b), so the perinatal lethality of RHIM mutant RIPK1 mice does not appear to be caused by impaired expression of prosurvival factors such as c-FLIP, A20, and c-IAP1/2. ZBP1/RIPK3/MLKL-dependent inflammation is detected as early as E17.5 in RIPK1 RHIM mutant mice (Newton et al. 2016b), so there is also the question of what ZBP1 might be sensing in utero. Given that interactions between RIPK1 and other RHIM-containing proteins are usually not detected without a stimulus, identification of the ZBP1 ligand is probably key to unraveling how RIPK1 prevents ZBP1-induced necroptosis.

The enzymatic activity of RIPK1 is dispensable for the prosurvival function of RIPK1 because mice expressing catalytically inactive RIPK1 are viable (Kaiser et al. 2014; Newton

et al. 2014; Polykratis et al. 2014). It may be that inhibition of RIPK1 enzymatic activity prevents cell death in certain contexts (e.g., necroptosis induced by TLR3 agonist plus z-VAD-FMK) by enforcing this ill-defined brake function of RIPK1. Therefore, a better understanding of how RIPK1 suppresses activation of caspase-8 and RIPK3 might help identify disease settings in which inhibitors of RIPK1 could offer a therapeutic benefit.

In humans, RIPK1 deficiency is associated with immunodeficiency, very early-onset inflammatory bowel disease, and arthritis (Cuchet-Lourenço et al 2018; Li et al 2019). Although inflammation appears less widespread than in RIPK1-deficient mice, cells from RIPK1-deficient patients display similar defects in signaling to cells from RIPK1-deficient mice. For example, fibroblasts fail to activate MAPK and NF-κB signaling normally in response to TNF or a TLR3 agonist, and they show increased cell death. Blood cells treated with LPS also show impaired cytokine production (Cuchet-Lourenço et al. 2018). Therefore, aberrant cell death owing to RIPK1 deficiency might drive disease in humans as well as mice.

PRECLINICAL DISEASE MODELS IN WHICH INHIBITION OF RIPK1 IS BENEFICIAL

The ability of catalytically inactive RIPK1 or small molecule inhibitors of RIPK1 to prevent inflammatory skin disease and multiorgan inflammation in mice lacking Sharpin (Berger et al. 2014), colitis in mice lacking NEMO in intestinal epithelial cells (Vlantis et al. 2016), or inflammation and neurodegeneration in mice heterozygous for both TAK1 and TBK1 in microglia (Xu et al. 2018b) is consistent with the notion that cell death can be a potent driver of inflammatory disease. These genetic disease models reduce or remove known brakes on the activation of RIPK1, but in what other settings might excessive activation of RIPK1 be deleterious and promote disease? Activation of RIPK1 appears to contribute to ischemia-reperfusion injury because mice expressing catalytically inactive RIPK1 show improved survival in a kidney ischemia-reperfusion injury model (Newton et al. 2016a), as do mice dosed with the RIPK1 inhibitor Nec-1 (Linkermann et al. 2012). Nec-1 also reduces reperfusion injury after myocardial infarction in mice and pigs (Oerlemans et al. 2012; Koudstaal et al. 2015). Which cell types activate RIPK1 in these ischemia-reperfusion injury models is unclear. Autophosphorylated RIPK1 is detected in endothelial cells in human heart samples showing acute myocardial infarction (Patel et al. 2019), so the benefits of RIPK1 inhibition may stem from its effects on the vasculature. Mice expressing catalytically inactive RIPK1 also have improved outcomes in an intracerebral hemorrhage stroke model (Lule et al. 2017), plus RIPK1 inhibitors can reduce infarct size in a major cerebral artery occlusion stroke model (Degterev et al. 2005).

Besides ischemia-reperfusion injury, inhibition of RIPK1 pharmacologically or genetically is reported to ameliorate disease in mouse models of retinitis pigmentosa (Murakami et al. 2012), multiple sclerosis (Ofengeim et al. 2015; Zhang et al. 2019b), acetaminophen hepatotoxicity (Dara et al. 2015), atherosclerosis (Karunakaran et al. 2016), amyotrophic lateral sclerosis (Ito et al. 2016), aortic aneurysm (Wang et al. 2017a), and Alzheimer's disease (Ofengeim et al. 2017). More recent studies have investigated the effect of inhibiting RIPK1 on tumor development, progression, and metastasis, but discrepant results do not produce a clear picture. One group has reported that inhibition of RIPK1 in a mouse model of pancreatic ductal adenocarcinoma slows tumor progression by reprogramming tumor-associated macrophages to promote tumor immunity (Seifert et al. 2016; Wang et al. 2018). An independent study, however, observed no benefit from inhibiting RIPK1 in an intervention setting, and the changes in gene expression that were reported for macrophages expressing catalytically inactive RIPK1 were not replicated (Patel et al. 2019). Contradictory results have also been reported regarding the ability of inhibited RIPK1 to limit tumor cell metastasis (Strilic et al. 2016; Hänggi et al. 2017; Patel et al. 2019), which purportedly acts by either blocking the death of endothelial cells (Strilic et al. 2016) or altering vascular permeability (Hänggi et al. 2017). Unexpectedly, in a mouse model of liver cancer, rather than limiting tumor development, MLKL deficiency or Nec-1 skewed tumor development toward hepatocellular carcinoma instead of intrahepatic cholangiocarcinoma (Seehawer et al. 2018). Necroptosis-associated cytokines were proposed to drive the development of cholangiocarcinoma, but exactly how these cytokines achieve this needs clarification. It is also unclear what triggers hepatocyte necroptosis in this setting.

EFFECT OF INHIBITING RIPK1 ON INNATE IMMUNITY

Cell death can be an important immune defense against infection, so there are studies that have examined whether inhibition of RIPK1 compromises pathogen clearance. Results have differed depending on the pathogen and the route of infection. Activation of RIPK1 appears beneficial to the host in some infections. For example, mice expressing catalytically inactive RIPK1 are less able to clear the Gram-negative bacteria *Salmonella typhimurium* (Shutinoski et al. 2016) and *Yersinia pseudotuberculosis* (Peterson et al. 2017). Inhibition of RIPK1 with Nec-1 has also been shown to increase the susceptibility of mice to the fungus *Candida albicans* (Cao et al. 2019). In other infections, however, activation of RIPK1 appears deleterious to the host. For example, inhibition of RIPK1 with Nec-1s is reported to reduce the number of bacteria in the bronchoalveolar lavage fluid of mice at 4 h after being dosed intranasally with Gram-positive *Staphylococcus aureus* (Kitur et al. 2015). In the same study, mice lacking RIPK3 were analyzed at 24 h after infection, and results were consistent with RIPK1/RIPK3-dependent proinflammatory signals contributing to lung damage and impeding bacterial clearance (Kitur et al. 2015). In contrast, in *Staphylococcus aureus* skin infection or sepsis models, inhibition of RIPK1 or MLKL deficiency increased bacterial counts and was deleterious (Kitur et al. 2016). Falling into yet a third category, some infections are unaffected by the inhibition of RIPK1. For example, neither Nec-1s nor MLKL deficiency altered bacterial burdens in a humanized mouse model of *Mycobacterium tuberculosis* infection (Stutz et al. 2018).

In terms of viral infections, mice expressing catalytically inactive RIPK1 are reported to have elevated virus titers in the spleen and liver after intraperitoneal injection of vaccinia virus (Polykratis et al. 2014). It was unclear, however, when after infection the mice were analyzed. Whether the mice eventually cleared the virus or succumbed to infection was not reported. As mentioned earlier, mice expressing catalytically inactive RIPK1 are reported to be more susceptible to infection with West Nile virus or Zika virus (Daniels et al. 2017, 2019). MEFs or mouse macrophages expressing catalytically inactive RIPK1 are killed by influenza A virus just like wild-type cells (Kuriakose et al. 2016; Nogusa et al. 2016), but whether mice clear the virus normally has not been reported. It will be important to explore the role of RIPK1 during infection further as inhibitors of RIPK1 are considered for therapeutic purposes (Harris et al. 2017).

CONCLUDING REMARKS

Experiments in preclinical disease models highlight the potential therapeutic benefit of inhibiting RIPK1 to limit inflammation and tissue injury, but still there is much we do not understand about this kinase. The kinase-independent prosurvival scaffolding functions of RIPK1, although clear from genetic experiments, are poorly understood at a biochemical level. We would also benefit from further insights into how activation of RIPK1 induces expression of proinflammatory genes independent of cell death. Other areas that merit further examination are the role of RIPK1 in signaling by RIG-I, the role of ubiquitination and proteins such as A20, ABIN1, and XIAP in modulating activation of RIPK1, and the role of RIPK1-dependent signaling in cancer.

ACKNOWLEDGMENTS

I thank Vishva M. Dixit and Domagoj Vucic for helpful discussions. K.N. is an employee of Genentech.

REFERENCES

Alvarez-Diaz S, Dillon CP, Lalaoui N, Tanzer MC, Rodriguez DA, Lin A, Lebois M, Hakem R, Josefsson EC, O'Reilly LA, et al. 2016. The pseudokinase MLKL and the kinase RIPK3 have distinct roles in autoimmune disease caused by loss of death-receptor-induced apoptosis. *Immunity* **45:** 513–526. doi:10.1016/j.immuni.2016.07.016

Anderton H, Bandala-Sanchez E, Simpson DS, Rickard JA, Ng AP, Di Rago L, Hall C, Vince JE, Silke J, Liccardi G, et al. 2019. RIPK1 prevents TRADD-driven, but TNFR1 independent, apoptosis during development. *Cell Death Differ* **26:** 877–889. doi:10.1038/s41418-018-0166-8

Annibaldi A, Wicky John S, Vanden Berghe T, Swatek KN, Ruan J, Liccardi G, Bianchi K, Elliot PR, Choi SM, Van Coillie S, et al. 2018. Ubiquitin-mediated regulation of RIPK1 kinase activity independent of IKK and MK2. *Mol Cell* **69:** 566–580.e5. doi:10.1016/j.molcel.2018.01.027

Bakshi S, Taylor J, Strickson S, McCartney T, Cohen P. 2017. Identification of TBK1 complexes required for the phosphorylation of IRF3 and the production of interferon β. *Biochemical J* **474:** 1163–1174. doi:10.1042/BCJ20160992

Balachandran S, Thomas E, Barber GN. 2004. A FADD-dependent innate immune mechanism in mammalian cells. *Nature* **432:** 401–405. doi:10.1038/nature03124

Berger SB, Kasparcova V, Hoffman S, Swift B, Dare L, Schaeffer M, Capriotti C, Cook M, Finger J, Hughes-Earle

Cite this article as *Cold Spring Harb Perspect Biol* doi: 10.1101/cshperspect.a036368

A, et al. 2014. Cutting edge: RIP1 kinase activity is dispensable for normal development but is a key regulator of inflammation in SHARPIN-deficient mice. *J Immunol* **192:** 5476–5480. doi:10.4049/jimmunol.1400499

Bertrand MJ, Milutinovic S, Dickson KM, Ho WC, Boudreault A, Durkin J, Gillard JW, Jaquith JB, Morris SJ, Barker PA. 2008. cIAP1 and cIAP2 facilitate cancer cell survival by functioning as E3 ligases that promote RIP1 ubiquitination. *Mol Cell* **30:** 689–700. doi:10.1016/j.molcel.2008.05.014

Bonnard M, Mirtsos C, Suzuki S, Graham K, Huang J, Ng M, Itie A, Wakeham A, Shahinian A, Henzel WJ, et al. 2000. Deficiency of T2K leads to apoptotic liver degeneration and impaired NF-κB-dependent gene transcription. *EMBO J* **19:** 4976–4985. doi:10.1093/emboj/19.18.4976

Bonnet MC, Preukschat D, Welz PS, van Loo G, Ermolaeva MA, Bloch W, Haase I, Pasparakis M. 2011. The adaptor protein FADD protects epidermal keratinocytes from necroptosis in vivo and prevents skin inflammation. *Immunity* **35:** 572–582. doi:10.1016/j.immuni.2011.08.014

Brault M, Olsen TM, Martinez J, Stetson DB, Oberst A. 2018. Intracellular nucleic acid sensing triggers necroptosis through synergistic type I IFN and TNF signaling. *J Immunol* **200:** 2748–2756. doi:10.4049/jimmunol.1701492

Buchrieser J, Oliva-Martin MJ, Moore MD, Long JCD, Cowley SA, Perez-Simón JA, James W, Venero JL. 2018. RIPK1 is a critical modulator of both tonic and TLR-responsive inflammatory and cell death pathways in human macrophage differentiation. *Cell Death Dis* **9:** 973. doi:10.1038/s41419-018-1053-4

Cai Z, Jitkaew S, Zhao J, Chiang HC, Choksi S, Liu J, Ward Y, Wu LG, Liu ZG. 2014. Plasma membrane translocation of trimerized MLKL protein is required for TNF-induced necroptosis. *Nat Cell Biol* **16:** 55–65. doi:10.1038/ncb2883

Callow MG, Watanabe C, Wickliffe KE, Bainer R, Kummerfield S, Weng J, Cuellar T, Janakiraman V, Chen H, Chih B, et al. 2018. CRISPR whole-genome screening identifies new necroptosis regulators and RIPK1 alternative splicing. *Cell Death Dis* **9:** 261. doi:10.1038/s41419-018-0301-y

Cao M, Wu Z, Lou Q, Lu W, Zhang J, Li Q, Zhang Y, Yao Y, Zhao Q, Li M, et al. 2019. Dectin-1-induced RIPK1 and RIPK3 activation protects host against Candida albicans infection. *Cell Death Differ* doi:10.1038/s41418-019-0323-8

Ch'en IL, Beisner DR, Degterev A, Yuan J, Hoffmann A, Hedrick SM. 2008. Antigen-mediated T cell expansion regulated by parallel pathways of death. *Proc Natl Acad Sci* **105:** 17463–17468. doi:10.1073/pnas.0808043105

Ch'en IL, Tsau JS, Molkentin JD, Komatsu M, Hedrick SM. 2011. Mechanisms of necroptosis in T cells. *J Exp Med* **208:** 633–641. doi:10.1084/jem.20110251

Chan FK, Shisler J, Bixby JG, Felices M, Zheng L, Appel M, Orenstein J, Moss B, Lenardo MJ. 2003. A role for tumor necrosis factor receptor-2 and receptor-interacting protein in programmed necrosis and antiviral responses. *J Biol Chem* **278:** 51613–51621. doi:10.1074/jbc.M305633200

Chang M, Jin W, Sun SC. 2009. Peli1 facilitates TRIF-dependent Toll-like receptor signaling and proinflammatory cytokine production. *Nat Immunol* **10:** 1089–1095. doi:10.1038/ni.1777

Chen NJ, Chio II, Lin WJ, Duncan G, Chau H, Katz D, Huang HL, Pike KA, Hao Z, Su YW, et al. 2008. Beyond tumor necrosis factor receptor: TRADD signaling in Toll-like receptors. *Proc Natl Acad Sci* **105:** 12429–12434. doi:10.1073/pnas.0806585105

Chen W, Zhou Z, Li L, Zhong CQ, Zheng X, Wu X, Zhang Y, Ma H, Huang D, Li W, et al. 2013. Diverse sequence determinants control human and mouse receptor interacting protein 3 (RIP3) and mixed lineage kinase domain-like (MLKL) interaction in necroptotic signaling. *J Biol Chem* **288:** 16247–16261. doi:10.1074/jbc.M112.435545

Cheung PC, Nebreda AR, Cohen P. 2004. TAB3, a new binding partner of the protein kinase TAK1. *Biochem J* **378:** 27–34. doi:10.1042/bj20031794

Cho YS, Challa S, Moquin D, Genga R, Ray TD, Guildford M, Chan FK. 2009. Phosphorylation-driven assembly of the RIP1–RIP3 complex regulates programmed necrosis and virus-induced inflammation. *Cell* **137:** 1112–1123. doi:10.1016/j.cell.2009.05.037

Clark K, Peggie M, Plater L, Sorcek RJ, Young ER, Madwed JB, Hough J, McIver EG, Cohen P. 2011. Novel cross-talk within the IKK family controls innate immunity. *Biochem J* **434:** 93–104. doi:10.1042/BJ20101701

Cuchet-Lourenço D, Eletto D, Wu C, Plagnol V, Papapietro O, Curtis J, Ceron-Gutierrez L, Bacon CM, Hackett S, Alsaleem B, et al. 2018. Biallelic *RIPK1* mutations in humans cause severe immunodeficiency, arthritis, and intestinal inflammation. *Science* **361:** 810–813. doi:10.1126/science.aar2641

Cusson-Hermance N, Khurana S, Lee TH, Fitzgerald KA, Kelliher MA. 2005. Rip1 mediates the Trif-dependent Toll-like receptor 3- and 4-induced NF-κB activation but does not contribute to interferon regulatory factor 3 activation. *J Biol Chem* **280:** 36560–36566. doi:10.1074/jbc.M506831200

Daniels BP, Snyder AG, Olsen TM, Orozco S, Oguin TH III, Tait SWG, Martinez J, Gale M Jr, Loo YM, Oberst A. 2017. RIPK3 restricts viral pathogenesis via cell death-independent neuroinflammation. *Cell* **169:** 301–313.e11. doi:10.1016/j.cell.2017.03.011

Daniels BP, Kofman SB, Smith JR, Norris GT, Snyder AG, Kolb JP, Gao X, Locasale JW, Martinez J, Gale M, et al. 2019. The nucleotide sensor ZBP1 and kinase RIPK3 induce the enzyme IRG1 to promote an antiviral metabolic state in neurons. *Immunity* **50:** 64–76.e4. doi:10.1016/j.immuni.2018.11.017

Dannappel M, Vlantis K, Kumari S, Polykratis A, Kim C, Wachsmuth L, Eftychi C, Lin J, Corona T, Hermance N, et al. 2014. RIPK1 maintains epithelial homeostasis by inhibiting apoptosis and necroptosis. *Nature* **513:** 90–94. doi:10.1038/nature13608

Dara L, Johnson H, Suda J, Win S, Gaarde W, Han D, Kaplowitz N. 2015. Receptor interacting protein kinase 1 mediates murine acetaminophen toxicity independent of the necrosome and not through necroptosis. *Hepatology* **62:** 1847–1857. doi:10.1002/hep.27939

Degterev A, Huang Z, Boyce M, Li Y, Jagtap P, Mizushima N, Cuny GD, Mitchison TJ, Moskowitz MA, Yuan J. 2005. Chemical inhibitor of nonapoptotic cell death with therapeutic potential for ischemic brain injury. *Nat Chem Biol* **1:** 112–119. doi:10.1038/nchembio711

Degterev A, Hitomi J, Germscheid M, Ch'en IL, Korkina O, Teng X, Abbott D, Cuny GD, Yuan C, Wagner G, et al. 2008. Identification of RIP1 kinase as a specific cellular target of necrostatins. *Nat Chem Biol* **4**: 313–321. doi:10.1038/nchembio.83

Dillon CP, Oberst A, Weinlich R, Janke LJ, Kang TB, Ben-Moshe T, Mak TW, Wallach D, Green DR. 2012. Survival function of the FADD-CASPASE-8-cFLIP_L complex. *Cell Rep* **1**: 401–407. doi:10.1016/j.celrep.2012.03.010

Dillon CP, Weinlich R, Rodriguez DA, Cripps JG, Quarato G, Gurung P, Verbist KC, Brewer TL, Llambi F, Gong YN, et al. 2014. RIPK1 blocks early postnatal lethality mediated by caspase-8 and RIPK3. *Cell* **157**: 1189–1202. doi:10.1016/j.cell.2014.04.018

Dondelinger Y, Aguileta MA, Goossens V, Dubuisson C, Grootjans S, Dejardin E, Vandenabeele P, Bertrand MJ. 2013. RIPK3 contributes to TNFR1-mediated RIPK1 kinase-dependent apoptosis in conditions of cIAP1/2 depletion or TAK1 kinase inhibition. *Cell Death Differ* **20**: 1381–1392. doi:10.1038/cdd.2013.94

Dondelinger Y, Jouan-Lanhouet S, Divert T, Theatre E, Bertin J, Gough PJ, Giansanti P, Heck AJ, Dejardin E, Vandenabeele P, et al. 2015. NF-κB-independent role of IKKα/IKKβ in preventing RIPK1 kinase-dependent apoptotic and necroptotic cell death during TNF signaling. *Mol Cell* **60**: 63–76. doi:10.1016/j.molcel.2015.07.032

Dondelinger Y, Delanghe T, Priem D, Wynosky-Dolfi MA, Dorobetea D, Rojas-Rivera D, Giansanti P, Roelandt R, Gropengiesser J, Ruckdeschel K, et al. 2019. Serine 25 phosphorylation inhibits RIPK1 kinase-dependent cell death in models of infection and inflammation. *Nat Commun* **10**: 1729. doi:10.1038/s41467-019-09690-0

Draber P, Kupka S, Reichert M, Draberova H, Lafont E, de Miguel D, Spilgies L, Surinova S, Taraborrelli L, Hartwig T, et al. 2015. LUBAC-recruited CYLD and A20 regulate gene activation and cell death by exerting opposing effects on linear ubiquitin in signaling complexes. *Cell Rep* **13**: 2258–2272. doi:10.1016/j.celrep.2015.11.009

Dynek JN, Goncharov T, Dueber EC, Fedorova AV, Izrael-Tomasevic A, Phu L, Helgason E, Fairbrother WJ, Deshayes K, Kirkpatrick DS, et al. 2010. c-IAP1 and UbcH5 promote K11-linked polyubiquitination of RIP1 in TNF signalling. *EMBO J* **29**: 4198–4209. doi:10.1038/emboj.2010.300

Dziedzic SA, Su Z, Jean Barrett V, Najafov A, Mookhtiar AK, Amin P, Pan H, Sun L, Zhu H, Ma A, et al. 2018. ABIN-1 regulates RIPK1 activation by linking Met1 ubiquitylation with Lys63 deubiquitylation in TNF-RSC. *Nat Cell Biol* **20**: 58–68. doi:10.1038/s41556-017-0003-1

Ea CK, Deng L, Xia ZP, Pineda G, Chen ZJ. 2006. Activation of IKK by TNFα requires site-specific ubiquitination of RIP1 and polyubiquitin binding by NEMO. *Mol Cell* **22**: 245–257. doi:10.1016/j.molcel.2006.03.026

Eliopoulos AG, Wang C, Dumitru CD, Tsichlis PN. 2003. Tpl2 transduces CD40 and TNF signals that activate ERK and regulates IgE induction by CD40. *EMBO J* **22**: 3855–3864. doi:10.1093/emboj/cdg386

Elliott PR, Leske D, Hrdinka M, Bagola K, Fiil BK, McLaughlin SH, Wagstaff J, Volkmar N, Christianson JC, Kessler BM, et al. 2016. SPATA2 links CYLD to LUBAC, activates CYLD, and controls LUBAC signaling. *Mol Cell* **63**: 990–1005. doi:10.1016/j.molcel.2016.08.001

Emmerich CH, Bakshi S, Kelsall IR, Ortiz-Guerrero J, Shpiro N, Cohen P. 2016. Lys63/Met1-hybrid ubiquitin chains are commonly formed during the activation of innate immune signalling. *Biochem Biophys Res Commun* **474**: 452–461. doi:10.1016/j.bbrc.2016.04.141

Enesa K, Ordureau A, Smith H, Barford D, Cheung PC, Patterson-Kane J, Arthur JS, Cohen P. 2012. Pellino1 is required for interferon production by viral double-stranded RNA. *J Biol Chem* **287**: 34825–34835. doi:10.1074/jbc.M112.367557

Ermolaeva MA, Michallet MC, Papadopoulou N, Utermöhlen O, Kranidioti K, Kollias G, Tschopp J, Pasparakis M. 2008. Function of TRADD in tumor necrosis factor receptor 1 signaling and in TRIF-dependent inflammatory responses. *Nat Immunol* **9**: 1037–1046. doi:10.1038/ni.1638

Fitzgerald KA, McWhirter SM, Faia KL, Rowe DC, Latz E, Golenbock DT, Coyle AJ, Liao SM, Maniatis T. 2003. IKKε and TBK1 are essential components of the IRF3 signaling pathway. *Nat Immunol* **4**: 491–496. doi:10.1038/ni921

Geng J, Ito Y, Shi L, Amin P, Chu J, Ouchida AT, Mookhtiar AK, Zhao H, Xu D, Shan B, et al. 2017. Regulation of RIPK1 activation by TAK1-mediated phosphorylation dictates apoptosis and necroptosis. *Nat Commun* **8**: 359. doi:10.1038/s41467-017-00406-w

Gerlach B, Cordier SM, Schmukle AC, Emmerich CH, Rieser E, Haas TL, Webb AI, Rickard JA, Anderton H, Wong WW, et al. 2011. Linear ubiquitination prevents inflammation and regulates immune signalling. *Nature* **471**: 591–596. doi:10.1038/nature09816

Guo H, Omoto S, Harris PA, Finger JN, Bertin J, Gough PJ, Kaiser WJ, Mocarski ES. 2015. Herpes simplex virus suppresses necroptosis in human cells. *Cell Host Microbe* **17**: 243–251. doi:10.1016/j.chom.2015.01.003

Guo H, Gilley RP, Fisher A, Lane R, Landsteiner VJ, Ragan KB, Dovey CM, Carette JE, Upton JW, Mocarski ES, et al. 2018. Species-independent contribution of ZBP1/DAI/DLM-1-triggered necroptosis in host defense against HSV1. *Cell Death Dis* **9**: 816. doi:10.1038/s41419-018-0868-3

Haas TL, Emmerich CH, Gerlach B, Schmukle AC, Cordier SM, Rieser E, Feltham R, Vince J, Warnken U, Wenger T, et al. 2009. Recruitment of the linear ubiquitin chain assembly complex stabilizes the TNF-R1 signaling complex and is required for TNF-mediated gene induction. *Mol Cell* **36**: 831–844. doi:10.1016/j.molcel.2009.10.013

Hänggi K, Vasilikos L, Valls AF, Yerbes R, Knop J, Spilgies LM, Rieck K, Misra T, Bertin J, Gough PJ, et al. 2017. RIPK1/RIPK3 promotes vascular permeability to allow tumor cell extravasation independent of its necroptotic function. *Cell Death Dis* **8**: e2588. doi:10.1038/cddis.2017.20

Harris PA, Berger SB, Jeong JU, Nagilla R, Bandyopadhyay D, Campobasso N, Capriotti CA, Cox JA, Dare L, Dong X, et al. 2017. Discovery of a first-in-class receptor interacting protein 1 (RIP1) kinase specific clinical candidate (GSK2982772) for the treatment of inflammatory diseases. *J Med Chem* **60**: 1247–1261. doi:10.1021/acs.jmedchem.6b01751

He S, Wang L, Miao L, Wang T, Du F, Zhao L, Wang X. 2009. Receptor interacting protein kinase-3 determines cellular necrotic response to TNF-α. *Cell* **137:** 1100–1111. doi:10.1016/j.cell.2009.05.021

He S, Liang Y, Shao F, Wang X. 2011. Toll-like receptors activate programmed necrosis in macrophages through a receptor-interacting kinase-3-mediated pathway. *Proc Natl Acad Sci* **108:** 20054–20059. doi:10.1073/pnas.1116302108

Heger K, Wickliffe KE, Ndoja A, Zhang J, Murthy A, Dugger DL, Maltzman A, de Sousa EMF, Hung J, Zeng Y, et al. 2018. OTULIN limits cell death and inflammation by deubiquitinating LUBAC. *Nature* **559:** 120–124. doi:10.1038/s41586-018-0256-2

Hemmi H, Takeuchi O, Sato S, Yamamoto M, Kaisho T, Sanjo H, Kawai T, Hoshino K, Takeda K, Akira S. 2004. The roles of two IκB kinase-related kinases in lipopolysaccharide and double stranded RNA signaling and viral infection. *J Exp Med* **199:** 1641–1650. doi:10.1084/jem.20040520

Hildebrand JM, Tanzer MC, Lucet IS, Young SN, Spall SK, Sharma P, Pierotti C, Garnier JM, Dobson RC, Webb AI, et al. 2014. Activation of the pseudokinase MLKL unleashes the four-helix bundle domain to induce membrane localization and necroptotic cell death. *Proc Natl Acad Sci* **111:** 15072–15077. doi:10.1073/pnas.1408987111

Hitomi J, Christofferson DE, Ng A, Yao J, Degterev A, Xavier RJ, Yuan J. 2008. Identification of a molecular signaling network that regulates a cellular necrotic cell death pathway. *Cell* **135:** 1311–1323. doi:10.1016/j.cell.2008.10.044

Holler N, Zaru R, Micheau O, Thome M, Attinger A, Valitutti S, Bodmer JL, Schneider P, Seed B, Tschopp J. 2000. Fas triggers an alternative, caspase-8-independent cell death pathway using the kinase RIP as effector molecule. *Nat Immunol* **1:** 489–495. doi:10.1038/82732

Hsu H, Huang J, Shu HB, Baichwal V, Goeddel DV. 1996a. TNF-dependent recruitment of the protein kinase RIP to the TNF receptor-1 signaling complex. *Immunity* **4:** 387–396. doi:10.1016/S1074-7613(00)80252-6

Hsu H, Shu HB, Pan MG, Goeddel DV. 1996b. TRADD-TRAF2 and TRADD-FADD interactions define two distinct TNF receptor 1 signal transduction pathways. *Cell* **84:** 299–308. doi:10.1016/S0092-8674(00)80984-8

Irmler M, Thome M, Hahne M, Schneider P, Hofmann K, Steiner V, Bodmer JL, Schroter M, Burns K, Mattmann C, et al. 1997. Inhibition of death receptor signals by cellular FLIP. *Nature* **388:** 190–195. doi:10.1038/40657

Ito Y, Ofengeim D, Najafov A, Das S, Saberi S, Li Y, Hitomi J, Zhu H, Chen H, Mayo L, et al. 2016. RIPK1 mediates axonal degeneration by promoting inflammation and necroptosis in ALS. *Science* **353:** 603–608. doi:10.1126/science.aaf6803

Jaco I, Annibaldi A, Lalaoui N, Wilson R, Tenev T, Laurien L, Kim C, Jamal K, Wicky John S, Liccardi G, et al. 2017. MK2 phosphorylates RIPK1 to prevent TNF-induced cell death. *Mol Cell* **66:** 698–710.e5. doi:10.1016/j.molcel.2017.05.003

Jiang Z, Zamanian-Daryoush M, Nie H, Silva AM, Williams BR, Li X. 2003. Poly(I-C)-induced Toll-like receptor 3 (TLR3)-mediated activation of NFκB and MAP kinase is through an interleukin-1 receptor-associated kinase (IRAK)-independent pathway employing the signaling

components TLR3-TRAF6-TAK1-TAB2-PKR. *J Biol Chem* **278:** 16713–16719. doi:10.1074/jbc.M300562200

Jiang Z, Mak TW, Sen G, Li X. 2004. Toll-like receptor 3-mediated activation of NF-κB and IRF3 diverges at Toll-IL-1 receptor domain-containing adapter inducing IFN-β. *Proc Natl Acad Sci* **101:** 3533–3538. doi:10.1073/pnas.0308496101

Kaiser WJ, Offermann MK. 2005. Apoptosis induced by the Toll-like receptor adaptor TRIF is dependent on its receptor interacting protein homotypic interaction motif. *J Immunol* **174:** 4942–4952. doi:10.4049/jimmunol.174.8.4942

Kaiser WJ, Upton JW, Mocarski ES. 2008. Receptor-interacting protein homotypic interaction motif-dependent control of NF-κB activation via the DNA-dependent activator of IFN regulatory factors. *J Immunol* **181:** 6427–6434. doi:10.4049/jimmunol.181.9.6427

Kaiser WJ, Upton JW, Long AB, Livingston-Rosanoff D, Daley-Bauer LP, Hakem R, Caspary T, Mocarski ES. 2011. RIP3 mediates the embryonic lethality of caspase-8-deficient mice. *Nature* **471:** 368–372. doi:10.1038/nature09857

Kaiser WJ, Sridharan H, Huang C, Mandal P, Upton JW, Gough PJ, Sehon CA, Marquis RW, Bertin J, Mocarski ES. 2013. Toll-like receptor 3-mediated necrosis via TRIF, RIP3, and MLKL. *J Biol Chem* **288:** 31268–31279. doi:10.1074/jbc.M113.462341

Kaiser WJ, Daley-Bauer LP, Thapa RJ, Mandal P, Berger SB, Huang C, Sundararajan A, Guo H, Roback L, Speck SH, et al. 2014. RIP1 suppresses innate immune necrotic as well as apoptotic cell death during mammalian parturition. *Proc Natl Acad Sci* **111:** 7753–7758. doi:10.1073/pnas.1401857111

Kanayama A, Seth RB, Sun L, Ea CK, Hong M, Shaito A, Chiu YH, Deng L, Chen ZJ. 2004. TAB2 and TAB3 activate the NF-κB pathway through binding to polyubiquitin chains. *Mol Cell* **15:** 535–548. doi:10.1016/j.molcel.2004.08.008

Kang TB, Ben-Moshe T, Varfolomeev EE, Pewzner-Jung Y, Yogev N, Jurewicz A, Waisman A, Brenner O, Haffner R, Gustafsson E, et al. 2004. Caspase-8 serves both apoptotic and nonapoptotic roles. *J Immunol* **173:** 2976–2984. doi:10.4049/jimmunol.173.5.2976

Kang TB, Oh GS, Scandella E, Bolinger B, Ludewig B, Kovalenko A, Wallach D. 2008. Mutation of a self-processing site in caspase-8 compromises its apoptotic but not its nonapoptotic functions in bacterial artificial chromosome-transgenic mice. *J Immunol* **181:** 2522–2532. doi:10.4049/jimmunol.181.4.2522

Kang TB, Jeong JS, Yang SH, Kovalenko A, Wallach D. 2018. Caspase-8 deficiency in mouse embryos triggers chronic RIPK1-dependent activation of inflammatory genes, independently of RIPK3. *Cell Death Differ* **25:** 1107–1117. doi:10.1038/s41418-018-0104-9

Karunakaran D, Geoffrion M, Wei L, Gan W, Richards L, Shangari P, DeKemp EM, Beanlands RA, Perisic L, Maegdefessel L, et al. 2016. Targeting macrophage necroptosis for therapeutic and diagnostic interventions in atherosclerosis. *Sci Adv* **2:** e1600224. doi:10.1126/sciadv.1600224

Kattah MG, Shao L, Rosli YY, Shimizu H, Whang MI, Advincula R, Achacoso P, Shah S, Duong BH, Onizawa M, et al. 2018. A20 and ABIN-1 synergistically preserve intestinal epithelial cell survival. *J Exp Med* **215**: 1839–1852.

Kelliher MA, Grimm S, Ishida Y, Kuo F, Stanger BZ, Leder P. 1998. The death domain kinase RIP mediates the TNF-induced NF-κB signal. *Immunity* **8**: 297–303. doi:10.1016/S1074-7613(00)80535-X

Kitur K, Parker D, Nieto P, Ahn DS, Cohen TS, Chung S, Wachtel S, Bueno S, Prince A. 2015. Toxin-induced necroptosis is a major mechanism of *Staphylococcus aureus* lung damage. *PLoS Pathog* **11**: e1004820. doi:10.1371/journal.ppat.1004820

Kitur K, Wachtel S, Brown A, Wickersham M, Paulino F, Peñaloza HF, Soong G, Bueno S, Parker D, Prince A. 2016. Necroptosis promotes *Staphylococcus aureus* clearance by inhibiting excessive inflammatory signaling. *Cell Rep* **16**: 2219–2230. doi:10.1016/j.celrep.2016.07.039

Kondylis V, Polykratis A, Ehlken H, Ochoa-Callejero L, Straub BK, Krishna-Subramanian S, Van TM, Curth HM, Heise N, Weih F, et al. 2015. NEMO prevents steatohepatitis and hepatocellular carcinoma by inhibiting RIPK1 kinase activity-mediated hepatocyte apoptosis. *Cancer Cell* **28**: 582–598. doi:10.1016/j.ccell.2015.10.001

Koudstaal S, Oerlemans MI, Van der Spoel TI, Janssen AW, Hoefer IE, Doevendans PA, Sluijter JP, Chamuleau SA. 2015. Necrostatin-1 alleviates reperfusion injury following acute myocardial infarction in pigs. *Eur J Clin Invest* **45**: 150–159. doi:10.1111/eci.12391

Kreuz S, Siegmund D, Scheurich P, Wajant H. 2001. NF-κB inducers upregulate cFLIP, a cycloheximide-sensitive inhibitor of death receptor signaling. *Mol Cell Biol* **21**: 3964–3973. doi:10.1128/MCB.21.12.3964-3973.2001

Krikos A, Laherty CD, Dixit VM. 1992. Transcriptional activation of the tumor necrosis factor α-inducible zinc finger protein, A20, is mediated by κB elements. *J Biol Chem* **267**: 17971–17976.

Kumari S, Redouane Y, Lopez-Mosqueda J, Shiraishi R, Romanowska M, Lutzmayer S, Kuiper J, Martinez C, Dikic I, Pasparakis M, et al. 2014. Sharpin prevents skin inflammation by inhibiting TNFR1-induced keratinocyte apoptosis. *eLife* **3**: e03422. doi:10.7554/eLife.03422

Kupka S, De Miguel D, Draber P, Martino L, Surinova S, Rittinger K, Walczak H. 2016. SPATA2-mediated binding of CYLD to HOIP enables CYLD recruitment to signaling complexes. *Cell Rep* **16**: 2271–2280. doi:10.1016/j.celrep.2016.07.086

Kuriakose T, Man SM, Malireddi RK, Karki R, Kesavardhana S, Place DE, Neale G, Vogel P, Kanneganti TD. 2016. ZBP1/DAI is an innate sensor of influenza virus triggering the NLRP3 inflammasome and programmed cell death pathways. *Sci Immunol* **1**: aag2045. doi:10.1126/sciimmunol.aag2045

Lafont E, Draber P, Rieser E, Reichert M, Kupka S, de Miguel D, Draberova H, von Massenhausen A, Bhamra A, Henderson S, et al. 2018. TBK1 and IKKε prevent TNF-induced cell death by RIPK1 phosphorylation. *Nat Cell Biol* **20**: 1389–1399. doi:10.1038/s41556-018-0229-6

Lamothe B, Lai Y, Xie M, Schneider MD, Darnay BG. 2013. TAK1 is essential for osteoclast differentiation and is an important modulator of cell death by apoptosis and necroptosis. *Mol Cell Biol* **33**: 582–595. doi:10.1128/MCB.01225-12

Lee TH, Shank J, Cusson N, Kelliher MA. 2004. The kinase activity of Rip1 is not required for tumor necrosis factor-α-induced IκB kinase or p38 MAP kinase activation or for the ubiquitination of Rip1 by Traf2. *J Biol Chem* **279**: 33185–33191. doi:10.1074/jbc.M404206200

Legarda D, Justus SJ, Ang RL, Rikhi N, Li W, Moran TM, Zhang J, Mizoguchi E, Zelic M, Kelliher MA, et al. 2016. CYLD proteolysis protects macrophages from TNF-mediated auto-necroptosis induced by LPS and licensed by type I IFN. *Cell Rep* **15**: 2449–2461. doi:10.1016/j.celrep.2016.05.032

Li Y, Fuhrer M, Bahrami E, Socha P, Dlaudel-Dreszler M, Bouzidi A, Liu Y, Lehle AS, Magg T, Hollizeck S, et al. 2019. Human RIPK1 deficiency causes combined immunodeficiency and inflammatory bowel diseases. *Proc Natl Acad Sci* **116**: 970–975. doi:10.1073/pnas.1813582116

Lin Y, Devin A, Rodriguez Y, Liu ZG. 1999. Cleavage of the death domain kinase RIP by caspase-8 prompts TNF-induced apoptosis. *Genes Dev* **13**: 2514–2526. doi:10.1101/gad.13.19.2514

Lin J, Kumari S, Kim C, Van TM, Wachsmuth L, Polykratis A, Pasparakis M. 2016. RIPK1 counteracts ZBP1-mediated necroptosis to inhibit inflammation. *Nature* **540**: 124–128. doi:10.1038/nature20558

Linkermann A, Bräsen JH, Himmerkus N, Liu S, Huber TB, Kunzendorf U, Krautwald S. 2012. Rip1 (receptor-interacting protein kinase 1) mediates necroptosis and contributes to renal ischemia/reperfusion injury. *Kidney Int* **81**: 751–761. doi:10.1038/ki.2011.450

Liu S, Chen J, Cai X, Wu J, Chen X, Wu YT, Sun L, Chen ZJ. 2013. MAVS recruits multiple ubiquitin E3 ligases to activate antiviral signaling cascades. *eLife* **2**: e00785. doi:10.7554/eLife.00785

Liu S, Cai X, Wu J, Cong Q, Chen X, Li T, Du F, Ren J, Wu YT, Grishin NV, et al. 2015. Phosphorylation of innate immune adaptor proteins MAVS, STING, and TRIF induces IRF3 activation. *Science* **347**: aaa2630. doi:10.1126/science.aaa2630

Lo YC, Lin SC, Rospigliosi CC, Conze DB, Wu CJ, Ashwell JD, Eliezer D, Wu H. 2009. Structural basis for recognition of diubiquitins by NEMO. *Mol Cell* **33**: 602–615. doi:10.1016/j.molcel.2009.01.012

Lule S, Wu L, McAllister LM, Edmiston WJ, Chung JY, Levy E, Zheng Y, Gough PJ, Bertin J, Degterev A, et al. 2017. Genetic inhibition of receptor interacting protein kinase-1 reduces cell death and improves functional outcome after intracerebral hemorrhage in mice. *Stroke* **48**: 2549–2556. doi:10.1161/STROKEAHA.117.017702

Maelfait J, Liverpool L, Bridgeman A, Ragan KB, Upton JW, Rehwinkel J. 2017. Sensing of viral and endogenous RNA by ZBP1/DAI induces necroptosis. *EMBO J* **36**: 2529–2543. doi:10.15252/embj.201796476

Mahoney DJ, Cheung HH, Mrad RL, Plenchette S, Simard C, Enwere E, Arora V, Mak TW, Lacasse EC, Waring J, et al. 2008. Both cIAP1 and cIAP2 regulate TNFα-mediated NF-κB activation. *Proc Natl Acad Sci* **105**: 11778–11783. doi:10.1073/pnas.0711122105

Matsui K, Kumagai Y, Kato H, Sato S, Kawagoe T, Uematsu S, Takeuchi O, Akira S. 2006. Cutting edge: Role of TANK-binding kinase 1 and inducible IκB kinase in IFN responses against viruses in innate immune cells. *J*

Cite this article as *Cold Spring Harb Perspect Biol* doi: 10.1101/cshperspect.a036368

Immunol 177: 5785–5789. doi:10.4049/jimmunol.177.9
.5785

McComb S, Cessford E, Alturki NA, Joseph J, Shutinoski B, Startek JB, Gamero AM, Mossman KL, Sad S. 2014. Type-I interferon signaling through ISGF3 complex is required for sustained Rip3 activation and necroptosis in macrophages. Proc Natl Acad Sci 111: E3206–E3213. doi:10.1073/pnas.1407068111

McWhirter SM, Fitzgerald KA, Rosains J, Rowe DC, Golenbock DT, Maniatis T. 2004. IFN-regulatory factor 3-dependent gene expression is defective in Tbk1-deficient mouse embryonic fibroblasts. Proc Natl Acad Sci 101: 233–238. doi:10.1073/pnas.2237236100

Meng H, Liu Z, Li X, Wang H, Jin T, Wu G, Shan B, Christofferson DE, Qi C, Yu Q, et al. 2018. Death-domain dimerization-mediated activation of RIPK1 controls necroptosis and RIPK1-dependent apoptosis. Proc Natl Acad Sci 115: E2001–E2009. doi:10.1073/pnas.1722013115

Menon MB, Gropengießer J, Fischer J, Novikova L, Deuretzbacher A, Lafera J, Schimmeck H, Czymmeck N, Ronkina N, Kotlyarov A, et al. 2017. p38^{MAPK}/MK2-dependent phosphorylation controls cytotoxic RIPK1 signalling in inflammation and infection. Nat Cell Biol 19: 1248–1259. doi:10.1038/ncb3614

Meylan E, Burns K, Hofmann K, Blancheteau V, Martinon F, Kelliher M, Tschopp J. 2004. RIP1 is an essential mediator of Toll-like receptor 3–induced NF-κB activation. Nat Immunol 5: 503–507. doi:10.1038/ni1061

Michallet MC, Meylan E, Ermolaeva MA, Vazquez J, Rebsamen M, Curran J, Poeck H, Bscheider M, Hartmann G, Konig M, et al. 2008. TRADD protein is an essential component of the RIG-like helicase antiviral pathway. Immunity 28: 651–661. doi:10.1016/j.immuni.2008.03.013

Micheau O, Tschopp J. 2003. Induction of TNF receptor I-mediated apoptosis via two sequential signaling complexes. Cell 114: 181–190. doi:10.1016/S0092-8674(03)00521-X

Mompeán M, Li W, Li J, Laage S, Siemer AB, Bozkurt G, Wu H, McDermott AE. 2018. The structure of the necrosome RIPK1–RIPK3 Core, a human hetero-amyloid signaling complex. Cell 173: 1244–1253.e10. doi:10.1016/j.cell.2018.03.032

Moquin DM, McQuade T, Chan FK. 2013. CYLD deubiquitinates RIP1 in the TNFα-induced necrosome to facilitate kinase activation and programmed necrosis. PLoS ONE 8: e76841. doi:10.1371/journal.pone.0076841

Moulin M, Anderton H, Voss AK, Thomas T, Wong WW, Bankovacki A, Feltham R, Chau D, Cook WD, Silke J, et al. 2012. IAPs limit activation of RIP kinases by TNF receptor 1 during development. EMBO J 31: 1679–1691. doi:10.1038/emboj.2012.18

Murakami Y, Matsumoto H, Roh M, Suzuki J, Hisatomi T, Ikeda Y, Miller JW, Vavvas DG. 2012. Receptor interacting protein kinase mediates necrotic cone but not rod cell death in a mouse model of inherited degeneration. Proc Natl Acad Sci 109: 14598–14603. doi:10.1073/pnas.1206937109

Murphy JM, Czabotar PE, Hildebrand JM, Lucet IS, Zhang JG, Alvarez-Diaz S, Lewis R, Lalaoui N, Metcalf D, Webb AI, et al. 2013. The pseudokinase MLKL mediates necroptosis via a molecular switch mechanism. Immunity 39: 443–453. doi:10.1016/j.immuni.2013.06.018

Najjar M, Saleh D, Zelic M, Nogusa S, Shah S, Tai A, Finger JN, Polykratis A, Gough PJ, Bertin J, et al. 2016. RIPK1 and RIPK3 kinases promote cell-death-independent inflammation by Toll-like receptor 4. Immunity 45: 46–59. doi:10.1016/j.immuni.2016.06.007

Newton K, Dugger DL, Wickliffe KE, Kapoor N, de Almagro MC, Vucic D, Komuves L, Ferrando RE, French DM, Webster J, et al. 2014. Activity of protein kinase RIPK3 determines whether cells die by necroptosis or apoptosis. Science 343: 1357–1360. doi:10.1126/science.1249361

Newton K, Dugger DL, Maltzman A, Greve JM, Hedehus M, Martin-McNulty B, Carano RA, Cao TC, van Bruggen N, Bernstein L, et al. 2016a. RIPK3 deficiency or catalytically inactive RIPK1 provides greater benefit than MLKL deficiency in mouse models of inflammation and tissue injury. Cell Death Differ 23: 1565–1576. doi:10.1038/cdd.2016.46

Newton K, Wickliffe KE, Maltzman A, Dugger DL, Strasser A, Pham VC, Lill JR, Roose-Girma M, Warming S, Solon M, et al. 2016b. RIPK1 inhibits ZBP1-driven necroptosis during development. Nature 540: 129–133. doi:10.1038/nature20559

Nogusa S, Thapa RJ, Dillon CP, Liedmann S, Oguin TH III, Ingram JP, Rodriguez DA, Kosoff R, Sharma S, Sturm O, et al. 2016. RIPK3 activates parallel pathways of MLKL-driven necroptosis and FADD-mediated apoptosis to protect against influenza A virus. Cell Host Microbe 20: 13–24. doi:10.1016/j.chom.2016.05.011

Oberst A, Dillon CP, Weinlich R, McCormick LL, Fitzgerald P, Pop C, Hakem R, Salvesen GS, Green DR. 2011. Catalytic activity of the caspase-8–FLIP$_L$ complex inhibits RIPK3-dependent necrosis. Nature 471: 363–367. doi:10.1038/nature09852

O'Donnell MA, Perez-Jimenez E, Oberst A, Ng A, Massoumi R, Xavier R, Green DR, Ting AT. 2011. Caspase 8 inhibits programmed necrosis by processing CYLD. Nat Cell Biol 13: 1437–1442. doi:10.1038/ncb2362

Oerlemans MI, Liu J, Arslan F, den Ouden K, van Middelaar BJ, Doevendans PA, Sluijter JP. 2012. Inhibition of RIP1-dependent necrosis prevents adverse cardiac remodeling after myocardial ischemia-reperfusion in vivo. Basic Res Cardiol 107: 270. doi:10.1007/s00395-012-0270-8

Ofengeim D, Ito Y, Najafov A, Zhang Y, Shan B, DeWitt JP, Ye J, Zhang X, Chang A, Vakifahmetoglu-Norberg H, et al. 2015. Activation of necroptosis in multiple sclerosis. Cell Rep 10: 1836–1849. doi:10.1016/j.celrep.2015.02.051

Ofengeim D, Mazzitelli S, Ito Y, DeWitt JP, Mifflin L, Zou C, Das S, Adiconis X, Chen H, Zhu H, et al. 2017. RIPK1 mediates a disease-associated microglial response in Alzheimer's disease. Proc Natl Acad Sci 114: E8788–E8797. doi:10.1073/pnas.1714175114

Onizawa M, Oshima S, Schulze-Topphoff U, Oses-Prieto JA, Lu T, Tavares R, Prodhomme T, Duong B, Whang MI, Advincula R, et al. 2015. The ubiquitin-modifying enzyme A20 restricts ubiquitination of the kinase RIPK3 and protects cells from necroptosis. Nat Immunol 16: 618–627. doi:10.1038/ni.3172

Orning P, Weng D, Starheim K, Ratner D, Best Z, Lee B, Brooks A, Xia S, Wu H, Kelliher MA, et al. 2018. Pathogen blockade of TAK1 triggers caspase-8-dependent cleavage of gasdermin D and cell death. *Science* **362:** 1064–1069. doi:10.1126/science.aau2818

Orozco S, Yatim N, Werner MR, Tran H, Gunja SY, Tait SW, Albert ML, Green DR, Oberst A. 2014. RIPK1 both positively and negatively regulates RIPK3 oligomerization and necroptosis. *Cell Death Differ* **21:** 1511–1521. doi:10.1038/cdd.2014.76

Oshima S, Turer EE, Callahan JA, Chai S, Advincula R, Barrera J, Shifrin N, Lee B, Benedict Yen TS, Woo T, et al. 2009. ABIN-1 is a ubiquitin sensor that restricts cell death and sustains embryonic development. *Nature* **457:** 906–909. doi:10.1038/nature07575

Park YH, Jeong MS, Park HH, Jang SB. 2013. Formation of the death domain complex between FADD and RIP1 proteins in vitro. *Biochim Biophys Acta* **1834:** 292–300. doi:10.1016/j.bbapap.2012.08.013

Patel S, Webster JD, Varfolomeev E, Kwon YC, Cheng JH, Zhang J, Dugger DL, Wickliffe KE, Maltzman A, Sujatha-Bhaskar S, et al. 2019. RIP1 inhibition blocks inflammatory diseases but not tumor growth or metastases. *Cell Death Differ* doi: 10.1038/s41418-019-0347-0.

Pattison MJ, Mitchell O, Flynn HR, Chen CS, Yang HT, Ben-Addi H, Boeing S, Snijders AP, Ley SC. 2016. TLR and TNF-R1 activation of the MKK3/MKK6-p38α axis in macrophages is mediated by TPL-2 kinase. *Biochem J* **473:** 2845–2861. doi:10.1042/BCJ20160502

Peltzer N, Rieser E, Taraborrelli L, Draber P, Darding M, Pernaute B, Shimizu Y, Sarr A, Draberova H, Montinaro A, et al. 2014. HOIP deficiency causes embryonic lethality by aberrant TNFR1-mediated endothelial cell death. *Cell Rep* **9:** 153–165. doi:10.1016/j.celrep.2014.08.066

Peltzer N, Darding M, Montinaro A, Draber P, Draberova H, Kupka S, Rieser E, Fisher A, Hutchinson C, Taraborrelli L, et al. 2018. LUBAC is essential for embryogenesis by preventing cell death and enabling haematopoiesis. *Nature* **557:** 112–117. doi:10.1038/s41586-018-0064-8

Perry AK, Chow EK, Goodnough JB, Yeh WC, Cheng G. 2004. Differential requirement for TANK-binding kinase-1 in type I interferon responses to Toll-like receptor activation and viral infection. *J Exp Med* **199:** 1651–1658. doi:10.1084/jem.20040528

Peterson LW, Philip NH, DeLaney A, Wynosky-Dolfi MA, Asklof K, Gray F, Choa R, Bjanes E, Buza EL, Hu B, et al. 2017. RIPK1-dependent apoptosis bypasses pathogen blockade of innate signaling to promote immune defense. *J Exp Med* **214:** 3171–3182. doi:10.1084/jem.20170347

Petrie EJ, Sandow JJ, Jacobsen AV, Smith BJ, Griffin MDW, Lucet IS, Dai W, Young SN, Tanzer MC, Wardak A, et al. 2018. Conformational switching of the pseudokinase domain promotes human MLKL tetramerization and cell death by necroptosis. *Nat Commun* **9:** 2422. doi:10.1038/s41467-018-04714-7

Pobezinskaya YL, Kim YS, Choksi S, Morgan MJ, Li T, Liu C, Liu Z. 2008. The function of TRADD in signaling through tumor necrosis factor receptor 1 and TRIF-dependent Toll-like receptors. *Nat Immunol* **9:** 1047–1054. doi:10.1038/ni.1639

Polykratis A, Hermance N, Zelic M, Roderick J, Kim C, Van TM, Lee TH, Chan FK, Pasparakis M, Kelliher MA. 2014. Cutting edge: RIPK1 Kinase inactive mice are viable and protected from TNF-induced necroptosis in vivo. *J Immunol* **193:** 1539–1543. doi:10.4049/jimmunol.1400590

Polykratis A, Martens A, Eren RO, Shirasaki Y, Yamagishi M, Yamaguchi Y, Uemura S, Miura M, Holzmann B, Kollias G, et al. 2019. A20 prevents inflammasome-dependent arthritis by inhibiting macrophage necroptosis through its ZnF7 ubiquitin-binding domain. *Nat Cell Biol* **21:** 731–742. doi:10.1038/s41556-019-0324-3

Rahighi S, Ikeda F, Kawasaki M, Akutsu M, Suzuki N, Kato R, Kensche T, Uejima T, Bloor S, Komander D, et al. 2009. Specific recognition of linear ubiquitin chains by NEMO is important for NF-κB activation. *Cell* **136:** 1098–1109. doi:10.1016/j.cell.2009.03.007

Rajput A, Kovalenko A, Bogdanov K, Yang SH, Kang TB, Kim JC, Du J, Wallach D. 2011. RIG-I RNA helicase activation of IRF3 transcription factor is negatively regulated by caspase-8-mediated cleavage of the RIP1 protein. *Immunity* **34:** 340–351. doi:10.1016/j.immuni.2010.12.018

Raju S, Whalen DM, Mengistu M, Swanson C, Quinn JG, Taylor SS, Webster JD, Newton K, Shaw AS. 2018. Kinase domain dimerization drives RIPK3-dependent necroptosis. *Sci Signal* **11:** eaar2188. doi: 10.1126/scisignal.aar2188

Rickard JA, Anderton H, Etemadi N, Nachbur U, Darding M, Peltzer N, Lalaoui N, Lawlor KE, Vanyai H, Hall C, et al. 2014a. TNFR1-dependent cell death drives inflammation in *Sharpin*-deficient mice. *eLife* **3:** e03464. doi:10.7554/eLife.03464

Rickard JA, O'Donnell JA, Evans JM, Lalaoui N, Poh AR, Rogers T, Vince JE, Lawlor KE, Ninnis RL, Anderton H, et al. 2014b. RIPK1 regulates RIPK3-MLKL-driven systemic inflammation and emergency hematopoiesis. *Cell* **157:** 1175–1188. doi:10.1016/j.cell.2014.04.019

Rijal D, Ariana A, Wight A, Kim K, Alturki NA, Aamir Z, Ametepe ES, Korneluk RG, Tiedje C, Menon MB, et al. 2018. Differentiated macrophages acquire a pro-inflammatory and cell death-resistant phenotype due to increasing XIAP and p38-mediated inhibition of RipK1. *J Biol Chem* **293:** 11913–11927. doi:10.1074/jbc.RA118.003614

Roderick JE, Hermance N, Zelic M, Simmons MJ, Polykratis A, Pasparakis M, Kelliher MA. 2014. Hematopoietic RIPK1 deficiency results in bone marrow failure caused by apoptosis and RIPK3-mediated necroptosis. *Proc Natl Acad Sci* **111:** 14436–14441. doi:10.1073/pnas.1409389111

Rodriguez DA, Weinlich R, Brown S, Guy C, Fitzgerald P, Dillon CP, Oberst A, Quarato G, Low J, Cripps JG, et al. 2016. Characterization of RIPK3-mediated phosphorylation of the activation loop of MLKL during necroptosis. *Cell Death Differ* **23:** 76–88. doi:10.1038/cdd.2015.70

Saleh D, Najjar M, Zelic M, Shah S, Nogusa S, Polykratis A, Paczosa MK, Gough PJ, Bertin J, Whalen M, et al. 2017. Kinase activities of RIPK1 and RIPK3 can direct IFN-β synthesis induced by lipopolysaccharide. *J Immunol* **198:** 4435–4447. doi:10.4049/jimmunol.1601717

Sato S, Sugiyama M, Yamamoto M, Watanabe Y, Kawai T, Takeda K, Akira S. 2003. Toll/IL-1 receptor domain-containing adaptor inducing IFN-β (TRIF) associates with TNF receptor-associated factor 6 and TANK-binding kinase 1, and activates two distinct transcription factors, NF-κB and IFN-regulatory factor-3, in the Toll-like receptor signaling. *J Immunol* **171:** 4304–4310. doi:10 .4049/jimmunol.171.8.4304

Schlicher L, Wissler M, Preiss F, Brauns-Schubert P, Jakob C, Dumit V, Borner C, Dengjel J, Maurer U. 2016. SPATA2 promotes CYLD activity and regulates TNF-induced NF-κB signaling and cell death. *EMBO Rep* **17:** 1485–1497. doi:10.15252/embr.201642592

Schock SN, Chandra NV, Sun Y, Irie T, Kitagawa Y, Gotoh B, Coscoy L, Winoto A. 2017. Induction of necroptotic cell death by viral activation of the RIG-I or STING pathway. *Cell Death Differ* **24:** 615–625. doi:10.1038/cdd.2016 .153

Seehawer M, Heinzmann F, D'Artista L, Harbig J, Roux PF, Hoenicke L, Dang H, Klotz S, Robinson L, Doré G, et al. 2018. Necroptosis microenvironment directs lineage commitment in liver cancer. *Nature* **562:** 69–75. doi:10 .1038/s41586-018-0519-y

Seifert L, Werba G, Tiwari S, Giao Ly NN, Alothman S, Alqunaibit D, Avanzi A, Barilla R, Daley D, Greco SH, et al. 2016. The necrosome promotes pancreatic oncogenesis via CXCL1 and Mincle-induced immune suppression. *Nature* **532:** 245–249. doi:10.1038/na ture17403

Shim JH, Xiao C, Paschal AE, Bailey ST, Rao P, Hayden MS, Lee KY, Bussey C, Steckel M, Tanaka N, et al. 2005. TAK1, but not TAB1 or TAB2, plays an essential role in multiple signaling pathways in vivo. *Genes Dev* **19:** 2668–2681. doi:10.1101/gad.1360605

Shu HB, Takeuchi M, Goeddel DV. 1996. The tumor necrosis factor receptor 2 signal transducers TRAF2 and c-IAP1 are components of the tumor necrosis factor receptor 1 signaling complex. *Proc Natl Acad Sci* **93:** 13973–13978. doi:10.1073/pnas.93.24.13973

Shutinoski B, Alturki NA, Rijal D, Bertin J, Gough PJ, Schlossmacher MG, Sad S. 2016. K45A mutation of RIPK1 results in poor necroptosis and cytokine signaling in macrophages, which impacts inflammatory responses in vivo. *Cell Death Differ* **23:** 1628–1637. doi:10.1038/cdd .2016.51

Stanger BZ, Leder P, Lee TH, Kim E, Seed B. 1995. RIP: A novel protein containing a death domain that interacts with Fas/APO-1 (CD95) in yeast and causes cell death. *Cell* **81:** 513–523. doi:10.1016/0092-8674(95)90072-1

Stehlik C, de Martin R, Kumabashiri I, Schmid JA, Binder BR, Lipp J. 1998. Nuclear factor (NF)-κB-regulated X-chromosome–linked *iap* gene expression protects endothelial cells from tumor necrosis factor α–induced apoptosis. *J Exp Med* **188:** 211–216. doi:10.1084/jem.188.1 .211

Strilic B, Yang L, Albarrán-Juárez J, Wachsmuth L, Han K, Müller UC, Pasparakis M, Offermanns S. 2016. Tumour-cell-induced endothelial cell necroptosis via death receptor 6 promotes metastasis. *Nature* **536:** 215–218. doi:10 .1038/nature19076

Stutz MD, Ojaimi S, Allison C, Preston S, Arandjelovic P, Hildebrand JM, Sandow JJ, Webb AI, Silke J, Alexander WS, et al. 2018. Necroptotic signaling is primed in *Mycobacterium tuberculosis*-infected macrophages, but its pathophysiological consequence in disease is restricted. *Cell Death Differ* **25:** 951–965. doi:10.1038/s41418-017-0031-1

Su L, Quade B, Wang H, Sun L, Wang X, Rizo J. 2014. A plug release mechanism for membrane permeation by MLKL. *Structure* **22:** 1489–1500. doi:10.1016/j.str.2014.07.014

Sun X, Yin J, Starovasnik MA, Fairbrother WJ, Dixit VM. 2002. Identification of a novel homotypic interaction motif required for the phosphorylation of receptor-interacting protein (RIP) by RIP3. *J Biol Chem* **277:** 9505–9511. doi:10.1074/jbc.M109488200

Sun L, Wang H, Wang Z, He S, Chen S, Liao D, Wang L, Yan J, Liu W, Lei X, et al. 2012. Mixed lineage kinase domain-like protein mediates necrosis signaling downstream of RIP3 kinase. *Cell* **148:** 213–227. doi:10.1016/j.cell.2011 .11.031

Takahashi N, Vereecke L, Bertrand MJ, Duprez L, Berger SB, Divert T, Gonçalves A, Sze M, Gilbert B, Kourula S, et al. 2014. RIPK1 ensures intestinal homeostasis by protecting the epithelium against apoptosis. *Nature* **513:** 95–99. doi:10.1038/nature13706

Thapa RJ, Ingram JP, Ragan KB, Nogusa S, Boyd DF, Benitez AA, Sridharan H, Kosoff R, Shubina M, Landsteiner VJ, et al. 2016. DAI senses influenza A virus genomic RNA and activates RIPK3-dependent cell death. *Cell Host Microbe* **20:** 674–681. doi:10.1016/j.chom.2016.09.014

Ting AT, Pimentel-Muiños FX, Seed B. 1996. RIP mediates tumor necrosis factor receptor 1 activation of NF-κB but not Fas/APO-1-initiated apoptosis. *EMBO J* **15:** 6189–6196. doi:10.1002/j.1460-2075.1996.tb01007.x

Tokunaga F, Nakagawa T, Nakahara M, Saeki Y, Taniguchi M, Sakata S, Tanaka K, Nakano H, Iwai K. 2011. SHARPIN is a component of the NF-κB-activating linear ubiquitin chain assembly complex. *Nature* **471:** 633–636. doi:10.1038/nature09815

Tournier C, Dong C, Turner TK, Jones SN, Flavell RA, Davis RJ. 2001. MKK7 is an essential component of the JNK signal transduction pathway activated by proinflammatory cytokines. *Genes Dev* **15:** 1419–1426. doi:10.1101/ gad.888501

Upton JW, Kaiser WJ, Mocarski ES. 2012. DAI/ZBP1/DLM-1 complexes with RIP3 to mediate virus-induced programmed necrosis that is targeted by murine cytomegalovirus vIRA. *Cell Host Microbe* **11:** 290–297. doi:10.1016/ j.chom.2012.01.016

Van TM, Polykratis A, Straub BK, Kondylis V, Papadopoulou N, Pasparakis M. 2017. Kinase-independent functions of RIPK1 regulate hepatocyte survival and liver carcinogenesis. *J Clin Invest* **127:** 2662–2677. doi:10.1172/JCI 92508

Vanlangenakker N, Vanden Berghe T, Bogaert P, Laukens B, Zobel K, Deshayes K, Vucic D, Fulda S, Vandenabeele P, Bertrand MJ. 2011. cIAP1 and TAK1 protect cells from TNF-induced necrosis by preventing RIP1/RIP3-dependent reactive oxygen species production. *Cell Death Differ* **18:** 656–665. doi:10.1038/cdd.2010.138

Varfolomeev EE, Schuchmann M, Luria V, Chiannilkulchai N, Beckmann JS, Mett IL, Rebrikov D, Brodianski VM, Kemper OC, Kollet O, et al. 1998. Targeted disruption of the mouse *Caspase 8* gene ablates cell death induction by the TNF receptors, Fas/Apo1, and DR3 and is lethal prenatally. *Immunity* **9:** 267–276. doi:10.1016/S1074-7613 (00)80609-3

Varfolomeev E, Goncharov T, Fedorova AV, Dynek JN, Zobel K, Deshayes K, Fairbrother WJ, Vucic D. 2008. c-IAP1 and c-IAP2 are critical mediators of tumor necrosis factor α (TNFα)-induced NF-κB activation. *J Biol Chem* **283**: 24295–24299. doi:10.1074/jbc.C800128200

Vince JE, Pantaki D, Feltham R, Mace PD, Cordier SM, Schmukle AC, Davidson AJ, Callus BA, Wong WW, Gentle IE, et al. 2009. TRAF2 must bind to cellular inhibitors of apoptosis for tumor necrosis factor (TNF) to efficiently activate NF-κB and to prevent TNF-induced apoptosis. *J Biol Chem* **284**: 35906–35915. doi:10.1074/jbc.M109.072256

Vlantis K, Wullaert A, Polykratis A, Kondylis V, Dannappel M, Schwarzer R, Welz P, Corona T, Walczak H, Weih F, et al. 2016. NEMO prevents RIP kinase 1-mediated epithelial cell death and chronic intestinal inflammation by NF-κB-dependent and -independent functions. *Immunity* **44**: 553–567. doi:10.1016/j.immuni.2016.02.020

Wagner SA, Satpathy S, Beli P, Choudhary C. 2016. SPATA2 links CYLD to the TNF-α receptor signaling complex and modulates the receptor signaling outcomes. *EMBO J* **35**: 1868–1884. doi:10.15252/embj.201694300

Wang CY, Mayo MW, Korneluk RG, Goeddel DV, Baldwin AS Jr. 1998. NF-κB antiapoptosis: Induction of TRAF1 and TRAF2 and c-IAP1 and c-IAP2 to suppress caspase-8 activation. *Science* **281**: 1680–1683. doi:10.1126/science.281.5383.1680

Wang L, Du F, Wang X. 2008. TNF-α induces two distinct caspase-8 activation pathways. *Cell* **133**: 693–703. doi:10.1016/j.cell.2008.03.036

Wang H, Sun L, Su L, Rizo J, Liu L, Wang LF, Wang FS, Wang X. 2014. Mixed lineage kinase domain-like protein MLKL causes necrotic membrane disruption upon phosphorylation by RIP3. *Mol Cell* **54**: 133–146. doi:10.1016/j.molcel.2014.03.003

Wang Q, Zhou T, Liu Z, Ren J, Phan N, Gupta K, Stewart DM, Morgan S, Assa C, Kent KC, et al. 2017a. Inhibition of receptor-interacting protein kinase 1 with necrostatin-1s ameliorates disease progression in elastase-induced mouse abdominal aortic aneurysm model. *Sci Rep* **7**: 42159. doi:10.1038/srep42159

Wang Y, Gao W, Shi X, Ding J, Liu W, He H, Wang K, Shao F. 2017b. Chemotherapy drugs induce pyroptosis through caspase-3 cleavage of a gasdermin. *Nature* **547**: 99–103. doi:10.1038/nature22393

Wang W, Marinis JM, Beal AM, Savadkar S, Wu Y, Khan M, Taunk PS, Wu N, Su W, Wu J, et al. 2018. RIP1 kinase drives macrophage-mediated adaptive immune tolerance in pancreatic cancer. *Cancer Cell* **34**: 757–774.e7. doi:10.1016/j.ccell.2018.10.006

Welz PS, Wullaert A, Vlantis K, Kondylis V, Fernández-Majada V, Ermolaeva M, Kirsch P, Sterner-Kock A, van Loo G, Pasparakis M. 2011. FADD prevents RIP3-mediated epithelial cell necrosis and chronic intestinal inflammation. *Nature* **477**: 330–334. doi:10.1038/nature10273

Wong WW, Vince JE, Lalaoui N, Lawlor KE, Chau D, Bankovacki A, Anderton H, Metcalf D, O'Reilly L, Jost PJ, et al. 2014. cIAPs and XIAP regulate myelopoiesis through cytokine production in an RIPK1- and RIPK3-dependent manner. *Blood* **123**: 2562–2572. doi:10.1182/blood-2013-06-510743

Wu CJ, Conze DB, Li T, Srinivasula SM, Ashwell JD. 2006. Sensing of Lys 63-linked polyubiquitination by NEMO is a key event in NF-κB activation. *Nat Cell Biol* **8**: 398–406. doi:10.1038/ncb1384

Wu J, Huang Z, Ren J, Zhang Z, He P, Li Y, Ma J, Chen W, Zhang Y, Zhou X, et al. 2013. Mlkl knockout mice demonstrate the indispensable role of Mlkl in necroptosis. *Cell Res* **23**: 994–1006. doi:10.1038/cr.2013.91

Wu XN, Yang ZH, Wang XK, Zhang Y, Wan H, Song Y, Chen X, Shao J, Han J. 2014. Distinct roles of RIP1–RIP3 hetero- and RIP3–RIP3 homo-interaction in mediating necroptosis. *Cell Death Differ* **21**: 1709–1720. doi:10.1038/cdd.2014.77

Xie T, Peng W, Yan C, Wu J, Gong X, Shi Y. 2013. Structural insights into RIP3-mediated necroptotic signaling. *Cell Rep* **5**: 70–78. doi:10.1016/j.celrep.2013.08.044

Xu C, Wu X, Zhang X, Xie Q, Fan C, Zhang H. 2018a. Embryonic lethality and host immunity of RelA-deficient mice are mediated by both apoptosis and necroptosis. *J Immunol* **200**: 271–285. doi:10.4049/jimmunol.1700859

Xu D, Jin T, Zhu H, Chen H, Ofengeim D, Zou C, Mifflin L, Pan L, Amin P, Li W, et al. 2018b. TBK1 suppresses RIPK1-driven apoptosis and inflammation during development and in aging. *Cell* **174**: 1477–1491.e19. doi:10.1016/j.cell.2018.07.041

Yamaguchi N, Yamaguchi N. 2015. The seventh zinc finger motif of A20 is required for the suppression of TNF-α-induced apoptosis. *FEBS Lett* **589**: 1369–1375. doi:10.1016/j.febslet.2015.04.022

Yeh WC, de la Pompa JL, McCurrach ME, Shu HB, Elia AJ, Shahinian A, Ng M, Wakeham A, Khoo W, Mitchell K, et al. 1998. FADD: Essential for embryo development and signaling from some, but not all, inducers of apoptosis. *Science* **279**: 1954–1958. doi:10.1126/science.279.5358.1954

Yeh WC, Itie A, Elia AJ, Ng M, Shu HB, Wakeham A, Mirtsos C, Suzuki N, Bonnard M, Goeddel DV, et al. 2000. Requirement for Casper (c-FLIP) in regulation of death receptor-induced apoptosis and embryonic development. *Immunity* **12**: 633–642. doi:10.1016/S1074-7613(00)80214-9

Zhang DW, Shao J, Lin J, Zhang N, Lu BJ, Lin SC, Dong MQ, Han J. 2009. RIP3, an energy metabolism regulator that switches TNF-induced cell death from apoptosis to necrosis. *Science* **325**: 332–336. doi:10.1126/science.1172308

Zhang H, Zhou X, McQuade T, Li J, Chan FK, Zhang J. 2011. Functional complementation between FADD and RIP1 in embryos and lymphocytes. *Nature* **471**: 373–376. doi:10.1038/nature09878

Zhang J, Clark K, Lawrence T, Peggie MW, Cohen P. 2014. An unexpected twist to the activation of IKKβ: TAK1 primes IKKβ for activation by autophosphorylation. *Biochemical J* **461**: 531–537. doi:10.1042/BJ20140444

Zhang X, Fan C, Zhang H, Zhao Q, Liu Y, Xu C, Xie Q, Wu X, Yu X, Zhang J, et al. 2016. MLKL and FADD are critical for suppressing progressive lymphoproliferative disease and activating the NLRP3 inflammasome. *Cell Rep* **16**: 3247–3259. doi:10.1016/j.celrep.2016.06.103

Cite this article as *Cold Spring Harb Perspect Biol* doi: 10.1101/cshperspect.a036368

Zhang Q, Lenardo MJ, Baltimore D. 2017. 30 years of NF-κB: A blossoming of relevance to human pathobiology. *Cell* **168:** 37–57. doi:10.1016/j.cell.2016.12.012

Zhang X, Dowling JP, Zhang J. 2019a. RIPK1 can mediate apoptosis in addition to necroptosis during embryonic development. *Cell Death Dis* **10:** 245. doi:10.1038/s41419-019-1490-8

Zhang S, Su Y, Ying Z, Guo D, Zou Z, Wang L, Zhang Z, Jiang Z, Zhang Z, Wang X. 2019b. RIP1 kinase inhibitor halts the progression of an immune-induced demyelination disease at the stage of monocyte elevation. *Proc*

Natl Acad Sci **116:** 5675–5680. doi:10.1073/pnas.1819917116

Zhao J, Jitkaew S, Cai Z, Choksi S, Li Q, Luo J, Liu ZG. 2012. Mixed lineage kinase domain-like is a key receptor interacting protein 3 downstream component of TNF-induced necrosis. *Proc Natl Acad Sci* **109:** 5322–5327. doi:10.1073/pnas.1200012109

Zinngrebe J, Rieser E, Taraborrelli L, Peltzer N, Hartwig T, Ren H, Kovács I, Endres C, Draber P, Darding M, et al. 2016. LUBAC deficiency perturbs TLR3 signaling to cause immunodeficiency and autoinflammation. *J Exp Med* **213:** 2671–2689. doi:10.1084/jem.20160041

The Killer Pseudokinase Mixed Lineage Kinase Domain-Like Protein (MLKL)

James M. Murphy[1,2]

[1]Walter and Eliza Hall Institute of Medical Research, Parkville, Victoria 3052, Australia
[2]Department of Medical Biology, University of Melbourne, Parkville, Victoria 3052, Australia

Correspondence: jamesm@wehi.edu.au

Whereas the apoptosis cell death pathway typically enables cells to undergo death in an immunologically silent manner, cell death by necroptosis induces cell lysis and release of cellular constituents known to elicit an immune response. Consequently, the origins of necroptosis likely originated in host defense against pathogens, although recently it has emerged that dysregulation of the pathway underlies many human pathologies. The past decade has seen a rapid advance in our understanding of the molecular mechanisms underlying necroptotic cell death, including the implication of the pseudokinase, mixed lineage kinase domain-like protein (MLKL), as the terminal effector in the pathway. Here, I review our current understanding of how MLKL is activated by the upstream receptor interacting protein kinase (RIPK)3, the proposed mechanism(s) by which MLKL kills cells, and recently described layers of regulation that tune MLKL's killing activity.

The programmed necrosis pathway was first described by Holler, Tschopp, and colleagues in 2000. This cell death mechanism relies on the protein kinase, receptor interacting protein kinase (RIPK)1, downstream of Fas, tumor necrosis factor (TNF), or TRAIL receptor activation, in the absence of the proteolytic activity of the apoptosis effector, caspase-8 (Holler et al. 2000). The name necroptosis was coined to describe the pathway in 2005 (Degterev et al. 2005) to convey that, like apoptosis, necroptosis is a form of regulated cell death. In 2009, the protein kinase RIPK3 was implicated in the pathway (Cho et al. 2009; He et al. 2009; Zhang et al. 2009), with the terminal effector pseudokinase, mixed lineage kinase domain-like protein (MLKL), implicated 3 years later (Sun et al. 2012; Zhao et al.

2012). In parallel with these advances, our knowledge of the upstream cues that induce necroptotic signaling has advanced. Stimulation of TNF receptor 1 is the best studied mode of necroptosis induction, with other death receptor ligands, Toll-like receptor (TLR) ligands (e.g., double-stranded RNA, lipopolysaccharide) (He et al. 2011; Kaiser et al. 2013), type I interferons (Thapa et al. 2011; McComb et al. 2014), and, most recently, oxidized low-density lipoprotein (oxLDL) (Karunakaran et al. 2016) known to induce necroptosis in various cellular contexts.

Physiologically, necroptosis signaling has been attributed roles in a wide range of human pathologies. Most of these attributions have arisen from studies of RIPK3-deficient mice (Newton et al. 2004), which, like MLKL-deficient mice

(Murphy et al. 2013; Wu et al. 2013), are viable, indicating that necroptosis is dispensable for development and not triggered in the absence of challenge. Recent revelations that RIPK3 performs additional functions beyond necroptosis signaling, such as directing inflammatory cytokine production and apoptotic signaling (Mandal et al. 2014; Newton et al. 2014; Wong et al. 2014; Alvarez-Diaz et al. 2016), have led to reevaluation of whether phenotypes observed in challenged RIPK3-knockout mice represent necroptosis. Because MLKL has only been credibly attributed functions in necroptosis signaling, studies of MLKL-deficient mice have enabled validation of dysregulated necroptosis contributing to the pathologies of several diseases, including kidney ischemic-reperfusion injury (Müller et al. 2017; Pefanis et al. 2019), pathogen clearance (Kitur et al. 2016), sterile sepsis (Newton et al. 2016a; Moerke et al. 2019), dermatitis (Dannappel et al. 2014), and multiorgan inflammation (Rickard et al. 2014a,b). One of the major challenges with ascribing roles for necroptosis in pathologies is that different death modalities can cooccur in diseased tissues, as evidenced by apoptosis and necroptosis cooccurring in mice with hyperactive TNF signaling (Dannappel et al. 2014; Rickard et al. 2014a), and necroptosis and ferroptosis cooccurring following acute kidney injury (Müller et al. 2017). Nevertheless, the recent discovery of a mouse strain harboring an activated MLKL mutant clearly implicates inflammation as an important contributor to necroptotic pathologies, as these animals succumbed to a lethal inflammatory syndrome within 6 days of birth (Hildebrand et al. 2019). Consequently, much interest has arisen in therapeutically targeting the necroptosis pathway. To this end, there has been a flurry of interest in the molecular mechanisms underlying MLKL activity, how MLKL kills cells and regulators of its lytic activity, which are reviewed below.

OVERVIEW OF THE NECROPTOSIS PATHWAY

Following ligation of cell surface receptors, such as TNF receptor 1, in contexts in which RIPK1 ubiquitylation by inhibitors of apoptosis proteins (IAPs) and caspase-8 activity are depleted or inhibited, a large molecular weight platform termed the "necrosome" assembles in the cytoplasm. Although the composition and magnitude of the necrosome complex likely differs depending on cellular context, at the core of this platform are the RIPK1 and RIPK3 protein kinases and the MLKL pseudokinase (Fig. 1). RIPK1 and RIPK3 contain an amino-terminal protein kinase domain and a short module termed the RIP homotypic interacting motif (RHIM) (Sun et al. 2002), responsible for oligomerization, carboxy-terminal to the kinase domain. RIPK1 contains a further protein interaction module, the death domain, at its carboxyl terminus, which is known to mediate oligomerization (Meng et al. 2018) and interaction with other death domain containing proteins, such as TNF receptor 1 (Hsu et al. 1996). Assembly of RIPK1-RIPK3 hetero-oligomers is facilitated by their RHIMs, which in isolation assemble into amyloid structures of indeterminate size (Li et al. 2012; Mompeán et al. 2018; Pham et al. 2019). Via its RHIM, RIPK1 is thought to seed assembly of RIPK3 into fibrils, where it is presumed recruitment of further RIPK3 protomers and their autophosphorylation are key events in necroptosis signal propagation. RIPK3 recruitment to other RIPK3 protomers within this assembly may be favored by allosteric interactions between their kinase domains and activation by phosphorylation of a site in the C-lobe of their kinase domains (Raju et al. 2018). It is from this complex that RIPK3 can interact with, and activate MLKL by phosphorylation to kill cells (described below; Figs. 1 and 2). The kinase activity of RIPK1, which is known only to phosphorylate its own activation loop, is required for necroptotic signaling (Berger et al. 2014; Newton et al. 2014; de Almagro et al. 2017; Newton 2019). Presumably this autophosphorylation leads to an electrostatic repulsion or conformational change that disfavors RIPK3 hetero-oligomer formation to allow RIPK3 to preferentially self-associate within the necrosome complex. Other RHIM proteins, such as ZBP1/DAI and TRIF, are thought to facilitate RIPK3 homo-oligomerization by sequestering RIPK1 or direct interaction with RIPK3, respectively (Kaiser

Cite this article as *Cold Spring Harb Perspect Biol* doi: 10.1101/cshperspect.a036376

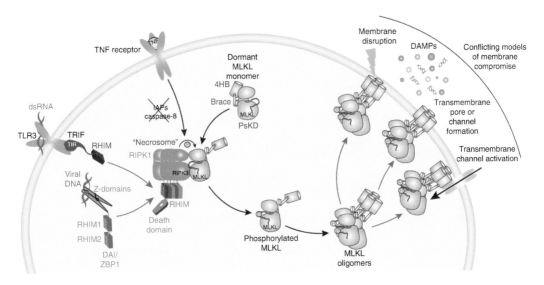

Figure 1. Simplified schematic overview of the necroptosis signaling pathway. Exogenous stimuli provoke assembly of the receptor interacting protein kinase (RIPK)1:RIPK3 oligomeric complex termed the necrosome, which acts as a platform for recruiting and activating mixed lineage kinase domain-like protein (MLKL). MLKL phosphorylation by RIPK3 induces assembly of killer complexes of unclear stoichiometry, which are trafficked to the plasma membrane where cell permeabilization occurs. Several conflicting models of permeabilization have been proposed (Cai et al. 2014; Chen et al. 2014; Dondelinger et al. 2014; Hildebrand et al. 2014; Su et al. 2014; Wang et al. 2014; Xia et al. 2016; Huang et al. 2017), and the precise mechanism remains to be deduced. (TIR) Toll-interleukin receptor, (4HB) four-helix bundle, (PsKD) pseudokinase domain, (RHIM) RIP homotypic interacting motif.

et al. 2013; Lin et al. 2016; Newton et al. 2016b). Although TRIF acts as a cytoplasmic adaptor for TLRs, whether ZBP1/DAI is activated by viral DNA, viral RNA, and/or RNA:protein complexes is incompletely understood. Full comprehension of the mechanism by which RHIM-containing proteins might activate RIPK3 and cues that prompt RIPK3 self-association await more complete structural examination.

MLKL: THE END OF THE LINE

MLKL is composed of an amino-terminal four-helix bundle (4HB) domain, a two-helix "brace" region, and a carboxy-terminal pseudokinase domain (Fig. 3A; Murphy et al. 2013). The 4HB domain functions as the executioner domain by virtue of its membrane permeabilization activity (Cai et al. 2014; Chen et al. 2014; Dondelinger et al. 2014; Hildebrand et al. 2014; Su et al. 2014; Wang et al. 2014; Tanzer et al. 2016). The 4HB domain fold is akin to the HeLo

domain, which exerts the membrane permeabilization function of the yeast Het-S protein (Daskalov et al. 2016) and the recently described plant MLKL ortholog, AtMLKL (Mahdi et al. 2019). The brace helices serve two principal functions within MLKL: as key contributors to MLKL oligomerization, and as levers to communicate signals from the pseudokinase domain to the 4HB domain (Su et al. 2014; Quarato et al. 2016; Davies et al. 2018). The carboxy-terminal pseudokinase domain serves a regulatory function by binding and restraining the executioner function of the 4HB domain in the dormant, nonactivated state (Murphy et al. 2013; Hildebrand et al. 2014; Tanzer et al. 2016; Davies et al. 2018; Petrie et al. 2018). Additionally, the pseudokinase domain serves as a signal integrator, in which phosphorylation of the activation loop by RIPK3 serves as a cue for activation. This event is thought to trigger a conformational change within the pseudokinase that promotes 4HB domain exposure, enabling

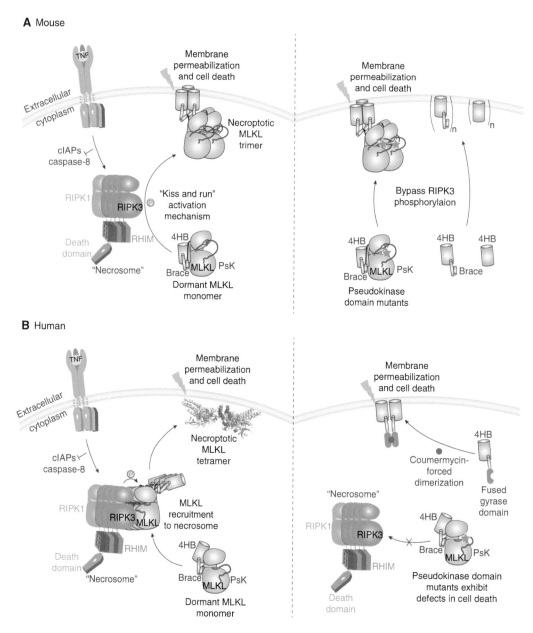

Figure 2. Mouse and human necroptosis pathways differ in mechanism. Although mouse mixed lineage kinase domain-like protein (MLKL) can be activated by receptor interacting protein kinase (RIPK)3 via a transient interaction that can be bypassed via mutations in the pseudoactive site or simply expression of the four-helix bundle (4HB) domain (A) (Murphy et al. 2013; Hildebrand et al. 2014), the same is not true of human MLKL (B) (Petrie et al. 2018). Human MLKL relies on recruitment to necrosomal RIPK3 for its activation (*left*) and it is not possible to bypass this activation step by introducing mutations in the human MLKL pseudoactive site (*right*). Furthermore, human MLKL 4HB domain expression is insufficient to induce necroptosis; forced dimerization via a fused domain is crucial to enabling the human MLKL 4HB domain to kill cells (Quarato et al. 2016; Tanzer et al. 2016). The concepts presented in this summary figure have been presented individually elsewhere previously (Tanzer et al. 2016; Petrie et al. 2018, 2019a).

Cite this article as *Cold Spring Harb Perspect Biol* doi: 10.1101/cshperspect.a036376

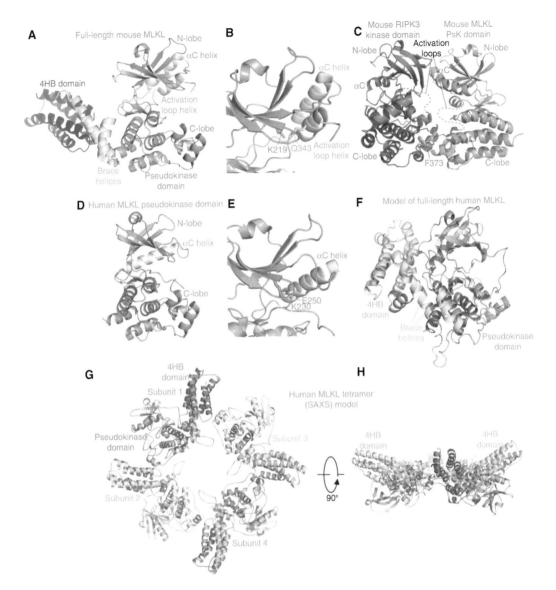

Figure 3. Structural studies of full-length mixed lineage kinase domain-like protein (MLKL) and the MLKL pseudokinase domain. (A) The structure of full-length mouse MLKL revealed a domain architecture comprising an amino-terminal four-helix bundle (4HB) domain, the connector "brace" helices, and the carboxy-terminal pseudokinase domain (PDB, 4BTF). (B) Zoomed-in view of the pseudokinase domain pseudoactive site within the structure shown in panel A. The atypical interacting residues K219 from the N-lobe β3 strand and the activation loop helix Q343 are shown as sticks. The αC helix is shown in gray in A, B, D, E, and F; the activation loop helix of mouse MLKL in cyan in A and B. (C) The mouse MLKL pseudokinase domain (blue):RIPK3 kinase domain (red) cocrystal structure identified a face-to-face interaction between the domains, with activation loops apposed (PDB, 4M69) (Xie et al. 2013). The MLKL F373 residue (blue sticks) is buried in the RIPK3 C-lobe, and was found to be a crucial determinant of human MLKL binding to RIPK3 in subsequent studies (Petrie et al. 2019b). (D) The human MLKL pseudokinase domain crystal structure revealed a more conventional kinase domain structure with a canonical β3 strand Lys salt bridge with the αC helix Glu (shown in sticks in zoomed image in panel E). This structure was drawn from PDB 4MWI (Murphy et al. 2014), but is comparable with PDB 4M67 solved in parallel (Xie et al. 2013). (F) The structure of full-length human MLKL was modeled using distance restraints derived from cross-linking mass spectrometry data (Petrie et al. 2018), which were consistent with the sequestration of the killer 4HB domain by the suppressor pseudokinase domain within the inactive human MLKL monomer. (G,H) Orthogonal views of the human MLKL tetramer modeled using small-angle X-ray scattering (SAXS) data (Petrie et al. 2018). Throughout the figure, mouse MLKL domains are shown in blue shades, and human MLKL domains in yellow-orange shades.

MLKL to oligomerize, translocate to, and permeabilize the plasma membrane (Fig. 2; Petrie et al. 2017, 2019a).

DIVERGENT MECHANISMS OF MOUSE AND HUMAN MLKL ACTIVATION

The necroptosis pathway is by definition a caspase-independent pathway relying on the core RIPK3-MLKL cassette for execution. Recent advances have illuminated important mechanistic differences between the mouse and human necroptotic signaling pathways (Fig. 2), with marked differences between RIPK3's engagement and phosphorylation of MLKL (Petrie et al. 2018, 2019a). Importantly, human and mouse MLKL are not interchangeable and do not complement necroptotic signaling in cells derived from the opposing species (Tanzer et al. 2016; Davies et al. 2018; Petrie et al. 2018). The crucial factor governing this break in communication is the capacity of MLKL to interact with, and be phosphorylated by, the RIPK3 endogenous to the species in which MLKL is expressed. Consequently, these data support the idea that RIPK3 and MLKL have coevolved and diverged in a species-specific manner.

At the heart of MLKL activation is its phosphorylation by RIPK3 (Sun et al. 2012; Murphy et al. 2013; Tanzer et al. 2015; Rodriguez et al. 2016). It is the RIPK3-mediated phosphorylation of the activation loop of the MLKL pseudokinase domain at T357/S358 in human MLKL and S345 in mouse MLKL that is considered a key hallmark of MLKL activation in both species, and is widely used to monitor MLKL activation experimentally. However, consistent with divergent evolution of the RIPK3-MLKL cassettes between mouse and human cells, recent studies suggest the functions these events serve in signaling are quite distinct between species (Fig. 2).

Mouse MLKL

In the case of mouse MLKL, mutations within the regulatory pseudokinase domain led to stimulus-independent cell death (Murphy et al. 2013; Hildebrand et al. 2014). Mutations within

the "pseudoactive site" (the ATP-binding site, such as K219M or Q343A; Fig. 3A,B) or mutation of the adjacent activation loop residue, S345, to a negatively charged residue to mimic RIPK3 phosphorylation, are known to trigger MLKL activation (Murphy et al. 2013). We proposed that this phosphorylation event triggers a conformational change in the pseudokinase domain, representing a molecular switch that converts dormant cytoplasmic MLKL into a killer protein that can form oligomers and translocate to, and permeabilize, the plasma membrane. Indeed, expression of the amino-terminal 4HB domain of mouse MLKL is sufficient to kill cells (Hildebrand et al. 2014), consistent with the role of the carboxy-terminal pseudokinase domain in suppressing the killer activity of the 4HB domain until this molecular switch is toggled by RIPK3 modification. Mutation of the pseudokinase domain within full-length mouse MLKL or expression of the 4HB executioner domain overrides this molecular switch (Murphy et al. 2013; Tanzer et al. 2015; Jacobsen et al. 2016; Rodriguez et al. 2016), because these constructs kill RIPK3-deficient cells in the absence of any further stimulus. It is highly likely that this is not a simple binary switch, however, because mutational studies of other phosphosites in mouse MLKL, including S158 in the interdomain "brace" helices, S228 and S248 in the pseudokinase domain, indicate other phosphorylation events within mouse MLKL likely tune the killing activity (Tanzer et al. 2015).

Human MLKL

In the case of human MLKL, however, activation is markedly more tightly regulated. Early studies suggested that mutation of the pseudokinase activation loop residues, T357 and S358, to mimic RIPK3 phosphorylation did not compromise cell killing by human MLKL, whereas mutation of these sites to Ala abrogated necroptosis signaling (Sun et al. 2012). Subsequent studies suggested that phosphomimetic mutant MLKL could modestly enhance cell death relative to the wild-type counterpart when oligomerized via a fused domain (Wang et al. 2014). Further studies, also in cell lines that do not typically undergo necroptosis and also involving knock-down, rather than knockout, of endogenous MLKL supported the idea that phosphomimetic mutation of T357/S358 in human MLKL could promote stimulus-independent death (Yoon et al. 2016), as observed in studies of mouse MLKL. However, subsequent studies in the human histiocytic lymphoma U937 and adeno-carcinoma HT29 cell lines, which are commonly used for studies of the necroptosis signaling pathway, suggested the opposite (Petrie et al. 2018). These cell lines were edited to delete endogenous human MLKL, and wild-type or mutant human MLKL constructs stably intro-duced and inducibly expressed. In both lines, although wild-type human MLKL could recon-stitute the signaling pathway, neither the T357E/S358E phosphomimic nor the T357A/S358A phospho-ablating human MLKL constructs sig-naled either in the absence or presence of nec-roptosis stimuli. Furthermore, introduction of constructs harboring mutations within the hu-man MLKL pseudoactive site, such as those observed in colon, lung, and endometrial carcinomas and melanoma specimens, into MLKL$^{-/-}$ U937 cells did not promote MLKL's killing activity, but rather delayed the kinetics of cell death following treatment with a necroptosis stimulus (Petrie et al. 2018). Biophysical data with recombinant proteins suggest the basis for these mutants showing defects in signaling is that they are locked in a monomeric confor-mation, which hampers assembly into higher order oligomers that are responsible for cell death. The basis for complete abrogation of sig-naling by human MLKL activation loop muta-tions was rationalized by binding studies using recombinant proteins. Although wild-type hu-man MLKL robustly bound human RIPK3 ki-nase domain, no binding was detected for the T357E/S358E human MLKL mutant (Petrie et al. 2018). These data support the idea that contrary to mouse MLKL activation, which can be achieved artificially in the absence of RIPK3, human MLKL relies on recruitment to human RIPK3 in cells as a precursor to its acti-vation (Petrie et al. 2019a).

The role of RIPK3-mediated phosphory-lation of human MLKL then becomes a conun-drum. Although it is possible that phosphomi-metic mutations fail to fully recapitulate the effects of MLKL activation loop phosphoryla-tion, it is clear that human MLKL activation is not triggered by such mutations and this differs from observations using mouse MLKL mutant counterparts. Instead, it is probable that human MLKL phosphorylation serves a different role in activation, such as by promoting higher order MLKL assemblies that are required for killing cells, prompting reorganization of MLKL sub-units with such assemblies, and/or promoting disengagement from RIPK3 to facilitate mem-brane translocation and permeabilization. The crucial role oligomerization plays in human MLKL activation is evident from studies of the 4HB domain or the 4HB + brace domain in human cells. Expression of these portions of hu-man MLKL do not measurably induce necro-ptosis in cell lines typically used to study the pathway. However, if these domains are fused to an inducible dimerization domain, forced oligomerization can provoke cell death, support-ing a key function for MLKL oligomerization in killing human cells (Quarato et al. 2016; Tanzer et al. 2016). The particulars of this choreography remain of enormous interest in the field, because each checkpoint represents a possible target for therapeutic intervention in the pathway.

FROM PROTEIN STRUCTURE TO MECHANISM

MLKL Full-Length and Pseudokinase Domain Structures

Structural studies of MLKL have played a cru-cial role in advancing our understanding of the mechanism underlying its activation and cell killing. To date, the only full-length protein structure among the necroptosis machinery is the X-ray crystal structure of mouse MLKL (Fig. 3A,B). This structure identified a number of defining features, including the aforemen-tioned amino-terminal 4HB domain (discussed further below), the two helix linker termed the "brace helices," and the carboxy-terminal pseu-dokinase domain (Murphy et al. 2013). MLKL was designated a pseudokinase because it has divergent sequences in place of the Mg^{2+}-binding DFG motif at the start of the activation loop (conserved as GFE throughout species), with variable sequences in place of the conventional catalytic loop HRD motif (e.g., HRN in mouse, HGK in human) (Manning et al. 2002; Jacobsen and Murphy 2017). The latter is consistent with a loss of selective pressure to retain a catalytically competent sequence and active site geometry that would be requisite for a catalytic enzyme. Instead, an absence of catalytic activity has allowed sequence divergence among species within these motifs. Intriguingly, MLKL has retained the N-lobe β3 strand lysine of the VAIK motif (K219 in mouse, Fig. 3B; K230 in human, Fig. 3E), which positions ATP for phosphoryl transfer in catalytically active protein kinases. This is the last remnant of a conventional protein kinase among the catalytic motifs in MLKL and, accordingly, biophysical studies have revealed the importance of this Lys in mediating ATP binding (Murphy et al. 2013, 2014, 2017; Petrie et al. 2018). The physiological function of ATP binding in MLKL's regulation has been somewhat of a conundrum because mutation of K219 in mouse MLKL not only abrogated ATP binding but also triggered constitutive necroptosis (Murphy et al. 2013). This was not a consequence of loss of ATP binding, however, but rather a result of perturbing the molecular switch mechanism by disrupting the hydrogen bond between K219 and a residue in the noncanonical activation loop helix, Q343. This helix is unusual and was observed to displace the typical N-lobe regulatory element, the αC helix, which conventionally contributes a Glu to a charged pair with the VAIK Lys. Furthermore, biophysical studies revealed that, contrary to conventional protein kinases, cations competed for nucleotide binding to MLKL (Murphy et al. 2013). Thus, the likelihood that ATP binding could serve a physiological role appeared unlikely.

The structure of the human MLKL pseudokinase domain revealed stark differences to its mouse counterpart (Fig. 3D,E; Xie et al. 2013; Murphy et al. 2014). Unlike the mouse MLKL pseudokinase domain, human MLKL shows a conventional "closed" conformation, in which the αC helix is in the position of a typical active

protein kinase with the key αC helix Glu (Glu250) engaged in an ion pair with the N-lobe β3 Lys (K230; Fig. 3D,E). Contrary to the mouse pseudokinase domain structure, a ladder of hydrophobic interactions termed the regulatory "R-spine" (Kornev et al. 2008), which is synonymous with the active conformations of conventional protein kinases, is intact in the human structure. The catalytic "C-spine" is not intact in either structure, but would typically rely on the adenine ring of ATP to complete the ladder. To date, no structures of MLKL bound to ATP or analogs have been reported, so it remains unclear whether such binding might promote toggling of the conformational switch. It also remains unclear whether there may be different functions conferred by ATP binding in mouse versus human MLKL. An intriguing observation from biophysical analyses of mutant proteins was that the determinants of ATP binding have diverged between mouse and human MLKL (Murphy et al. 2014). This supports the concept that without the selective pressure of maintaining active site geometry for catalysis that is imposed on conventional enzymes, the pseudoactive sites of pseudoenzymes have greater liberty to diverge (Ribeiro et al. 2019). In the case of human MLKL, it was observed that ATP binding destabilized a tetrameric form to favor a monomer, and human MLKL mutations that abrogated ATP binding led to delayed kinetics in necroptotic signaling when expressed in $MLKL^{-/-}$ cells (Petrie et al. 2018). Collectively, these data suggest that ATP binding is likely to be synonymous with the capacity of MLKL to adopt different conformers, thereby enabling the pseudokinase domain to perform a molecular switch function, rather than a cue for activation or attenuation of MLKL activation per se.

The pseudokinase domain of MLKL is known to engage the kinase domain of RIPK3, stably in the case of the human system (Sun et al. 2012; Zhao et al. 2012; Petrie et al. 2018), but transiently in the mouse system (Murphy et al. 2013). The only structural information on this interaction to date arise from a cocrystal structure of the mouse MLKL pseudokinase and RIPK3 kinase domains, where the proteins were crystallized in a face-to-face orientation

Cite this article as *Cold Spring Harb Perspect Biol* doi: 10.1101/cshperspect.a036376

with activation loops apposed (Fig. 3C; Xie et al. 2013). Notably, the pseudokinase domain conformation observed in the full-length mouse MLKL crystal structure (Fig. 3A,B) was preserved in the context of this complex, supporting the idea that the activation loop helix and displaced αC helix are not artifacts arising from crystallization conditions or contacts. Interestingly, recombinant mouse RIPK3 kinase domain and MLKL pseudokinase domains do not form a stable complex when mixed in vitro, but were copurified in complex when coexpressed in insect cells (Xie et al. 2013). This implicates cotranslational binding or the phosphorylation state of either protein as possible determinants of complex formation, which are currently poorly understood. Nonetheless, in this complex, it was observed that F373 of mouse MLKL projects from the αF-αG loop into a cavity adjacent to αG in RIPK3 (Fig. 3C). Although the importance of this interaction to mouse MLKL interaction with RIPK3 has not been examined, recent data suggest a similar mode of interaction occurs in the human system (Petrie et al. 2019b). Ala substitution of the equivalent human MLKL residue, F386, abrogated reconstitution of necroptotic signaling in $MLKL^{-/-}$ U937 cells, suggesting more broadly that this C-lobe:C-lobe interaction underpins RIPK3 engagement by MLKL (Petrie et al. 2019b). Even though this Phe is conserved in human and mouse MLKL, additional interactions must govern the RIPK3–MLKL interaction because mouse RIPK3 was not able to activate human MLKL expressed in $Mlkl^{-/-}$ mouse fibroblasts.

In the structure of full-length mouse MLKL, the killer 4HB domain was crystallized in a conformation extended away from the pseudokinase domain (Fig. 3A). This contrasts a conformation in which the 4HB domain is sequestered by the pseudokinase domain to prevent killing, which was predicted from biochemical studies (Hildebrand et al. 2014). In the absence of an experimental human MLKL full-length structure, structural mass spectrometry was used to deduce the organization of domains in the dormant monomer form (Petrie et al. 2018). In this model, the 4HB domain is closely associated with the pseudokinase domain (Fig. 3F),

consistent with the suppressor function of the latter. Using small-angle scattering, it was possible to model the higher order, tetrameric assembly of human MLKL at low resolution (Fig. 3G,H). Coupled with cross-linking mass spectrometry data, this revealed a torus structure where the 4HB domains are directed toward a common face (Petrie et al. 2018), which is consistent with a role for positively charged residues on this surface in lipid association and membrane permeabilization (discussed below). Our current knowledge is insufficient to deduce whether this is the assembly that permeabilizes plasma membranes or whether higher order assembly is a prerequisite for cell killing.

The Killer 4HB Domain

The function of the amino-terminal 4HB domain as the cellular executioner was revealed from truncation and overexpression studies (Chen et al. 2014; Dondelinger et al. 2014; Hildebrand et al. 2014; Wang et al. 2014). The mouse MLKL 4HB domain is a potent killer of cells, as mentioned above, and recombinant protein potently permeabilizes lipid bilayers in vitro (Tanzer et al. 2016). In contrast, the human counterpart shows much less of these activities, and instead relies on forced oligomerization to induce cell death in human cells (Tanzer et al. 2016). The basis for these differences is not yet clear, and because structural studies of MLKL amino-termini have focused on the human 4HB domain, much of our understanding is based on the human system. Small-angle X-ray scattering studies of the mouse MLKL 4HB domain + brace support the idea that mouse MLKL assembles into trimers (Fig. 4A; Davies et al. 2018), unlike the tetramers observed for full-length human MLKL (Fig. 3G,H; Petrie et al. 2018), and that trimerization is centered on the brace helices in mouse MLKL. Validation of this organization and stoichiometry in the context of full-length mouse MLKL awaits further experimentation. Although the residues that facilitate lipid binding and membrane permeabilization have not yet been formally defined for the mouse MLKL 4HB domain, scanning mutagenesis identified two clusters of residues centered on

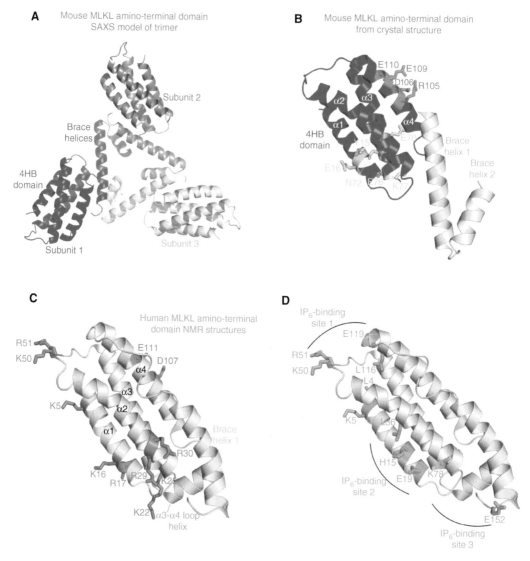

Figure 4. Comparison of four-helix bundle (4HB) domain structures between mouse and human mixed lineage kinase domain-like protein (MLKL). (*A*) The mouse MLKL 4HB + brace trimer structures were modeled based on small-angle X-ray scattering (SAXS) studies (Davies et al. 2018). (*B*) The mouse MLKL 4HB + brace region from the full-length structure (PDB 4BTF). Two sites identified in alanine scanning experiments as key for necroptotic signaling are shown as magenta and salmon sticks (Hildebrand et al. 2014). (*C*) The nuclear magnetic resonance (NMR) structure of human MLKL 4HB domain with first brace helix (PDB 2MSV). Distinct from the mouse 4HB structure, an additional loop was observed between the α3 and α4 helices (Su et al. 2014). Residues identified as key to necroptotic signaling from studies of mutants in cells are shown in magenta. Those involved in phospholipid binding are shown as red sticks. Most residues were identified by Quarato and colleagues (2016), but K22 and K25 were identified by Su et al. (2014). (*D*) The human MLKL 4HB domain and first brace helix structure shown in panel *C* with residues identified as inositol phosphate interactors shown as cyan sticks (Dovey et al. 2018; McNamara et al. 2019).

Cite this article as *Cold Spring Harb Perspect Biol* doi: 10.1101/cshperspect.a036376

the α1–α2 helix junction and the α4 helix (salmon and magenta sticks, respectively, Fig. 4B) that were required for MLKL to induce cell death (Hildebrand et al. 2014). The most deleterious were alanine substitutions of the α4 helix, E109/E110 and R105/D106 (Fig. 4B), which completely abrogated cell death. Interestingly, forced oligomerization of full-length mouse MLKL harboring these mutations was insufficient to hurdle this signaling checkpoint (Tanzer et al. 2016). Because many of these residues possess negatively charged side chains, it is likely that they serve functions distinct from directly engaging the negatively charged phospholipids of the plasma membrane.

The solution structure of the human 4HB domain and first brace helix topologically resembled the mouse counterpart, albeit with the additional structural feature of an additional helix in the loop connecting the α3 and α4 helices of the 4HB domain (Su et al. 2014; McNamara et al. 2019), which was not ordered in the mouse crystal structure. Several basic amino acids have been identified as mediators of lipid engagement in nuclear magnetic resonance (NMR) spectroscopy studies (red sticks, Fig. 4C; Su et al. 2014; Quarato et al. 2016). Consistent with a role in lipid binding, Ala substitution of K17/R18 led to delays in necroptotic signaling (Quarato et al. 2016; Petrie et al. 2018), although curiously did not abrogate cell death altogether. This suggests that others of the lipid-binding residues enable lipid engagement to proceed, although in the absence of a lipid bound structure, the precise details remain unknown. Mutational analyses identified D107/E111 (magenta sticks, Fig. 4C) as essential for necroptosis signaling (Petrie et al. 2018). This site is spatially analogous to the aforementioned key residues in mouse MLKL, D105, and E109 (Fig. 4B), which suggests that both human and mouse MLKL rely on nonphospholipid interactions with currently unknown coeffectors to induce necroptotic cell death. This possibility remains of enormous ongoing interest. The recent discovery of inositol phosphates as additional activators of human 4HB domain activation (interaction sites shown as cyan sticks, Fig. 4D) indicates that not all regulators of MLKL

activation need be proteins (Dovey et al. 2018; McNamara et al. 2019).

In the absence of a membrane-bound 4HB domain structure, the underlying mechanism of membrane permeabilization remains unclear. The membrane aperture is referred to loosely as a "pore" although the precise composition, including the stoichiometry of MLKL and the inclusion of other components, remain unclear. The aperture size has been estimated as ∼4 nm diameter indirectly (Ros et al. 2017), although whether it is determinate in size remains to be established. One study has proposed the 4HB domain might reorganize in membranes to assemble into ion channels (Xia et al. 2016); however, this remains to be fully explored in cellular contexts and structurally. Earlier studies implicated MLKL in engaging mitochondrial membranes to provoke mitochondrial fission or promote ion channel activity (Cai et al. 2014), although subsequent studies have discounted these possibilities (Murphy et al. 2013; Tait et al. 2013; Moujalled et al. 2014; Remijsen et al. 2014; Wang et al. 2014). Based on studies showing that the 4HB domain can permeabilize membranes in vitro (Dondelinger et al. 2014; Su et al. 2014; Wang et al. 2014; Tanzer et al. 2016; Petrie et al. 2018), it is thought that MLKL kills cells via direct action on the plasma membrane. Such a function has previously been attributed to an ancient 4HB domain initially characterized in yeast termed the HeLo domain (Daskalov et al. 2016). While HeLo domain-containing proteins are abundant in yeast, with similar domains present in plants (Jubic et al. 2019; Mahdi et al. 2019), the only example in animals is that of MLKL. Whether the domain has a common ancestor or arose by convergent evolution is an interesting conundrum.

TUNING NECROPTOSIS SIGNALING

Although recombinant MLKL or the 4HB domain alone can permeabilize artificial membranes, it is evident that there are many layers of regulation in cells that prevent unbridled cell killing that would lead to inflammatory diseases (Fig. 5). MLKL is present in most cell types throughout animals as a dormant protein (Mur-

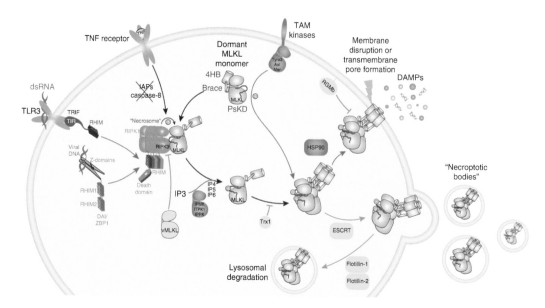

Figure 5. Multiple effectors influence the necroptosis signaling pathway. Numerous auxiliary interactions with mixed lineage kinase domain-like protein (MLKL) have been reported to promote (green arrows, Bigenzahn et al. 2016; Jacobsen et al. 2016; Zhao et al. 2016; Dovey et al. 2018; McNamara et al. 2019; Najafov et al. 2019) or counteract (red arrows, Gong et al. 2017a,b; Reynoso et al. 2017; Yoon et al. 2017; Zargarian et al. 2017; Fan et al. 2019; Petrie et al. 2019b) necroptosis.

phy et al. 2013; Wu et al. 2013), and cues are necessary to toggle MLKL to an activated form that kills cells. Many checkpoints have been reported to date, which collectively likely dictate the relatively slow kinetics of necroptotic death. Induction of necroptosis leads to early changes in the plasma membrane, such as phosphatidyl-serine (PS) exposure, within minutes (Gong et al. 2017b; Zargarian et al. 2017), but only after 3–6 hours (depending on strength of stimulus) is necroptotic death observed (Tanzer et al. 2015; Gong et al. 2017b; Zargarian et al. 2017). Recent efforts have been directed toward understanding whether there are additional steps involved in facilitating MLKL to translocate and assemble into killer complexes and, conversely, whether additional mechanisms exist to suppress activated MLKL and limit its killing capacity. These studies have identified several processes, of which our understanding is only emerging.

Posttranslational Modifications

The most definitive step in the conversion of MLKL from a dormant form to a killer is phos-phorylation of the MLKL pseudokinase domain activation loop by RIPK3. The observation of additional posttranslational modifications has led to a picture of MLKL activation as not a binary switch, but rather a graded switch, where multiple signals can be integrated to tune MLKL activity. Other phosphorylation events in human MLKL have been identified in mass spectrometry screens, such as in a cell-cycle-dependent manner in the 4HB domain (S125) and pseudokinase domain (T377) (Dephoure et al. 2008), but the responsible kinases and the impact of these events on signaling is yet to be determined. Most recently, the PS-binding TAM (Tyro3, Axl, Mer) tyrosine kinase family were reported to phosphorylate MLKL at Y376 and proposed to promote MLKL activation by directing MLKL oligomerization (Najafov et al. 2019). Because Y376 is best known for its role as a structural residue in MLKL, the precise mechanism by which its phosphorylation might trigger MLKL oligomerization remains to be completely elucidated. To date, no phosphatases have been reported to dephosphorylate MLKL phosphosites, although it is possible that an "off switch" of this nature exists. Furthermore, MLKL has been shown to undergo ubiquitylation on induction of necroptosis signaling (Lawlor et al. 2015). Whether such a signal consigns MLKL to proteasomal degradation to defuse necroptosis signaling, or whether it serves to regulate MLKL localization and thus positively or negatively impact necroptosis remains of outstanding interest. Recently, RIPK3 was reported to undergo O-GlcNAcylation, which limited its capacity to engage in RIPK1 hetero-oligomers and RIPK3 homo-oligomers (Li et al. 2019) and attenuated necroptotic signaling. Whether similar modifications might occur in MLKL remain to be determined.

Positive Regulators of MLKL Activation

Recent data suggest a role for second messengers, such as inositol phosphates generated by the IPMK, ITPK1, and IPPK metabolic kinases, in direct binding of the human MLKL 4HB domain (Fig. 4D) to perform an auxiliary role in promoting 4HB domain unmasking to kill cells (Fig. 5; Dovey et al. 2018; McNamara et al. 2019). The ubiquity of the "inositol phosphate code," the relative contributions of each metabolite and their cognate kinases in directing necroptosis, and whether inhibition of IPMK, ITPK1, and/ or IPPK could be targeted to block necroptosis therapeutically, remain to be further explored.

Beyond RIPK3, remarkably few proteins have been identified as MLKL interactors, with the cochaperones, HSP90-Cdc37, among the few identified as playing a key role in MLKL activation (Fig. 5; Bigenzahn et al. 2016; Jacobsen et al. 2016; Zhao et al. 2016). These chaperones contribute to activation of most protein kinases, including RIPK1 and RIPK3 (Lewis et al. 2000; Cho et al. 2009), and with their role in regulating MLKL activation, they are established as important coeffectors in the necroptosis pathway. Whether HSP90 exerts its effects on MLKL folding, oligomerization, or translocation to membranes has not been precisely determined, although it is plausible HSP90 impacts each of these activation steps. It is highly likely that additional positive regulators of MLKL activation exist and await discovery.

Negative Regulators of MLKL Activation

Over the past few years, a number of proteins have emerged as candidate attenuators of MLKL activity (Fig. 5, red arrows), although it remains to be determined whether they act directly on MLKL or by proxy. Further, it remains to be established how universally these proteins are involved in necroptosis, because they have been studied only in limited contexts to date. For instance, repulsive guidance molecule-b (RGMb) was implicated in inhibiting necroptosis in kidney tubular cells, and by overexpression was observed to block human MLKL membrane translocation and therefore necroptotic cell death (Liu et al. 2018). Whether this protein functions more broadly in necroptotic signaling, and at endogenous levels, requires further investigation. The putative interactor, thioredoxin-1 (Trx1), was proposed to limit MLKL oxidation and assembly into higher order disulfide cross-linked species (Reynoso et al. 2017), which have been widely used as a readout of MLKL activation without complete knowledge of whether they represent killer assemblies or are a consequence of membrane compromise and a dysregulated redox environment in the cytoplasm.

Recent studies have suggested exocytosis and endocytosis of membrane-associated MLKL can extinguish the necroptotic signal by respectively shedding or degrading activated (phosphorylated) MLKL from membranes (Fig. 5). Exocytosis of phospho-MLKL via the ESCRT machinery in vesicles or exosomes referred to as "necroptotic bodies" was identified by three independent groups recently (Gong et al. 2017a,b; Yoon et al. 2017; Zargarian et al. 2017). Further studies have implicated phospho-MLKL interaction with ALIX/syntenin-1 in the exocytosis process, and interaction with flotillin-1 and flotillin-2 within lipid rafts in endocytosis-mediated lysosomal degradation (Fan et al. 2019). Whether these interactions are directly with phospho-MLKL, or membrane-scaffolded, remains of enormous interest, but they illustrate the breadth of mechanisms used by cells to counteract a death signal and prolong longevity of the cell.

In a similar vein, pathogens have evolved mechanisms to prolong cell longevity by counteracting death pathways, including necroptosis. The best characterized of these are RHIM-targeting pathogens (Pearson and Murphy 2017), which either act to inhibit RIPK3 oligomerization by competition in the case of viral proteins (Upton et al. 2010, 2012) or by cleaving RHIM domains in the case of the enteropathogenic *Escherichia coli* EspL protease (Pearson et al. 2017). Very recently, an additional mechanism to disarm necroptosis by pathogens was uncovered. MLKL xenologs encoded in the genomes of some poxviruses were identified to function as mimics of cellular MLKL by sequestering RIPK3 in the host cell to prevent activation of cellular MLKL (Fig. 5; Petrie et al. 2019b). One of the curiosities of necroptosis signaling is that some organisms, like carnivores, lack RIPK3 and MLKL, and others like birds and rabbits, lack RIPK3 (Dondelinger et al. 2016; Newton and Manning 2016), suggesting that the pathway was negatively selected against in evolution of some species. In addition, the amino acid sequences of MLKL are highly divergent among species, with identity as low as 68% between human and mouse MLKL, which contributes to their inability to complement signaling in cells derived from the respective other species. The forces driving such divergence have remained unclear. The recent discovery of virally encoded MLKL homolog pseudokinase domains indicates that pathogens have contributed to, if not driven, RIPK3-MLKL evolution. MLKL orthologs encoded by two related rodent poxviruses were found to inhibit necroptosis in human and mouse cells, while an ortholog from swinepox did not (Petrie et al. 2019b). Interestingly, although RIPK3 is the target in mouse and human cells, the mechanism of inhibition differed. In human cells, a highly phosphorylated form of RIPK3 was stabilized but could not activate cellular MLKL; in mouse cells, RIPK3 phosphorylation was attenuated. Curiously, birds and rabbit do not express RIPK3, which raises the question about whether viral MLKL in avian poxes and myxomas have evolved a distinct target and, more generally, whether other non-RIPK3 mediated signals might exist that can trigger MLKL activation. Like animal MLKL sequences, viral MLKL sequences are highly divergent, with exemplars from each poxviral MLKL ortholog family sharing only ~30% sequence identity. This is consistent with the idea that coevolution of viral MLKL with the host RIPK3-MLKL has led to divergence of both systems from those of other species.

CONCLUDING REMARKS

Our current understanding of how the terminal necroptosis effector, MLKL, is activated and kills cells has emerged from only 7 years of intensive study. Consequently, there are many gaps in our understanding. Precisely how MLKL permeabilizes membranes, whether this is a determinate stoichiometry (like with the pore-forming pyroptosis effector gasdermin proteins) (Ruan et al. 2018; Xia et al. 2019), and whether this is a highly organized structure await further structural studies. It is a curiosity that to date very few MLKL interactors have been identified. Thus, it remains of enormous interest to understand which proteins or metabolites are essential for MLKL activation and the range of mechanisms by which necroptosis can be defused. Finally, obtaining a thorough understanding of differences in the necroptosis pathway between species is essential as endeavors to therapeutically target this pathway advance.

ACKNOWLEDGMENTS

I thank members of my laboratory and colleagues for discussions, Emma Petrie for critical reading of the manuscript, and collaborators who have contributed to many of the studies described herein. I thank the National Health and Medical Research Council of Australia for funding support (1105754, 1124735, 1124737, and 9000433) and the Victorian State Government Operational Infrastructure Support scheme.

REFERENCES

*Reference is also in this collection.

Alvarez-Diaz S, Dillon CP, Lalaoui N, Tanzer MC, Rodriguez DA, Lin A, Lebois M, Hakem R, Josefsson EC, O'Reilly LA, et al. 2016. The pseudokinase MLKL and the kinase RIPK3 have distinct roles in autoimmune disease caused by loss of death-receptor-induced apoptosis. *Immunity* **45**: 513–526. doi:10.1016/j.immuni.2016.07.016

Berger SB, Kasparcova V, Hoffman S, Swift B, Dare L, Schaeffer M, Capriotti C, Cook M, Finger J, Hughes-Earle A, et al. 2014. Cutting edge: RIP1 kinase activity is dispensable for normal development but is a key regulator of inflammation in SHARPIN-deficient mice. *J Immunol* **192**: 5476–5480. doi:10.4049/jimmunol.1400499

Bigenzahn JW, Fauster A, Rebsamen M, Kandasamy RK, Scorzoni S, Vladimer GI, Müller AC, Gstaiger M, Zuber J, Bennett KL, et al. 2016. An inducible retroviral expression system for tandem affinity purification mass-spectrometry-based proteomics identifies mixed lineage kinase domain-like protein (MLKL) as a heat shock protein 90 (HSP90) client. *Mol Cell Proteomics* **15**: 1139–1150. doi:10.1074/mcp.O115.055350

Cai Z, Jitkaew S, Zhao J, Chiang HC, Choksi S, Liu J, Ward Y, Wu LG, Liu ZG. 2014. Plasma membrane translocation of trimerized MLKL protein is required for TNF-induced necroptosis. *Nat Cell Biol* **16**: 55–65. doi:10.1038/ncb2883

Chen X, Li W, Ren J, Huang D, He WT, Song Y, Yang C, Li W, Zheng X, Chen P, et al. 2014. Translocation of mixed lineage kinase domain-like protein to plasma membrane leads to necrotic cell death. *Cell Res* **24**: 105–121. doi:10.1038/cr.2013.171

Cho YS, Challa S, Moquin D, Genga R, Ray TD, Guildford M, Chan FK. 2009. Phosphorylation-driven assembly of the RIP1-RIP3 complex regulates programmed necrosis and virus-induced inflammation. *Cell* **137**: 1112–1123. doi:10.1016/j.cell.2009.05.037

Dannappel M, Vlantis K, Kumari S, Polykratis A, Kim C, Wachsmuth L, Eftychi C, Lin J, Corona T, Hermance N, et al. 2014. RIPK1 maintains epithelial homeostasis by inhibiting apoptosis and necroptosis. *Nature* **513**: 90–94. doi:10.1038/nature13608

Daskalov A, Habenstein B, Sabaté R, Berbon M, Martinez D, Chaignepain S, Coulary-Salin B, Hofmann K, Loquet A, Saupe SJ. 2016. Identification of a novel cell death-inducing domain reveals that fungal amyloid-controlled programmed cell death is related to necroptosis. *Proc Natl Acad Sci* **113**: 2720–2725. doi:10.1073/pnas.1522361113

Davies KA, Tanzer MC, Griffin MDW, Mok YF, Young SN, Qin R, Petrie EJ, Czabotar PE, Silke J, Murphy JM. 2018. The brace helices of MLKL mediate interdomain communication and oligomerisation to regulate cell death by necroptosis. *Cell Death Differ* **25**: 1567–1580. doi:10.1038/s41418-018-0061-3

de Almagro MC, Goncharov T, Izrael-Tomasevic A, Duttler S, Kist M, Varfolomeev E, Wu X, Lee WP, Murray J, Webster JD, et al. 2017. Coordinated ubiquitination and phosphorylation of RIP1 regulates necroptotic cell death. *Cell Death Differ* **24**: 26–37. doi:10.1038/cdd.2016.78

Degterev A, Huang Z, Boyce M, Li Y, Jagtap P, Mizushima N, Cuny GD, Mitchison TJ, Moskowitz MA, Yuan J. 2005. Chemical inhibitor of nonapoptotic cell death with therapeutic potential for ischemic brain injury. *Nat Chem Biol* **1**: 112–119. doi:10.1038/nchembio711

Dephoure N, Zhou C, Villen J, Beausoleil SA, Bakalarski CE, Elledge SJ, Gygi SP. 2008. A quantitative atlas of mitotic phosphorylation. *Proc Natl Acad Sci* **105**: 10762–10767. doi:10.1073/pnas.0805139105

Dondelinger Y, Declercq W, Montessuit S, Roelandt R, Goncalves A, Bruggeman I, Hulpiau P, Weber K, Sehon CA, Marquis RW, et al. 2014. MLKL compromises plasma membrane integrity by binding to phosphatidylinositol phosphates. *Cell Rep* **7**: 971–981. doi:10.1016/j.celrep.2014.04.026

Dondelinger Y, Hulpiau P, Saeys Y, Bertrand MJM, Vandenabeele P. 2016. An evolutionary perspective on the necroptotic pathway. *Trends Cell Biol* **26**: 721–732. doi:10.1016/j.tcb.2016.06.004

Dovey CM, Diep J, Clarke BP, Hale AT, McNamara DE, Guo H, Brown NW Jr, Cao JY, Grace CR, Gough PJ, et al. 2018. MLKL requires the inositol phosphate code to execute necroptosis. *Mol Cell* **70**: 936–948.e7. doi:10.1016/j.molcel.2018.05.010

Fan W, Guo J, Gao B, Zhang W, Ling L, Xu T, Pan C, Li L, Chen S, Wang H, et al. 2019. Flotillin-mediated endocytosis and ALIX-syntenin-1-mediated exocytosis protect the cell membrane from damage caused by necroptosis. *Sci Signal* **12**: eaaw3423. doi:10.1126/scisignal.aaw3423

Gong YN, Guy C, Crawford JC, Green DR. 2017a. Biological events and molecular signaling following MLKL activation during necroptosis. *Cell Cycle* **16**: 1748–1760. doi:10.1080/15384101.2017.1371889

Gong YN, Guy C, Olauson H, Becker JU, Yang M, Fitzgerald P, Linkermann A, Green DR. 2017b. ESCRT-III acts downstream of MLKL to regulate necroptotic cell death and its consequences. *Cell* **169**: 286–300.e16. doi:10.1016/j.cell.2017.03.020

He S, Wang L, Miao L, Wang T, Du F, Zhao L, Wang X. 2009. Receptor interacting protein kinase-3 determines cellular necrotic response to TNF-α. *Cell* **137**: 1100–1111. doi:10.1016/j.cell.2009.05.021

He S, Liang Y, Shao F, Wang X. 2011. Toll-like receptors activate programmed necrosis in macrophages through a receptor-interacting kinase-3-mediated pathway. *Proc Natl Acad Sci* **108**: 20054–20059. doi:10.1073/pnas.1116302108

Hildebrand JM, Tanzer MC, Lucet IS, Young SN, Spall SK, Sharma P, Pierotti C, Garnier JM, Dobson RC, Webb AI, et al. 2014. Activation of the pseudokinase MLKL unleashes the four-helix bundle domain to induce membrane localization and necroptotic cell death. *Proc Natl Acad Sci* **111**: 15072–15077. doi:10.1073/pnas.1408987111

Hildebrand JM, Kauppi M, Majewski IJ, Liu Z, Cox A, Miyake S, Petrie EJ, Silk MA, Li Z, Tanzer MC, et al. 2019. Missense mutations in the MLKL "brace" region lead to lethal neonatal inflammation in mice and are present in dependent frequency in humans. bioRxiv 628370.

Holler N, Zaru R, Micheau O, Thome M, Attinger A, Valitutti S, Bodmer JL, Schneider P, Seed B, Tschopp J. 2000. Fas triggers an alternative, caspase-8-independent cell death pathway using the kinase RIP as effector molecule. *Nat Immunol* **1**: 489–495. doi:10.1038/82732

Hsu H, Huang J, Shu HB, Baichwal V, Goeddel DV. 1996. TNF-dependent recruitment of the protein kinase RIP to the TNF receptor-1 signaling complex. *Immunity* **4**: 387–396. doi:10.1016/S1074-7613(00)80252-6

Huang D, Zheng X, Wang ZA, Chen X, He WT, Zhang Y, Xu JG, Zhao H, Shi W, Wang X, et al. 2017. The MLKL channel in necroptosis is an octamer formed by tetramers in a dyadic process. *Mol Cell Biol* **37**: e00497–e00416.

Jacobsen AV, Murphy JM. 2017. The secret life of kinases: insights into non-catalytic signalling functions from pseudokinases. *Biochem Soc Trans* **45**: 665–681. doi:10 .1042/BST20160331

Jacobsen AV, Lowes KN, Tanzer MC, Lucet IS, Hildebrand JM, Petrie EJ, van Delft MF, Liu Z, Conos SA, Zhang JG, et al. 2016. HSP90 activity is required for MLKL oligomerisation and membrane translocation and the induction of necroptotic cell death. *Cell Death Dis* **7**: e2051. doi:10 .1038/cddis.2015.386

Jubic LM, Saile S, Furzer OJ, El Kasmi F, Dangl JL. 2019. Help wanted: helper NLRs and plant immune responses. *Curr Opin Plant Biol* **50**: 82–94. doi:10.1016/j.pbi.2019.03 .013

Kaiser WJ, Sridharan H, Huang C, Mandal P, Upton JW, Gough PJ, Sehon CA, Marquis RW, Bertin J, Mocarski ES. 2013. Toll-like receptor 3-mediated necrosis via TRIF, RIP3, and MLKL. *J Biol Chem* **288**: 31268–31279. doi:10.1074/jbc.M113.462341

Karunakaran D, Geoffrion M, Wei L, Gan W, Richards L, Shangari P, DeKemp EM, Beanlands RA, Perisic L, Maegdefessel L, et al. 2016. Targeting macrophage necroptosis for therapeutic and diagnostic interventions in atherosclerosis. *Sci Adv* **2**: e1600224. doi:10.1126/sciadv.160 0224

Kitur K, Wachtel S, Brown A, Wickersham M, Paulino F, Peñaloza HF, Soong G, Bueno S, Parker D, Prince A. 2016. Necroptosis promotes *Staphylococcus aureus* clearance by inhibiting excessive inflammatory signaling. *Cell Rep* **16**: 2219–2230. doi:10.1016/j.celrep.2016.07.039

Kornev AP, Taylor SS, Ten Eyck LF. 2008. A helix scaffold for the assembly of active protein kinases. *Proc Natl Acad Sci* **105**: 14377–14382. doi:10.1073/pnas.0807988105

Lawlor KE, Khan N, Mildenhall A, Gerlic M, Croker BA, D'Cruz AA, Hall C, Kaur Spall S, Anderton H, Masters SL, et al. 2015. RIPK3 promotes cell death and NLRP3 inflammasome activation in the absence of MLKL. *Nat Commun* **6**: 6282. doi:10.1038/ncomms7282

Lewis J, Devin A, Miller A, Lin Y, Rodriguez Y, Neckers L, Liu ZG. 2000. Disruption of hsp90 function results in degradation of the death domain kinase, receptor-interacting protein (RIP), and blockage of tumor necrosis factor-induced nuclear factor-κB activation. *J Biol Chem* **275**: 10519–10526. doi:10.1074/jbc.275.14.10519

Li J, McQuade T, Siemer AB, Napetschnig J, Moriwaki K, Hsiao YS, Damko E, Moquin D, Walz T, McDermott A, et al. 2012. The RIP1/RIP3 necrosome forms a functional amyloid signaling complex required for programmed necrosis. *Cell* **150**: 339–350. doi:10.1016/j.cell.2012.06.019

Li X, Gong W, Wang H, Li T, Attri KS, Lewis RE, Kalil AC, Bhinderwala F, Powers R, Yin G, et al. 2019. O-GlcNAc transferase suppresses inflammation and necroptosis by targeting receptor-interacting serine/threonine-protein kinase 3. *Immunity* **50**: 576–590.E6. doi:10.1016/j .immuni.2019.01.007

Lin J, Kumari S, Kim C, Van TM, Wachsmuth L, Polykratis A, Pasparakis M. 2016. RIPK1 counteracts ZBP1-mediated necroptosis to inhibit inflammation. *Nature* **540**: 124–128. doi:10.1038/nature20558

Liu W, Chen B, Wang Y, Meng C, Huang H, Huang XR, Qin J, Mulay SR, Anders HJ, Qiu A, et al. 2018. RGMb protects against acute kidney injury by inhibiting tubular cell necroptosis via an MLKL-dependent mechanism. *Proc Natl Acad Sci* **115**: E1475–E1484. doi:10.1073/pnas .1716959115

Mahdi L, Huang M, Zhang X, Nakano RT, Kopp LB, Saur IML, Jacob F, Kovacova V, Lapin D, Parker JE, et al. 2019. Plant mixed lineage kinase domain-like proteins limit biotrophic pathogen growth. bioRxiv 681015. doi:10 .1101/681015

Mandal P, Berger SB, Pillay S, Moriwaki K, Huang C, Guo H, Lich JD, Finger J, Kasparcova V, Votta B, et al. 2014. RIP3 induces apoptosis independent of pronecrotic kinase activity. *Mol Cell* **56**: 481–495. doi:10.1016/j.molcel.2014.10 .021

Manning G, Whyte DB, Martinez R, Hunter T, Sudarsanam S. 2002. The protein kinase complement of the human genome. *Science* **298**: 1912–1934. doi:10.1126/science .1075762

McComb S, Cessford E, Alturki NA, Joseph J, Shutinoski B, Startek JB, Gamero AM, Mossman KL, Sad S. 2014. Type-I interferon signaling through ISGF3 complex is required for sustained Rip3 activation and necroptosis in macrophages. *Proc Natl Acad Sci* **111**: E3206–E3213. doi:10 .1073/pnas.1407068111

McNamara DE, Dovey CM, Hale AT, Quarato G, Grace CR, Guibao CD, Diep J, Nourse A, Cai CR, Wu H, et al. 2019. Direct activation of human MLKL by a select repertoire of inositol phosphate metabolites. *Cell Chem Biol* **26**: 863–877.E7. doi:10.1016/j.chembiol.2019.03.010

Meng H, Liu Z, Li X, Wang H, Jin T, Wu G, Shan B, Christofferson DE, Qi C, Yu Q, et al. 2018. Death-domain dimerization-mediated activation of RIPK1 controls necroptosis and RIPK1-dependent apoptosis. *Proc Natl Acad Sci* **115**: E2001–E2009. doi:10.1073/pnas.1722013115

Moerke C, Bleibaum F, Kunzendorf U, Krautwald S. 2019. Combined knockout of RIPK3 and MLKL reveals unexpected outcome on tissue injury and inflammation. *Front Cell Dev Biol* **7**: 19. doi:10.3389/fcell.2019.00019

Mompeán M, Li W, Li J, Laage S, Siemer AB, Bozkurt G, Wu H, McDermott AE. 2018. The structure of the necrosome RIPK1-RIPK3 core, a human hetero-amyloid signaling complex. *Cell* **173**: 1244–1253.e10. doi:10.1016/j.cell .2018.03.032

Moujalled DM, Cook WD, Murphy JM, Vaux DL. 2014. Necroptosis induced by RIPK3 requires MLKL but not Drp1. *Cell Death Dis* **5**: e1086. doi:10.1038/cddis.2014.18

Müller T, Dewitz C, Schmitz J, Schroder AS, Bräsen JH, Stockwell BR, Murphy JM, Kunzendorf U, Krautwald S. 2017. Necroptosis and ferroptosis are alternative cell death pathways that operate in acute kidney failure. *Cell Mol Life Sci* **74**: 3631–3645. doi:10.1007/s00018-017-2547-4

Murphy JM, Czabotar PE, Hildebrand JM, Lucet IS, Zhang JG, Alvarez-Diaz S, Lewis R, Lalaoui N, Metcalf D, Webb AI, et al. 2013. The pseudokinase MLKL mediates necroptosis via a molecular switch mechanism. *Immunity* **39**: 443–453. doi:10.1016/j.immuni.2013.06.018

Murphy JM, Lucet IS, Hildebrand JM, Tanzer MC, Young SN, Sharma P, Lessene G, Alexander WS, Babon JJ, Silke J, et al. 2014. Insights into the evolution of divergent nucleotide-binding mechanisms among pseudokinases revealed by crystal structures of human and mouse MLKL. *Biochem J* **457**: 369–377. doi:10.1042/BJ20131270

Murphy JM, Zhang Q, Young SN, Reese ML, Bailey FP, Eyers PA, Ungureanu D, Hammaren H, Silvennoinen O, Varghese LN, et al. 2017. A robust methodology to subclassify pseudokinases based on their nucleotide binding-properties. *Biochem J* **457**: 323–334. doi:10.1042/BJ20131174

Najafov A, Mookhtiar AK, Luu HS, Ordureau A, Pan H, Amin PP, Li Y, Lu Q, Yuan J. 2019. TAM kinases promote necroptosis by regulating oligomerization of MLKL. *Mol Cell* **75**: 457–468.E4. doi:10.1016/j.molcel.2019.05.022

* Newton K. 2019. Multitasking kinase RIPK1 regulates cell death and inflammation. *Cold Spring Harb Perspect Biol* doi:10.1101/cshperspect.a036368

Newton K, Manning G. 2016. Necroptosis and inflammation. *Annu Rev Biochem* **85**: 743–763. doi:10.1146/annurev-biochem-060815-014830

Newton K, Sun X, Dixit VM. 2004. Kinase RIP3 is dispensable for normal NF-κBs, signaling by the B-cell and T-cell receptors, tumor necrosis factor receptor 1, and Toll-like receptors 2 and 4. *Mol Cell Biol* **24**: 1464–1469. doi:10.1128/MCB.24.4.1464-1469.2004

Newton K, Dugger DL, Wickliffe KE, Kapoor N, de Almagro MC, Vucic D, Komuves L, Ferrando RE, French DM, Webster J, et al. 2014. Activity of protein kinase RIPK3 determines whether cells die by necroptosis or apoptosis. *Science* **343**: 1357–1360. doi:10.1126/science.1249361

Newton K, Dugger DL, Maltzman A, Greve JM, Hedehus M, Martin-McNulty B, Carano RA, Cao TC, van Bruggen N, Bernstein L, et al. 2016a. RIPK3 deficiency or catalytically inactive RIPK1 provides greater benefit than MLKL deficiency in mouse models of inflammation and tissue injury. *Cell Death Differ* **23**: 1565–1576. doi:10.1038/cdd.2016.46

Newton K, Wickliffe KE, Maltzman A, Dugger DL, Strasser A, Pham VC, Lill JR, Roose-Girma M, Warming S, Solon M, et al. 2016b. RIPK1 inhibits ZBP1-driven necroptosis during development. *Nature* **540**: 129–133. doi:10.1038/nature20559

Pearson JS, Murphy JM. 2017. Down the rabbit hole: is necroptosis truly an innate response to infection? *Cell Microbiol* **19**: e12750. doi:10.1111/cmi.12750

Pearson JS, Giogha C, Mühlen S, Nachbur U, Pham CL, Zhang Y, Hildebrand JM, Oates CV, Lung TW, Ingle D, et al. 2017. EspL is a bacterial cysteine protease effector that cleaves RHIM proteins to block necroptosis and inflammation. *Nat Microbiol* **2**: 16258. doi:10.1038/nmicrobiol.2016.258

Pefanis A, Ierino FL, Murphy JM, Cowan PJ. 2019. Regulated necrosis in kidney ischemia-reperfusion injury. *Kidney Int* **96**: 291–301. doi:10.1016/j.kint.2019.02.009

Petrie EJ, Hildebrand JM, Murphy JM. 2017. Insane in the membrane: a structural perspective of MLKL function in necroptosis. *Immunol Cell Biol* **95**: 152–159. doi:10.1038/icb.2016.125

Petrie EJ, Sandow JJ, Jacobsen AV, Smith BJ, Griffin MDW, Lucet IS, Dai W, Young SN, Tanzer MC, Wardak A, et al. 2018. Conformational switching of the pseudokinase domain promotes human MLKL tetramerization and cell death by necroptosis. *Nat Commun* **9**: 2422. doi:10.1038/s41467-018-04714-7

Petrie EJ, Czabotar PE, Murphy JM. 2019a. The structural basis of necroptotic cell death signaling. *Trends Biochem Sci* **44**: 53–63. doi:10.1016/j.tibs.2018.11.002

Petrie EJ, Sandow JJ, Lehmann WIL, Liang LY, Coursier D, Young SN, Kersten WJA, Fitzgibbon C, Samson AL, Jacobsen AV, et al. 2019b. Viral MLKL homologs subvert necroptotic cell death by sequestering cellular RIPK3. *Cell Rep* **28**: 3309–3319.

Pham CL, Shanmugam N, Strange M, O'Carroll A, Brown JW, Sierecki E, Gambin Y, Steain M, Sunde M. 2019. Viral M45 and necroptosis-associated proteins form heteromeric amyloid assemblies. *EMBO Rep* **20**: e46518.

Quarato G, Guy CS, Grace CR, Llambi F, Nourse A, Rodriguez DA, Wakefield R, Frase S, Moldoveanu T, Green DR. 2016. Sequential engagement of distinct MLKL phosphatidylinositol-binding sites executes necroptosis. *Mol Cell* **61**: 589–601. doi:10.1016/j.molcel.2016.01.011

Raju S, Whalen DM, Mengistu M, Swanson C, Quinn JG, Taylor SS, Webster JD, Newton K, Shaw AS. 2018. Kinase domain dimerization drives RIPK3-dependent necroptosis. *Sci Signal* **11**: eaar2188. doi:10.1126/scisignal.aar2188

Ribeiro AJM, Das S, Dawson N, Zaru R, Orchard S, Thornton JM, Orengo C, Zeqiraj E, Murphy JM, Eyers PA. 2019. Emerging concepts in pseudoenzyme classification, evolution, and signaling. *Sci Signal* **12**: eaat9797.

Remijsen Q, Goossens V, Grootjans S, Van den Haute C, Vanlangenakker N, Dondelinger Y, Roelandt R, Bruggeman I, Goncalves A, Bertrand MJ, et al. 2014. Depletion of RIPK3 or MLKL blocks TNF-driven necroptosis and switches towards a delayed RIPK1 kinase-dependent apoptosis. *Cell Death Dis* **5**: e1004. doi:10.1038/cddis.2013.531

Reynoso E, Liu H, Li L, Yuan AL, Chen S, Wang Z. 2017. Thioredoxin-1 actively maintains the pseudokinase MLKL in a reduced state to suppress disulfide bond-dependent MLKL polymer formation and necroptosis. *J Biol Chem* **292**: 17514–17524. doi:10.1074/jbc.M117.799353

Rickard JA, Anderton H, Etemadi N, Nachbur U, Darding M, Peltzer N, Lalaoui N, Lawlor KE, Vanyai H, Hall C, et al. 2014a. TNFR1-dependent cell death drives inflammation in Sharpin-deficient mice. *eLife* **3**: e03464. doi:10.7554/eLife.03464

Rickard JA, O'Donnell JA, Evans JM, Lalaoui N, Poh AR, Rogers TW, Vince JE, Lawlor KE, Ninnis RL, Anderton H, et al. 2014b. RIPK1 regulates RIPK3-MLKL-driven systemic inflammation and emergency hematopoiesis. *Cell* **157**: 1175–1188. doi:10.1016/j.cell.2014.04.019

Rodriguez DA, Weinlich R, Brown S, Guy C, Fitzgerald P, Dillon CP, Oberst A, Quarato G, Low J, Cripps JG, et al. 2016. Characterization of RIPK3-mediated phosphorylation of the activation loop of MLKL during necroptosis. *Cell Death Differ* **23**: 76–88. doi:10.1038/cdd.2015.70

Ros U, Peña-Blanco A, Hänggi K, Kunzendorf U, Krautwald S, Wong WW, García-Sáez AJ. 2017. Necroptosis execution is mediated by plasma membrane nanopores independent of calcium. *Cell Rep* **19**: 175–187. doi:10.1016/j.celrep.2017.03.024

Ruan J, Xia S, Liu X, Lieberman J, Wu H. 2018. Cryo-EM structure of the gasdermin A3 membrane pore. *Nature* **557**: 62–67. doi:10.1038/s41586-018-0058-6

Su L, Quade B, Wang H, Sun L, Wang X, Rizo J. 2014. A plug release mechanism for membrane permeation by MLKL. *Structure* **22:** 1489–1500. doi:10.1016/j.str.2014.07.014

Sun X, Yin J, Starovasnik MA, Fairbrother WJ, Dixit VM. 2002. Identification of a novel homotypic interaction motif required for the phosphorylation of receptor-interacting protein (RIP) by RIP3. *J Biol Chem* **277:** 9505–9511. doi:10.1074/jbc.M109488200

Sun L, Wang H, Wang Z, He S, Chen S, Liao D, Wang L, Yan J, Liu W, Lei X, et al. 2012. Mixed lineage kinase domain-like protein mediates necrosis signaling downstream of RIP3 kinase. *Cell* **148:** 213–227. doi:10.1016/j.cell.2011.11.031

Tait SW, Oberst A, Quarato G, Milasta S, Haller M, Wang R, Karvela M, Ichim G, Yatim N, Albert ML, et al. 2013. Widespread mitochondrial depletion via mitophagy does not compromise necroptosis. *Cell Rep* **5:** 878–885. doi:10.1016/j.celrep.2013.10.034

Tanzer MC, Tripaydonis A, Webb AI, Young SN, Varghese LN, Hall C, Alexander WS, Hildebrand JM, Silke J, Murphy JM. 2015. Necroptosis signalling is tuned by phosphorylation of MLKL residues outside the pseudokinase domain activation loop. *Biochem J* **471:** 255–265. doi:10.1042/BJ20150678

Tanzer MC, Matti I, Hildebrand JM, Young SN, Wardak A, Tripaydonis A, Petrie EJ, Mildenhall AL, Vaux DL, Vince JE, et al. 2016. Evolutionary divergence of the necroptosis effector MLKL. *Cell Death Differ* **23:** 1185–1197. doi:10.1038/cdd.2015.169

Thapa RJ, Basagoudanavar SH, Nogusa S, Irrinki K, Mallilankaraman K, Slifker MJ, Beg AA, Madesh M, Balachandran S. 2011. NF-κB protects cells from gamma interferon-induced RIP1-dependent necroptosis. *Mol Cell Biol* **31:** 2934–2946. doi:10.1128/MCB.05445-11

Upton JW, Kaiser WJ, Mocarski ES. 2010. Virus inhibition of RIP3-dependent necrosis. *Cell Host Microbe* **7:** 302–313. doi:10.1016/j.chom.2010.03.006

Upton JW, Kaiser WJ, Mocarski ES. 2012. DAI/ZBP1/DLM-1 complexes with RIP3 to mediate virus-induced programmed necrosis that is targeted by murine cytomegalovirus vIRA. *Cell Host Microbe* **11:** 290–297. doi:10.1016/j.chom.2012.01.016

Wang H, Sun L, Su L, Rizo J, Liu L, Wang LF, Wang FS, Wang X. 2014. Mixed lineage kinase domain-like protein MLKL causes necrotic membrane disruption upon phosphorylation by RIP3. *Mol Cell* **54:** 133–146. doi:10.1016/j.molcel.2014.03.003

Wong WW, Vince JE, Lalaoui N, Lawlor KE, Chau D, Bankovacki A, Anderton H, Metcalf D, O'Reilly L, Jost PJ, et al. 2014. cIAPs and XIAP regulate myelopoiesis through cytokine production in an RIPK1- and RIPK3-dependent manner. *Blood* **123:** 2562–2572. doi:10.1182/blood-2013-06-510743

Wu J, Huang Z, Ren J, Zhang Z, He P, Li Y, Ma J, Chen W, Zhang Y, Zhou X, et al. 2013. Mlkl knockout mice demonstrate the indispensable role of Mlkl in necroptosis. *Cell Res* **23:** 994–1006. doi:10.1038/cr.2013.91

Xia B, Fang S, Chen X, Hu H, Chen P, Wang H, Gao Z. 2016. MLKL forms cation channels. *Cell Res* **26:** 517–528. doi:10.1038/cr.2016.26

* Xia S, Hollingsworth LR IV, Wu H. 2019. Mechanism and regulation of gasdermin-mediated cell death. *Cold Spring Harb Perspect Biol* doi:10.1101/cshperspect.a036400

Xie T, Peng W, Yan C, Wu J, Gong X, Shi Y. 2013. Structural insights into RIP3-mediated necroptotic signaling. *Cell Rep* **5:** 70–78. doi:10.1016/j.celrep.2013.08.044

Yoon S, Bogdanov K, Kovalenko A, Wallach D. 2016. Necroptosis is preceded by nuclear translocation of the signaling proteins that induce it. *Cell Death Differ* **23:** 253–260. doi:10.1038/cdd.2015.92

Yoon S, Kovalenko A, Bogdanov K, Wallach D. 2017. MLKL, the protein that mediates necroptosis, also regulates endosomal trafficking and extracellular vesicle generation. *Immunity* **47:** 51–65.e7. doi:10.1016/j.immuni.2017.06.001

Zargarian S, Shlomovitz I, Erlich Z, Hourizadeh A, Ofir-Birin Y, Croker BA, Regev-Rudzki N, Edry-Botzer L, Gerlic M. 2017. Phosphatidylserine externalization, "necroptotic bodies" release, and phagocytosis during necroptosis. *PLoS Biol* **15:** e2002711. doi:10.1371/journal.pbio.2002711

Zhang DW, Shao J, Lin J, Zhang N, Lu BJ, Lin SC, Dong MQ, Han J. 2009. RIP3, an energy metabolism regulator that switches TNF-induced cell death from apoptosis to necrosis. *Science* **325:** 332–336. doi:10.1126/science.1172308

Zhao J, Jitkaew S, Cai Z, Choksi S, Li Q, Luo J, Liu ZG. 2012. Mixed lineage kinase domain-like is a key receptor interacting protein 3 downstream component of TNF-induced necrosis. *Proc Natl Acad Sci* **109:** 5322–5327. doi:10.1073/pnas.1200012109

Zhao XM, Chen Z, Zhao JB, Zhang PP, Pu YF, Jiang SH, Hou JJ, Cui YM, Jia XL, Zhang SQ. 2016. Hsp90 modulates the stability of MLKL and is required for TNF-induced necroptosis. *Cell Death Dis* **7:** e2089. doi:10.1038/cddis.2015.390

Cite this article as *Cold Spring Harb Perspect Biol* doi: 10.1101/cshperspect.a036376

Death Receptors and Their Ligands in Inflammatory Disease and Cancer

Alessandro Annibaldi[1] and Henning Walczak[2,3,4]

[1]Center for Molecular Medicine Cologne, University of Cologne, 50931 Cologne, Germany

[2]Center for Biochemistry, University of Cologne, 50931 Cologne, Germany

[3]Cologne Excellence Cluster on Cellular Stress Responses in Aging-Associated Diseases (CECAD), University of Cologne, 50931 Cologne, Germany

[4]Centre for Cell Death, Cancer, and Inflammation (CCCI), UCL Cancer Institute, University College, London WC1E 6BT, United Kingdom

Correspondence: a.annibaldi@uni-koeln.de; h.walczak@uni-koeln.de; h.walczak@ucl.ac.uk

On binding to their cognate ligands, death receptors can initiate a cascade of events that can result in two distinct outcomes: gene expression and cell death. The study of three different death receptor–ligand systems, the tumor necrosis factor (TNF)–TNF receptor 1 (TNFR1), the CD95L–CD95, and the TNF-related apoptosis-inducing ligand (TRAIL)–TRAIL-R1/2 system, has drawn the attention of generations of scientists over the past 50 years. This scientific journey, as often happens in science, has been anything but a straight line to success and discoveries in this field were often made by serendipity, catching the scientists by surprise. However, as Louis Pasteur pointed out, luck prefers the prepared mind. It is therefore not surprising that the most impactful discovery of the field to date, the fact that TNF inhibition serves as an effective treatment for several inflammatory and autoimmune diseases, has been like this. Luckily, the scientists who made this discovery were prepared and, most importantly, determined to harness their discovery for therapeutic benefit. Today's research on these death receptor–ligand systems has led to the discovery of a causal link between cell death induced by a variety of these systems and inflammation. In this review, we explain why we predict that therapeutic exploitation of this discovery may profoundly impact the future treatment of inflammatory disease and cancer.

SIGNALING BY DEATH RECEPTORS AND THEIR LIGANDS

Before we discuss the role of death receptor (DR)–ligand systems in inflammation-associated diseases and cancer, we will give an update on the current understanding of ligand-stimulated signaling by these receptors.

DRs are a class of cell surface–expressed type I transmembrane receptors that form part of the tumor necrosis factor (TNF) receptor superfamily (TNFRSF) (Walczak 2013). The defining characteristic of a DR is the presence of the so-called death domain (DD) within the cytoplasmic portion. This class of receptors includes TNF receptor 1 (TNFR1) (Loetscher et al. 1990; Schall et al. 1990), CD95 (Fas/APO-1) (Oehm et al. 1992; Itoh and Nagata 1993), TNF-related apoptosis-inducing ligand (TRAIL)-R1 (DR4) (Pan et al. 1997b), TRAIL-R2 (DR5, APO-2/TRICK/DR5/KILLER) (Pan et al. 1997a; Screaton et al. 1997a; Sheridan et al. 1997; Walczak et al. 1997; Wu et al. 1997), and DR3 (TRAMP) (Chinnaiyan et al. 1996). On cross-linking by their respective cognate ligands, DRs have the capacity to induce the death of the cell on which they are expressed. However, despite the concept implicit in their name, cell death is by no means the only possible functional signaling outcome of DR stimulation and, at least for some of them, it is also not the default outcome (Fig. 1).

TNF binds to TNFR1 and TNFR2. However, only TNFR1 contains a DD and therefore forms part of the DR subfamily (Medler and Wajant 2019; Wajant and Siegmund 2019). The TNFR1 signaling pathway is the best studied of all DR signaling pathways and we will begin by providing an in-depth explanation of this pathway (Fig. 1). TNF binding to TNFR1 triggers receptor trimerization and the formation of the TNFR1 signaling complex (TNFR1-SC; also referred to as TNF–RSC or complex I of TNFR1 signaling) (Micheau and Tschopp 2003; Annibaldi and Meier 2018). TNFR1-SC formation is initiated by DD-dependent recruitment of TRADD and RIPK1 to the receptor (Kelliher et al. 1998). Subsequent binding of the adaptor TRAF2 to TRADD, in turn, mediates recruitment of the E3 ligases cIAP1 and cIAP2 (Rothe et al. 1995; Hsu et al. 1996; Micheau and Tschopp 2003; Ermolaeva et al. 2008). These two cIAPs place ubiquitin chains of various topologies (i.e., K11, K48, and K63 linkages) on RIPK1 and other components of the TNFR1-SC (Dynek et al. 2010; Annibaldi et al. 2018). These ubiquitin chains serve as scaffolds to recruit the linear ubiquitin chain assembly complex (LUBAC) (Haas et al. 2009). LUBAC is a tripartite E3 ligase complex that consists of the central catalytic component HOIP (also called RNF31), HOIL-1 (also called RBCK1), and SHARPIN (also called SIPL1) (Gerlach et al. 2011; Ikeda et al. 2011). LUBAC conjugates linear ubiquitin chains (also referred to as Met1 or M1 chains) to several TNFR1-SC components, including

RIPK1, NEMO, TRADD, and TNFR1 itself (Gerlach et al. 2011; Draber et al. 2015). To date, LUBAC is the only E3 ligase known to generate linear ubiquitin chains de novo (Haas et al. 2009; Gerlach et al. 2011). This activity is crucial in the context of various signaling pathways, including in TNFR1 signaling (Zinngrebe et al. 2014; Peltzer et al. 2016). The ubiquitin chains generated by cIAP1/2 and LUBAC recruit different kinase-containing subcomplexes: the TAK1/TAB2/TAB3 (Ori et al. 2013), NEMO/IKKα/IKKβ (Rahighi et al. 2009), NEMO/TANK/TBK1/IKKε, and NEMO/NAP1/TBK1 complexes (Lafont et al. 2018). Although the TAK1 and IKKα/β complexes are necessary for gene activation via MAPKs and NF-κB, the different NEMO-containing kinase complexes are required to phosphorylate RIPK1 in TNFR1-SC at distinct sites, and this inhibits TNF-induced cell death (Dondelinger et al. 2015, 2019; Jaco et al. 2017; Lafont et al. 2018). TNF-induced gene expression leads to the production of cytokines and prosurvival proteins, which are required to mount an innate immune response.

TNF signaling is tightly regulated by a series of checkpoints that are dependent on ubiquitination, phosphorylation, gene expression, and protein cleavage events (Ting and Bertrand 2016; Annibaldi and Meier 2018). Circumstances that compromise these checkpoints often result in the formation of a secondary, cytoplasmic complex, referred to as complex II (Micheau and Tschopp 2003). The core components of complex II are RIPK1, FADD, caspase-8, cFLIP and, if expressed in the cell, RIPK3 (Wang et al. 2008; Zhang et al. 2009; Feoktistova et al. 2011; Tenev et al. 2011). The signals that induce cell death by apoptosis and necroptosis emanate from this complex. Apoptosis is initiated by RIPK1/FADD-mediated activation of caspase-8, which in turn activates the executioner caspase-3 and caspase-7. Necroptosis is mediated by the RIPK1/RIPK3/MLKL axis on genetic deletion or pharmacological inhibition of caspase-8. It requires the kinase activities of both RIPK1 and RIPK3 (Pasparakis and Vandenabeele 2015; Peltzer and Walczak 2019). Both ubiquitination and phosphorylation of RIPK1

Cite this article as *Cold Spring Harb Perspect Biol* doi: 10.1101/cshperspect.a036384

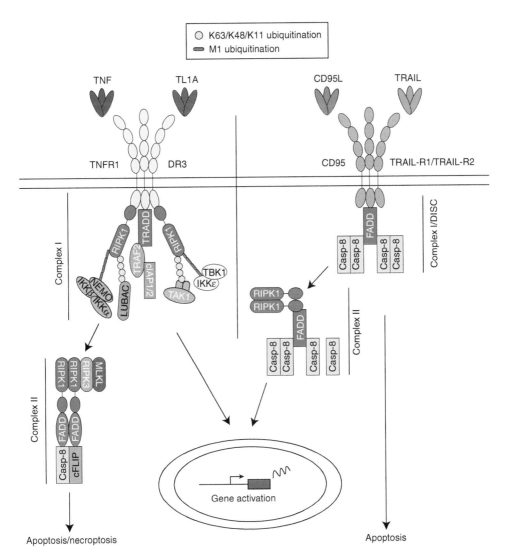

Figure 1. Tumor necrosis factor (TNF) receptor 1 (TNFR1)/DR3 and CD95/TNF-related apoptosis-inducing ligand (TRAIL)-R1/2 signaling pathways. Binding of TNF and TL1A to TNR-R1 and DR3, respectively, and CD95L and TRAIL to CD95 and TRAIL-R1/2, respectively, induces formation of a membrane-bound complex referred to as complex I or death-inducing signaling complex (DISC) in the case of CD95 and TRAIL-R1/2. TNFR1- and DR3-mediated complex I triggers gene expression via NF-κB and MAPKs, whereas the DISC has the potential to induce cell death by apoptosis or necroptosis. These two primary complexes dissociate from the respective receptors and incorporate additional proteins to form a secondary cytosolic complex called complex II. In the case of TNFR1 and TL1A, this complex induces caspase-8-mediated apoptosis or RIPK3/MLKL-mediated necroptosis. CD95 and TRAIL-R1/2 complex II triggers gene activation via NF-κB and MAPKs.

are required to prevent it from leaving TNFR1-SC and nucleating a death-promoting complex II.

Until a few years ago, the prevailing concept was that abnormally high TNF-induced gene expression was the (only) cause of TNF-induced inflammation. TNF-induced cell death was re-garded as less relevant, perhaps even irrelevant, to the chronic inflammatory and autoimmune disorders known to be driven by TNF. However, this view has dramatically changed in recent years as explained in the next section of this review.

DR3 and its ligand TL1A appear to signal in essentially the same way as TNF and TNFR1 (Fig. 1). Expression of TL1A and DR3 is mainly restricted to immune cells (Marsters et al. 1996; Screaton et al. 1997b). This system has been reviewed thoroughly elsewhere (Richard et al. 2015), but suffice to state that little is known regarding the relative contribution of TL1A-mediated gene activation versus TL1A-induced cell death to inflammation. This topic likely represents an interesting field for future study of TL1A and DR3.

The TRAIL and CD95L (FasL/APO-1L) signaling pathways share many components with the TNF pathways. Both ligands trigger the formation of membrane-bound and cytoplasmic complexes, but their roles are reversed in comparison to TNFR1 complexes (Walczak 2013). Although TNFR1-SC mediates gene activation and TNFR1 complex II triggers cell death, TRAIL and CD95L assemble a primary death-inducing signaling complex (DISC). Secondly, cytoplasmic complexes activate gene expression. TRAIL binding to TRAIL-R1 and/or TRAIL-R2 (or CD95L binding to CD95) oligomerizes the receptor and the intracellular DDs adopt a conformation that enables FADD recruitment (Chinnaiyan et al. 1995; Shirley et al. 2011). It has been shown that receptor hexamerization is the optimal assembly for cell death induction by these two death ligands, which explains why hexameric receptor agonists are potent inducers of cell death by TRAIL-R1/2 and CD95 (Holler et al. 2003; Valley et al. 2012). FADD in turn recruits caspase-8 and caspase-10 via death effector domain (DED) interactions. Caspase-8 undergoes ubiquitination by cullin-3, increasing its clustering and activation (Jin et al. 2009). Besides inducing cell death, TRAIL and CD95L can also activate NF-κB and MAPKs through a secondary complex, complex II, whose core components are RIPK1, NEMO, TRAF2, caspase-8, and FADD (von Karstedt et al. 2017). Recently, it was shown that the TRAIL DISC can also elicit gene activation through NF-κB (Hartwig et al. 2017; Henry and Martin 2017) using LUBAC (Lafont et al. 2017). Interestingly, whereas RIPK1 is only important for TNF-induced gene activation in some instances (Wong et al. 2010), RIPK1 is required for TRAIL-induced gene expression (Hartwig et al. 2017; Henry and Martin 2017).

DEATH RECEPTORS AND THEIR LIGANDS IN INFLAMMATORY DISEASES

TNF in Chronic Inflammation and Autoimmunity

TNF is a potent inflammatory cytokine that coordinates immune responses following tissue damage or pathogen infection. Initially, however, TNF was thought to induce necrosis in tumor cells, a concept that obviously inspired the naming of this cytokine. Although the purified activity then referred to as TNF did induce tumor necrosis, most strikingly in certain sarcomas (Carswell et al. 1975), this proved the exception rather than the rule after TNF was cloned. Most cell types exposed to TNF induce the expression of a plethora of proinflammatory cytokines and chemokines, rather than dying (Beutler 1999). It is therefore not surprising that the study of TNF-induced cell death became the wallflower of cell death research for many years. The pathway appeared obscure at a time when the field was focused on deciphering the canonical apoptosis program. A few scientists kept studying this form of cell death, inspired by the fact that it could clearly be triggered, and thus occurred in a programmed manner (Vanden Berghe et al. 2004; Vandenabeele et al. 2006).

The concept of TNF being a potent proinflammatory factor became entrenched in the scientific community in the late 1980s and early 1990s, which is when scientists were trying to understand how inhibition of cytokines could be used in the treatment of inflammatory diseases. According to one school of thought, therapeutic intervention in such multifactorial diseases would be impossible because of redundancy among the cytokines instigating inflammation. However, the optimists' view was that an apical cytokine triggered the inflammatory cascade involving a plethora of cytokines. The first indication that TNF might be that apical cytokine, and consequently a worthy therapeutic target, came from cultures of dissociated rheumatoid synovial membranes obtained from rheumatoid arthritis (RA) patients. In this study, blocking TNF prevented the production of many inflammatory cytokines, including interleukin (IL)-1, IL-6, and IL-8, which had also been considered candidate apical cytokines (Brennan et al. 1989). Further support for this concept came from studies showing that TNF and TNFR1 are up-regulated in inflamed tissues of RA patients (Chu et al. 1991; Deleuran et al. 1992). The ultimate proof for this groundbreaking discovery came when TNF inhibitors afforded therapeutic benefit to patients suffering from RA (Elliott et al. 1993; Moreland et al. 1999; Weinblatt et al. 1999; Bathon et al. 2000; Lovell et al. 2000). Shortly after the approval of the first TNF inhibitor (enbrel/etanercept) in 1997, other biotherapeutic inhibitors of TNF were approved for the treatment of various chronic inflammatory and autoimmune diseases. Besides RA and other arthritides, the most prominent indications were psoriasis, and the two most prevalent forms of inflammatory bowel disease (IBD) in Crohn's disease (CD) and ulcerative colitis (UC) (Monaco et al. 2015). The therapeutic success of TNF inhibition has been tremendous, indeed unprecedented in the treatment of chronic inflammatory and autoimmune diseases (Brenner et al. 2015; Kalliolias and Ivashkiv 2016). Since their introduction in 1997, TNF blockers have helped millions of patients live a life in which their disease is controlled, albeit not cured.

Death Ligands beyond TNF in Inflammation and Autoimmunity

Unfortunately, anti-TNF therapy does not help all patients suffering from one of the aforementioned disorders. In about half the patients suffering from RA, long-lasting benefits can be achieved through the inhibition of TNF, whereas the other half of RA patients do not benefit from this treatment. The same is true for ~35% of patients with psoriasis and between 20% and 40% of patients suffering from the different forms of IBD (Cho and Feldman 2015; Lopetuso et al. 2017). TNF inhibition also provided no therapeutic benefit in patients with the autoim-

mune disorders multiple sclerosis (MS) and amyotropic lateral sclerosis (ALS), despite a clear up-regulation of the TNF system in these disorders (Lenercept Multiple Sclerosis Study Group 1999). One interpretation of these failures is that TNF is not the apical regulator of inflammation in nonresponding patients, and inflammation is instead instigated by different means.

This view was challenged by the realization that TNF can mediate inflammation not only by activating gene expression, but also by inducing aberrant cell death (Fig. 2). For example, mutation of the *Sharpin* gene in mice causes chronic proliferative dermatitis (cpdm) (Seymour et al. 2007). In cells of these mice, TNF-induced gene activation is attenuated, whereas the induction of cell death is increased (Gerlach et al. 2011). At the time, this result provided a conundrum: how did the mouse develop an inflammatory disease if proinflammatory gene expression in response to TNF was compromised? One could argue that an inflammatory mediator other than TNF drove inflammation. However, gene activation by other inflammatory mediators, including CD40L, was also compromised in *Sharpin*-deficient cells (Gerlach et al. 2011). Given that TNF-induced cell death was aberrantly increased, we reasoned that genetic ablation of *Tnf* could address two questions: (1) does aberrant cell death cause the inflammatory syndrome in *cpdm* mice, and (2) is TNF the intrinsic *agent provocateur* of this cell death? Indeed, genetic deletion of *Tnf*, even heterozygous deletion of *Tnf*, prevented both cell death in the skin and inflammatory disease in *cpdm* mice. Therefore, we concluded that TNF-induced cell death caused inflammation in these mice (Gerlach et al. 2011). This discovery was genetically confirmed when it was shown that codeletion of *Casp8* and *Mlkl*, or *Fadd* and *Ripk3*, also prevented cell death in the skin and inflammatory disease in *cpdm* mice (Kumari et al. 2014; Rickard et al. 2014; Peltzer et al. 2018). Other genetic mouse models of inflammatory disease have also been shown to be driven by excessive cell death (Welz et al. 2011; Dannappel et al. 2014; Kumari et al. 2014; Vlantis et al. 2016). Important-

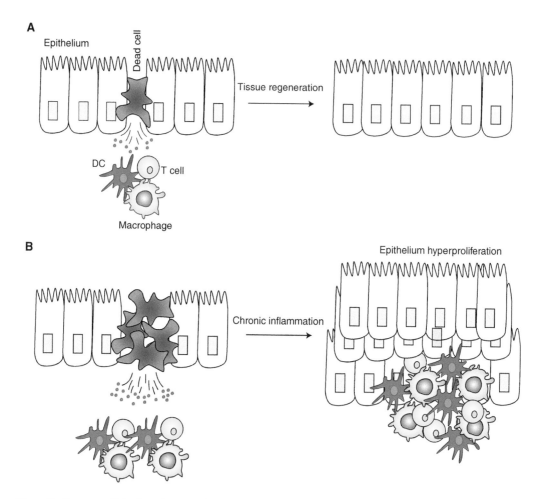

Figure 2. Aberrant cell death in chronic inflammation. (*A*) Cell death is important for the repair and regeneration programs of different types of tissues, such as the epithelium of the skin and the intestine. Dying cells release factors that trigger the activation of an inflammatory program, whose ultimate purpose is to restore tissue integrity. (*B*) Deregulated cell death determines the persistence of tissue repair programs, which in turn leads to chronic inflammation, epithelial cell hyperproliferation, and, ultimately, to an autoinflammatory or autoimmune disorder. (DC) Dendritic cell.

ly, several of these genetic alterations have also been found in humans, although these alterations appear to be rare. These individuals often suffer from a combination of autoinflammation and immunodeficiency (Fusco et al. 2008; Cuchet-Lourenco et al. 2018; Oda et al. 2019).

A further advance in the understanding of the etiology of inflammatory diseases came when it was shown that pharmacologic or genetic inhibition of RIPK1 prevented dermatitis in *cpdm* mice (Berger et al. 2014). Thus, by blocking the kinase activity of RIPK1, TNF-induced

cell death and skin inflammation was prevented. These findings raised the possibility that inhibition of RIPK1 might suffice to block aberrant TNF-induced cell death and consequent inflammation. If this were the case, then TNF blockade could potentially be replaced by inhibition of RIPK1. Additional preclinical studies showed that inhibition of RIPK1 could also ameliorate TNF-induced septic shock, anticollagen antibody-induced arthritis, and inflammation and tissue injury caused by A20 deficiency, but it provided no benefit in other models, such as

Cite this article as *Cold Spring Harb Perspect Biol* doi: 10.1101/cshperspect.a036384

chemically induced pancreatitis (Newton et al. 2016; Patel et al. 2020).

Given that several biotherapeutic inhibitors of TNF figure in the list of the world's top 10 best-selling drugs, it is not surprising that this concept garnered the attention of the biotech and pharmaceutical industries. Many companies are currently bringing RIPK1 kinase inhibitors into clinical testing. However, the first phase II clinical trials, which were performed in RA, psoriasis, and IBD patients, did not appear to reveal a benefit and the inhibitor has recently "moved back to research" (see gsk.com/media/5745/q3-2019-results-slides.pdf). Could it be that the notion of inflammatory disease being caused by aberrant TNF-induced RIPK1-kinase-dependent cell death is just too simple? Recent evidence suggests this might indeed be the case. For example, lethal dermatitis induced by keratinocyte-specific genetic deletion of *Hoip* or *Hoil-1* (*Hoip*$^{E-KO}$ or *Hoil-1*$^{E-KO}$) is considerably delayed by codeletion of *Tnfr1*, whereas inhibition of RIPK1 only delays lethal dermatitis by a few days (Taraborrelli et al. 2018). Intriguingly, however, when *Hoip*$^{E-KO}$;*Tnfr1*$^{-/-}$ or *Hoil-1*$^{E-KO}$;*Tnfr1*$^{-/-}$ mice are given a RIPK1 kinase inhibitor, lethal dermatitis is prevented (Taraborrelli et al. 2018). Thus, at least in this genetic model, TNFR1 ablation and RIPK1 kinase inhibition are synergistic at inhibiting cell death and consequent inflammation. Nevertheless, this discovery suggests that combining a TNF inhibitor and a RIPK1-specific inhibitor might benefit certain autoimmune or autoinflammatory disease patients. For patients with rare germline mutations impacting LUBAC components, this combination may represent an alternative treatment strategy if their current therapies do not work or if they develop resistance to them. An outstanding question, however, is what drives RIPK1-dependent cell death and disease when the TNF–TNFR1 system is absent.

The idea that TNF superfamily (TNFSF) cytokines beyond TNF itself could act individually or in a concerted manner to promote inflammation has been proposed, albeit with cell-death-independent mechanisms and disease etiologies in mind and, hence, without considering the coinhibition of other cell death–inducing ligands (Croft and Siegel 2017). At present, several TNFSF proteins are under evaluation in preclinical and clinical studies as potential targets in various rheumatic and chronic inflammatory diseases. Especially noteworthy in the context of this review are TRAIL and CD95L. The primary signaling output of these two TNFSF members is cell death, which is in contrast to that of TNF and TL1A (Fig. 1). Nevertheless, as described earlier, both the TRAIL–TRAIL-R1/R2 and CD95L–CD95 systems can induce gene activation (Hartwig et al. 2017; Henry and Martin 2017). The CD95L–CD95 system drives activation-induced T-cell death, which is instrumental in immune homeostasis (Alderson et al. 1995; Brunner et al. 1995; Dhein et al. 1995; Ju et al. 1995). Indeed, mutations in *CD95* cause the accumulation of aberrant T cells, which ultimately results in autoimmunity (Watanabe-Fukunaga et al. 1992). The TRAIL–TRAIL-R system has mainly been characterized for its ability to selectively kill cancerous, but not essential normal cells. In addition to being expressed in cancer cells, TRAIL-R1 and TRAIL-R2 are also expressed on different subpopulations of T cells, whose apoptosis they can promote (Roberts et al. 2003; Zhang et al. 2003). Hence, the immunological "day job" of the CD95 and TRAIL systems appears to be the proper termination of an immune response and the prevention of autoimmunity. Indeed, the killing function of CD95 and TRAIL-R1/R2 in activated or autoreactive T cells has led to the idea that autoreactive T cells might be eliminated in the context of autoimmunity using CD95L, TRAIL, or other DR agonists. The severe hepatotoxicity of CD95 agonists (Ogasawara et al. 1993) excludes their use, but TRAIL and other agonists of the TRAIL DRs may yet be an option, although this awaits therapeutic validation in patients.

Given that caspase-8-dependent cell death causes lethal inflammation in *Hoip*$^{E-KO}$ and *Hoil-1*$^{E-KO}$ mice, as well as in *Hoil-1*$^{E-KO}$; *Tnfr1*$^{-/-}$ and *Hoip*$^{-KO}$;*Tnfr1*$^{-/-}$ mice, the role of other DRs in inflammation must be considered (Taraborrelli et al. 2018). Indeed, cells deficient for HOIL-1 or HOIP are more sensitive to killing by TNF, TRAIL, or CD95L. Although neither constitutive deletion of *Trail-r* nor keratinocyte-specific deletion of the DD of CD95 ameliorated dermatitis in *Hoil-1^{E-KO};Tnfr1* mice, eliminating both CD95 and TRAIL-R signaling significantly ameliorated dermatitis (Taraborrelli et al. 2018). Hence, cell death driven by CD95L and TRAIL also contributes to inflammation. The therapeutic potential this discovery may unleash is tremendous because it suggests that the prevention of disease-causing inflammation may require the simultaneous neutralization of cell death induction by multiple DR–ligand systems. Thus, for patients with a cell death–dependent disease etiology, blocking TNF may not be sufficient to achieve a lasting therapeutic benefit, whereas blocking TNF and these additional death ligands or their cell death–inducing signaling pathways, may well afford such benefit.

In summary, during the past decade, we learned that (1) TNF-driven inflammatory and autoimmune disorders can also stem from TNF-induced cell death and not only from TNF-induced gene activation; (2) failure of TNF inhibitors to provide therapeutic benefit in all patients with inflammation-associated disease does not necessarily mean that TNF is not a valuable target in nonresponders; it might simply mean that blocking TNF is not sufficient and other death ligands—or their downstream effectors—have to be blocked in addition; and (3) these other death ligands can be TRAIL and CD95L, and the downstream effector essential for mediating cell death by them and not TNF could be the kinase activity of RIPK1. This last point leaves us with an intriguing biochemical question because, if anything, TNF-induced cell death would have been expected to require RIPK1 activity but not cell death induced by TRAIL and CD95L. This unexpected result leaves us with the realization that there is more plasticity between cell death pathways triggered by the different DRs than previously thought. Extending this concept, two recent studies showed that when the proteolytic activity of caspase-8 is genetically ablated and necroptosis is inhibited, the system is rewired to activate yet a third modality of programmed cell death, cas-

pase-1-dependent pyroptosis (Fritsch et al. 2019; Newton et al. 2019). This discovery exposes a previously unappreciated plasticity between the three major programmed cell death pathways. This rewiring likely represents an evolutionary necessity to survive the challenge of infection. The other, pathological side of the coin is something we are only beginning to unearth. It may have a major impact on our understanding of neuroinflammatory and neurodegenerative diseases (Ising et al. 2019), possibly guiding how we treat these diseases effectively in the future.

In conclusion, combining a TNF inhibitor with a RIPK1 kinase inhibitor or with inhibitors of CD95L, TRAIL, and/or other death ligands, might extend the reach of TNF-inhibitory therapies to patients who currently do not benefit from them. For example, patients with a disorder in which TNF inhibitors provide therapeutic benefit in many, but not all patients, such as RA, IBD (CD and UC), or psoriasis. However, equally, or perhaps even more, excitingly, this concept could extend to chronic inflammatory, autoimmune, and neuroinflammatory disorders in which TNF inhibition so far failed. Only future clinical trials testing the therapeutic concepts summarized here will reveal whether the insights gained from studying mouse models are also relevant to human patients.

DEATH RECEPTORS AND THEIR LIGANDS IN CANCER

The TNF System in Cancer

TNF owes its name to the fact that it was identified as the agent responsible for the tumor-necrotizing activity of "Coley's toxins." Discovered and used at the end of the 19th century by William Coley in New York (Coley 1893), Coley's toxins consisted of a mixture of bacterial lysate products. It induced regressions of certain tumors, especially sarcomas (Nauts et al. 1946). However, enthusiasm for the identification of a soluble factor that could revolutionize cancer treatment did not persist in an era when the discoveries of chemo- and radiotherapy dominated cancer research and therapy. Nevertheless, scientists kept chasing this ominous activity and TNF was cloned with the advent of modern biochemical and genetic techniques (Gray et al. 1984; Pennica et al. 1984; Marmenout et al. 1985). The promise held by TNF for cancer treatment proved short-lived because systemic treatment with TNF caused a lethal inflammatory shock syndrome with massive cytokine induction (Tracey et al. 1988). For a long time, this was thought to be the result of increased TNF-induced gene expression rather than cell death (Balkwill 2009). Intriguingly, however, the cytokine storm responsible for TNF-induced shock was recently shown to indeed be a consequence of TNF-induced cell death (Newton et al. 2016). Today, TNF is exploited to treat cancer locally by isolated limb perfusion (ILP). In ILP, TNF is used in combination with chemotherapy (melphalan) for the treatment of locally advanced extremity soft tissue sarcoma (STS) (Neuwirth et al. 2017). In this case, the toxicity linked to systemic administration of TNF is circumvented by local treatment. At present, it is not entirely clear why some types of cancer, especially sarcomas, are susceptible to TNF, whereas the vast majority of cancers are not. TNF killing of tumor endothelial cells is likely involved in the observed tumor necrosis (Robaye et al. 1991).

Further discouraging the use of TNF as a cancer drug, a study published in 1999 showed that $Tnf^{-/-}$ mice challenged with the skin carcinogen DMBA and the tumor promoter TPA developed fewer tumors than wild-type mice (Moore et al. 1999). TNF also promotes the growth of syngeneic and carcinogen-induced tumors of the skin, pancreas, colon, and ovary (Suganuma et al. 1999; Kulbe et al. 2007; Zins et al. 2007; Egberts et al. 2008). These findings, together with evidence indicating that TNF is produced by both malignant cells and cells in the tumor microenvironment, suggest that TNF is a crucial, if not the central, regulator of cancer-related inflammation (Mantovani et al. 2008; Balkwill and Mantovani 2012). Mechanistically, we now know that cancer cells promote a pro-inflammatory microenvironment that supports tumor cell survival and proliferation, angiogenesis, and metastasis, although dampening adaptive antitumor immune responses. The ability of cancer cells to promote such an inflammatory microenvironment is often mediated by the production of TNF. Collectively, these discoveries led to a paradigm shift with TNF being regarded as an endogenous tumor-promoting agent. Consequently, inhibition of TNF started to be considered in cancer therapy.

Our knowledge of the role of TNF in inflammation-related disorders might give us clues on how to best harness the biology of TNF in cancer therapy. As explained in the previous section, TNF-dependent autoimmune and inflammatory disease can be caused by aberrant TNF-induced cell death. Given the abundance of TNF in the microenvironment of many different cancers, the question arises as to whether we can rewire TNF signaling from its prosurvival and proliferative function into a cell death–inducing function. In other words, can we harness endogenous TNF in the microenvironment and use it to kill cancer cells? Several reports indicate this might indeed be possible. In these studies, some of the checkpoints that prevent TNF-induced cell death were manipulated for TNF to trigger cell death. For example, SMAC mimetics, which degrade cIAP1/2 and promote TNFR1 complex II formation, rendered cancer cells susceptible to TNF-induced cell death (Beug et al. 2014, 2017; Lalaoui et al. 2016). Pan-caspase inhibitors, like zVAD or emricasan, can also convert the TNF-induced signal in cancer cells into a necroptosis-inducing one (Brumatti et al. 2016). Intriguingly, cell death induced by TNF, especially necroptosis, appears to be highly immunogenic, facilitating tumor immunity (Yatim et al. 2015, 2017; Snyder et al. 2019; Aaes et al. 2020). A recent study reported that lowering the threshold of TNF cytotoxicity increases the efficacy of cancer immunotherapy (Vredevoogd et al. 2019). Thus, if we could manipulate TNF signaling in cancer cells to trigger an immunogenic cell death, we might kill a fraction of the cells directly and evoke an antitumor immune response to target the escapees.

In summary, more than a century of biomedical research on TNF has been highly stimulating from a scientific perspective. Realization of the extraordinary potential of TNF inhibition in the treatment of autoimmune and chronic inflammatory diseases represents a striking advance of modern medicine. Future opportunities related to its role in cancer-related inflammation await further preclinical research followed by clinical validation. Efforts focused on harnessing endogenous TNF in the tumor microenvironment and rendering it capable of killing cancer cells in a manner that stimulates antitumor immunity represents a fascinating new challenge. Clearly, TNF's exciting journey in tumor biology and immunology is far from over.

The CD95 and TRAIL Systems in Cancer

Coincident with hope fading for TNF as a viable cancer therapeutic, two antibodies, anti-APO-1 and anti-Fas, drew attention owing to their ability to directly kill cancer cells (Trauth et al. 1989; Yonehara et al. 1989). Their targets, Fas and APO-1, were found to belong to the TNFRSF (Itoh et al. 1991; Oehm et al. 1992). Disappointingly, however, both antibody and ligand-derived CD95 agonists were highly toxic, causing massive hepatocyte death and fulminant hepatitis (Ogasawara et al. 1993).

Subsequently, the CD95L–CD95 system was shown to stimulate cancer cell proliferation, migration, and invasion (Owen-Schaub et al. 1993; Barnhart et al. 2004; Chen et al. 2010), including in glioblastoma (Kleber et al. 2008). Subsequent clinical testing of CD95-Fc (APG101/asunercept), a CD95L antagonist, showed that glioblastoma patients treated with this inhibitor plus radiotherapy benefited when compared with radiotherapy alone (Wick et al. 2014). Thus, based on our current understanding of the cancer biology of the CD95L–CD95 system, its inhibition rather than activation appears to provide benefit for cancer patients.

Soon after it was realized that CD95 agonists, like TNF, would not be a "magic bullet" for killing cancer cells, a new member of the TNFSF was identified. It most resembled CD95L, which was also known as FasL or APO-1L. Given that it could induce apoptosis similar to FasL, this new TNFSF member was named TRAIL or Apo2L (Wiley et al. 1995; Pitti et al. 1996). Importantly, TRAIL selectively killed cancer cells, without killing any essential normal cells, a unique com-

bination of characteristics, which holds true both in vitro and in vivo (Walczak et al. 1997). This finding provided the scientific basis for developing TRAIL-receptor agonists (TRAs) as novel cancer therapeutics. However, TRAIL-R1- and TRAIL-R2-specific antibodies as well as a first recombinant form of TRAIL showed very limited anticancer activity in patients in clinical trials. As recently thoroughly reviewed elsewhere (von Karstedt et al. 2017), various strategies are currently underway to overcome the two issues which, together, are likely responsible for the failure of these first-generation TRAs, that is, (1) poor agonistic activity, and (2) resistance of most primary cancers to TRAIL-induced apoptosis.

Additionally, however, it recently emerged that the TRAIL–TRAIL-R system can be "hijacked" by certain cancers. Rather than killing cancer cells, recombinant TRAIL was found to act in a tumor-supportive manner (Trauzold et al. 2006), particularly in KRAS-mutated cancers (Hoogwater et al. 2010). It was then realized that endogenous TRAIL can be recruited by KRAS-mutated cancers to act in this manner. The mechanism whereby this is achieved is two-pronged: (1) by enhancing tumor cell proliferation, invasion, and metastasis in a TRAIL-R2 DD- and FADD-independent manner (von Karstedt et al. 2015), and (2) by creating a TRAIL-R2 DD- and FADD-dependent cytokine-rich micromilieu in which monocytes are polarized to become myeloid-derived suppressor cells (MDSCs) or alternatively activated (M2) macrophages (Fig. 3; Hartwig et al. 2017). Interestingly, the latter mechanism not only supports tumor growth, it also acts via immunosuppression. Therefore, TRAIL may represent a previously unrecognized immune checkpoint whose inhibition may unleash tumor immunity. It is interesting to note that TRAIL, albeit mostly in non-cancer-related contexts, has been shown to inhibit T helper type 1 (Th1) cells (Ikeda et al. 2010), promote regulatory T cells (Pillai et al. 2011), and suppress T-cell activation and proliferation by interfering with proximal T-cell receptor (TCR) signaling (Lehnert et al. 2014). It can also kill immature dendritic cells (Leverkus et al. 2000). Thus, TRAIL inhibition may act via tu-

Cite this article as *Cold Spring Harb Perspect Biol* doi: 10.1101/cshperspect.a036384

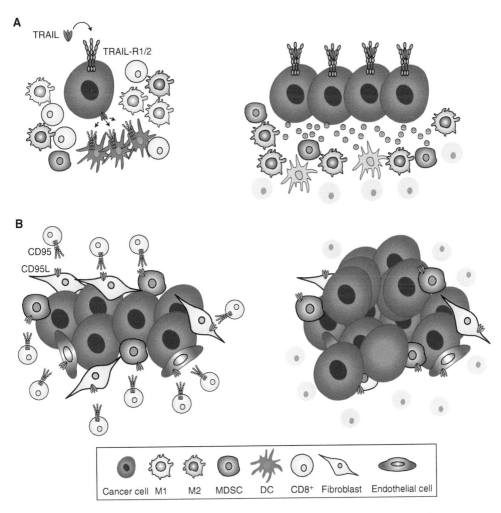

Figure 3. Mechanisms of death receptor (DR)-mediated immunosuppression in cancer. (*A*) The tumor necrosis factor (TNF)-related apoptosis-inducing ligand (TRAIL)/TRAIL-R1/2 system contributes to immunosuppression via distinct mechanisms. The activation of TRAIL-R1/2 expressed by cancer cells induces the secretion of cytokines that polarize monocytes into myeloid-derived suppressor cells (MDSCs) and alternatively activated (M2) macrophages. These cell types in turn create an immunosuppressive tumor microenvironment that prevents CD8$^+$ T-cell-mediated antitumor immune responses. In addition, TRAIL expressed by cancer cells or noncancer cells in the tumor microenvironment can directly kill TRAIL-R1/2-expressing immune cells, which are required for mounting an adaptive immune response, such as dendritic cells (DCs) or T cells, and thereby interfere with adaptive antitumor immunity. (*B*) Cancer cells have the potential to induce the expression of CD95L on different cell types of the tumor microenvironment, such as endothelial cells, MDSCs, and fibroblasts. By cross-linking CD95 on the surface of T cells, CD95L induces their demise by apoptosis, thereby facilitating tumor cell evasion from an adaptive immune attack.

mor-cell-centered suppressive as well as immunity-enabling antitumor effects.

Intriguingly, CD95L, arguably the best T-cell killer encoded by our genome and the physiological mediator of activation-induced T-cell death (Alderson et al. 1995; Brunner et al. 1995; Dhein et al. 1995; Ju et al. 1995), can also be expressed by various cells in the tumor microenvironment, including endothelial cells (Motz et al. 2014), MDSCs (Zhu et al. 2017), and cancer-associated fibroblasts (Lakins et al. 2018). Thus, CD95L inhibition may also act by a combination of tumor-cell-centered suppressive and immunity-enabling antitumor effects (Fig. 3).

In summary, both TRAIL and CD95L may represent the ultimate form of an immune checkpoint. Rather than suppressing tumor immunity in the subtle manner of CTLA4 and PD-1/PD-L1, they can remove cells required for adaptive antitumor immunity simply by killing them. Whether or not these different means are co-opted by a given cancer, or represent alternative ways by which particular cancers circumvent immune recognition, will determine whether TRAIL and/or CD95L inhibitors can serve as immune checkpoint blockers in their own right or whether they may (only) act in synergy with CTLA4- and/or PD-1/L1-targeting therapeutics. It will be exciting to answer these eminent questions in cancer immunotherapy.

CONCLUDING REMARKS

From today's perspective, it may appear surprising that the tremendous success of TNF inhibitors did not spur the pharmaceutical world to follow their usual strategy, which is to move to the next family member and develop an inhibitor for that one. Although this appears the obvious strategy with the hindsight we have today, the initial concept that TNF blockers worked because they blocked TNF-induced gene expression, and not TNF-induced cell death, blurred the picture. Now we know that in many cases it is the inhibition of TNF-induced cell death that affords therapeutic benefit. Consequently, there is no excuse for the pharmaceutical industry not to tackle TNF's closest relatives. Targeting the other death ligands may provide therapies for patients with autoimmune or chronic inflammatory diseases, perhaps also neuroinflammatory and neurodegenerative diseases, who do not benefit from the inhibition of TNF(-induced cell death), or at least the inhibition of TNF alone. Given the newly appreciated role of DRs and their ligands in cancer-related inflammation, tumor promotion, and tumor immune suppression and evasion, the therapeutic principle of death ligand inhibition may also extend to cancer therapy.

ACKNOWLEDGMENTS

This work was funded by a Wellcome Trust Investigator Award (214342/Z/18/Z), a Medical Research Council Grant (MR/S00811X/1), a Cancer Research UK Programme Grant (A17341), a Collaborative Research Centre (SFB 1399) from the Deutsche Forschungsgemeinschaft (DFG) and the Alexander von Humboldt Foundation to H.W. and a CMMC Junior Research Group Program to A.A.

REFERENCES

Aaes TL, Verschuere H, Kaczmarek A, Heyndrickx L, Bartosz Wiernicki B, Delrue I, De Craene B, Taminau J, Delvaeye T, Bertrand MJM, et al. 2020. Immunodominant AH1 antigen-deficient necroptotic, but not apoptotic, murine cancer cells induce antitumor protection. *J Immunol* doi:10.4049/jimmunol.1900072

Alderson MR, Tough TW, Davis-Smith T, Braddy S, Falk B, Schooley KA, Goodwin RG, Smith CA, Ramsdell F, Lynch DH. 1995. Fas ligand mediates activation-induced cell death in human T lymphocytes. *J Exp Med* 181: 71–77. doi:10.1084/jem.181.1.71

Annibaldi A, Meier P. 2018. Checkpoints in TNF-induced cell death: implications in inflammation and cancer. *Trends Mol Med* 24: 49–65. doi:10.1016/j.molmed.2017.11.002

Annibaldi A, Wicky John S, Vanden Berghe T, Swatek KN, Ruan J, Liccardi G, Bianchi K, Elliott PR, Choi SM, Van Coillie S, et al. 2018. Ubiquitin-mediated regulation of RIPK1 kinase activity independent of IKK and MK2. *Mol Cell* 69: 566–580.e5. doi:10.1016/j.molcel.2018.01.027

Balkwill F. 2009. Tumour necrosis factor and cancer. *Nat Rev Cancer* 9: 361–371. doi:10.1038/nrc2628

Balkwill FR, Mantovani A. 2012. Cancer-related inflammation: common themes and therapeutic opportunities. *Semin Cancer Biol* 22: 33–40. doi:10.1016/j.semcancer.2011.12.005

Barnhart BC, Legembre P, Pietras E, Bubici C, Franzoso G, Peter ME. 2004. CD95 ligand induces motility and invasiveness of apoptosis-resistant tumor cells. *EMBO J* 23: 3175–3185. doi:10.1038/sj.emboj.7600325

Bathon JM, Martin RW, Fleischmann RM, Tesser JR, Schiff MH, Keystone EC, Genovese MC, Wasko MC, Moreland LW, Weaver AL, et al. 2000. A comparison of etanercept and methotrexate in patients with early rheumatoid arthritis. *N Engl J Med* 343: 1586–1593. doi:10.1056/NEJM200011303432201

Berger SB, Kasparcova V, Hoffman S, Swift B, Dare L, Schaeffer M, Capriotti C, Cook M, Finger J, Hughes-Earle A, et al. 2014. Cutting edge: RIP1 kinase activity is dispensable for normal development but is a key regulator of inflammation in SHARPIN-deficient mice. *J Immunol* 192: 5476–5480. doi:10.4049/jimmunol.1400499

Beug ST, Tang VA, LaCasse EC, Cheung HH, Beauregard CE, Brun J, Nuyens JP, Earl N, St-Jean M, Holbrook J, et

Cite this article as *Cold Spring Harb Perspect Biol* doi: 10.1101/cshperspect.a036384

al. 2014. Smac mimetics and innate immune stimuli synergize to promote tumor death. *Nat Biotechnol* **32:** 182–190. doi:10.1038/nbt.2806

Beug ST, Beauregard CE, Healy C, Sanda T, St-Jean M, Chabot J, Walker DE, Mohan A, Earl N, Lun X, et al. 2017. Smac mimetics synergize with immune checkpoint inhibitors to promote tumour immunity against glioblastoma. *Nat Commun* **8:** 14278. doi:10.1038/ncomms14278

Beutler BA. 1999. The role of tumor necrosis factor in health and disease. *J Rheumatol Suppl* **57:** 16–21.

Brennan FM, Chantry D, Jackson A, Maini R, Feldmann M. 1989. Inhibitory effect of TNFα antibodies on synovial cell interleukin-1 production in rheumatoid arthritis. *Lancet* **334:** 244–247. doi:10.1016/S0140-6736(89)90430-3

Brenner D, Blaser H, Mak TW. 2015. Regulation of tumour necrosis factor signalling: live or let die. *Nat Rev Immunol* **15:** 362–374. doi:10.1038/nri3834

Brumatti G, Ma C, Lalaoui N, Nguyen NY, Navarro M, Tanzer MC, Richmond J, Ghisi M, Salmon JM, Silke N, et al. 2016. The caspase-8 inhibitor emricasan combines with the SMAC mimetic birinapant to induce necroptosis and treat acute myeloid leukemia. *Sci Transl Med* **8:** 339ra369. doi:10.1126/scitranslmed.aad3099

Brunner T, Mogil RJ, LaFace D, Yoo NJ, Mahboubi A, Echeverri F, Martin SJ, Force WR, Lynch DH, Ware CF, et al. 1995. Cell-autonomous Fas (CD95)/Fas-ligand interaction mediates activation-induced apoptosis in T-cell hybridomas. *Nature* **373:** 441–444. doi:10.1038/373441a0

Carswell EA, Old LJ, Kassel RL, Green S, Fiore N, Williamson B. 1975. An endotoxin-induced serum factor that causes necrosis of tumors. *Proc Natl Acad Sci* **72:** 3666–3670. doi:10.1073/pnas.72.9.3666

Chen L, Park SM, Tumanov AV, Hau A, Sawada K, Feig C, Turner JR, Fu YX, Romero IL, Lengyel E, et al. 2010. CD95 promotes tumour growth. *Nature* **465:** 492–496. doi:10.1038/nature09075

Chinnaiyan AM, O'Rourke K, Tewari M, Dixit VM. 1995. FADD, a novel death domain-containing protein, interacts with the death domain of Fas and initiates apoptosis. *Cell* **81:** 505–512. doi:10.1016/0092-8674(95)90071-3

Chinnaiyan AM, O'Rourke K, Yu GL, Lyons RH, Garg M, Duan DR, Xing L, Gentz R, Ni J, Dixit VM. 1996. Signal transduction by DR3, a death domain-containing receptor related to TNFR-1 and CD95. *Science* **274:** 990–992. doi:10.1126/science.274.5289.990

Cho JH, Feldman M. 2015. Heterogeneity of autoimmune diseases: pathophysiologic insights from genetics and implications for new therapies. *Nat Med* **21:** 730–738. doi:10.1038/nm.3897

Chu CQ, Field M, Feldmann M, Maini RN. 1991. Localization of tumor necrosis factor α in synovial tissues and at the cartilage–pannus junction in patients with rheumatoid arthritis. *Arthritis Rheum* **34:** 1125–1132. doi:10.1002/art.1780340908

Coley WB. 1893. The treatment of malignant tumors by repeated inoculations of erysipelas: with a report of ten original cases. 1893. *Clin Orthop Relat Res* **262:** 3–11.

Croft M, Siegel RM. 2017. Beyond TNF: TNF superfamily cytokines as targets for the treatment of rheumatic diseases. *Nat Rev Rheumatol* **13:** 217–233. doi:10.1038/nrrheum.2017.22

Cuchet-Lourenco D, Eletto D, Wu C, Plagnol V, Papapietro O, Curtis J, Ceron-Gutierrez L, Bacon CM, Hackett S, Alsaleem B, et al. 2018. Biallelic *RIPK1* mutations in humans cause severe immunodeficiency, arthritis, and intestinal inflammation. *Science* **361:** 810–813. doi:10.1126/science.aar2641

Dannappel M, Vlantis K, Kumari S, Polykratis A, Kim C, Wachsmuth L, Eftychi C, Lin J, Corona T, Hermance N, et al. 2014. RIPK1 maintains epithelial homeostasis by inhibiting apoptosis and necroptosis. *Nature* **513:** 90–94. doi:10.1038/nature13608

Deleuran BW, Chu CQ, Field M, Brennan FM, Mitchell T, Feldmann M, Maini RN. 1992. Localization of tumor necrosis factor receptors in the synovial tissue and cartilage-pannus junction in patients with rheumatoid arthritis. Implications for local actions of tumor necrosis factor α. *Arthritis Rheum* **35:** 1170–1178. doi:10.1002/art.1780351009

Dhein J, Walczak H, Bäumler C, Debatin KM, Krammer PH. 1995. Autocrine T-cell suicide mediated by APO-1/(Fas/CD95). *Nature* **373:** 438–441. doi:10.1038/373438a0

Dondelinger Y, Jouan-Lanhouet S, Divert T, Theatre E, Bertin J, Gough PJ, Giansanti P, Heck AJ, Dejardin E, Vandenabeele P, et al. 2015. NF-κB-independent role of IKKα/IKKβ in preventing RIPK1 kinase-dependent apoptotic and necroptotic cell death during TNF signaling. *Mol Cell* **60:** 63–76. doi:10.1016/j.molcel.2015.07.032

Dondelinger Y, Delanghe T, Priem D, Wynosky-Dolfi MA, Sorobetea D, Rojas-Rivera D, Giansanti P, Roelandt R, Gropengiesser J, Ruckdeschel K, et al. 2019. Serine 25 phosphorylation inhibits RIPK1 kinase-dependent cell death in models of infection and inflammation. *Nat Commun* **10:** 1729. doi:10.1038/s41467-019-09690-0

Draber P, Kupka S, Reichert M, Draberova H, Lafont E, de Miguel D, Spilgies L, Surinova S, Taraborrelli L, Hartwig T, et al. 2015. LUBAC-recruited CYLD and A20 regulate gene activation and cell death by exerting opposing effects on linear ubiquitin in signaling complexes. *Cell Rep* **13:** 2258–2272. doi:10.1016/j.celrep.2015.11.009

Dynek JN, Goncharov T, Dueber EC, Fedorova AV, Izrael-Tomasevic A, Phu L, Helgason E, Fairbrother WJ, Deshayes K, Kirkpatrick DS, et al. 2010. c-IAP1 and UbcH5 promote K11-linked polyubiquitination of RIP1 in TNF signalling. *EMBO J* **29:** 4198–4209. doi:10.1038/emboj.2010.300

Egberts JH, Cloosters V, Noack A, Schniewind B, Thon L, Klose S, Kettler B, von Forstner C, Kneitz C, Tepel J, et al. 2008. Anti-tumor necrosis factor therapy inhibits pancreatic tumor growth and metastasis. *Cancer Res* **68:** 1443–1450. doi:10.1158/0008-5472.CAN-07-5704

Elliott MJ, Maini RN, Feldmann M, Long-Fox A, Charles P, Katsikis P, Brennan FM, Walker J, Bijl H, Ghrayeb J, et al. 1993. Treatment of rheumatoid arthritis with chimeric monoclonal antibodies to tumor necrosis factor α. *Arthritis Rheum* **36:** 1681–1690. doi:10.1002/art.1780361206

Ermolaeva MA, Michallet MC, Papadopoulou N, Utermöhlen O, Kranidioti K, Kollias G, Tschopp J, Pasparakis M. 2008. Function of TRADD in tumor necrosis factor receptor 1 signaling and in TRIF-dependent inflammatory responses. *Nat Immunol* **9:** 1037–1046. doi:10.1038/ni.1638

Feoktistova M, Geserick P, Kellert B, Dimitrova DP, Langlais C, Hupe M, Cain K, MacFarlane M, Häcker G, Leverkus M. 2011. cIAPs block ripoptosome formation, a RIP1/caspase-8 containing intracellular cell death complex differentially regulated by cFLIP isoforms. *Mol Cell* **43:** 449–463. doi:10.1016/j.molcel.2011.06.011

Fritsch M, Günther SD, Schwarzer R, Albert MC, Schorn F, Werthenbach JP, Schiffmann LM, Stair N, Stocks H, Seeger JM, et al. 2019. Caspase-8 is the molecular switch for apoptosis, necroptosis and pyroptosis. *Nature* **575:** 683–687. doi:10.1038/s41586-019-1770-6

Fusco F, Pescatore A, Bal E, Ghoul A, Paciolla M, Lioi MB, D'Urso M, Rabia SH, Bodemer C, Bonnefont JP, et al. 2008. Alterations of the *IKBKG* locus and diseases: an update and a report of 13 novel mutations. *Hum Mutat* **29:** 595–604. doi:10.1002/humu.20739

Gerlach B, Cordier SM, Schmukle AC, Emmerich CH, Rieser E, Haas TL, Webb AI, Rickard JA, Anderton H, Wong WW, et al. 2011. Linear ubiquitination prevents inflammation and regulates immune signalling. *Nature* **471:** 591–596. doi:10.1038/nature09816

Gray PW, Aggarwal BB, Benton CV, Bringman TS, Henzel WJ, Jarrett JA, Leung DW, Moffat B, Ng P, Svedersky LP, et al. 1984. Cloning and expression of cDNA for human lymphotoxin, a lymphokine with tumour necrosis activity. *Nature* **312:** 721–724. doi:10.1038/312721a0

Haas TL, Emmerich CH, Gerlach B, Schmukle AC, Cordier SM, Rieser E, Feltham R, Vince J, Warnken U, Wenger T, et al. 2009. Recruitment of the linear ubiquitin chain assembly complex stabilizes the TNF-R1 signaling complex and is required for TNF-mediated gene induction. *Mol Cell* **36:** 831–844. doi:10.1016/j.molcel.2009.10.013

Hartwig T, Montinaro A, von Karstedt S, Sevko A, Surinova S, Chakravarthy A, Taraborrelli L, Draber P, Lafont E, Arce Vargas F, et al. 2017. The TRAIL-induced cancer secretome promotes a tumor-supportive immune microenvironment via CCR2. *Mol Cell* **65:** 730–742.e5. doi:10.1016/j.molcel.2017.01.021

Henry CM, Martin SJ. 2017. Caspase-8 acts in a non-enzymatic role as a scaffold for assembly of a pro-inflammatory "FADDosome" complex upon TRAIL stimulation. *Mol Cell* **65:** 715–729.e5. doi:10.1016/j.molcel.2017.01.022

Holler N, Tardivel A, Kovacsovics-Bankowski M, Hertig S, Gaide O, Martinon F, Tinel A, Deperthes D, Calderara S, Schulthess T, et al. 2003. Two adjacent trimeric Fas ligands are required for Fas signaling and formation of a death-inducing signaling complex. *Mol Cell Biol* **23:** 1428–1440. doi:10.1128/MCB.23.4.1428-1440.2003

Hoogwater FJ, Nijkamp MW, Smakman N, Steller EJ, Emmink BL, Westendorp BF, Raats DA, Sprick MR, Schaefer U, Van Houdt WJ, et al. 2010. Oncogenic K-Ras turns death receptors into metastasis-promoting receptors in human and mouse colorectal cancer cells. *Gastroenterology* **138:** 2357–2367. doi:10.1053/j.gastro.2010.02.046

Hsu H, Huang J, Shu HB, Baichwal V, Goeddel DV. 1996. TNF-dependent recruitment of the protein kinase RIP to the TNF receptor-1 signaling complex. *Immunity* **4:** 387–396. doi:10.1016/S1074-7613(00)80252-6

Ikeda T, Hirata S, Fukushima S, Matsunaga Y, Ito T, Uchino M, Nishimura Y, Senju S. 2010. Dual effects of TRAIL in suppression of autoimmunity: the inhibition of Th1 cells and the promotion of regulatory T cells. *J Immunol* **185:** 5259–5267. doi:10.4049/jimmunol.0902797

Ikeda F, Deribe YL, Skånland SS, Stieglitz B, Grabbe C, Franz-Wachtel M, van Wijk SJ, Goswami P, Nagy V, Terzic J, et al. 2011. SHARPIN forms a linear ubiquitin ligase complex regulating NF-κB activity and apoptosis. *Nature* **471:** 637–641. doi:10.1038/nature09814

Ising C, Venegas C, Zhang S, Scheiblich H, Schmidt SV, Vieira-Saecker A, Schwartz S, Albasset S, McManus RM, Tejera D, et al. 2019. NLRP3 inflammasome activation drives tau pathology. *Nature* **575:** 669–673. doi:10.1038/s41586-019-1769-z

Itoh N, Nagata S. 1993. A novel protein domain required for apoptosis. Mutational analysis of human Fas antigen. *J Biol Chem* **268:** 10932–10937.

Itoh N, Yonehara S, Ishii A, Yonehara M, Mizushima S, Sameshima M, Hase A, Seto Y, Nagata S. 1991. The polypeptide encoded by the cDNA for human cell surface antigen Fas can mediate apoptosis. *Cell* **66:** 233–243. doi:10.1016/0092-8674(91)90614-5

Jaco I, Annibaldi A, Lalaoui N, Wilson R, Tenev T, Laurien L, Kim C, Jamal K, Wicky John S, Liccardi G, et al. 2017. MK2 phosphorylates RIPK1 to prevent TNF-induced cell death. *Mol Cell* **66:** 698–710.e5. doi:10.1016/j.molcel.2017.05.003

Jin Z, Li Y, Pitti R, Lawrence D, Pham VC, Lill JR, Ashkenazi A. 2009. Cullin3-based polyubiquitination and p62-dependent aggregation of caspase-8 mediate extrinsic apoptosis signaling. *Cell* **137:** 721–735. doi:10.1016/j.cell.2009.03.015

Ju ST, Panka DJ, Cui H, Ettinger R, El-Khatib M, Sherr DH, Stanger BZ, Marshak-Rothstein A. 1995. Fas(CD95)/FasL interactions required for programmed cell death after T-cell activation. *Nature* **373:** 444–448. doi:10.1038/373444a0

Kalliolias GD, Ivashkiv LB. 2016. TNF biology, pathogenic mechanisms and emerging therapeutic strategies. *Nat Rev Rheumatol* **12:** 49–62. doi:10.1038/nrrheum.2015.169

Kelliher MA, Grimm S, Ishida Y, Kuo F, Stanger BZ, Leder P. 1998. The death domain kinase RIP mediates the TNF-induced NF-κB signal. *Immunity* **8:** 297–303. doi:10.1016/S1074-7613(00)80535-X

Kleber S, Sancho-Martinez I, Wiestler B, Beisel A, Gieffers C, Hill O, Thiemann M, Mueller W, Sykora J, Kuhn A, et al. 2008. Yes and PI3K bind CD95 to signal invasion of glioblastoma. *Cancer Cell* **13:** 235–248. doi:10.1016/j.ccr.2008.02.003

Kulbe H, Thompson R, Wilson JL, Robinson S, Hagemann T, Fatah R, Gould D, Ayhan A, Balkwill F. 2007. The inflammatory cytokine tumor necrosis factor-α generates an autocrine tumor-promoting network in epithelial ovarian cancer cells. *Cancer Res* **67:** 585–592. doi:10.1158/0008-5472.CAN-06-2941

Kumari S, Redouane Y, Lopez-Mosqueda J, Shiraishi R, Romanowska M, Lutzmayer S, Kuiper J, Martinez C, Dikic I, Pasparakis M, et al. 2014. Sharpin prevents skin inflammation by inhibiting TNFR1-induced keratinocyte apoptosis. *eLife* **3:** e03422. doi:10.7554/eLife.03422

Lafont E, Kantari-Mimoun C, Draber P, De Miguel D, Hartwig T, Reichert M, Kupka S, Shimizu Y, Taraborrelli L, Spit M, et al. 2017. The linear ubiquitin chain assembly complex regulates TRAIL-induced gene activation and

Cite this article as *Cold Spring Harb Perspect Biol* doi: 10.1101/cshperspect.a036384

cell death. *EMBO J* **36:** 1147–1166. doi:10.15252/embj .201695699

Lafont E, Draber P, Rieser E, Reichert M, Kupka S, de Miguel D, Draberova H, von Mässenhausen A, Bhamra A, Henderson S, et al. 2018. TBK1 and IKKε prevent TNF-induced cell death by RIPK1 phosphorylation. *Nat Cell Biol* **20:** 1389–1399. doi:10.1038/s41556-018-0229-6

Lakins MA, Ghorani E, Munir H, Martins CP, Shields JD. 2018. Cancer-associated fibroblasts induce antigen-specific deletion of CD8⁺ T cells to protect tumour cells. *Nat Commun* **9:** 948. doi:10.1038/s41467-018-03347-0

Lalaoui N, Hänggi K, Brumatti G, Chau D, Nguyen NY, Vasilikos L, Spilgies LM, Heckmann DA, Ma C, Ghisi M, et al. 2016. Targeting p38 or MK2 enhances the anti-leukemic activity of Smac-mimetics. *Cancer Cell* **29:** 145–158. doi:10.1016/j.ccell.2016.01.006

Lehnert C, Weiswange M, Jeremias I, Bayer C, Grunert M, Debatin KM, Strauss G. 2014. TRAIL-receptor costimulation inhibits proximal TCR signaling and suppresses human T cell activation and proliferation. *J Immunol* **193:** 4021–4031. doi:10.4049/jimmunol.1303242

Lenercept Multiple Sclerosis Study Group. 1999. TNF neutralization in MS: results of a randomized, placebo-controlled multicenter study. The Lenercept Multiple Sclerosis Study Group and The University of British Columbia MS/MRI Analysis Group. *Neurology* **53:** 457–465. doi:10.1212/WNL.53.3.457

Leverkus M, Walczak H, McLellan A, Fries HW, Terbeck G, Bröcker EB, Kämpgen E. 2000. Maturation of dendritic cells leads to up-regulation of cellular FLICE-inhibitory protein and concomitant down-regulation of death ligand-mediated apoptosis. *Blood* **96:** 2628–2631. doi:10 .1182/blood.V96.7.2628

Loetscher H, Pan YC, Lahm HW, Gentz R, Brockhaus M, Tabuchi H, Lesslauer W. 1990. Molecular cloning and expression of the human 55 kd tumor necrosis factor receptor. *Cell* **61:** 351–359. doi:10.1016/0092-8674(90) 90815-V

Lopetuso LR, Gerardi V, Papa V, Scaldaferri F, Rapaccini GL, Gasbarrini A, Papa A. 2017. Can we predict the efficacy of anti-TNF-α agents? *Int J Mol Sci* **18:** 1973. doi:10.3390/ ijms18091973

Lovell DJ, Giannini EH, Reiff A, Cawkwell GD, Silverman ED, Nocton JJ, Stein LD, Gedalia A, Ilowite NT, Wallace CA, et al. 2000. Etanercept in children with polyarticular juvenile rheumatoid arthritis. *N Engl J Med* **342:** 763–769. doi:10.1056/NEJM200003163421103

Mantovani A, Allavena P, Sica A, Balkwill F. 2008. Cancer-related inflammation. *Nature* **454:** 436–444. doi:10.1038/ nature07205

Marmenout A, Fransen L, Tavernier J, Van der Heyden J, Tizard R, Kawashima E, Shaw A, Johnson MJ, Semon D, Muller R, et al. 1985. Molecular cloning and expression of human tumor necrosis factor and comparison with mouse tumor necrosis factor. *Eur J Biochem* **152:** 515–522. doi:10.1111/j.1432-1033.1985.tb09226.x

Marsters SA, Sheridan JP, Donahue CJ, Pitti RM, Gray CL, Goddard AD, Bauer KD, Ashkenazi A. 1996. Apo-3, a new member of the tumor necrosis factor receptor family, contains a death domain and activates apoptosis and NF-κB. *Curr Biol* **6:** 1669–1676. doi:10.1016/S0960-9822(02) 70791-4

Medler J, Wajant H. 2019. Tumor necrosis factor receptor-2 (TNFR2): an overview of an emerging drug target. *Expert Opin Ther Targets* **23:** 295–307. doi:10.1080/14728222 .2019.1586886

Micheau O, Tschopp J. 2003. Induction of TNF receptor I-mediated apoptosis via two sequential signaling complexes. *Cell* **114:** 181–190. doi:10.1016/S0092-8674(03) 00521-X

Monaco C, Nanchahal J, Taylor P, Feldmann M. 2015. Anti-TNF therapy: past, present and future. *Int Immunol* **27:** 55–62. doi:10.1093/intimm/dxu102

Moore RJ, Owens DM, Stamp G, Arnott C, Burke F, East N, Holdsworth H, Turner L, Rollins B, Pasparakis M, et al. 1999. Mice deficient in tumor necrosis factor-α are resistant to skin carcinogenesis. *Nat Med* **5:** 828–831. doi:10 .1038/10552

Moreland LW, Schiff MH, Baumgartner SW, Tindall EA, Fleischmann RM, Bulpitt KJ, Weaver AL, Keystone EC, Furst DE, Mease PJ, et al. 1999. Etanercept therapy in rheumatoid arthritis: a randomized, controlled trial. *Ann Intern Med* **130:** 478–486. doi:10.7326/0003-4819-130-6-199903160-00004

Motz GT, Santoro SP, Wang LP, Garrabrant T, Lastra RR, Hagemann IS, Lal P, Feldman MD, Benencia F, Coukos G. 2014. Tumor endothelium FasL establishes a selective immune barrier promoting tolerance in tumors. *Nat Med* **20:** 607–615. doi:10.1038/nm.3541

Nauts HC, Swift WE, Coley BL. 1946. The treatment of malignant tumors by bacterial toxins as developed by the late William B. Coley, M.D., reviewed in the light of modern research. *Cancer Res* **6:** 205–216.

Neuwirth MG, Song Y, Sinnamon AJ, Fraker DL, Zager JS, Karakousis GC. 2017. Isolated limb perfusion and infusion for extremity soft tissue sarcoma: a contemporary systematic review and meta-analysis. *Ann Surg Oncol* **24:** 3803–3810. doi:10.1245/s10434-017-6109-7

Newton K, Dugger DL, Maltzman A, Greve JM, Hedehus M, Martin-McNulty B, Carano RA, Cao TC, van Bruggen N, Bernstein L, et al. 2016. RIPK3 deficiency or catalytically inactive RIPK1 provides greater benefit than MLKL deficiency in mouse models of inflammation and tissue injury. *Cell Death Differ* **23:** 1565–1576. doi:10.1038/cdd.2016 .46

Newton K, Wickliffe KE, Maltzman A, Dugger DL, Reja R, Zhang Y, Roose-Girma M, Modrusan Z, Sagolla MS, Webster JD, et al. 2019. Activity of caspase-8 determines plasticity between cell death pathways. *Nature* **575:** 679–682. doi:10.1038/s41586-019-1752-8

Oda H, Beck DB, Kuehn HS, Sampaio Moura N, Hoffmann P, Ibarra M, Stoddard J, Tsai WL, Gutierrez-Cruz G, Gadina M, et al. 2019. Second case of HOIP deficiency expands clinical features and defines inflammatory transcriptome regulated by LUBAC. *Front Immunol* **10:** 479. doi:10.3389/fimmu.2019.00479

Oehm A, Behrmann I, Falk W, Pawlita M, Maier G, Klas C, Li-Weber M, Richards S, Dhein J, Trauth BC, et al. 1992. Purification and molecular cloning of the APO-1 cell surface antigen, a member of the tumor necrosis factor/nerve growth factor receptor superfamily. Sequence identity with the Fas antigen. *J Biol Chem* **267:** 10709–10715.

Ogasawara J, Watanabe-Fukunaga R, Adachi M, Matsuzawa A, Kasugai T, Kitamura Y, Itoh N, Suda T, Nagata S. 1993.

Lethal effect of the anti-Fas antibody in mice. *Nature* **364:** 806–809. doi:10.1038/364806a0

Ori D, Kato H, Sanjo H, Tartey S, Mino T, Akira S, Takeuchi O. 2013. Essential roles of K63-linked polyubiquitin-binding proteins TAB2 and TAB3 in B cell activation via MAPKs. *J Immunol* **190:** 4037–4045. doi:10.4049/jimmunol.1300173

Owen-Schaub LB, Meterissian S, Ford RJ. 1993. Fas/APO-1 expression and function on malignant cells of hematologic and nonhematologic origin. *J Immunother Emphasis Tumor Immunol* **14:** 234–241. doi:10.1097/00002371-199310000-00011

Pan G, Ni J, Wei YF, Yu G, Gentz R, Dixit VM. 1997a. An antagonist decoy receptor and a death domain-containing receptor for TRAIL. *Science* **277:** 815–818. doi:10.1126/science.277.5327.815

Pan G, O'Rourke K, Chinnaiyan AM, Gentz R, Ebner R, Ni J, Dixit VM. 1997b. The receptor for the cytotoxic ligand TRAIL. *Science* **276:** 111–113. doi:10.1126/science.276.5309.111

Pasparakis M, Vandenabeele P. 2015. Necroptosis and its role in inflammation. *Nature* **517:** 311–320. doi:10.1038/nature14191

Patel S, Webster JD, Varfolomeev E, Kwon YC, Cheng JH, Zhang J, Dugger DL, Wickliffe KE, Maltzman A, Sujatha-Bhaskar S, et al. 2020. RIP1 inhibition blocks inflammatory diseases but not tumor growth or metastases. *Cell Death Differ* **27:** 161–175. doi: 10.1038/s41418-019-0347-0.

Peltzer N, Walczak H. 2019. Cell death and inflammation—a vital but dangerous liaison. *Trends Immunol* **40:** 387–402. doi:10.1016/j.it.2019.03.006

Peltzer N, Darding M, Walczak H. 2016. Holding RIPK1 on the ubiquitin leash in TNFR1 signaling. *Trends Cell Biol* **26:** 445–461. doi:10.1016/j.tcb.2016.01.006

Peltzer N, Darding M, Montinaro A, Draber P, Draberova H, Kupka S, Rieser E, Fisher A, Hutchinson C, Taraborrelli L, et al. 2018. LUBAC is essential for embryogenesis by preventing cell death and enabling haematopoiesis. *Nature* **557:** 112–117. doi:10.1038/s41586-018-0064-8

Pennica D, Nedwin GE, Hayflick JS, Seeburg PH, Derynck R, Palladino MA, Kohr WJ, Aggarwal BB, Goeddel DV. 1984. Human tumour necrosis factor: precursor structure, expression and homology to lymphotoxin. *Nature* **312:** 724–729. doi:10.1038/312724a0

Pillai MR, Collison LW, Wang X, Finkelstein D, Rehg JE, Boyd K, Szymczak-Workman AL, Doggett T, Griffith TS, Ferguson TA, et al. 2011. The plasticity of regulatory T cell function. *J Immunol* **187:** 4987–4997. doi:10.4049/jimmunol.1102173

Pitti RM, Marsters SA, Ruppert S, Donahue CJ, Moore A, Ashkenazi A. 1996. Induction of apoptosis by Apo-2 ligand, a new member of the tumor necrosis factor cytokine family. *J Biol Chem* **271:** 12687–12690. doi:10.1074/jbc.271.22.12687

Rahighi S, Ikeda F, Kawasaki M, Akutsu M, Suzuki N, Kato R, Kensche T, Uejima T, Bloor S, Komander D, et al. 2009. Specific recognition of linear ubiquitin chains by NEMO is important for NF-κB activation. *Cell* **136:** 1098–1109. doi:10.1016/j.cell.2009.03.007

Richard AC, Ferdinand JR, Meylan F, Hayes ET, Gabay O, Siegel RM. 2015. The TNF-family cytokine TL1A: from lymphocyte costimulator to disease co-conspirator. *J Leukoc Biol* **98:** 333–345. doi:10.1189/jlb.3RI0315-095R

Rickard JA, Anderton H, Etemadi N, Nachbur U, Darding M, Peltzer N, Lalaoui N, Lawlor KE, Vanyai H, Hall C, et al. 2014. TNFR1-dependent cell death drives inflammation in *Sharpin*-deficient mice. *eLife* **3:** e03464. doi:10.7554/eLife.03464

Robaye B, Mosselmans R, Fiers W, Dumont JE, Galand P. 1991. Tumor necrosis factor induces apoptosis (programmed cell death) in normal endothelial cells in vitro. *Am J Pathol* **138:** 447–453.

Roberts AI, Devadas S, Zhang X, Zhang L, Keegan A, Greeneltch K, Solomon J, Wei L, Das J, Sun E, et al. 2003. The role of activation-induced cell death in the differentiation of T-helper-cell subsets. *Immunol Res* **28:** 285–293. doi:10.1385/IR:28:3:285

Rothe M, Pan MG, Henzel WJ, Ayres TM, Goeddel DV. 1995. The TNFR2–TRAF signaling complex contains two novel proteins related to baculoviral inhibitor of apoptosis proteins. *Cell* **83:** 1243–1252. doi:10.1016/0092-8674(95)90149-3

Schall TJ, Lewis M, Koller KJ, Lee A, Rice GC, Wong GH, Gatanaga T, Granger GA, Lentz R, Raab H, et al. 1990. Molecular cloning and expression of a receptor for human tumor necrosis factor. *Cell* **61:** 361–370. doi:10.1016/0092-8674(90)90816-W

Screaton GR, Mongkolsapaya J, Xu XN, Cowper AE, McMichael AJ, Bell JI. 1997a. TRICK2, a new alternatively spliced receptor that transduces the cytotoxic signal from TRAIL. *Curr Biol* **7:** 693–696. doi:10.1016/S0960-9822(06)00297-1

Screaton GR, Xu XN, Olsen AL, Cowper AE, Tan R, McMichael AJ, Bell JI. 1997b. LARD: a new lymphoid-specific death domain containing receptor regulated by alternative pre-mRNA splicing. *Proc Natl Acad Sci* **94:** 4615–4619. doi:10.1073/pnas.94.9.4615

Seymour RE, Hasham MG, Cox GA, Shultz LD, Hogenesch H, Roopenian DC, Sundberg JP. 2007. Spontaneous mutations in the mouse *Sharpin* gene result in multiorgan inflammation, immune system dysregulation and dermatitis. *Genes Immun* **8:** 416–421. doi:10.1038/sj.gene.6364403

Sheridan JP, Marsters SA, Pitti RM, Gurney A, Skubatch M, Baldwin D, Ramakrishnan L, Gray CL, Baker K, Wood WI, et al. 1997. Control of TRAIL-induced apoptosis by a family of signaling and decoy receptors. *Science* **277:** 818–821. doi:10.1126/science.277.5327.818

Shirley S, Morizot A, Micheau O. 2011. Regulating TRAIL receptor-induced cell death at the membrane: a deadly discussion. *Recent Pat Anticancer Drug Discov* **6:** 311–323. doi:10.2174/157489211796957757

Snyder AG, Hubbard NW, Messmer MN, Kofman SB, Hagan CE, Orozco SL, Chiang K, Daniels BP, Baker D, Oberst A. 2019. Intratumoral activation of the necroptotic pathway components RIPK1 and RIPK3 potentiates antitumor immunity. *Sci Immunol* **4:** eaaw2004.

Suganuma M, Okabe S, Marino MW, Sakai A, Sueoka E, Fujiki H. 1999. Essential role of tumor necrosis factor α (TNF-α) in tumor promotion as revealed by TNF-α-deficient mice. *Cancer Res* **59:** 4516–4518.

Taraborrelli L, Peltzer N, Montinaro A, Kupka S, Rieser E, Hartwig T, Sarr A, Darding M, Draber P, Haas TL, et al.

2018. LUBAC prevents lethal dermatitis by inhibiting cell death induced by TNF, TRAIL and CD95L. *Nat Commun* **9:** 3910. doi:10.1038/s41467-018-06155-8

Tenev T, Bianchi K, Darding M, Broemer M, Langlais C, Wallberg F, Zachariou A, Lopez J, MacFarlane M, Cain K, et al. 2011. The ripoptosome, a signaling platform that assembles in response to genotoxic stress and loss of IAPs. *Mol Cell* **43:** 432–448. doi:10.1016/j.molcel.2011.06.006

Ting AT, Bertrand MJ. 2016. More to life than NF-κB in TNFR1 signaling. *Trends Immunol* **37:** 535–545. doi:10.1016/j.it.2016.06.002

Tracey KJ, Lowry SF, Cerami A. 1988. Cachetin/TNF-α in septic shock and septic adult respiratory distress syndrome. *Am Rev Respir Dis* **138:** 1377–1379. doi:10.1164/ajrccm/138.6.1377

Trauth BC, Klas C, Peters AM, Matzku S, Moller P, Falk W, Debatin KM, Krammer PH. 1989. Monoclonal antibody-mediated tumor regression by induction of apoptosis. *Science* **245:** 301–305. doi:10.1126/science.2787530

Trauzold A, Siegmund D, Schniewind B, Sipos B, Egberts J, Zorenkov D, Emme D, Röder C, Kalthoff H, Wajant H. 2006. TRAIL promotes metastasis of human pancreatic ductal adenocarcinoma. *Oncogene* **25:** 7434–7439. doi:10.1038/sj.onc.1209719

Valley CC, Lewis AK, Mudaliar DJ, Perlmutter JD, Braun AR, Karim CB, Thomas DD, Brody JR, Sachs JN. 2012. Tumor necrosis factor-related apoptosis-inducing ligand (TRAIL) induces death receptor 5 networks that are highly organized. *J Biol Chem* **287:** 21265–21278. doi:10.1074/jbc.M111.306480

Vandenabeele P, Vanden Berghe T, Festjens N. 2006. Caspase inhibitors promote alternative cell death pathways. *Sci STKE* **2006:** pe44. doi:10.1126/stke.3582006pe44

Vanden Berghe T, Denecker G, Brouckaert G, Vadimovisch Krysko D, D'Herde K, Vandenabeele P. 2004. More than one way to die: methods to determine TNF-induced apoptosis and necrosis. *Methods Mol Med* **98:** 101–126.

Vlantis K, Wullaert A, Polykratis A, Kondylis V, Dannappel M, Schwarzer R, Welz P, Corona T, Walczak H, Weih F, et al. 2016. NEMO prevents RIP kinase 1-mediated epithelial cell death and chronic intestinal inflammation by NF-κB-dependent and -independent functions. *Immunity* **44:** 553–567. doi:10.1016/j.immuni.2016.02.020

von Karstedt S, Conti A, Nobis M, Montinaro A, Hartwig T, Lemke J, Legler K, Annewanter F, Campbell AD, Taraborrelli L, et al. 2015. Cancer cell-autonomous TRAIL-R signaling promotes KRAS-driven cancer progression, invasion, and metastasis. *Cancer Cell* **27:** 561–573. doi:10.1016/j.ccell.2015.02.014

von Karstedt S, Montinaro A, Walczak H. 2017. Exploring the TRAILs less travelled: TRAIL in cancer biology and therapy. *Nat Rev Cancer* **17:** 352–366. doi:10.1038/nrc.2017.28

Vredevoogd DW, Kuilman T, Ligtenberg MA, Boshuizen J, Stecker KE, de Bruijn B, Krijgsman O, Huang X, Kenski JCN, Lacroix R, et al. 2019. Augmenting immunotherapy impact by lowering tumor TNF cytotoxicity threshold. *Cell* **178:** 585–599.e15. doi:10.1016/j.cell.2019.06.014

Wajant H, Siegmund D. 2019. TNFR1 and TNFR2 in the control of the life and death balance of macrophages. *Front Cell Dev Biol* **7:** 91. doi:10.3389/fcell.2019.00091

Walczak H. 2013. Death receptor-ligand systems in cancer, cell death, and inflammation. *Cold Spring Harb Perspect Biol* **5:** a008698. doi:10.1101/cshperspect.a008698

Walczak H, Degli-Esposti MA, Johnson RS, Smolak PJ, Waugh JY, Boiani N, Timour MS, Gerhart MJ, Schooley KA, Smith CA, et al. 1997. TRAIL-R2: a novel apoptosis-mediating receptor for TRAIL. *EMBO J* **16:** 5386–5397. doi:10.1093/emboj/16.17.5386

Wang L, Du F, Wang X. 2008. TNF-α induces two distinct caspase-8 activation pathways. *Cell* **133:** 693–703. doi:10.1016/j.cell.2008.03.036

Watanabe-Fukunaga R, Brannan CI, Copeland NG, Jenkins NA, Nagata S. 1992. Lymphoproliferation disorder in mice explained by defects in Fas antigen that mediates apoptosis. *Nature* **356:** 314–317. doi:10.1038/356314a0

Weinblatt ME, Kremer JM, Bankhurst AD, Bulpitt KJ, Fleischmann RM, Fox RI, Jackson CG, Lange M, Burge DJ. 1999. A trial of etanercept, a recombinant tumor necrosis factor receptor:Fc fusion protein, in patients with rheumatoid arthritis receiving methotrexate. *N Engl J Med* **340:** 253–259. doi:10.1056/NEJM199901283400401

Welz PS, Wullaert A, Vlantis K, Kondylis V, Fernández-Majada V, Ermolaeva M, Kirsch P, Sterner-Kock A, van Loo G, Pasparakis M. 2011. FADD prevents RIP3-mediated epithelial cell necrosis and chronic intestinal inflammation. *Nature* **477:** 330–334. doi:10.1038/nature10273

Wick W, Fricke H, Junge K, Kobyakov G, Martens T, Heese O, Wiestler B, Schliesser MG, von Deimling A, Pichler J, et al. 2014. A phase II, randomized, study of weekly APG101+reirradiation versus reirradiation in progressive glioblastoma. *Clin Cancer Res* **20:** 6304–6313. doi:10.1158/1078-0432.CCR-14-0951-T

Wiley SR, Schooley K, Smolak PJ, Din WS, Huang CP, Nicholl JK, Sutherland GR, Smith TD, Rauch C, Smith CA, et al. 1995. Identification and characterization of a new member of the TNF family that induces apoptosis. *Immunity* **3:** 673–682. doi:10.1016/1074-7613(95)90057-8

Wong WW, Gentle IE, Nachbur U, Anderton H, Vaux DL, Silke J. 2010. RIPK1 is not essential for TNFR1-induced activation of NF-κB. *Cell Death Differ* **17:** 482–487. doi:10.1038/cdd.2009.178

Wu GS, Burns TF, McDonald ER III, Jiang W, Meng R, Krantz ID, Kao G, Gan DD, Zhou JY, Muschel R, et al. 1997. KILLER/DR5 is a DNA damage-inducible p53-regulated death receptor gene. *Nat Genet* **17:** 141–143. doi:10.1038/ng1097-141

Yatim N, Jusforgues-Saklani H, Orozco S, Schulz O, Barreira da Silva R, Reis e Sousa C, Green DR, Oberst A, Albert ML. 2015. RIPK1 and NF-κB signaling in dying cells determines cross-priming of CD8[+] T cells. *Science* **350:** 328–334.

Yatim N, Cullen S, Albert ML. 2017. Dying cells actively regulate adaptive immune responses. *Nat Rev Immunol* **17:** 262–275.

Yonehara S, Ishii A, Yonehara M. 1989. A cell-killing monoclonal antibody (anti-Fas) to a cell surface antigen co-downregulated with the receptor of tumor necrosis factor. *J Exp Med* **169:** 1747–1756. doi:10.1084/jem.169.5.1747

Zhang XR, Zhang LY, Devadas S, Li L, Keegan AD, Shi YF. 2003. Reciprocal expression of TRAIL and CD95L in Th1 and Th2 cells: role of apoptosis in T helper subset differ-

entiation. *Cell Death Differ* **10:** 203–210. doi:10.1038/sj
.cdd.4401138

Zhang DW, Shao J, Lin J, Zhang N, Lu BJ, Lin SC, Dong MQ, Han J. 2009. RIP3, an energy metabolism regulator that switches TNF-induced cell death from apoptosis to necrosis. *Science* **325:** 332–336. doi:10.1126/science.117 2308

Zhu J, de Tenbossche CGP, Cané S, Colau D, van Baren N, Lurquin C, Schmitt-Verhulst AM, Liljeström P, Uyttenhove C, Van den Eynde BJ. 2017. Resistance to cancer immunotherapy mediated by apoptosis of tumor-infil-

trating lymphocytes. *Nat Commun* **8:** 1404. doi:10.1038/s41467-017-00784-1

Zinngrebe J, Montinaro A, Peltzer N, Walczak H. 2014. Ubiquitin in the immune system. *EMBO Rep* **15:** 28–45. doi:10.1002/embr.201338025

Zins K, Abraham D, Sioud M, Aharinejad S. 2007. Colon cancer cell–derived tumor necrosis factor-α mediates the tumor growth-promoting response in macrophages by up-regulating the colony-stimulating factor-1 pathway. *Cancer Res* **67:** 1038–1045. doi:10.1158/0008-5472 .CAN-06-2295

Recent Insights on Inflammasomes, Gasdermin Pores, and Pyroptosis

Nathalia M. de Vasconcelos[1,2,4] and Mohamed Lamkanfi[1,3]

[1]Department of Internal Medicine and Pediatrics, Ghent University, B-9000 Ghent, Belgium

[2]VIB-UGhent Center for Inflammation Research, VIB, B-9052 Ghent, Belgium

[3]Janssen Immunosciences, World Without Disease Accelerator, Pharmaceutical Companies of Johnson & Johnson, B-2340 Beerse, Belgium

Correspondence: mlamkanf@its.jnj.com

Inflammasomes assemble in the cytosol of myeloid and epithelial cells on sensing of cellular stress and pathogen-associated molecular patterns and serve as scaffolds for recruitment and activation of inflammatory caspases. Inflammasomes play beneficial roles in host and immune responses against diverse pathogens but may also promote inflammatory tissue damage if uncontrolled. Gasdermin D (GSDMD) is a recently identified substrate of murine caspase-1 and caspase-11, and human caspases-1, -4, and -5 that mediates a regulated lytic cell death mode termed pyroptosis. Recent studies have identified pyroptosis as a critical inflammasome effector mechanism that controls inflammasome-dependent cytokine secretion and contributes to antimicrobial defense and inflammasome-mediated autoinflammatory diseases. Here, we review recent developments on inflammasome-associated effector functions with an emphasis on the emerging roles of gasdermin pores and pyroptosis.

CANONICAL AND NONCANONICAL INFLAMMASOME PATHWAYS

Although frequently providing protection against bacterial, viral, and fungal pathogens, inflammasome activation is often detrimental in a diverse range of (auto)inflammatory, autoimmune, metabolic, neurodegenerative, and malignant diseases (Mangan et al. 2018; Van Gorp et al. 2019; Voet et al. 2019).

Inflammasome pathways are traditionally divided into the canonical and noncanonical pathways, referring to which inflammatory caspase initiates signaling that critically contributes to inflammation and host–defense responses (Fig. 1). The canonical inflammasomes by definition assemble a multiprotein complex that activates caspase-1, and their signaling is initiated by engagement of cytosolic pattern recognition receptors (PRRs) mostly from the nucleotide-binding domain, leucine-rich repeat containing (NLR) protein family (Lamkanfi and Dixit 2014; Broz and Dixit 2016). NLRP3 scavenges the cytosol of myeloid and epithelial cells for a plethora of danger- and pathogen-associated molecular patterns (DAMPs and PAMPs); thus, it is speculated that this inflammasome senses cellular stress and imbalance. Other

---

[4]Present address: The Francis Crick Institute, London, United Kingdom.

Cite this article as *Cold Spring Harb Perspect Biol* doi: 10.1101/cshperspect.a036392

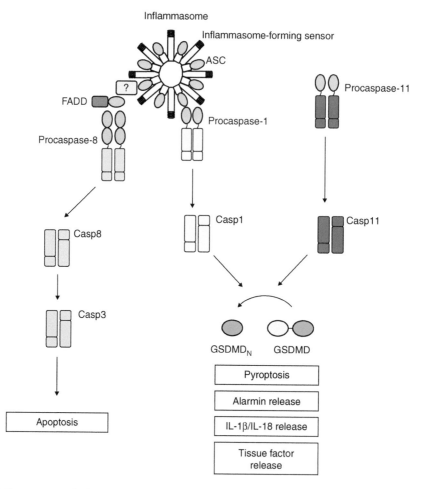

Figure 1. Inflammasome platforms activate caspase-1, -11, and -8 for multiple downstream effector functions. Intracellular lipopolysaccharide (LPS) precipitates the oligomerization of caspase-11, in the so-called noncanonical inflammasome (*right*), and its automaturation. Cytosolic presence of danger- and pathogen-associated molecular patterns (DAMPs or PAMPs) can also trigger the canonical inflammasome pathway, which recruits procaspase-1 for proximity-induced autoactivation (*left*). Caspase-1 and caspase-11 cleave the common substrate, gasdermin D (GSDMD), releasing an amino-terminal fragment (GSDMD$_N$), which initiates cell death by pyroptosis. This lytic cell death mode is associated with alarmin release to the extracellular space and discharge of tissue factor–containing microvesicles. Inflammasomes also mediate the activation and secretion of the proinflammatory cytokines interleukin (IL)-1β and IL-18, either through caspase-1-directed cleavage of the procytokines or through indirect NLRP3 activation. At the canonical inflammasome, ASC also recruits FADD and caspase-8, through a still-undefined mechanism. Caspase-8 activation at inflammasomes leads to apoptosis in caspase-1-deficient cells.

well-defined canonical inflammasome-forming NLRs include NLRC4, human NLRP1, and murine Nlrp1a and Nlrp1b, which respond to stimuli such as flagellin, inhibition of dipeptidyl peptidases (DPPs)8/9, and *Bacillus anthracis*–derived lethal toxin (LeTx) (Franchi et al. 2006; Miao et al. 2006; Okondo et al. 2018; Zhong et al. 2018; de Vasconcelos et al. 2019b). Akin to NLRs, the HIN200 family member AIM2 and the TRIM protein family member pyrin are well-defined inflammasome-forming PRRs. AIM2 directly binds to cytosolic DNA via its carboxy-terminal HIN200 domain, and pyrin indirectly senses inactivating RhoA GTPase modifications by microbial pathogens (Bürck-stümmer et al. 2009; Fernandes-Alnemri et al. 2009; Xu et al. 2014). Once active, these inflammasome receptors recruit the bipartite inflammasome adaptor ASC, which bridges the interaction with caspase-1, although direct engagement of caspase-1 is also possible in the context of the caspase recruitment domain (CARD)-containing NLRs including NLRP1, Nlrp1a, Nlrp1b, and NLRC4 (Broz et al. 2010; Masters et al. 2012; Van Opdenbosch et al. 2014; Broz and Dixit 2016). The noncanonical inflammasome is engaged on sensing of cytosolic lipopolysaccharide (LPS) by murine caspase-11 and its human orthologs, caspase-4 and -5 (Kayagaki et al. 2011; Shi et al. 2014; Schmid-Burgk et al. 2015). Inflammasomes perform several cell-intrinsic and extrinsic functions, and highlighting these are complex machineries integrating several cellular signaling pathways (Fig. 1). Yet, defining the contribution of the different effector functions initiated on inflammasome activation is only now emerging. This review will discuss recent progress in understanding of inflammasome functions and how these may affect disease outcomes.

INFLAMMASOME EFFECTOR FUNCTIONS: PYROPTOSIS AND MATURATION OF INTERLEUKIN-1 CYTOKINES

Both canonical and noncanonical inflammasome pathways converge on secretion of proinflammatory cytokines and initiation of a lytic cell death mode named pyroptosis, both of which are unequivocally linked to inflammasome-driven antimicrobial functions (Fig. 1). Interleukin (IL)-1β and IL-18 are key proinflammatory cytokines of the IL-1 family that are produced and stored in the cytosol as functionally inactive proteins (Dinarello 2009). Caspase-1, originally named IL-1β converting enzyme (ICE), cleaves both IL-1β and IL-18 after specific aspartate residues in their linker regions, allowing the mature cytokines to bind their cognate receptors on their extracellular release (Black et al. 1989; Kostura et al. 1989; Ghayur et al.

1997; Gu et al. 1997). IL-1β is a pyrogenic cytokine, and its binding to IL-1 receptor 1 (IL-1R1) triggers assembly of a tertiary receptor complex with the IL-1 receptor accessory protein (IL-1RAcP) on effector cells that promotes secretion of inflammatory mediators, local infiltration of immune cells, antibody production by B cells, and development of a T helper (Th)17 response. IL-18 is a costimulatory factor for interferon γ (IFN-γ) production and consequently a stimulator of natural killer (NK) cell function, polarizing the T-cell response to either a Th1 or Th2 pattern according to the context (Dinarello 2009). Protein expression of pro-IL-1β is mainly confined to cells of the myeloid lineage and further regulated by the requirement of NF-κB and MAPK signaling for its transcriptional up-regulation (often named signal 1), whereas pro-IL-18 is constitutively present in cells throughout the body (Van Opdenbosch et al. 2017; Van Gorp et al. 2019). Only caspase-1 can directly cleave IL-1β and IL-18; meanwhile, caspase-11 must activate the NLRP3 inflammasome for cytokine secretion in the context of noncanonical inflammasome signaling (Kayagaki et al. 2011, 2015).

In contrast to maturing cytokines, murine caspase-1 and -11 (and human caspases -1, -4, and -5) are equally capable of autonomously inducing pyroptosis (Kayagaki et al. 2011). This is accomplished through caspase-1/-11-directed cleavage of a common pyroptosis substrate, gasdermin D (GSDMD), at Asp275 (corresponding to Asp276 in human GSDMD), releasing a cytotoxic GSDMD amino-terminal fragment ($GSDMD_N$) (Kayagaki et al. 2015; Shi et al. 2015). $GSDMD_N$ assembles pores in the plasma membrane for execution of pyroptosis (Aglietti et al. 2016; Ding et al. 2016; Liu et al. 2016; Sborgi et al. 2016).

In addition to pyroptosis, inflammasomes were also described to initiate an alternative form of cell death, namely, NETosis. The term NETosis refers to a neutrophil-associated cell death mode, accompanied by extrusion of DNA webs, called NETs (Papayannopoulos 2018). Histones and antimicrobial proteins aggregate in the decondensed DNA structure, serving as a scaffold to hold pathogens and facilitate antimicrobial responses. In the context of noncanonical inflammasome activation, cas-pase-11 was shown to control DNA extrusion and cell death by cleaving and activating GSDMD, after sensing LPS or *Citrobacter roden-tium* in the cytosol of neutrophils (Chen et al. 2018). In vivo, DNase treatment, known to dis-solve NETs, prevented the host response to the vacuole destabilizing and caspase-11-activating Δ*sifA* mutant of *Salmonella enterica* serovar Typhimurium (*S.* Typhimurium) strain, but it had no effect in the absence of caspase-11 or GSDMD. These data suggest that inflamma-somes have an ability to regulate the initiation of cell death, which would take on different mor-phological characteristics depending on the cell type. However, understanding the specific contri-butions of each form of inflammasome-mediated cell death during antimicrobial defenses awaits the use of targeted conditional knockouts.

Pyroptotic cell lysis releases intracellular contents that act in neighboring cells for initia-tion of inflammatory responses (Fig. 1). One of these components, called alarmins, is IL-1α, an IL-1 family member that, akin to IL-1β, binds to IL-1R1 (Dinarello 2009). However, different from IL-1β, IL-1α is present in the nucleus of resting cells and is biologically active both as a full-length or matured cytokine (Mosley et al. 1987; Dinarello 2009). Caspase-1/-11-initiated pyroptosis controls IL-1α release on several in-flammasome triggers (Kayagaki et al. 2011). An-other alarmin, HMGB1, is an abundant nuclear protein involved in chromatin maintenance at steady-state conditions. Yet, extracellular HMGB1 signals through RAGE to initiate a che-motactic response (Bertheloot and Latz 2017). It may also engage TLRs when bound to dsDNA or LPS to promote proinflammatory signaling. Studies have shown that HMGB1 is released on inflammasome activation both in vitro and in vivo in mice challenged with LPS or LeTx (Lam-kanfi et al. 2010; Kayagaki et al. 2011; Lu et al. 2012; Van Opdenbosch et al. 2014). Notably, recent studies have clarified the role of hepato-cyte-derived HMGB1 in promoting caspase-11-dependent lethality during LPS-induced endotoxemia, revealing that initial contact be-tween hepatocytes and LPS triggers the observed systemic HMGB1 release seen in this model

(Deng et al. 2018). In a second step, HMGB1 binding to extracellular LPS promotes RAGE-dependent translocation to the cytosol of the HMGB1–LPS complex, where LPS activates cas-pase-11. Indeed, impairing HMGB1 signaling confers protection to septic shock and endotox-emia, similar to the resistance seen in caspase-11-deficient mice (Lamkanfi et al. 2010; Deng et al. 2018). Pyroptotic cells also release micro-vesicles containing tissue factor (TF), which pro-motes intravascular blood clotting and further instigates LPS-induced lethality in vivo (Wu et al. 2019). Initiation of an eicosanoid storm is another inflammasome effector mechanism. Different from pyroptosis induction and IL-1β/IL-18 release; however, resident peritoneal mac-rophages are selectively capable of releasing ei-cosanoids, which contributed to hypothermia on systemic Nlrp1b and NLRC4 inflammasome activation in vivo (von Moltke et al. 2012).

Although caspase-1 is an absolute require-ment for initiation of pyroptosis during canon-ical inflammasome signaling, deletion of cas-pase-1 in murine macrophages and epithelial cells does not prevent cell death per se after triggering of the NLRP3, AIM2, NLRC4, and Nlrp1b inflammasomes (Puri et al. 2012; Sagu-lenko et al. 2013; Rauch et al. 2017; Schneider et al. 2017; Van Opdenbosch et al. 2017; Lee et al. 2018a). Instead, caspase-1 knockout cells switch to an apoptotic phenotype that is driven by ASC-mediated caspase-8 activation, albeit frequently delayed by a few hours (Sagulenko et al. 2013; Rauch et al. 2017; Van Opdenbosch et al. 2017; Lee et al. 2018a). Accordingly, knockin of the C284A amino acid substitution in the catalytic site of caspase-1 inactivates not only its protease activity, but also suffices for triggering ASC- and caspase-8-mediated apo-ptosis on activation of the NLRC4 and Nlrp1b inflammasome pathways (Van Opdenbosch et al. 2017; Lee et al. 2018a). Moreover, activa-tion of caspase-8 by inflammasomes was fully dependent on ASC (Man et al. 2013; Sagulenko et al. 2013; Rauch et al. 2017; Van Opdenbosch et al. 2017; Lee et al. 2018a), and genetic deletion of ASC in caspase-1-deficient macrophages fully rescued these cells from inflammasome-initi-ated apoptosis (Van Opdenbosch et al. 2017;

Cite this article as *Cold Spring Harb Perspect Biol* doi: 10.1101/cshperspect.a036392

Lee et al. 2018a). Furthermore, inflammasome-induced caspase-8 activation and apoptosis by the Nlrp1b and NLRC4 inflammasomes was negatively regulated by TLR-induced up-regulation of the caspase-8-regulatory protein cFLIP (Van Opdenbosch et al. 2017; Lee et al. 2018a). Whether cFLIP is also incorporated in ASC specks remains unknown. Overall, these findings suggest that inflammasome assembly and ASC specks might function as a critical checkpoint that integrates cellular signals committing an infected cell to die by redundant regulated cell death pathways.

Interestingly, recruitment of caspase-8 to ASC specks is also observed in wild-type macrophages, suggesting that it is a native aspect of inflammasome signaling and does not only occur in the absence of caspase-1 signaling (Fig. 1; Man et al. 2013; Gurung et al. 2014; Van Opdenbosch et al. 2017). Ultrastructure modeling of the NLRP3 and AIM2 inflammasomes suggests these receptors oligomerize in star-like structures to expose their pyrin (PYD) domains for ASC recruitment (Lu et al. 2014). Additionally, a recent cryo-electron microscopic (cryo-EM) analysis of NLRP3 complexed with NEK7 suggests that the latter stabilizes NLRP3 complexes by bridging the interaction between leucine-rich repeats (LRRs) of adjacent NLRP3 molecules in the inflammasome (Sharif et al. 2019). $PYD_{AIM2/NLRP3}/PYD_{ASC}$ interactions act as a seeding mechanism for rapidly assembling ASC molecules and forming PYD-based filaments, which are cross-linked by $CARD_{ASC}$ to form the speck and recruit caspase-1 (Lu et al. 2014; Dick et al. 2016; Schmidt et al. 2016). Although the mechanisms governing caspase-8 recruitment to the ASC speck are not fully clarified, FADD, a death effector domain (DED)-containing protein that recruits caspase-8 to death receptors during extrinsic apoptosis (Hughes et al. 2016), colocalizes with ASC following NLRC4 activation, which suggests that this adaptor protein may mediate caspase-8 recruitment to the ASC speck (Fig. 1; Van Opdenbosch et al. 2017). Consistently, FADD deficiency was shown to hamper caspase-8 maturation on canonical and noncanonical NLRP3 inflammasome activation (Gurung et al. 2014).

Because FADD does not have either a CARD or a PYD, its interaction with ASC may be indirect. Thus, more work is needed to clarify how FADD is recruited to ASC specks.

Remarkably, genetic deletion of GSDMD in murine bone marrow–derived macrophages (BMDMs) elicits apoptosis in canonical inflammasome-triggered cells (He et al. 2015; de Vasconcelos et al. 2019a). A recent report suggests that caspase-1-mediated cleavage of the pro-apoptotic Bcl-2 family member BID contributed to caspase-1-induced apoptosis induction in GSDMD-deficient BMDMs (Tsuchiya et al. 2019). Notably, caspase-1-dependent apoptosis might drive cell death in primary cortical neurons on oxygen/glucose deprivation (Zhang et al. 2003) because these cells express caspase-1 and have low levels of GSDMD protein expression (Tsuchiya et al. 2019).

A recent study has shown that full autoprocessing of caspase-1, which releases the soluble p20/p10 form from the complex, inactivates the protease, thus suggesting that only caspase-1 associated with the inflammasome is active and that the inflammasome platform together with the polymerized caspase functions as a holoenzyme (Boucher et al. 2018). In agreement, inflammasome receptors that contain a CARD domain, such as NLRC4 and Nlrp1b, were previously shown to potently induce pyroptosis in the absence of ASC, by activating caspase-1 in the apparent absence of caspase-1 autoprocessing (Broz et al. 2010; Guey et al. 2014; Van Opdenbosch et al. 2014). Consistently, other studies show that the full-length form of caspase-1 becomes catalytically active and cleaves GSDMD during NLRC4 activation in $Asc^{-/-}$ cells (Dick et al. 2016; Boucher et al. 2018). However, maturation of IL-1β is less effective when CARD-containing inflammasome receptors are engaged in ASC-deficient macrophages (Broz et al. 2010; Van Opdenbosch et al. 2014). This might suggest that GSDMD is a better caspase-1 substrate than pro-IL-1β and/or that it can be efficiently recruited to complexes composed of NLRC4/Nlrp1b and caspase-1, whereas pro-IL-1β recruitment would occur more efficiently to ASC-containing complexes such as the ASC speck. Notably, also GSDMD and pro-

IL-1β (Man et al. 2013; He et al. 2015) have been localized to inflammasome platforms alongside caspase-8 and FADD (Man et al. 2013; Van Opdenbosch et al. 2017).

Although alternative downstream cell death pathways have not been described for the noncanonical inflammasome, substantial progress has been made in understanding its upstream signaling requirements. Guanylate-binding proteins (GBPs)—small GTPases that are transcriptionally up-regulated in response to IFN stimulation—act upstream of caspase-11 in activation of the noncanonical inflammasome pathway by vacuolar pathogens (Meunier et al. 2014; Pilla et al. 2014; Santos et al. 2018). Several GBP family members accumulate at endosomal membranes and bacterial outer membrane vesicles, in which GBP2 and GBP5 are prominently involved in the release of PAMPs from the endolysosomal compartment or unmasking intramembrane domains of LPS for detection by caspase-11 (Meunier et al. 2014; Pilla et al. 2014; Shi et al. 2014; Santos et al. 2018). This process promotes caspase-11 dimerization and automaturation that is essential for it to gain catalytic activity and cleave GSDMD (Lee et al. 2018b; Ross et al. 2018). However, GBPs appear to be dispensable for caspase-11-mediated host defense and in vivo sensing of cytosol-invasive bacterial pathogens that deliberately access the cytosol in their course of infection such as *Burkholderia thailandensis* (Aachoui et al. 2015).

The noncanonical inflammasome has long been recognized to rely on the NLRP3 inflammasome for cytokine secretion, a process in which caspase-11-mediated plasma membrane damage allows K^+ efflux to activate NLRP3 (Kayagaki et al. 2011; Rühl and Broz 2015). In agreement, $Gsdmd^{-/-}$ macrophages autoprocess caspase-11 following LPS transfection, but fail to undergo pyroptosis, activate caspase-1, or mature IL-1β/IL-18 (Kayagaki et al. 2015; Shi et al. 2015; Gonzalez Ramirez et al. 2018). NLRP3 activation downstream from GSDMD has also been shown in the context of caspase-8-mediated GSDMD cleavage, and this mechanism promotes IL-1β secretion in parallel to direct caspase-8-mediated IL-1β maturation in the context of extrinsic and intrinsic apoptosis

induction in macrophages (Yabal et al. 2014; Lawlor et al. 2015; Chauhan et al. 2018; Malireddi et al. 2018; Orning et al. 2018; Sarhan et al. 2018). Caspase-8-mediated control of the NLRP3 inflammasome was shown in various conditions, such as TLR4 or TNFR1 signaling combined with inhibition of IAPs, after TAK1 inhibition and following *Yersinia pestis* infection (Yabal et al. 2014; Lawlor et al. 2015; Malireddi et al. 2018; Orning et al. 2018; Sarhan et al. 2018), and thus suggesting a role in diverse pathophysiological settings.

Gasdermin Proteins: A Conserved Family of Pore-Forming Proteins

Gasdermin proteins were initially identified as proteins with high expression levels in skin and the upper gastrointestinal tract (Saeki et al. 2000). There are four gasdermin family members in humans, namely, gasdermin A (GSDMA), gasdermin B (GSDMB), gasdermin C (GSDMC), and gasdermin D (GSDMD) (Tamura et al. 2007). The mouse genome does not encode homologs of *GSDMB*, but *GSDMA* and *GSDMC* are represented by three (*Gsdma1-3*) and four (*Gsdmc1-4*) paralogs, respectively. Mice further express a single *GSDMD* ortholog. All gasdermin family members are able to induce necrotic cell death through their amino-terminal domains (Fig. 2; Ding et al. 2016). Interestingly, gasdermin-induced cytotoxicity was shown to be shared even with the distant gasdermin relative, deafness-associated tumor suppressor (DFNA5). Identified as the gene bearing a mutation causative of nonsyndromic hearing impairment, DFNA5 was initially described as a separate branch of the gasdermin phylogenetic tree, but recently renamed gasdermin E (GSDME) (Van Laer et al. 1998; Tamura et al. 2007; Ding et al. 2016; Wang et al. 2017).

The amino-terminal domains of GSDMD, GSDMA, GSDMA3, and DFNA5/GSDME associate with phospholipids in vitro and perforate liposomes (Fig. 2). There seems to be a level of specialization because GSDMD, GSDMA3, DFNA5/GSDME, and GSDMB all bind to phosphoinositides, but only GSDMD, GSDMA3, and GSDMA efficiently interact with cardioli-

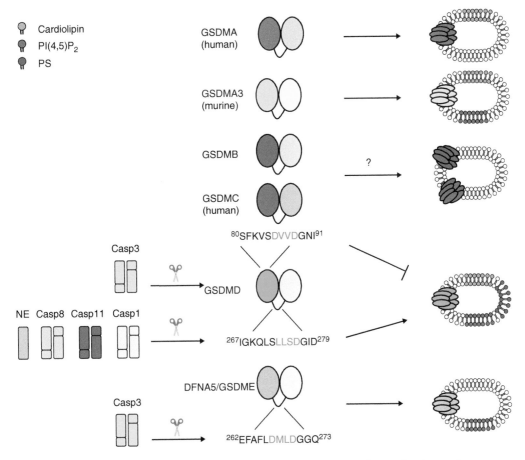

Figure 2. The gasdermin family encompasses pore-forming proteins. All gasdermin family members are composed of a pore-forming amino-terminal domain, which is kept autoinhibited by the carboxy-terminal domain, both connected by an unstructured flexible linker. Caspase-1 (Casp1), caspase-8 (Casp8), caspase-11 (Casp11), and neutrophil elastase (NE) cleave gasdermin D (GSDMD) in the linker region (the caspase recognition sequences are marked in red), releasing the pore forming amino-terminal fragment (GSDMD$_N$). GSDMD$_N$ is able to perforate liposomes containing complex mixtures, which include phosphatidylserine (PS), phosphatidylinositol 4,5-bisphosphate (PI(4,5)P$_2$) and cardiolipin-containing preparations. Conversely, caspase-3 (Casp3) cleavage of deafness-associated tumor suppressor/gasdermin E (DFNA5/GSDME) in the linker region releases an amino-terminal fragment that perforates liposomes containing PI(4,5)P$_2$. Caspase-3 (Casp3) can also cleave GSDMD, generating an amino-terminal fragment that is unable to perforate plasma membranes. Although amino-terminal fragments of human gasdermin A (GSDMA) and murine gasdermin A3 (GSDMA3) forms pores in lipososomes containing cardiolipin and/or PI(4,5)P$_2$, their activation mechanism remains unknown. It is also unclear how gasdermin B (GSDMB) and gasdermin C (GSDMC) could be activated and, although overexpression of their amino-terminal fragments triggers a lytic form of cell death, pore formation by these proteins has not yet been confirmed.

pin, a lipid found in the inner leaflet of the mitochondrial membrane (Aglietti et al. 2016; Ding et al. 2016; Liu et al. 2016; Chao et al. 2017; Wang et al. 2017). Uniquely, GSDMB efficiently binds to sulfatide, a lipid present in specialized areas such as the apical face of

epithelial cells (Chao et al. 2017). Together with their differential tissue distribution profiles (Tamura et al. 2007), these findings suggest that perforation of cellular membranes is a common mechanism of gasdermin-induced cytotoxicity (Aglietti et al. 2016; Ding et al. 2016;

Liu et al. 2016; Sborgi et al. 2016; Wang et al. 2017).

In all gasdermins, the amino-terminal pore-forming domain is kept inhibited by intramolecular interactions with the regulatory carboxy-terminal domain (Ding et al. 2016). However, other aspects of gasdermin biochemistry may be different between family members. For GSDMD, GSDMA3, GSDME, and GSDMA, the carboxy-terminal domain also prevents binding to lipids, in addition to preventing cell death (Shi et al. 2015; Ding et al. 2016). Conversely, GSDMB full-length protein is already able to bind lipids, but toxicity remains prevented by the presence of the carboxy-terminal domain (Ding et al. 2016; Chao et al. 2017).The regulatory interfaces in GSDMD and GSDMA3 were mapped to two main points of contact: (interface 1) the amino-terminal α1 helix and β1-β2 hairpin that intramolecularly interact with the carboxy-terminal domain; and (interface 2) the amino-terminal α4 helix, which interacts with the carboxy-terminal α9 and α11 helices at the upper part of the carboxy-terminal structure (Shi et al. 2015; Ding et al. 2016; Liu et al. 2018). Interface 1 is the lipid-binding component of GSDMA3, GSDMD, and DFNA5/GSDME, and this region is kept hidden by the carboxy-terminal domain in the full-length GSDMA3 and GSDMD proteins (Rogers et al. 2017; Liu et al. 2018; Ruan et al. 2018). Mutations that distress these molecular interfaces, particularly at interface 1, or that shorten the linker region destabilize the interdomain inhibitory interaction and are sufficient to promote cell death by full-length GSDMD and GSDMA3 (Ding et al. 2016; Kuang et al. 2017; Liu et al. 2018). Although the tertiary structure of DFNA5/GSDME has not been solved, mutations that cause hearing impairment in patients drive production of a truncated protein that lacks the last exon (Op de Beeck et al. 2012). Overexpression in HEK293T cells of these deafness-associated DFNA5/GSDME mutants triggers cell death (Wang et al. 2017). Therefore, it is likely that DFNA5/GSDME pathogenicity is associated with conformational changes that weaken the interface between the amino- and carboxy-terminal domains, al-though it remains unresolved why only cochlea cells are affected in patients expressing these mutant forms of DFNA5/GSDME.

Notwithstanding the considerable insights that have been reported on the secondary and tertiary structure of gasdermin family members (Ding et al. 2016), the mechanisms that govern dissociation of the amino- and carboxy-terminal domains are still unclear for most gasdermin proteins. GSDMD was initially discovered as a substrate of the inflammatory caspases -1, -11, -4, and -5, and the long-awaited executor of plasma membrane lysis during inflammasome-mediated pyroptosis (Fig. 2; Kayagaki et al. 2015; Shi et al. 2015). The cleavage site of inflammatory caspases lies at an unstructured loop that segregates the amino- and carboxy-terminal domains. Caspase-3 also cleaves GSDMD at Asp87, which destroys the amino-terminal domain and inactivates GSDMD (Rogers et al. 2017; Taabazuing et al. 2017; Orning et al. 2018), but further analysis is required to show that this regulation mechanism operates under physiological conditions. Furthermore, recent studies suggest that regulation of GSDMD activity is not restricted to caspase-3 and the inflammatory caspases because neutrophil elastase and caspase-8 also cleave GSDMD at the caspase-1/-11 cleavage site or at adjacent sites in the same linker (Fig. 2; Kambara et al. 2018; Orning et al. 2018; Sollberger et al. 2018). Notably, gasdermins have substantially less sequence similarity in the linker region, suggesting that each gasdermin may potentially use a different mechanism to disrupt the inhibitory interface (Tamura et al. 2007). Indeed, DFNA5/GSDME is cleaved in this same region by caspase-3 (Fig. 2; Rogers et al. 2017; Wang et al. 2017), a process that has been suggested to promote secondary necrosis of cells that follows apoptosis in cell culture settings (Rogers et al. 2017), although this has been contested by others (Lee et al. 2018a; Tixeira et al. 2018; Vince et al. 2018). Interestingly, cell types with high expression level of DFNA5/GSDME do not display features of apoptosis, such as cell shrinkage and blebbing, but instead undergo lytic cell death on activation of death receptors (Wang et al. 2017). This form of DFNA5/GSDME-mediated regulated necrosis was suggested to drive adverse inflammation in the lung and intestine in response to chemotherapy (Wang et al. 2017).

Overall, despite great recent advances in understanding the function of several gasdermin proteins, the field is still left with major questions. Most studies so far have relied on overexpression systems, which might overlook fine-tuned regulations and functions of these proteins. Therefore, there should be a focus in establishing in vivo physiological roles of gasdermins going forward. Further studies on the tissue and cell type–associated expression of gasdermin proteins and their regulation and activation mechanisms will allow a better understanding of different necrotic cell death pathways and their potential contributions to homeostasis and pathology.

Gasdermin Pores: Structural and Biochemical Features

Plasma membrane damage is one of the most outspoken features of pyroptosis. Yet, mechanistic understanding of pyroptotic cell lysis only gained significant traction with the discovery of GSDMD. In the context of inflammasome activation, cleavage of GSDMD by caspase-1, -11, -4, and -5 releases the pore-forming amino-terminal domain (GSDMD$_N$) from its inhibitory carboxy-terminal fragment (Kayagaki et al. 2015; Shi et al. 2015). GSDMD$_N$ monomers oligomerize in part through disulfide bonds, precipitated mainly by Cys39 and Cys192 of mouse GSDMD (Liu et al. 2016; Rathkey et al. 2018), and integration of GSDMD units into membranes promotes GSDMD$_N$-mediated pyroptotic cell lysis (Aglietti et al. 2016; Ding et al. 2016; Liu et al. 2016; Sborgi et al. 2016). Cryo-EM-based structural analysis indicated that GSDMA3 pores are formed by 26–28 monomers (Ruan et al. 2018). This ultrastructure suggests that, on membrane insertion, GSDMA3$_N$ undergoes significant conformational changes in which its unstructured loops reorganize into four β-strands. The final pore is composed of the adjoining of several β-strands from GSDMA3$_N$ subunits to form a β-barrel domain inserted in the lipid bilayer, with their

globular domains emerging at the surface of the membrane.

The observation of a GSDMA3$_N$ oligomer formed outside of the lipid bilayer, in which GSDMA3$_N$ still adopted its monomeric conformation, instigated the suggestion that this was a soluble prepore assembly, normally formed before membrane insertion (Ruan et al. 2018). Interestingly, in this same study, an attempt to analyze GSDMD pores was not successful, owing to the fact that these formed a heterogeneous population. Indeed, prior observations of the GSDMD$_N$ pore in overexpression systems had shown that its size ranges from 10 to 20 nm in internal diameter (Fig. 3; Ding et al. 2016; Liu et al. 2016; Sborgi et al. 2016). This pore assembly could potentially allow even large proteins to flux through, which appears to conflict with the observation that pyroptotic cells are subjected to osmotic pressure that drives cell lysis. However, further work is needed to understand how plasma membrane insertion of endogenous GSDMD$_N$ units proceeds in macrophages and other cell types that are triggered to undergo pyroptosis.

Early studies in this regard have shown that PEG-based osmoprotectants of only 2.4 nm prevented release of soluble cytosolic factors, including lactate dehydrogenase (LDH), from pyroptotic J774A.1 macrophages (Fink and Cookson 2006). Accordingly, osmoprotection with glycine allowed pyroptotic cells to be stained by a nuclear dye impermeable to intact membranes, named propidium iodide (PI), while bigger contents, tracked with LDH, were still retained in the cytoplasm (Russo et al. 2016). From this, the hypothesis emerged that caspase-1 initially triggers small pores at the plasma membrane that selectively allow ions to flux, with the associated osmotic pressure accounting for the ensuing total cell lysis. Consistent herewith, time-lapse analysis by atomic force microscopy of GSDMD pore formation showed that recombinant GSDMD$_N$ can initially form arc- or slit-shaped structures in liposomes, which grow over time into ring-shaped assemblies through the insertion of additional GSDMD$_N$ units at one end of the expanding oligomer (Mulvihill et al. 2018). Furthermore,

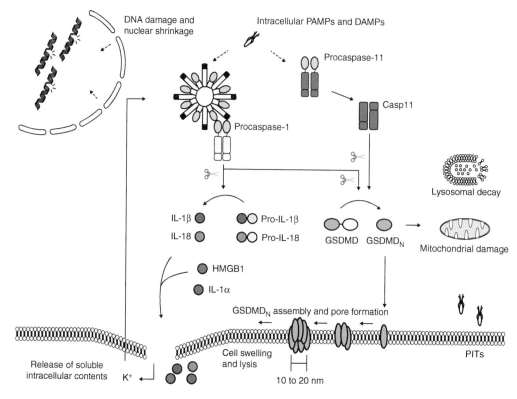

Figure 3. Amino-terminal gasdermin D (GSDMD) determines several morphological and biochemical features during pyroptosis. Presence of intracellular danger- and pathogen-associated molecular patterns (DAMPs and PAMPs) initiates signaling through canonical and noncanonical inflammasomes. Both caspase-1 and caspase-11 direct the cleavage of GSDMD, releasing its pore-forming amino-terminal fragment (GSDMD$_N$). Evidence suggests that GSDMD$_N$ inserts into the cellular plasma membrane in lower magnitude structures and assembles the pore through oligomerization within membranes until a higher magnitude final pore is formed. Finally, pyroptotic cellular swelling and plasma membrane rupture release soluble intracellular contents, including the proinflammatory cytokines IL-1β and IL-18, matured by caspase-1. During the noncanonical signaling, pyroptotic-mediated K$^+$ efflux causes cell-intrinsic NLRP3 inflammasome activation, accounting for cytokine maturation. GSDMD$_N$ also damages mitochondria during execution of cell death, and might be responsible for the organellar damage seen in lysosomes and nuclei. The pyroptotic cell corpse further maintains intracellular pathogens, in the pore-induced intracellular trap (PIT), facilitating their clearance by infiltrating immune cells.

GSDMD$_N$ monomers seem to require membranes to oligomerize, as the investigators could not detect any pore formation in the absence of lipids. Accordingly, a single-cell analysis of pyroptosis induction in primary macrophages showed that Ca^{2+} fluxes and osmotic increase in cell volume is detected before uptake of membrane-impermeable dyes (de Vasconcelos et al. 2019a). Furthermore, small plasma membrane–impermeant DNA dyes (molecular weight ~400 Da) accessed the nucleus of pyroptotic cells earlier than slightly larger ones (molecular weight ~600 Da). GSDMD-deficiency prevented these responses, suggesting that GSDMD$_N$ pores in the plasma membrane may gradually expand to promote plasma membrane permeability to larger cytosolic factors under physiological conditions (de Vasconcelos et al. 2019a). The concept that GSDMD$_N$ oligomerizes within the plasma membrane is further supported by the fact that GSDMD$_N$ higher magnitude assemblies are only found in membrane extracts (Ding et al. 2016; Liu et al. 2016). Therefore, the large pore GSDMD assemblies seen in cryo-EM and atomic force microscopy studies might be favored by the high concentrations of recombinant GSDMD$_N$ required for these analyses and may correspond to the final stages of in vivo pore formation, which may proceed more gradually at physiological concentrations (Fig. 3).

Pyroptosis: Morphological and Biochemical Hallmarks

In addition to the plasma membrane alterations and leakage promoted by GSDMD, pyroptotic cells undergo a series of intracellular changes during cell death, which is thought to be directed by inflammatory caspase activity. Of note, caspases were first recognized as mediators of apoptotic cell death, and their ability to cleave a large set of substrates generates the morphological and biochemical characteristics of this program of cellular demise (Taylor et al. 2008). Shigella- and Salmonella-induced cell death of infected macrophages was initially postulated to proceed through apoptosis (Zychlinsky et al. 1992; Chen et al. 1996; Monack et al. 1996; Hilbi et al. 1998; Hersh et al. 1999). However, these conclusions were debunked by the observation that caspase-1-initiated cell death triggered by these intracellular pathogens was accompanied by the release of cytosolic proteins, and the term pyroptosis—alluding to the inflammatory nature of this cell death—was coined (Cookson and Brennan 2001).

Although pyroptotic and apoptotic cells are morphologically distinct, they nevertheless have several biochemical features in common, which might be associated with their shared dependency on caspases. During canonical inflammasome activation, caspase-1 cleaves and activates caspase-7 (Lamkanfi et al. 2008). Caspase-7 activation does not contribute to inflammasome-mediated plasma membrane permeabilization or cytokine release, but it assists caspase-1 in the cleavage of PARP1, a nuclear protein involved in DNA repair (Lamkanfi et al. 2008; Malireddi et al. 2010). Interestingly, PARP1 cleavage during pyroptosis has been linked to an increased NF-κB response (Erener et al. 2012). Moreover, caspase-7 activation downstream from NLRC4 was shown to aid bacterial

control by promoting phagolysosomal transport of bacteria-containing vacuoles in infected cells (Akhter et al. 2009). Activation of inflammasomes was also found to promote cleavage of several glycolytic enzymes, such as GAPDH, α-enolase, and pyruvate kinase (Shao et al. 2007), although future studies should investigate how these proteolytic events impact inflammasome responses. Fewer proteins have been reported to be cleaved during noncanonical inflammasome signaling. Caspase-11 cleaves TRPC1, a component of cation-permeable pores, in vitro (Py et al. 2014). Absence of TRPC1 facilitates IL-1β release, but whether and how this protein is involved in pyroptosis is still unclear. The difference between the extensive list of caspase-1 putative substrates and the few proteins directly cleaved by caspase-11 might be related to enzyme substrate specificity. Indeed, in vitro comparisons of these two inflammatory caspases towards tetrapeptidic substrates suggested caspase-1 has a broader recognition spectrum than caspase-11 (Gonzalez Ramirez et al. 2018). Overall, the findings on alternative inflammatory substrates suggest that there might exist unidentified signaling pathways that contribute to the execution or consequences of pyroptosis.

Both apoptosis and pyroptosis are associated with DNA fragmentation, as evidenced by terminal deoxynucleotidyl transferase dUTP nick end (TUNEL) labeling, although a different staining pattern distinguishes these distinct regulated cell death modes (Fig. 3; Brennan and Cookson 2000). The mechanism for pyroptotic-initiated DNA damage is unclear, but the apoptotic caspase-3–ICAD/CAD pathway seems dispensable for pyroptosis as ICAD is not degraded in Salmonella-infected macrophages (Fink and Cookson 2006). Intriguingly, in addition to DNA fragmentation, the whole nuclear compartment changes during pyroptosis, taking a more roundish appearance (Fig. 3; de Vasconcelos et al. 2019a).

Mitochondrial damage has been observed on activation of canonical and noncanonical inflammasomes, suggesting this is a common feature of pyroptosis (Fig. 3; Allam et al. 2014; Yu et al. 2014; de Vasconcelos et al. 2019a).

BID, a member of the proapoptotic Bcl2 family, is also cleaved during pyroptosis (Yu et al. 2014; de Vasconcelos et al. 2019a). However, transgenic overexpression of antiapoptotic Bcl2 and genetic deletion of mitochondrial Bax/Bak pores does not impair permeabilization of the mitochondrial outer membrane (MOMP) and pyroptosis-associated cell permeabilization by the Nlrp1b and NLRP3 inflammasomes (Allam et al. 2014; de Vasconcelos et al. 2019a). Therefore, pyroptotic mitochondrial damage does not seem to rely on Bax/Bak pores. Instead, mitochondrial depolarization in the context of pyroptosis driven by the noncanonical inflammasome pathway is abolished in GSDMD-deficient macrophages (de Vasconcelos et al. 2019a), suggesting that GSDMD may also target intracellular membrane-bound organelles in addition to the plasma membrane (Aglietti et al. 2016; Ding et al. 2016; Liu et al. 2016; Sborgi et al. 2016). Consistently, pyroptotic organelle damage is not restricted to mitochondria, but also affects lysosomes and possibly other organelles (Fig. 3; de Vasconcelos et al. 2019a).

Release of the proinflammatory cytokines, IL-1β and IL-18, has long been associated with pyroptotic cell death (Fig. 3; Lamkanfi 2011). Several other soluble intracellular proteins, whether cytosolic or organellar, are also retrieved extracellularly following activation of the Nlrp1b and NLRC4 inflammasomes (de Vasconcelos et al. 2019a). Contrastingly, pyroptosis-induced membrane permeabilization does not allow the passage of organelles and phagocytosed pathogens (Jorgensen et al. 2016), suggesting there is a size limit for intracellular components that can be released during pyroptosis. Biologically, keeping larger intracellular agents, such as pathogenic bacteria, inside the pyroptotic corpse, which has been named the pore-induced intracellular trap (PIT), prevents the spread of such pathogens and could facilitate their clearance by infiltrating neutrophils and other professional phagocytes (Fig. 3; Jorgensen et al. 2016). Conversely, unspecific leakage in the extracellular milieu of soluble intracellular proteins and DAMPs may contribute to chemotactic and proinflammatory responses.

The Role of GSDMD in Inflammasome Effector Responses

IL-1β and IL-18 are cytosolic proteins that lack targeting sequences for secretion through the classical ER–Golgi secretion pathway (Rubartelli et al. 1990; Gu et al. 1997). Thus, a longstanding central question in the field has been how these cytokines are released from myeloid cells when their biological activity as a cytokine is required. $Casp1^{-/-}$ mice are impaired in both cytokine maturation and secretion (Kuida et al. 1995). Because caspase-1 also controls pyroptosis initiation, cell death has long been postulated to be the secretion mechanism for IL-1β and IL-18, although several alternative modes of IL-1β secretion from live cells have also been put forward. In support of a key role for pyroptosis in secretion of IL-1β and DAMPs, single-cell studies in the myeloid leukemia cell line THP-1 and in murine peritoneal macrophages have shown that extracellular release of IL-1β was only detected in the vicinity of dying cells (Liu et al. 2014; Cullen et al. 2015). In mouse BMDMs—the gold standard for inflammasome studies—and epithelial cells, most inflammasome triggers promote cell death coupled to IL-1β secretion (Kayagaki et al. 2011, 2015; Gurung et al. 2014; Zhong et al. 2016; Rauch et al. 2017; Van Opdenbosch et al. 2017). Accordingly, $Gsdmd^{-/-}$ BMDMs are impaired in LDH release and maintain matured IL-1β in their cytosol, failing to secrete the cytokine when shortly stimulated (He et al. 2015; Kayagaki et al. 2015; Shi et al. 2015). More importantly, GSDMD seems to be an essential controller of cytokine release in vivo, as $Gsdmd^{-/-}$ animals are rescued from the increase in circulating IL-1β caused by familial Mediterranean fever (FMF) and neonatal-onset multisystem inflammatory disease (NOMID)-associated mutations in mice (Kanneganti et al. 2018; Xiao et al. 2018). Thus, GSDMD is a critical determinant of inflammasome-mediated cytokine release in vitro and in vivo.

Nonetheless, several reports have shown a dissociation between cell death and cytokine secretion on inflammasome activation with specific stimuli and/or in different cell types. For example, activation of BMDMs with certain

Cite this article as *Cold Spring Harb Perspect Biol* doi: 10.1101/cshperspect.a036392

lipids of the mixture of oxidized phospholipids, oxPAPC, commonly released by necrotic cells, and with peptidoglycan from *O*-acetyltransferase A–deficient *Staphylococcus aureus* triggers IL-1β release in the absence of detectable LDH release (Shimada et al. 2010; Zanoni et al. 2017). In human primary monocytic cells, TLR4 signaling is sufficient to initiate IL-1β release, which is not accompanied by cell death above background levels (Gaidt et al. 2016). Similarly, IL-1β, but not LDH, was recently shown to be secreted by oxPAPC-triggered dendritic cells (DCs), although a recent publication shows that bone marrow–derived DCs are contaminated with BMDMs prompts a reevaluation of these conclusions (Zanoni et al. 2016; Erlich et al. 2019). Finally, a similar dissociation between cell death and cytokine release has been reported for *S.* Typhimurium–infected neutrophils (Chen et al. 2014; Heilig et al. 2018).

However, the molecular mechanisms driving cytokine secretion from live cells have so far remained unresolved. Alternative mechanisms for the active secretion of IL-1β could play a role in the absence of pyroptosis, as reviewed elsewhere (Monteleone et al. 2015). In some situations, macrophages and neutrophils were suggested to rely on GSDMD for IL-1β release without inducing LDH release (Evavold et al. 2018; Heilig et al. 2018), which could be explained by the rapid shedding of GSDMD pores in the plasma membrane by the endosomal sorting complexes required for transport (ESCRT) machinery (Rühl et al. 2018). It remains unclear whether this membrane recycling mechanism is sufficiently efficient and rapid to shed all GSDMD pores that appear in the plasma membrane to prevent cell death, and its relevance for controlling IL-1β secretion from activated monocytes and macrophages in vivo also warrants further analysis. Notably, delayed IL-1β release is detected in $Gsdmd^{-/-}$ BMDMs cultured ex vivo (Kayagaki et al. 2015). Careful analysis of GSDMD-deficient macrophages has revealed that they undergo apoptosis on activation of caspase-1, with morphological alterations appearing within the same time frame as pyroptosis is detected in wild-type cells through a pathway recently suggested to involve cleavage

of BID by caspase-1 (He et al. 2015; de Vasconcelos et al. 2019a; Tsuchiya et al. 2019).

ROLE OF PYROPTOSIS IN INFECTIONS AND INFLAMMATORY DISEASE

Inflammasomes have long been implicated in host defense against microbial pathogens. Studies in knockout mice have firmly established the critical role of inflammasome activation in protection against in vivo infection with *B. anthracis* spores (Moayeri et al. 2010; Terra et al. 2010), *Burkholderia* spp. (Ceballos-Olvera et al. 2011; Aachoui et al. 2015), *S.* Typhimurium (Karki et al. 2018), *Francisella tularensis* (Fernandes-Alnemri et al. 2010; Jones et al. 2010), murine cytomegalovirus (Rathinam et al. 2010), *Candida albicans* (Joly et al. 2009), and many other pathogens. Remarkably, certain effector functions of inflammasomes—that is, IL-1β or IL-18 secretion, pyroptosis, and DAMP release—may each exert specific functions in in vivo models of inflammasome activation that together contribute to host and immune responses during infection. As a result, genetic ablation of the inflammasome sensor not always phenocopies animals that lack expression of specific inflammasome effector proteins. $Il18^{-/-}$ mice are more susceptible to lethality when infected with *Burkholderia pseudomallei* or *B. thailandensis* compared with wild-type mice, whereas deletion of IL-1β did not alter host defense to the infections (Ceballos-Olvera et al. 2011; Aachoui et al. 2015). Additionally, IL-18 was shown to be essential for clearing *Chromobacterium violaceum* from the liver of infected mice (Maltez et al. 2015). Distinctively, IL-1-mediated signaling is important for resistance to *B. anthracis* infection, and deletion of IL-1R1 renders inbred mouse strains that express the LeTx-sensitive Nlrp1b allele susceptible to *B. anthracis* infection (Moayeri et al. 2010).

Another prominent example is that co-ablation of IL-1β and IL-18 signaling prevents lethality to lower LPS doses in the mouse endotoxemia model of sepsis, but this protective effect is not sustained at higher LPS doses (Lamkanfi et al. 2010; Berghe et al. 2014). However,

$Casp11^{-/-}$ and $Gsdmd^{-/-}$ mice are highly resistant to LPS-induced lethality (Lamkanfi et al. 2010; Kayagaki et al. 2011, 2015; Berghe et al. 2014). This decisive role for pyroptosis in LPS-induced endotoxic shock is at least partially related to the release of microvesicles containing TF by dying cells, thus initiating an overt systemic coagulation response (Wu et al. 2019). Similarly, on *C. violaceum* infection, IL-1β/IL-18 signaling contributes partially to host defense against this pathogen, whereas caspase-1/-11 double knockout mice succumb to infection (Maltez et al. 2015). Furthermore, caspase-1 and -11 are essential to control bacterial loads in the spleen, with no role found for inflammasome-produced cytokines. In an in vivo sterile model of inflammasome activation, intracellular flagellin delivery causes fast hypothermia in wild-type animals that is fully rescued by ablation of NLRC4 (von Moltke et al. 2012; Rauch et al. 2017). In contrast, $Casp1^{-/-}$ and $Gsdmd^{-/-}$ animals experience an intermediate phenotype. The hematocrit increases seen in wild-type animals was only observed in $Casp1^{-/-}$ at later time points, whereas $Nlrc4^{-/-}$ animals remained protected (von Moltke et al. 2012). A similar intermediate phenotype is seen on *Legionella* spp. infection, in which caspase-1/-11-deficient mice present with higher colony-forming unit (CFU) counts than wild-type controls, and $Nlrc4^{-/-}$ mice are even more susceptible (Mascarenhas et al. 2017).

These midway results obtained with $Casp1^{-/-}$ mice have been associated with caspase-8 activation at the ASC speck. Co-ablation of either ASC or caspase-8 (in RIPK3 knockout animals to prevent caspase-8-driven embryonic lethality [Kaiser et al. 2011; Oberst et al. 2011]) in the caspase-1/-11-deficient background phenocopies NLRC4 deletion on flagellin challenge in vivo or infection with *Legionella* spp. (Mascarenhas et al. 2017; Rauch et al. 2017). Therefore, in some disease settings, inflammasome signaling to caspase-8-dependent apoptosis suffices to clear the infected cell, resulting in a partial containment of the bacteria. Cell death also plays a role in the in vivo eicosanoid storm observed on NLRC4 triggering in epithelial cells, as ablation of both caspase-1

and -8 is required to protect mice from eicosanoid increase (Rauch et al. 2017).

In conclusion, it appears that there is an intricate relationship between inflammasome effector functions within several tissue compartments, which might determine susceptibility to specific pathogens. In *B. thailandensis* infection, IL-18 is required to up-regulate caspase-11 through IFN-γ production, and caspase-11-driven signaling is essential for mice to survive the infection (Aachoui et al. 2015). In addition, absence of caspase-11 does not alter *S.* Typhimurium bacterial loads in vivo, but $Casp1^{-/-}$ $Casp11^{-/-}$ mice in which caspase-11 expression was reconstituted from a C57BL/6 bacterial artificial chromosome (BAC) have higher CFUs in their organs than caspase-1/-11-codeleted controls (Broz et al. 2012). This surprisingly shows that caspase-11-mediated pyroptosis might be essential in the absence of caspase-1-driven signaling, and suggests proinflammatory cytokines and cell death may be needed simultaneously to promote full innate immune activation and protection.

Although the role of GSDMD in inflammasome-associated diseases remains less understood, much progress has been made in recent years. GSDMD was defined as an important effector to clear *Francisella novicida* in vivo infection (Banerjee et al. 2018; Zhu et al. 2018). However, similar to what has been observed with $Casp1^{-/-}$ mice, GSDMD-deficient animals have an intermediate phenotype on in vivo challenge with flagellin (Rauch et al. 2017). Defining the molecular mechanisms by which GSDMD-deficient cells switch to an apoptotic cell death mode on canonical inflammasome stimulation would help elucidate why $Gsdmd^{-/-}$ mice are not fully rescued on flagellin challenge. Remarkably, $Gsdmd^{-/-}$ animals were shown to clear *Escherichia coli* infection better than wild-type counterparts (Kambara et al. 2018). *E. coli* is sensed by caspase-11 in macrophages, triggering GSDMD-mediated pyroptosis (Kayagaki et al. 2011, 2015). Mice require caspase-11 to mount an immune response and survive in vivo delivery of *E. coli* outer membrane vesicles, showed to mediate LPS intracellular delivery (Vanaja et al. 2016; Santos et al. 2018). Although the protection observed in $Gsdmd^{-/-}$ mice was associated with lower cell death induction by neutrophils, this was not shown to be the case in vivo.

Inflammasome-initiated effector functions may play specific roles not only for infectious disease resolution, but also during development of sterile inflammasomopathies. These inflammasome-driven autoinflammatory diseases are caused by activating mutations in inflammasome-coding genes, and most of their symptoms recollect features of generalized inflammation, such as recurrent fever and body rash (Van Gorp et al. 2019). CAPS collectively represents three diseases—NOMID, Muckle–Wells syndrome (MWS), and familial cold autoinflammatory syndrome (FCAS)—which represent a spectrum of severity and are all caused by mutations in *NLRP3* (Kuemmerle-Deschner 2015). Deletion of the IL-18R in mice carrying the *Nlrp3* A350V allele, associated with MWS, is sufficient to rescue their survival (Brydges et al. 2013). In agreement with an important role for IL-18 in NLRC4-mediated autoinflammatory disease progression, a patient carrying the autoinflammatory *NLRC4* variant V341A was successfully treated with recombinant IL-18-binding protein, which prevents IL-18 binding to its receptor, while blocking IL-1 alone did not provide a benefit to this patient (Romberg et al. 2014; Canna et al. 2017). Development of FMF is in most patients linked to mutations in the *MEFV* gene, which encodes the inflammasome receptor pyrin. A knockin mouse model that expresses a chimeric humanized version of *Pyrin* with an FMF-associated mutation, V726A, develops systemic autoinflammatory pathology that is fully dependent on IL-1β signaling (Sharma et al. 2017). Interestingly, deletion of GSDMD completely rescued circulating IL-1β levels and provided the same level of protection as seen in $Il1b^{-/-}$ mice, highlighting the role of pyroptosis for cytokine secretion in vivo (Sharma et al. 2017; Kanneganti et al. 2018). GSDMD-deletion also rescued all signs of autoinflammation seen in the *Nlrp3* D301N mice, a mutation that causes NOMID in humans (Xiao et al. 2018). As deletion of GSDMD in these mice abolished circulating IL-1β levels, it would be of interest to determine the relative contribution of this cytokine to inflammatory pathology in *Nlrp3* D301N mice. A critical role for cell death in driving NLRP3-mediated pathology has also been suggested in mice bearing the FCAS-linked L351P mutation in *Nlrp3* (Brydges et al. 2013). Post-birth lethality seen in these animals is fully rescued by the absence of caspase-1, whereas co-ablation of IL-1R/IL-18 is only partially protective. Thus, pyroptosis appears to contribute to inflammasome-dependent cytokine secretion and other inflammasome-associated effector pathways. Modulating GSDMD activation may therefore hold promise for the treatment of inflammasome-driven diseases in the clinic.

CONCLUSIONS

Inflammasome activation is emerging as an increasingly complex mechanism for regulated activation of inflammatory caspases and induction of inflammatory and host defense responses. In addition to canonical proinflammatory cytokine maturation and secretion, GSDMD-driven pyroptosis is rapidly emerging as another key function associated with inflammasome activation that intercrosses with cytokine and DAMP release and other immune effector mechanisms. Functional analysis of additional caspase substrates that are cleaved during pyroptosis may reveal unexpected and diverse roles in intracellular signaling that contribute to how pyroptotic cells regulate immune activation and host defense. A better understanding of the diverse roles pyroptosis plays in the immune system could aid in understanding how inflammasome activation signals to neighbor cells to control pathogen spread and potentiates sterile tissue damage in inflammatory diseases. Furthermore, addressing these mechanisms could reveal novel therapeutic approaches to benefit patients suffering from a broad suite of inflammasome-mediated diseases.

ACKNOWLEDGMENTS

We apologize to colleagues whose work was not cited because of space constraints. This work was supported by European Research Council Grant 683144 (PyroPop) to M.L.

REFERENCES

Aachoui Y, Kajiwara Y, Leaf IA, Mao D, Ting JP, Coers J, Aderem A, Buxbaum JD, Miao EA. 2015. Canonical inflammasomes drive IFN-γ to prime caspase-11 in defense against a cytosol-invasive bacterium. *Cell Host Microbe* **18:** 320–332. doi:10.1016/j.chom.2015.07.016

Aglietti RA, Estevez A, Gupta A, Ramirez MG, Liu PS, Kayagaki N, Ciferri C, Dixit VM, Dueber EC. 2016. GsdmD p30 elicited by caspase-11 during pyroptosis forms pores in membranes. *Proc Natl Acad Sci* **113:** 7858–7863. doi:10.1073/pnas.1607769113

Akhter A, Gavrilin MA, Frantz L, Washington S, Ditty C, Limoli D, Day C, Sarkar A, Newland C, Butchar J, et al. 2009. Caspase-7 activation by the Nlrc4/Ipaf inflammasome restricts *Legionella pneumophila* infection. *PLoS Pathog* **5:** e1000361. doi:10.1371/journal.ppat.1000361

Allam R, Lawlor KE, Yu EC, Mildenhall AL, Moujalled DM, Lewis RS, Ke F, Mason KD, White MJ, Stacey KJ, et al. 2014. Mitochondrial apoptosis is dispensable for NLRP3 inflammasome activation but non-apoptotic caspase-8 is required for inflammasome priming. *EMBO Rep* **15:** 982–990. doi:10.15252/embr.201438463

Banerjee I, Behl B, Mendonca M, Shrivastava G, Russo AJ, Menoret A, Ghosh A, Vella AT, Vanaja SK, Sarkar SN, et al. 2018. Gasdermin D restrains type I interferon response to cytosolic DNA by disrupting ionic homeostasis. *Immunity* **49:** 413–426.e5. doi:10.1016/j.immuni.2018.07.006

Berghe TV, Demon D, Bogaert P, Vandendriessche B, Goethals A, Depuydt B, Vuylsteke M, Roelandt R, Van Wonterghem E, Vandenbroecke J, et al. 2014. Simultaneous targeting of IL-1 and IL-18 is required for protection against inflammatory and septic shock. *Am J Respir Crit Care Med* **189:** 282–291. doi:10.1164/rccm.201308-1535OC

Bertheloot D, Latz E. 2017. HMGB1, IL-1α, IL-33 and S100 proteins: dual-function alarmins. *Cell Mol Immunol* **14:** 43–64. doi:10.1038/cmi.2016.34

Black RA, Kronheim SR, Merriam JE, March CJ, Hopp TP. 1989. A pre-aspartate-specific protease from human leukocytes that cleaves pro-interleukin-1β. *J Biol Chem* **264:** 5323–5326.

Boucher D, Monteleone M, Coll RC, Chen KW, Ross CM, Teo JL, Gomez GA, Holley CL, Bierschenk D, Stacey KJ, et al. 2018. Caspase-1 self-cleavage is an intrinsic mechanism to terminate inflammasome activity. *J Exp Med* **215:** 827–840. doi:10.1084/jem.20172222

Brennan MA, Cookson BT. 2000. *Salmonella* induces macrophage death by caspase-1-dependent necrosis. *Mol Microbiol* **38:** 31–40. doi:10.1046/j.1365-2958.2000.02103.x

Broz P, Dixit VM. 2016. Inflammasomes: mechanism of assembly, regulation and signalling. *Nat Rev Immunol* **16:** 407–420. doi:10.1038/nri.2016.58

Broz P, von Moltke J, Jones JW, Vance RE, Monack DM. 2010. Differential requirement for caspase-1 autoproteolysis in pathogen-induced cell death and cytokine processing. *Cell Host Microbe* **8:** 471–483. doi:10.1016/j.chom.2010.11.007

Broz P, Ruby T, Belhocine K, Bouley DM, Kayagaki N, Dixit VM, Monack DM. 2012. Caspase-11 increases suscepti-bility to *Salmonella* infection in the absence of caspase-1. *Nature* **490:** 288–291. doi:10.1038/nature11419

Brydges SD, Broderick L, McGeough MD, Pena CA, Mueller JL, Hoffman HM. 2013. Divergence of IL-1, IL-18, and cell death in NLRP3 inflammasomopathies. *J Clin Invest* **123:** 4695–4705. doi:10.1172/JCI71543

Bürckstümmer T, Baumann C, Blüml S, Dixit E, Dürnberger G, Jahn H, Planyavsky M, Bilban M, Colinge J, Bennett KL, et al. 2009. An orthogonal proteomic-genomic screen identifies AIM2 as a cytoplasmic DNA sensor for the inflammasome. *Nat Immunol* **10:** 266–272. doi:10.1038/ni.1702

Canna SW, Girard C, Malle L, de Jesus A, Romberg N, Kelsen J, Surrey LF, Russo P, Sleight A, Schiffrin E, et al. 2017. Life-threatening NLRC4-associated hyperinflammation successfully treated with IL-18 inhibition. *J Allergy Clin Immunol* **139:** 1698–1701. doi:10.1016/j.jaci.2016.10.022

Ceballos-Olvera I, Sahoo M, Miller MA, Barrio L, Re F. 2011. Inflammasome-dependent pyroptosis and IL-18 protect against *Burkholderia pseudomallei* lung infection while IL-1β is deleterious. *PLoS Pathog* **7:** e1002452. doi:10.1371/journal.ppat.1002452

Chao KL, Kulakova L, Herzberg O. 2017. Gene polymorphism linked to increased asthma and IBD risk alters gasdermin-B structure, a sulfatide and phosphoinositide binding protein. *Proc Natl Acad Sci* **114:** E1128–E1137. doi:10.1073/pnas.1616783114

Chauhan D, Bartok E, Gaidt MM, Bock FJ, Herrmann J, Seeger JM, Broz P, Beckmann R, Kashkar H, Tait SWG, et al. 2018. BAX/BAK-induced apoptosis results in caspase-8-dependent IL-1β maturation in macrophages. *Cell Rep* **25:** 2354–2368.e5. doi:10.1016/j.celrep.2018.10.087

Chen LM, Kaniga K, Galan JE. 1996. *Salmonella* spp. are cytotoxic for cultured macrophages. *Mol Microbiol* **21:** 1101–1115. doi:10.1046/j.1365-2958.1996.471410.x

Chen KW, Groß CJ, Sotomayor FV, Stacey KJ, Tschopp J, Sweet MJ, Schroder K. 2014. The neutrophil NLRC4 inflammasome selectively promotes IL-1β maturation without pyroptosis during acute *Salmonella* challenge. *Cell Rep* **8:** 570–582. doi:10.1016/j.celrep.2014.06.028

Chen KW, Monteleone M, Boucher D, Sollberger G, Ramnath D, Condon ND, von Pein JB, Broz P, Sweet MJ, Schroder K. 2018. Noncanonical inflammasome signaling elicits gasdermin D-dependent neutrophil extracellular traps. *Sci Immunol* **3:** eaar6676. doi:10.1126/sciimmunol.aar6676

Cookson BT, Brennan MA. 2001. Pro-inflammatory programmed cell death. *Trends Microbiol* **9:** 113–114. doi:10.1016/S0966-842X(00)01936-3

Cullen SP, Kearney CJ, Clancy DM, Martin SJ. 2015. Diverse activators of the NLRP3 inflammasome promote IL-1β secretion by triggering necrosis. *Cell Rep* **11:** 1535–1548. doi:10.1016/j.celrep.2015.05.003

Deng M, Tang Y, Li W, Wang X, Zhang R, Zhang X, Zhao X, Liu J, Tang C, Liu Z, et al. 2018. The endotoxin delivery protein HMGB1 mediates caspase-11-dependent lethality in sepsis. *Immunity* **49:** 740–753.e7. doi:10.1016/j.immuni.2018.08.016

de Vasconcelos NM, Van Opdenbosch N, Van Gorp H, Parthoens E, Lamkanfi M. 2019a. Single-cell analysis of pyroptosis dynamics reveals conserved GSDMD-mediated subcellular events that precede plasma membrane rupture. *Cell Death Differ* **26:** 146–161. doi:10.1038/s41418-018-0106-7

de Vasconcelos NM, Vliegen G, Gonçalves A, De Hert E, Martín-Pérez R, Van Opdenbosch N, Jallapally A, Geiss-Friedlander R, Lambeir A-M, Augustyns K, et al. 2019b. DPP8/DPP9 inhibition elicits canonical Nlrp1b inflammasome hallmarks in murine macrophages. *Life Sci Alliance* **2:** e201900313. doi:10.26508/lsa.201900313

Dick MS, Sborgi L, Rühl S, Hiller S, Broz P. 2016. ASC filament formation serves as a signal amplification mechanism for inflammasomes. *Nat Commun* **7:** 11929. doi:10.1038/ncomms11929

Dinarello CA. 2009. Immunological and inflammatory functions of the interleukin-1 family. *Annu Rev Immunol* **27:** 519–550. doi:10.1146/annurev.immunol.021908.132612

Ding J, Wang K, Liu W, She Y, Sun Q, Shi J, Sun H, Wang DC, Shao F. 2016. Pore-forming activity and structural autoinhibition of the gasdermin family. *Nature* **535:** 111–116. doi:10.1038/nature18590

Erener S, Pétrilli V, Kassner I, Minotti R, Castillo R, Santoro R, Hassa Paul O, Tschopp J, Hottiger Michael O. 2012. Inflammasome-activated caspase 7 cleaves PARP1 to enhance the expression of a subset of NF-κB target genes. *Mol Cell* **46:** 200–211. doi:10.1016/j.molcel.2012.02.016

Erlich Z, Shlomovitz I, Edry-Botzer L, Cohen H, Frank D, Wang H, Lew AM, Lawlor KE, Zhan Y, Vince JE, et al. 2019. Macrophages, rather than DCs, are responsible for inflammasome activity in the GM-CSF BMDC model. *Nat Immunol* **20:** 397–406. doi:10.1038/s41590-019-0313-5

Evavold CL, Ruan J, Tan Y, Xia S, Wu H, Kagan JC. 2018. The pore-forming protein gasdermin D regulates interleukin-1 secretion from living macrophages. *Immunity* **48:** 35–44.e6. doi:10.1016/j.immuni.2017.11.013

Fernandes-Alnemri T, Yu JW, Datta P, Wu J, Alnemri ES. 2009. AIM2 activates the inflammasome and cell death in response to cytoplasmic DNA. *Nature* **458:** 509–513. doi:10.1038/nature07710

Fernandes-Alnemri T, Yu JW, Juliana C, Solorzano L, Kang S, Wu J, Datta P, McCormick M, Huang L, McDermott E, et al. 2010. The AIM2 inflammasome is critical for innate immunity to *Francisella tularensis*. *Nat Immunol* **11:** 385–393. doi:10.1038/ni.1859

Fink SL, Cookson BT. 2006. Caspase-1-dependent pore formation during pyroptosis leads to osmotic lysis of infected host macrophages. *Cell Microbiol* **8:** 1812–1825. doi:10.1111/j.1462-5822.2006.00751.x

Franchi L, Amer A, Body-Malapel M, Kanneganti TD, Özören N, Jagirdar R, Inohara N, Vandenabeele P, Bertin J, Coyle A, et al. 2006. Cytosolic flagellin requires Ipaf for activation of caspase-1 and interleukin 1β in *Salmonella*-infected macrophages. *Nat Immunol* **7:** 576–582. doi:10.1038/ni1346

Gaidt MM, Ebert TS, Chauhan D, Schmidt T, Schmid-Burgk JL, Rapino F, Robertson AA, Cooper MA, Graf T, Hornung V. 2016. Human monocytes engage an alternative inflammasome pathway. *Immunity* **44:** 833–846. doi:10.1016/j.immuni.2016.01.012

Ghayur T, Banerjee S, Hugunin M, Butler D, Herzog L, Carter A, Quintal L, Sekut L, Talanian R, Paskind M, et al. 1997. Caspase-1 processes IFN-γ-inducing factor and

regulates LPS-induced IFN-γ production. *Nature* **386:** 619–623. doi:10.1038/386619a0

Gonzalez Ramirez ML, Poreba M, Snipas SJ, Groborz K, Drag M, Salvesen GS. 2018. Extensive peptide and natural protein substrate screens reveal that mouse caspase-11 has much narrower substrate specificity than caspase-1. *J Biol Chem* **293:** 7058–7067. doi:10.1074/jbc.RA117.001329

Gu Y, Kuida K, Tsutsui H, Ku G, Hsiao K, Fleming MA, Hayashi N, Higashino K, Okamura H, Nakanishi K, et al. 1997. Activation of interferon-γ inducing factor mediated by interleukin-1β converting enzyme. *Science* **275:** 206–209. doi:10.1126/science.275.5297.206

Guey B, Bodnar M, Manié SN, Tardivel A, Petrilli V. 2014. Caspase-1 autoproteolysis is differentially required for NLRP1b and NLRP3 inflammasome function. *Proc Natl Acad Sci* **111:** 17254–17259. doi:10.1073/pnas.1415756111

Gurung P, Anand PK, Malireddi RKS, Vande Walle L, Van Opdenbosch N, Dillon CP, Weinlich R, Green DR, Lamkanfi M, Kanneganti TD. 2014. FADD and caspase-8 mediate priming and activation of the canonical and noncanonical Nlrp3 inflammasomes. *J Immunol* **192:** 1835–1846. doi:10.4049/jimmunol.1302839

He WT, Wan H, Hu L, Chen P, Wang X, Huang Z, Yang ZH, Zhong CQ, Han J. 2015. Gasdermin D is an executor of pyroptosis and required for interleukin-1β secretion. *Cell Res* **25:** 1285–1298. doi:10.1038/cr.2015.139

Heilig R, Dick MS, Sborgi L, Meunier E, Hiller S, Broz P. 2018. The gasdermin-D pore acts as a conduit for IL-1β secretion in mice. *European J Immunol* **48:** 584–592. doi:10.1002/eji.201747404

Hersh D, Monack DM, Smith MR, Ghori N, Falkow S, Zychlinsky A. 1999. The *Salmonella* invasin SipB induces macrophage apoptosis by binding to caspase-1. *Proc Natl Acad Sci* **96:** 2396–2401. doi:10.1073/pnas.96.5.2396

Hilbi H, Moss JE, Hersh D, Chen Y, Arondel J, Banerjee S, Flavell RA, Yuan J, Sansonetti PJ, Zychlinsky A. 1998. Shigella-induced apoptosis is dependent on caspase-1 which binds to IpaB. *J Biol Chem* **273:** 32895–32900. doi:10.1074/jbc.273.49.32895

Hughes MA, Powley IR, Jukes-Jones R, Horn S, Feoktistova M, Fairall L, Schwabe JW, Leverkus M, Cain K, MacFarlane M. 2016. Co-operative and hierarchical binding of c-FLIP and caspase-8: a unified model defines how c-FLIP isoforms differentially control cell fate. *Mol Cell* **61:** 834–849. doi:10.1016/j.molcel.2016.02.023

Joly S, Ma N, Sadler JJ, Soll DR, Cassel SL, Sutterwala FS. 2009. Cutting edge: *Candida albicans hyphae* formation triggers activation of the Nlrp3 inflammasome. *J Immunol* **183:** 3578–3581. doi:10.4049/jimmunol.0901323

Jones JW, Kayagaki N, Broz P, Henry T, Newton K, O'Rourke K, Chan S, Dong J, Qu Y, Roose-Girma M, et al. 2010. Absent in melanoma 2 is required for innate immune recognition of *Francisella tularensis*. *Proc Natl Acad Sci* **107:** 9771–9776. doi:10.1073/pnas.1003738107

Jorgensen I, Zhang Y, Krantz BA, Miao EA. 2016. Pyroptosis triggers pore-induced intracellular traps (PITs) that capture bacteria and lead to their clearance by efferocytosis. *J Exp Med* **213:** 2113–2128. doi:10.1084/jem.20151613

Kaiser WJ, Upton JW, Long AB, Livingston-Rosanoff D, Daley-Bauer LP, Hakem R, Caspary T, Mocarski ES. 2011. RIP3 mediates the embryonic lethality of caspase-

8-deficient mice. *Nature* **471**: 368–372. doi:10.1038/nature09857

Kambara H, Liu F, Zhang X, Liu P, Bajrami B, Teng Y, Zhao L, Zhou S, Yu H, Zhou W, et al. 2018. Gasdermin D exerts anti-inflammatory effects by promoting neutrophil death. *Cell Rep* **22**: 2924–2936. doi:10.1016/j.celrep.2018.02.067

Kanneganti A, Malireddi RKS, Saavedra PHV, Vande Walle L, Van Gorp H, Kambara H, Tillman H, Vogel P, Luo HR, Xavier RJ, et al. 2018. GSDMD is critical for autoinflammatory pathology in a mouse model of familial Mediterranean fever. *J Exp Med* **215**: 1519–1529. doi:10.1084/jem.20172060

Karki R, Lee E, Place D, Samir P, Mavuluri J, Sharma BR, Balakrishnan A, Malireddi RKS, Geiger R, Zhu Q, et al. 2018. IRF8 regulates transcription of *Naips* for NLRC4 inflammasome activation. *Cell* **173**: 920–933.e13. doi:10.1016/j.cell.2018.02.055

Kayagaki N, Warming S, Lamkanfi M, Vande Walle L, Louie S, Dong J, Newton K, Qu Y, Liu J, Heldens S, et al. 2011. Non-canonical inflammasome activation targets caspase-11. *Nature* **479**: 117–121. doi:10.1038/nature10558

Kayagaki N, Stowe IB, Lee BL, O'Rourke K, Anderson K, Warming S, Cuellar T, Haley B, Roose-Girma M, Phung QT, et al. 2015. Caspase-11 cleaves gasdermin D for non-canonical inflammasome signalling. *Nature* **526**: 666–671. doi:10.1038/nature15541

Kostura MJ, Tocci MJ, Limjuco G, Chin J, Cameron P, Hillman AG, Chartrain NA, Schmidt JA. 1989. Identification of a monocyte specific pre-interleukin 1β convertase activity. *Proc Natl Acad Sci* **86**: 5227–5231. doi:10.1073/pnas.86.14.5227

Kuang S, Zheng J, Yang H, Li S, Duan S, Shen Y, Ji C, Gan J, Xu XW, Li J. 2017. Structure insight of GSDMD reveals the basis of GSDMD autoinhibition in cell pyroptosis. *Proc Natl Acad Sci* **114**: 10642–10647. doi:10.1073/pnas.1708194114

Kuemmerle-Deschner JB. 2015. CAPS—pathogenesis, presentation and treatment of an autoinflammatory disease. *Semin Immunopathol* **37**: 377–385. doi:10.1007/s00281-015-0491-7

Kuida K, Lippke JA, Ku G, Harding MW, Livingston DJ, Su MS, Flavell RA. 1995. Altered cytokine export and apoptosis in mice deficient in interleukin-1β converting enzyme. *Science* **267**: 2000–2003. doi:10.1126/science.7535475

Lamkanfi M. 2011. Emerging inflammasome effector mechanisms. *Nat Rev Immunol* **11**: 213–220. doi:10.1038/nri2936

Lamkanfi M, Dixit VM. 2014. Mechanisms and functions of inflammasomes. *Cell* **157**: 1013–1022. doi:10.1016/j.cell.2014.04.007

Lamkanfi M, Kanneganti TD, Van Damme P, Vanden Berghe T, Vanoverberghe I, Vandekerckhove J, Vandenabeele P, Gevaert K, Núñez G. 2008. Targeted peptidecentric proteomics reveals caspase-7 as a substrate of the caspase-1 inflammasomes. *Mol Cell Proteomics* **7**: 2350–2363. doi:10.1074/mcp.M800132-MCP200

Lamkanfi M, Sarkar A, Vande Walle L, Vitari AC, Amer AO, Wewers MD, Tracey KJ, Kanneganti TD, Dixit VM. 2010. Inflammasome-dependent release of the alarmin HMGB1 in endotoxemia. *J Immunol* **185**: 4385–4392. doi:10.4049/jimmunol.1000803

Lawlor KE, Khan N, Mildenhall A, Gerlic M, Croker BA, D'Cruz AA, Hall C, Kaur Spall S, Anderton H, Masters SL, et al. 2015. RIPK3 promotes cell death and NLRP3 inflammasome activation in the absence of MLKL. *Nat Commun* **6**: 6282. doi:10.1038/ncomms7282

Lee BL, Mirrashidi KM, Stowe IB, Kummerfeld SK, Watanabe C, Haley B, Cuellar TL, Reichelt M, Kayagaki N. 2018a. ASC- and caspase-8-dependent apoptotic pathway diverges from the NLRC4 inflammasome in macrophages. *Sci Rep* **8**: 3788. doi:10.1038/s41598-018-21998-3

Lee BL, Stowe IB, Gupta A, Kornfeld OS, Roose-Girma M, Anderson K, Warming S, Zhang J, Lee WP, Kayagaki N. 2018b. Caspase-11 auto-proteolysis is crucial for noncanonical inflammasome activation. *J Exp Med* **215**: 2279–2288. doi:10.1084/jem.20180589

Liu T, Yamaguchi Y, Shirasaki Y, Shikada K, Yamagishi M, Hoshino K, Kaisho T, Takemoto K, Suzuki T, Kuranaga E, et al. 2014. Single-cell imaging of caspase-1 dynamics reveals an all-or-none inflammasome signaling response. *Cell Rep* **8**: 974–982. doi:10.1016/j.celrep.2014.07.012

Liu X, Zhang Z, Ruan J, Pan Y, Magupalli VG, Wu H, Lieberman J. 2016. Inflammasome-activated gasdermin D causes pyroptosis by forming membrane pores. *Nature* **535**: 153–158. doi:10.1038/nature18629

Liu Z, Wang C, Rathkey JK, Yang J, Dubyak GR, Abbott DW, Xiao TS. 2018. Structures of the gasdermin D C-terminal domains reveal mechanisms of autoinhibition. *Structure* **26**: 778–784.e3. doi:10.1016/j.str.2018.03.002

Lu B, Nakamura T, Inouye K, Li J, Tang Y, Lundbäck P, Valdes-Ferrer SI, Olofsson PS, Kalb T, Roth J, et al. 2012. Novel role of PKR in inflammasome activation and HMGB1 release. *Nature* **488**: 670–674. doi:10.1038/nature11290

Lu A, Magupalli VG, Ruan J, Yin Q, Atianand MK, Vos MR, Schröder GF, Fitzgerald KA, Wu H, Egelman EH. 2014. Unified polymerization mechanism for the assembly of ASC-dependent inflammasomes. *Cell* **156**: 1193–1206. doi:10.1016/j.cell.2014.02.008

Malireddi RK, Ippagunta S, Lamkanfi M, Kanneganti TD. 2010. Cutting edge: proteolytic inactivation of poly(ADP-ribose) polymerase 1 by the Nlrp3 and Nlrc4 inflammasomes. *J Immunol* **185**: 3127–3130. doi:10.4049/jimmunol.1001512

Malireddi RKS, Gurung P, Mavuluri J, Dasari TK, Klco JM, Chi H, Kanneganti T-D. 2018. TAK1 restricts spontaneous NLRP3 activation and cell death to control myeloid proliferation. *J Exp Med* **215**: 1023–1034. doi:10.1084/jem.20171922

Maltez VI, Tubbs AL, Cook KD, Aachoui Y, Falcone EL, Holland SM, Whitmire JK, Miao EA. 2015. Inflammasomes coordinate pyroptosis and natural killer cell cytotoxicity to clear infection by a ubiquitous environmental bacterium. *Immunity* **43**: 987–997. doi:10.1016/j.immuni.2015.10.010

Man SM, Tourlomousis P, Hopkins L, Monie TP, Fitzgerald KA, Bryant CE. 2013. *Salmonella* infection induces recruitment of caspase-8 to the inflammasome to modulate IL-1β production. *J Immunol* **191**: 5239–5246. doi:10.4049/jimmunol.1301581

Mangan MSJ, Olhava EJ, Roush WR, Seidel HM, Glick GD, Latz E. 2018. Targeting the NLRP3 inflammasome in inflammatory diseases. *Nat Rev Drug Discov* **17:** 688. doi:10.1038/nrd.2018.149

Mascarenhas DPA, Cerqueira DM, Pereira MSF, Castanheira FVS, Fernandes TD, Manin GZ, Cunha LD, Zamboni DS. 2017. Inhibition of caspase-1 or gasdermin-D enable caspase-8 activation in the Naip5/NLRC4/ASC inflammasome. *PLoS Pathog* **13:** e1006502. doi:10.1371/journal .ppat.1006502

Masters SL, Gerlic M, Metcalf D, Preston S, Pellegrini M, O'Donnell JA, McArthur K, Baldwin TM, Chevrier S, Nowell CJ, et al. 2012. NLRP1 inflammasome activation induces pyroptosis of hematopoietic progenitor cells. *Immunity* **37:** 1009–1023. doi:10.1016/j.immuni.2012.08 .027

Meunier E, Dick MS, Dreier RF, Schürmann N, Kenzelmann Broz D, Warming S, Roose-Girma M, Bumann D, Kayagaki N, Takeda K, et al. 2014. Caspase-11 activation requires lysis of pathogen-containing vacuoles by IFN-induced GTPases. *Nature* **509:** 366–370. doi:10.1038/ nature13157

Miao EA, Alpuche-Aranda CM, Dors M, Clark AE, Bader MW, Miller SI, Aderem A. 2006. Cytoplasmic flagellin activates caspase-1 and secretion of interleukin 1β via Ipaf. *Nat Immunol* **7:** 569–575. doi:10.1038/ni1344

Moayeri M, Crown D, Newman ZL, Okugawa S, Eckhaus M, Cataisson C, Liu S, Sastalla I, Leppla SH. 2010. Inflammasome sensor Nlrp1b-dependent resistance to anthrax is mediated by caspase-1, IL-1 signaling and neutrophil recruitment. *PLoS Pathog* **6:** e1001222. doi:10.1371/jour nal.ppat.1001222

Monack DM, Raupach B, Hromockyj AE, Falkow S. 1996. *Salmonella typhimurium* invasion induces apoptosis in infected macrophages. *Proc Natl Acad Sci* **93:** 9833–9838. doi:10.1073/pnas.93.18.9833

Monteleone M, Stow JL, Schroder K. 2015. Mechanisms of unconventional secretion of IL-1 family cytokines. *Cytokine* **74:** 213–218. doi:10.1016/j.cyto.2015.03.022

Mosley B, Urdal DL, Prickett KS, Larsen A, Cosman D, Conlon PJ, Gillis S, Dower SK. 1987. The interleukin-1 receptor binds the human interleukin-1α precursor but not the interleukin-1β precursor. *J Biol Chem* **262:** 2941–2944.

Mulvihill E, Sborgi L, Mari SA, Pfreundschuh M, Hiller S, Müller DJ. 2018. Mechanism of membrane pore formation by human gasdermin-D. *EMBO J* **37:** e98321. doi:10 .15252/embj.201798321

Oberst A, Dillon CP, Weinlich R, McCormick LL, Fitzgerald P, Pop C, Hakem R, Salvesen GS, Green DR. 2011. Catalytic activity of the caspase-8-FLIP$_L$ complex inhibits RIPK3-dependent necrosis. *Nature* **471:** 363–367. doi:10.1038/nature09852

Okondo MC, Rao SD, Taabazuing CY, Chui AJ, Poplawski SE, Johnson DC, Bachovchin DA. 2018. Inhibition of Dpp8/9 activates the Nlrp1b inflammasome. *Cell Chem Biol* **25:** 262–267.e5. doi:10.1016/j.chembiol.2017.12.013

Op de Beeck K, Van Laer L, Van Camp G. 2012. *DFNA5*, a gene involved in hearing loss and cancer: a review. *Ann Otol Rhinol Laryngol* **121:** 197–207. doi:10.1177/000 348941212100310

Orning P, Weng D, Starheim K, Ratner D, Best Z, Lee B, Brooks A, Xia S, Wu H, Kelliher MA, et al. 2018. Pathogen blockade of TAK1 triggers caspase-8–dependent cleavage

of gasdermin D and cell death. *Science* **362:** 1064–1069. doi:10.1126/science.aau2818

Papayannopoulos V. 2018. Neutrophil extracellular traps in immunity and disease. *Nat Rev Immunol* **18:** 134–147. doi:10.1038/nri.2017.105

Pilla DM, Hagar JA, Haldar AK, Mason AK, Degrandi D, Pfeffer K, Ernst RK, Yamamoto M, Miao EA, Coers J. 2014. Guanylate binding proteins promote caspase-11-dependent pyroptosis in response to cytoplasmic LPS. *Proc Natl Acad Sci* **111:** 6046–6051. doi:10.1073/pnas .1321700111

Puri AW, Broz P, Shen A, Monack DM, Bogyo M. 2012. Caspase-1 activity is required to bypass macrophage apoptosis upon *Salmonella* infection. *Nat Chem Biol* **8:** 745–747. doi:10.1038/nchembio.1023

Py BF, Jin M, Desai BN, Penumaka A, Zhu H, Kober M, Dietrich A, Lipinski MM, Henry T, Clapham DE, et al. 2014. Caspase-11 controls interleukin-1β release through degradation of TRPC1. *Cell Rep* **6:** 1122–1128. doi:10 .1016/j.celrep.2014.02.015

Rathinam VA, Jiang Z, Waggoner SN, Sharma S, Cole LE, Waggoner L, Vanaja SK, Monks BG, Ganesan S, Latz E, et al. 2010. The AIM2 inflammasome is essential for host defense against cytosolic bacteria and DNA viruses. *Nat Immunol* **11:** 395–402. doi:10.1038/ni.1864

Rathkey JK, Zhao J, Liu Z, Chen Y, Yang J, Kondolf HC, Benson BL, Chirieleison SM, Huang AY, Dubyak GR, et al. 2018. Chemical disruption of the pyroptotic pore-forming protein gasdermin D inhibits inflammatory cell death and sepsis. *Sci Immunol* **3:** eaat2738. doi:10.1126/ sciimmunol.aat2738

Rauch I, Deets KA, Ji DX, von Moltke J, Tenthorey JL, Lee AY, Philip NH, Ayres JS, Brodsky IE, Gronert K, et al. 2017. NAIP-NLRC4 inflammasomes coordinate intestinal epithelial cell expulsion with eicosanoid and IL-18 release via activation of caspase-1 and -8. *Immunity* **46:** 649–659. doi:10.1016/j.immuni.2017.03.016

Rogers C, Fernandes-Alnemri T, Mayes L, Alnemri D, Cingolani G, Alnemri ES. 2017. Cleavage of DFNA5 by caspase-3 during apoptosis mediates progression to secondary necrotic/pyroptotic cell death. *Nat Commun* **8:** 14128. doi:10.1038/ncomms14128

Romberg N, Al Moussawi K, Nelson-Williams C, Stiegler AL, Loring E, Choi M, Overton J, Meffre E, Khokha MK, Huttner AJ, et al. 2014. Mutation of NLRC4 causes a syndrome of enterocolitis and autoinflammation. *Nat Genet* **46:** 1135–1139. doi:10.1038/ng.3066

Ross C, Chan AH, Von Pein J, Boucher D, Schroder K. 2018. Dimerization and auto-processing induce caspase-11 protease activation within the non-canonical inflammasome. *Life Sci Alliance* **1:** e201800237. doi:10.26508/lsa .201800237

Ruan J, Xia S, Liu X, Lieberman J, Wu H. 2018. Cryo-EM structure of the gasdermin A3 membrane pore. *Nature* **557:** 62–67. doi:10.1038/s41586-018-0058-6

Rubartelli A, Cozzolino F, Talio M, Sitia R. 1990. A novel secretory pathway for interleukin-1β, a protein lacking a signal sequence. *EMBO J* **9:** 1503–1510. doi:10.1002/j .1460-2075.1990.tb08268.x

Rühl S, Broz P. 2015. Caspase-11 activates a canonical NLRP3 inflammasome by promoting K$^+$ efflux. *Eur J Immunol* **45:** 2927–2936. doi:10.1002/eji.201545772

Rühl S, Shkarina K, Demarco B, Heilig R, Santos JC, Broz P. 2018. ESCRT-dependent membrane repair negatively regulates pyroptosis downstream of GSDMD activation. *Science* **362**: 956–960. doi:10.1126/science.aar7607

Russo HM, Rathkey J, Boyd-Tressler A, Katsnelson MA, Abbott DW, Dubyak GR. 2016. Active caspase-1 induces plasma membrane pores that precede pyroptotic lysis and are blocked by lanthanides. *J Immunol* **197**: 1353–1367. doi:10.4049/jimmunol.1600699

Saeki N, Kuwahara Y, Sasaki H, Satoh H, Shiroishi T. 2000. Gasdermin (Gsdm) localizing to mouse chromosome 11 is predominantly expressed in upper gastrointestinal tract but significantly suppressed in human gastric cancer cells. *Mamm Genome* **11**: 718–724. doi:10.1007/s003350010138

Sagulenko V, Thygesen SJ, Sester DP, Idris A, Cridland JA, Vajjhala PR, Roberts TL, Schroder K, Vince JE, Hill JM, et al. 2013. AIM2 and NLRP3 inflammasomes activate both apoptotic and pyroptotic death pathways via ASC. *Cell Death Differ* **20**: 1149–1160. doi:10.1038/cdd.2013.37

Santos JC, Dick MS, Lagrange B, Degrandi D, Pfeffer K, Yamamoto M, Meunier E, Pelczar P, Henry T, Broz P. 2018. LPS targets host guanylate-binding proteins to the bacterial outer membrane for non-canonical inflammasome activation. *EMBO J* **37**: e98089. doi:10.15252/embj.201798089

Sarhan J, Liu BC, Muendlein HI, Li P, Nilson R, Tang AY, Rongvaux A, Bunnell SC, Shao F, Green DR, et al. 2018. Caspase-8 induces cleavage of gasdermin D to elicit pyroptosis during *Yersinia* infection. *Proc Natl Acad Sci* **115**: E10888. doi:10.1073/pnas.1809548115

Sborgi L, Rühl S, Mulvihill E, Pipercevic J, Heilig R, Stahlberg H, Farady CJ, Müller DJ, Broz P, Hiller S. 2016. GSDMD membrane pore formation constitutes the mechanism of pyroptotic cell death. *EMBO J* **35**: 1766–1778. doi:10.15252/embj.201694696

Schmid-Burgk JL, Gaidt MM, Schmidt T, Ebert TS, Bartok E, Hornung V. 2015. Caspase-4 mediates non-canonical activation of the NLRP3 inflammasome in human myeloid cells. *Eur J Immunol* **45**: 2911–2917. doi:10.1002/eji.201545523

Schmidt FI, Lu A, Chen JW, Ruan J, Tang C, Wu H, Ploegh HL. 2016. A single domain antibody fragment that recognizes the adaptor ASC defines the role of ASC domains in inflammasome assembly. *J Exp Med* **213**: 771–790. doi:10.1084/jem.20151790

Schneider KS, Groß CJ, Dreier RF, Saller BS, Mishra R, Gorka O, Heilig R, Meunier E, Dick MS, Ćiković T, et al. 2017. The Inflammasome drives GSDMD-independent secondary pyroptosis and IL-1 release in the absence of caspase-1 protease activity. *Cell Rep* **21**: 3846–3859. doi:10.1016/j.celrep.2017.12.018

Shao W, Yeretssian G, Doiron K, Hussain SN, Saleh M. 2007. The caspase-1 digestome identifies the glycolysis pathway as a target during infection and septic shock. *J Biol Chem* **282**: 36321–36329. doi:10.1074/jbc.M708182200

Sharif H, Wang L, Wang WL, Magupalli VG, Andreeva L, Qiao Q, Hauenstein AV, Wu Z, Núñez G, Mao Y, et al. 2019. Structural mechanism for NEK7-licensed activation of NLRP3 inflammasome. *Nature* **570**: 338–343. doi:10.1038/s41586-019-1295-z

Sharma D, Sharma BR, Vogel P, Kanneganti TD. 2017. IL-1β and caspase-1 drive autoinflammatory disease independently of IL-1α or caspase-8 in a mouse model of familial Mediterranean fever. *Am J Pathol* **187**: 236–244. doi:10.1016/j.ajpath.2016.10.015

Shi J, Zhao Y, Wang Y, Gao W, Ding J, Li P, Hu L, Shao F. 2014. Inflammatory caspases are innate immune receptors for intracellular LPS. *Nature* **514**: 187–192. doi:10.1038/nature13683

Shi J, Zhao Y, Wang K, Shi X, Wang Y, Huang H, Zhuang Y, Cai T, Wang F, Shao F. 2015. Cleavage of GSDMD by inflammatory caspases determines pyroptotic cell death. *Nature* **526**: 660–665. doi:10.1038/nature15514

Shimada T, Park BG, Wolf AJ, Brikos C, Goodridge HS, Becker CA, Reyes CN, Miao EA, Aderem A, Götz F, et al. 2010. *Staphylococcus aureus* evades lysozyme-based peptidoglycan digestion that links phagocytosis, inflammasome activation, and IL-1β secretion. *Cell Host Microbe* **7**: 38–49. doi:10.1016/j.chom.2009.12.008

Sollberger G, Choidas A, Burn GL, Habenberger P, Di Lucrezia R, Kordes S, Menninger S, Eickhoff J, Nussbaumer P, Klebl B, et al. 2018. Gasdermin D plays a vital role in the generation of neutrophil extracellular traps. *Sci Immunol* **3**: eaar6689. doi:10.1126/sciimmunol.aar6689

Taabazuing CY, Okondo MC, Bachovchin DA. 2017. Pyroptosis and apoptosis pathways engage in bidirectional crosstalk in monocytes and macrophages. *Cell Chem Biol* **24**: 507–514.e4. doi:10.1016/j.chembiol.2017.03.009

Tamura M, Tanaka S, Fujii T, Aoki A, Komiyama H, Ezawa K, Sumiyama K, Sagai T, Shiroishi T. 2007. Members of a novel gene family, Gsdm, are expressed exclusively in the epithelium of the skin and gastrointestinal tract in a highly tissue-specific manner. *Genomics* **89**: 618–629. doi:10.1016/j.ygeno.2007.01.003

Taylor RC, Cullen SP, Martin SJ. 2008. Apoptosis: controlled demolition at the cellular level. *Nat Rev Mol Cell Biol* **9**: 231–241. doi:10.1038/nrm2312

Terra JK, Cote CK, France B, Jenkins AL, Bozue JA, Welkos SL, LeVine SM, Bradley KA. 2010. Cutting edge: resistance to *Bacillus anthracis* infection mediated by a lethal toxin sensitive allele of *Nalp1b/Nlrp1b*. *J Immunol* **184**: 17–20. doi:10.4049/jimmunol.0903114

Tixeira R, Shi B, Parkes MAF, Hodge AL, Caruso S, Hulett MD, Baxter AA, Phan TK, Poon IKH. 2018. Gasdermin E does not limit apoptotic cell disassembly by promoting early onset of secondary necrosis in Jurkat T cells and THP-1 monocytes. *Front Immunol* **9**: 2842. doi:10.3389/fimmu.2018.02842

Tsuchiya K, Nakajima S, Hosojima S, Thi Nguyen D, Hattori T, Manh Le T, Hori O, Mahib MR, Yamaguchi Y, Miura M, et al. 2019. Caspase-1 initiates apoptosis in the absence of gasdermin D. *Nat Commun* **10**: 2091. doi:10.1038/s41467-019-09753-2

Vanaja SK, Russo AJ, Behl B, Banerjee I, Yankova M, Deshmukh SD, Rathinam VAK. 2016. Bacterial outer membrane vesicles mediate cytosolic localization of LPS and caspase-11 activation. *Cell* **165**: 1106–1119. doi:10.1016/j.cell.2016.04.015

Van Gorp H, Van Opdenbosch N, Lamkanfi M. 2019. Inflammasome-dependent cytokines at the crossroads of health and autoinflammatory disease. *Cold Spring Harb Perspect Biol* **11**: a028563. doi:10.1101/cshperspect.a028563

Van Laer L, Huizing EH, Verstreken M, van Zuijlen D, Wauters JG, Bossuyt PJ, Van de Heyning P, McGuirt WT, Smith RJ, Willems PJ, et al. 1998. Nonsyndromic hearing impairment is associated with a mutation in *DFNA5*. *Nat Genet* **20:** 194–197. doi:10.1038/2503

Van Opdenbosch N, Gurung P, Vande Walle L, Fossoul A, Kanneganti TD, Lamkanfi M. 2014. Activation of the NLRP1b inflammasome independently of ASC-mediated caspase-1 autoproteolysis and speck formation. *Nat Commun* **5:** 3209. doi:10.1038/ncomms4209

Van Opdenbosch N, Van Gorp H, Verdonckt M, Saavedra PHV, de Vasconcelos NM, Goncalves A, Vande Walle L, Demon D, Matusiak M, Van Hauwermeiren F, et al. 2017. Caspase-1 engagement and TLR-induced c-FLIP expression suppress ASC/caspase-8-dependent apoptosis by inflammasome sensors NLRP1b and NLRC4. *Cell Rep* **21:** 3427–3444. doi:10.1016/j.celrep.2017.11.088

Vince JE, De Nardo D, Gao W, Vince AJ, Hall C, McArthur K, Simpson D, Vijayaraj S, Lindqvist LM, Bouillet P, et al. 2018. The mitochondrial apoptotic effectors BAX/BAK activate caspase-3 and -7 to trigger NLRP3 inflammasome and caspase-8 driven IL-1β activation. *Cell Rep* **25:** 2339–2353.e4. doi:10.1016/j.celrep.2018.10.103

Voet S, Srinivasan S, Lamkanfi M, van Loo G. 2019. Inflammasomes in neuroinflammatory and neurodegenerative diseases. *EMBO Mol Med* **11:** e10248. doi:10.15252/emmm.201810248

von Moltke J, Trinidad NJ, Moayeri M, Kintzer AF, Wang SB, van Rooijen N, Brown CR, Krantz BA, Leppla SH, Gronert K, et al. 2012. Rapid induction of inflammatory lipid mediators by the inflammasome in vivo. *Nature* **490:** 107–111. doi:10.1038/nature11351

Wang Y, Gao W, Shi X, Ding J, Liu W, He H, Wang K, Shao F. 2017. Chemotherapy drugs induce pyroptosis through caspase-3 cleavage of a gasdermin. *Nature* **547:** 99–103. doi:10.1038/nature22393

Willingham SB, Bergstralh DT, O'Connor W, Morrison AC, Taxman DJ, Duncan JA, Barnoy S, Venkatesan MM, Flavell RA, Deshmukh M, et al. 2007. Microbial pathogen-induced necrotic cell death mediated by the inflammasome components CIAS1/cryopyrin/NLRP3 and ASC. *Cell Host Microbe* **2:** 147–159. doi:10.1016/j.chom.2007.07.009

Wu C, Lu W, Zhang Y, Zhang G, Shi X, Hisada Y, Grover SP, Zhang X, Li L, Xiang B, et al. 2019. Inflammasome activation triggers blood clotting and host death through pyroptosis. *Immunity* **50:** 1401–1411.e4. doi:10.1016/j.immuni.2019.04.003

Xiao J, Wang C, Yao JC, Alippe Y, Xu C, Kress D, Civitelli R, Abu-Amer Y, Kanneganti TD, Link DC, et al. 2018. Gasdermin D mediates the pathogenesis of neonatal-onset multisystem inflammatory disease in mice. *PLoS Biol* **16:** e3000047. doi:10.1371/journal.pbio.3000047

Xu H, Yang J, Gao W, Li L, Li P, Zhang L, Gong YN, Peng X, Xi JJ, Chen S, et al. 2014. Innate immune sensing of bacterial modifications of Rho GTPases by the Pyrin inflammasome. *Nature* **513:** 237–241. doi:10.1038/nature13449

Yabal M, Müller N, Adler H, Knies N, Groß Christina J, Damgaard Rune B, Kanegane H, Ringelhan M, Kaufmann T, Heikenwälder M, et al. 2014. XIAP restricts TNF- and RIP3-dependent cell death and inflammasome activation. *Cell Rep* **7:** 1796–1808. doi:10.1016/j.celrep.2014.05.008

Yu J, Nagasu H, Murakami T, Hoang H, Broderick L, Hoffman HM, Horng T. 2014. Inflammasome activation leads to caspase-1-dependent mitochondrial damage and block of mitophagy. *Proc Natl Acad Sci* **111:** 15514–15519. doi:10.1073/pnas.1414859111

Zanoni I, Tan Y, Di Gioia M, Broggi A, Ruan J, Shi J, Donado CA, Shao F, Wu H, Springstead JR, et al. 2016. An endogenous caspase-11 ligand elicits interleukin-1 release from living dendritic cells. *Science* **352:** 1232–1236. doi:10.1126/science.aaf3036

Zanoni I, Tan Y, Di Gioia M, Springstead JR, Kagan JC. 2017. By capturing inflammatory lipids released from dying cells, the receptor CD14 induces inflammasome-dependent phagocyte hyperactivation. *Immunity* **47:** 697–709.e3. doi:10.1016/j.immuni.2017.09.010

Zhang WH, Wang X, Narayanan M, Zhang Y, Huo C, Reed JC, Friedlander RM. 2003. Fundamental role of the Rip2/caspase-1 pathway in hypoxia and ischemia-induced neuronal cell death. *Proc Natl Acad Sci* **100:** 16012–16017. doi:10.1073/pnas.2534856100

Zhong FL, Mamaï O, Sborgi L, Boussofara L, Hopkins R, Robinson K, Szeverényi I, Takeichi T, Balaji R, Lau A, et al. 2016. Germline NLRP1 mutations cause skin inflammatory and cancer susceptibility syndromes via inflammasome activation. *Cell* **167:** 187–202.e17. doi:10.1016/j.cell.2016.09.001

Zhong FL, Robinson K, Teo DET, Tan KY, Lim C, Harapas CR, Yu CH, Xie WH, Sobota RM, Au VB, et al. 2018. Human DPP9 represses NLRP1 inflammasome and protects against autoinflammatory diseases via both peptidase activity and FIIND domain binding. *J Biol Chem* **293:** 18864–18878. doi:10.1074/jbc.RA118.004350

Zhu Q, Zheng M, Balakrishnan A, Karki R, Kanneganti T-D. 2018. Gasdermin D promotes AIM2 inflammasome activation and is required for host protection against *Francisella novicida*. *J Immunol* **201:** 3662–3668. doi:10.4049/jimmunol.1800788

Zychlinsky A, Prevost MC, Sansonetti PJ. 1992. *Shigella flexneri* induces apoptosis in infected macrophages. *Nature* **358:** 167–169. doi:10.1038/358167a0

Mechanism and Regulation of Gasdermin-Mediated Cell Death

Shiyu Xia,[1] Louis Robert Hollingsworth IV,[1] and Hao Wu

Program in Cellular and Molecular Medicine, Boston Children's Hospital, and Department of Biological Chemistry and Molecular Pharmacology, Harvard Medical School, Boston, Massachusetts 02115, USA

Correspondence: wu@crystal.harvard.edu

The innate immune system senses and responds to pathogens and endogenous damage through supramolecular protein complexes known as inflammasomes. Cytosolic inflammasome sensor proteins trigger inflammasome assembly on detection of infection and danger. Assembled inflammasomes activate a cascade of inflammatory caspases, which process procytokines and gasdermin D (GSDMD). Cleaved GSDMD forms membrane pores that lead to cytokine release and/or programmed lytic cell death, called pyroptosis. In this review, we provide a primer on pyroptosis and focus on its executioner, the GSDM protein family. In addition to inflammasome-mediated GSDMD pore formation, we describe recently discovered GSDMD activation by caspase-8 and elastase in *Yersinia*-infected macrophages and aging neutrophils, respectively, and GSDME activation by apoptotic caspases. Finally, we discuss strategies that host cells and pathogens use to restrict GSDMD pore formation, in addition to therapeutics targeting the GSDM family.

PYROPTOSIS AND GASDERMIN D ACTIVATION

Canonical and Noncanonical Inflammasome Pathways

The innate immune system recognizes a broad array of exogenous pathogens and endogenous damage to elicit an inflammatory response. To accomplish this, cytosolic sensor proteins (de Vasconcelos and Lamkanfi 2019) rapidly detect a diverse set of pathogen- and danger-associated molecular patterns, PAMPs and DAMPs, respectively (Bergsbaken et al. 2009). In one pathway, PAMPs and DAMPs such as bacterial flagellin, lipopolysaccharides (LPSs) of Gram-negative bacteria, bacterial toxins, uric acid crystals, cytosolic double-stranded DNA (dsDNA), and many others trigger the assembly of supramolecular signaling complexes called inflammasomes (Fig. 1A; Martinon et al. 2006; Pétrilli et al. 2007; Hornung et al. 2009; Zhao et al. 2011; Zhou et al. 2011; Hagar et al. 2013; Kayagaki et al. 2013; de Vasconcelos and Lamkanfi 2019). Inflammasomes activate inflammatory caspases, which include human and mouse caspase-1, human caspase-4 and -5, and mouse caspase-11.

Canonical inflammasomes are comprised of a pattern recognition receptor (PRR) such as NLRP3 or AIM2, the adaptor protein ASC, and the effector caspase-1 (Martinon et al. 2002; Lamkanfi and Dixit 2014; Man and Kanneganti 2015). Different PAMPs and DAMPs

---

[1]These authors contributed equally to this work.

Cite this article as *Cold Spring Harb Perspect Biol* doi: 10.1101/cshperspect.a036400

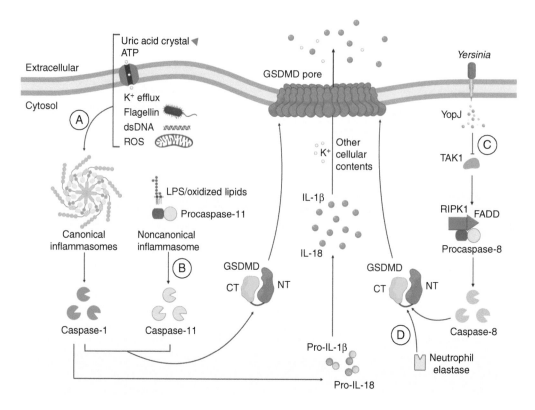

Figure 1. Biological pathways of gasdermin D (GSDMD) activation. (*A*) Canonical inflammasomes triggered by a wide array of pathogen- and danger-associated molecular patterns (DAMPs and PAMPs) activate caspase-1, which cleaves GSDMD and liberates GSDMD-NT for pore formation, cell content release, and pyroptosis. Caspase-1 also cleaves certain interleukin 1 (IL-1) family cytokines into their mature forms. (*B*) A complex of cytosolic lipopolysaccharides (LPSs) and caspase-11 known as the noncanonical inflammasome cleaves GSDMD to generate the pore-forming GSDMD-NT. (*C*) YopJ of *Yersinia pestis* inhibits transforming growth factor β–activated kinase 1 (TAK1) for caspase-8-mediated cleavage of GSDMD and pyroptosis. (*D*) In aging neutrophils, an elastase cleaves and activates GSDMD.

activate unique sensor proteins, leading to their homo-oligomerization. Sensor protein oligomers recruit ASC, which scaffolds procaspase-1 polymerization (Lu et al. 2014). The locally high concentration of procaspase-1 at the inflammasome facilitates its homodimerization, autoprocessing, and activation (Lamkanfi and Dixit 2014; Lu et al. 2014; Man and Kanneganti 2015). Activated caspase-1 subsequently cleaves the procytokines pro-IL-1β and pro-IL-18 into their bioactive forms, IL-1β and IL-18, respectively (Martinon et al. 2002; Faustin et al. 2007), and leads to pyroptosis, a lytic form of programmed cell death (Fink and Cookson 2005; Chen et al. 2016; Zhang et al. 2018).

The molecular platform that leads to the activation of murine caspase-11 or its human homologs caspases-4 and -5 is known as the noncanonical inflammasome (Fig. 1B; Kayagaki et al. 2011; Broz et al. 2012). Intracellular LPSs from Gram-negative bacteria and self-encoded oxidized phospholipids directly bind these caspases, resulting in their oligomerization, autoproteolysis, and activation (Hagar et al. 2013; Kayagaki et al. 2013; Shi et al. 2014; Zanoni et al. 2016; Lee et al. 2018). Although all inflammatory caspases can lead to pyroptosis, the molecular player that mediates this effector function and the mechanism for inflammasome-induced cytokine release were largely unknown until recently.

Cite this article as *Cold Spring Harb Perspect Biol* doi: 10.1101/cshperspect.a036400

GSDMD Identification, Autoinhibition, and Activation by Inflammasomes

To identify a molecule that is responsible for cytokine secretion and pyroptosis, Kayagaki et al. (2015) conducted an *N*-ethyl-*N*-nitrosourea (ENU)-based forward genetic screen in mice to identify mutations that impaired IL-1β secretion in bone marrow macrophages following LPS challenge. Simultaneously, Shi et al. (2015) performed a genome-wide CRISPR-Cas9 knockout screen in murine macrophages for genes involved in caspase-1- and caspase-11-dependent pyroptosis, and He et al. (2015) used quantitative mass spectrometry to discover new inflammasome component proteins. All studies identified gasdermin D (GSDMD) as a direct substrate of the inflammatory caspases downstream from canonical and noncanonical inflammasomes, which is required for both cytokine release and pyroptosis.

GSDMD features a two-domain architecture, with a pore-forming amino-terminal fragment (GSDMD-NT) of ~30 kDa (p30 fragment) and a repressive carboxy-terminal fragment (GSDMD-CT) of ~20 kDa (p20 fragment). These domains are separated by a long disordered interdomain linker cleavable by the inflammatory caspases after Asp276 (D276) (Kayagaki et al. 2015; Shi et al. 2015). On cleavage, GSDMD-NTs translocate to the plasma and mitochondrial membranes to bind acidic lipids, such as phosphatidylinositol phosphates (PIPs), phosphatidylserine (PS), and cardiolipin (CL), oligomerize into a ring-like structure, and insert into the lipid bilayer to form large transmembrane (TM) pores (Fig. 1; Aglietti et al. 2016; Chen et al. 2016; Ding et al. 2016; Liu et al. 2016; Sborgi et al. 2016; Platnich et al. 2018). These GSDMD pores release cytosolic contents, including the inflammatory cytokines IL-1β and IL-18, cause membrane rupture by disrupting osmotic potentials, and eventually result in pyroptotic cell death (Fink and Cookson 2006; Evavold et al. 2018; Heilig et al. 2018). Cleaved GSDMD can also form pores on bacterial membranes, directly killing intracellular bacteria (Liu et al. 2016). Thus, GSDMD facilitates cytokine release, pyroptotic

cell death, and pathogen destruction in innate immunity.

Because of its inflammatory role, pyroptosis was long regarded as exclusive to monocytes, macrophages, and dendritic cells. Recently, however, pyroptosis was shown to occur in various other cell types including epithelial cells, keratinocytes, and peripheral blood mononuclear cells (Shi et al. 2014; Zhong et al. 2016). GSDMD and caspase-4/5/11 are expressed in a wide range of tissues and cell types (Kayagaki et al. 2011; Broz et al. 2012; Saeki and Sasaki 2012; Hagar et al. 2013; Uhlen et al. 2015). These data are consistent with the finding that pyroptosis is not limited to a few subgroups of immune cells.

PORE-FORMING MECHANISM OF THE GSDM FAMILY

GSDMD belongs to the GSDM family, which is comprised of five other members in humans including GSDMA, GSDMB, GSDMC, DFNA5/GSDME, and DFNB59/GSDMF. Mice have three GSDMAs (GSDMA1–3) and four GSDMCs (GSDMC1–4) generated by gene duplication events, and they lack GSDMB (Tamura et al. 2007; Tanaka et al. 2013). Human GSDMB has multiple transcript variants that might show differences in their mechanisms of autoinhibition (Ding et al. 2016; Chao et al. 2017; Chen et al. 2018b; Panganiban et al. 2018). Because the GSDM family shares the common pore-forming activity of GSDMD, the Nomenclature Committee on Cell Death recently redefined pyroptosis as a type of regulated lytic cell death mediated by GSDM pore formation on the plasma membrane, often but not always as a result of inflammatory caspase activation (Kovacs and Miao 2017; Galluzzi et al. 2018).

Mechanism of GSDM Autoinhibition

GSDMs contain an amino-terminal pore-forming domain (GSDM-NT) and a carboxy-terminal autoinhibitory domain (GSDM-CT), with the exception of GSDMF, which lacks homology in the GSDM-CT and might have different regulatory mechanisms (Feng et al.

2018). In the autoinhibited state, GSDM-CT folds back onto GSDM-NT to repress its activity. The basis for this autoinhibitory mechanism was elucidated by the X-ray crystal structures of full-length (FL) mouse GSDMA3 as well as both human and mouse GSDMD (Ding et al. 2016; Kuang et al. 2017; Liu et al. 2018, 2019). GSDM-NT contains both α-helices and extended β-strands, whereas GSDM-CT is almost exclusively α-helical. The long linker between these domains is disordered in these crystal structures, indicating flexibility that underlies its accessibility to caspases and perhaps other unknown activating proteases (Fig. 2A,B). The α1 and α4 helices of GSDM-NT closely interact with GSDM-CT in the autoinhibited state via electrostatic and hydrophobic interactions (Fig. 2A). GSDM-NT and GSDM-CT remain noncovalently associated in vitro following cleavage of the interdomain linker, suggesting that a lipidic environment drives complex separation, GSDM-NT oligomerization, and pore formation (Ding et al. 2016; Ruan et al. 2018).

Mechanism of GSDM Pore Formation

Several GSDMs have been shown to form large TM pores following proteolytic activation of their NTs, including GSDMD, GSDMA3, and GSDME (Aglietti et al. 2016; Chen et al. 2016; Ding et al. 2016; Liu et al. 2016; Wang et al. 2017). The cryo-electron microscopic (cryo-EM) structure of the mouse GSDMA3 pore elucidated the mechanism of pore formation in the GSDM family (Ruan et al. 2018). The pore is a 27-fold symmetric, ~0.8-MDa oligomeric assembly, spanning 18 nm in inner diameter (Fig. 2C). The size of the inner diameter permits the passage of many soluble cytosolic contents to the extracellular space including proinflammatory cytokines IL-1β and IL-18 (Fig. 2C). However, lactate dehydrogenase (LDH), whose release is often used as an indication of lytic cell death, is too large to pass through the GSDMD pore, consistent with the use of LDH release as a hallmark of membrane rupture and cell death. In addition to the 27-subunit structure, 26- and 28-fold symmetric GSDMA3 pores were also reported, indicating heterogeneity and

plasticity in GSDM-NT oligomerization (Mulvihill et al. 2018; Ruan et al. 2018).

The 27-subunit GSDMA3 pore features a complete 108-strand antiparallel β-barrel formed by two β-hairpins from each GSDMA3-NT subunit and a rim adjacent to the cytoplasmic side of the lipid bilayer (Fig. 2C). Whereas the β-barrel with a height of ~7 nm serves as the TM region, the rim is soluble and does not integrate into the membrane. In the cryo-EM structure, the positively charged α1 helix lies in close proximity to the negative lipid headgroup of CL, consistent with the observation that GSDMs bind acidic lipids (Ding et al. 2016; Liu et al. 2016). It should be noted that acidic lipids are not exclusive to the plasma membrane, and the localization of CL in mitochondrial and bacterial membranes raises the possibility of GSDM pore formation elsewhere. The α1 helix is occluded by GSDM-CT in the autoinhibited conformation, and therefore FL GSDMs do not associate with lipids. Superimposition of the GSDMA3 autoinhibited NT and pore structures reveal that large conformational changes occur at the TM β-strands, whereas the globular domain that constitutes the soluble rim remains relatively unaltered (Fig. 2D). On membrane insertion, the entire β3-β4-β5 region of GSDMA3-NT reaches into the membrane to form the first β-hairpin, and the β7-α4-β8 region straightens into the second β-hairpin.

Although other protein families are known to form large β-barrel TM pores, including the membrane attack complex component/perforin (MACPF) and cholesterol-dependent cytolysin (CDC) superfamily, which include pneumolysin and hemolysin (Song et al. 1996; van Pee et al. 2017), GSDMs likely represent a novel class of pore-forming proteins given their unique membrane insertion mechanism and divergence at the amino acid sequence level (Ruan et al. 2018).

ADDITIONAL PATHWAYS REGULATING GSDM ACTIVATION

GSDMD Activation by Other Proteases

In addition to the inflammasome pathways, several other mechanisms for GSDMD activation

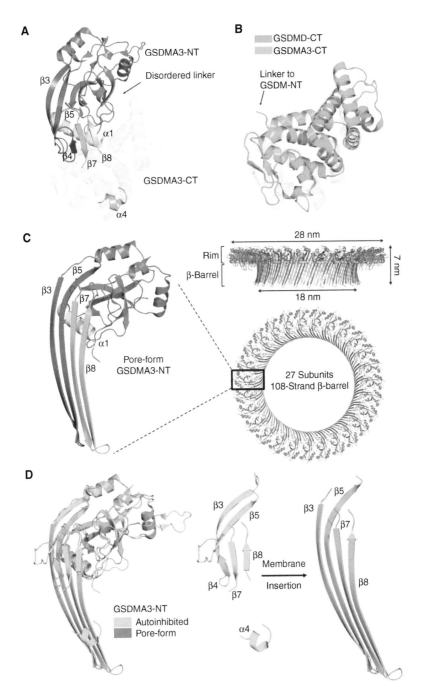

Figure 2. Mechanisms of gasdermin (GSDM) autoinhibition and pore formation. (*A*) Crystal structure of full-length GSDMA3 (PDB ID: 5B5R) shows a two-domain organization of the GSDM family. GSDMA3-CT is colored in orange and key secondary elements in GSDMA3-NT are labeled. (*B*) Structural overlay of mouse GSDMD-CT (PDB ID: 6AO3) and GSDMA3-CT suggests a conserved mechanism of autoinhibition in the GSDM family. (*C*) Cryo-electron microspcopic (cryo-EM) structure of the GSDMA3 pore (PDB ID: 6CB8) shows a large 27-subunit ring with a 108-strand β-barrel as the TM region. Each subunit contributes an α1 helix that directly interacts with acidic lipid heads and two β-hairpins that line the β-barrel. (*D*) Structural overlay of GSDMA3-NT in autoinhibited and membrane-inserted conformations reveals drastic conformational changes in the β3-β4-β5 and β7-α4-β8 regions, which form the two TM β-hairpins. (PyMOL session files are available at osf.io/x7dv8.)

have recently been elucidated. One such activation pathway occurs through transforming growth factor β–activated kinase 1 (TAK1) and caspase-8 in response to *Yersinia pestis* infection (Fig. 1C). Under normal conditions, TAK1 restricts the NLRP3 inflammasome and phosphorylation of receptor-interacting protein 1 (RIPK1) (Geng et al. 2017; Malireddi et al. 2018). YopJ, a *Y. pestis* effector molecule, binds TAK1 to abolish its repression of RIPK1 (Orning et al. 2018; Sarhan et al. 2018). Active RIPK1 then recruits FADD and procaspase-8 to form a structure called the RIPoptosome, which activates caspase-8 through autoprocessing. Active caspase-8 then cleaves GSDMD to promote pyroptosis (Fig. 1C; Orning et al. 2018; Sarhan et al. 2018).

Although the majority of GSDMD-activating enzymes are caspases, a GSDMD activator in aging neutrophils, an inflammatory cell type (Mayadas et al. 2014), is the neutrophil elastase (ELANE). ELANE cleaves GSDMD at a location upstream of the caspase cleavage site to generate functional GSDMD-NTs that form pores and result in neutrophil death (Fig. 1D; Kambara et al. 2018). ELANE-mediated cleavage of GSDMD also facilitates the formation of neutrophil extracellular traps (NETs) in response to noncanonical inflammasome stimuli (Sollberger et al. 2018; Chen et al. 2018a). Neutrophils expel these NETs, antimicrobial weblike structures composed of DNA and proteins, under certain pathogenic conditions. GSDMD facilitates the formation of NETs by forming pores in organelles, as well as their extrusion into the surrounding extracellular milieu by causing cell rupture (Chen et al. 2018a; Sollberger et al. 2018). Thus, the function of the GSDM family likely extends beyond pyroptosis and can be modulated through different upstream activating enzymes.

GSDME Activation by Caspase-3

GSDME was recently shown to form pores to mediate pyroptosis following its cleavage and activation by caspase-3 (Wang et al. 2017). Interestingly, caspase-3 is traditionally defined as an apoptotic caspase, highlighting a role of GSDME in cross talk between multiple cell death pathways. Additionally, GSDME functions as a tumor suppressor by potentiating pyroptosis in response to chemotherapeutics that activate caspase-3 and becomes silenced in many tumors. Cancer cells that express high levels of GSDME undergo pyroptosis in response to chemotherapies that generally result in apoptosis (Rogers et al. 2017; Wang et al. 2018). Further research into the regulation of GSDME expression and activation will elucidate the complex circuitry underlying its cross talk among multiple cell death pathways.

Proteolytic Restriction of GSDMD by Caspase-3/7 and Enterovirus71

Before pore formation, cleavage of GSDMD at noncanonical amino-terminal sites (amino-terminal to the caspase-1/11 cleavage site) can abolish its ability to form functional pores. One such mechanism involves proteolysis by apoptotic caspases (MacFarlane 2019), particularly caspases-3 and -7 (Fig. 3B; Taabazuing et al. 2017). In murine monocytes, multiple apoptotic stimuli resulted in a restricted "p43" GSDMD fragment, rather than the canonical p30 (GSDMD-NT) and p20 (GSDMD-CT) fragments, suggesting that apoptotic caspases might inactivate GSDMD. In vitro cleavage reactions showed that caspase-3 and caspase-7 indeed cleaved GSDMD at a different site than inflammatory caspases, and the cleavage site was mapped to aspartate 87 (D87). Neither the 1–87 nor 88–484 GSDMD fragment was toxic to HEK293T cells (Taabazuing et al. 2017), confirming that caspase-3 and caspase-7 inhibit pore formation by cleaving GSDMD within its NT domain. Conversely, however, a related GSDM family member, GSDME, can be activated by caspase-3 (Rogers et al. 2017; Wang et al. 2017). Cross talk between apoptosis and pyroptosis likely prevents an inflammatory response during apoptosis, which could otherwise lead to chronic inflammatory diseases (Nagata 2018). Failure to execute apoptosis might necessitate other pathways such as pyroptosis to ensure cell death.

Pathogens would also benefit from preventing pyroptosis by silencing pore formation, thus

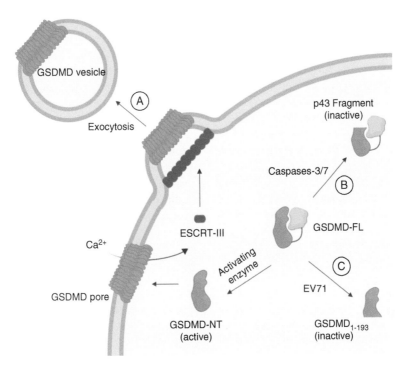

Figure 3. Pyroptotic death evasion through gasdermin D (GSDMD). (*A*) GSDMD pores are exocytosed from the cell through the endosomal sorting complexes required for transport (ESCRT)-III machinery. (*B*) Caspase-3 and caspase-7 cleave the GSDMD-NT after D87, generating a nonfunctional protein fragment. (*C*) The 3C protease of enterovirus 71 cleaves the GSDMD-NT at a noncanonical site, generating an inactive NT fragment.

protecting their niche and avoiding the innate immune response. One such case is enterovirus 71 (EV71), the causative agent of hand-foot-and-mouth disease. The 3C protease of EV71 cleaves GSDMD at an amino-terminal site, inhibiting GSDMD (Fig. 3C; Lei et al. 2017). EV71 infection resulted in this GSDMD cleavage in multiple cell lines (Lei et al. 2017). The same GSDMD cleavage bands were observed by co-expressing EV71 3C protease with GSDMD, and its catalytic activity was necessary for GSDMD cleavage. By mutating a 3C-like consensus site, glutamate 193 (Q193) of GSDMD was identified as the EV71 3C protease cleavage site, which is amino-terminal to the conventional caspase cleavage site (D276). The truncated GSDMD-NT, residues 1–193, could not cause cell death (Lei et al. 2017). Although EV71 remains the only known pathogen to cleave GSDMD

directly, other pathogens have likely evolved such mechanisms.

Mechanisms of Cytokine Release without Pyroptosis

Although GSDMD pore formation generally leads to pyroptotic cell death and membrane rupture, recent discoveries call into question the finality of eliciting pyroptosis. Several studies showed uncoupling of inflammasome activation and IL-1β release from cell death in multiple cell types, suggesting that live cells might release cytokines through GSDMD pores and perhaps even survive pore formation. The first line of such evidence showed that murine neutrophils undergo NLRC4-dependent cytokine maturation in response to *Salmonella* stimulus but not concomitant loss of membrane integrity and cell death (Chen et al. 2014). Macrophages underwent cell death under the same stimuli (Chen et al. 2014) but were also recently shown capable of releasing cytokines while maintaining membrane integrity when buffered with glycine, an osmoprotectant (Evavold et al. 2018). Murine macrophages evade cell death in the presence of Gram-negative peptidoglycan, which results in cytokine release with unconventional signaling through the metabolic enzyme hexokinase and the NLRP3 inflammasome (Wolf et al. 2016). In dendritic cells, NLRP3 stimuli such as LPS and ATP resulted in pyroptotic cell death, whereas oxidized phospholipids, which often accompany tissue cell death (Chang et al. 2004), triggered cytokine release without eliciting pyroptotic cell death (Zanoni et al. 2016). The NLRP3 inflammasome also mediates living IL-1β release in peripheral blood mononuclear cells (Gaidt et al. 2016). Thus, several cell types were shown capable of releasing inflammatory cytokines through GSDMD pores without concomitant cell death.

Of note, many of these studies used bulk cell assays, in which it is difficult to distinguish cytokine release from a small population of dying cells if IL-1β effect sizes are not sufficiently large. Furthermore, cell death (cessation of cell movement/mitochondrial integrity) likely precedes GSDMD-mediated cell lysis (DiPeso et al. 2017), confounding the ability to experimentally decouple pore formation (PI uptake), cell death, and cell lysis (LDH release). Nonetheless, prolonged cytokine maturation and release in living cells would be a reasonable mechanism for a more robust recruitment of immune effector cells. Indeed, several mechanisms to evade GSDMD-mediated cell death have been discovered, such as the exocytosis of GSDMD pore-containing membrane vesicles (Rühl et al. 2018) as described below.

GSDMD cleavage leads to its oligomerization and membrane insertion, which is likely an energetically irreversible process. How, then, can cells repair these GSDMD pores once they have been inserted into the plasma membrane? The first and only known mechanism of such repair involves exocytosis of vesicles containing GSDMD pores (Fig. 3A). Removal of GSDMD

pores from the cell membrane prevents or prolongs membrane lysis (Rühl et al. 2018). This pathway occurs through the endosomal sorting complexes required for transport (ESCRT)-III machinery (Rühl et al. 2018), which mediates membrane fission and subsequent exocytosis of membrane vesicles in many cellular processes (McCullough et al. 2018), including exocytosis of mixed lineage kinase domain like pseudokinase (MLKL) pores in necroptosis (Gong et al. 2017).

Similar to MLKL pores (Gong et al. 2017), GSDMD pores cause rapid calcium ion (Ca^{2+}) influx (Rühl et al. 2018), which is known to activate ESCRT-III (Scheffer et al. 2014). Inhibition of ESCRT-III through Ca^{2+} chelation or knockdowns of ESCRT-III components led to increased cell death in response to several inflammasome stimuli, including those that activate NLRC4, AIM2, and noncanonical activation of NLRP3 through caspase-11. Additionally, fluorescently labeled CHMP4, a component of the ESCRT-III machinery, formed membrane-localized puncta in response to inflammasome stimulus (Rühl et al. 2018), consistent with other ESCRT-III-mediated repair processes (Jimenez et al. 2014). Thus, cells can shed GSDMD pores to prevent or prolong pyroptotic cell death, enabling survival and/or increased cytokine maturation (Fig. 3). It remains unclear whether additional cellular mechanisms disassemble or destroy GSDMD pores, and the functional outcomes of these mechanisms in vivo.

Pyroptosis in Disease

Precise discrimination between healthy and infected host cells is an important challenge of innate immunity, as unchecked activity can lead to autoimmune disorders. Thus, although phagocytes, including neutrophils and macrophages, efficiently detect pathogens in the extracellular space, pathogens that enter host cells might escape surveillance. Pyroptosis serves a crucial protective function of alerting the immune system to intracellular pathogenic insult. Pyroptosis directly kills compromised host cells, removing the niche for invading pathogens. Additionally, release of cytokines through GSDMD

Cite this article as *Cold Spring Harb Perspect Biol* doi: 10.1101/cshperspect.a036400

pores establishes a cytokine gradient that directs phagocytes to the site of infection. Moreover, tears on the membranes of postlytic pyroptotic cells are small enough to retain large organelles and bacteria by forming an intracellular trap, driving the recruitment of phagocytes to engulf the trapped bacteria and cell debris through a process termed efferocytosis (Jorgensen et al. 2016; Davidson and Wood 2019; Moldoveanu and Czabotar 2019).

Despite its importance to pathogen defense, inappropriate or hyperactive pyroptosis can be highly pathological, causing damaging inflammation that contributes to diseases such as arthritis, inflammatory bowel disease, Alzheimer's, and sepsis (Lamkanfi and Dixit 2012). Thus, inhibiting pyroptosis might improve autoimmune and autoinflammatory disease prognoses.

Inhibiting Pyroptosis with Small Molecules

Because GSDMD integrates a multitude of endogenous and exogenous upstream signals (de Vasconcelos and Lamkanfi 2019) by executing pyroptosis, it presents a highly desirable molecular target for therapeutic intervention. Mechanistically, strategies to directly inhibit GSDMD could involve preventing its cleavage, lipid binding, oligomerization, or membrane insertion (Fig. 4A). Recently, four small molecules were identified as targeting GSDMD: necrosulfonamide (Rathkey et al. 2018), LDC7559 (Sollberger et al. 2018), Bay 11-7082, and disulfiram (Hu et al. 2018). Necrosulfonamide was originally identified as a cysteine reactive drug that targeted MLKL to inhibit necroptosis (Sun et al. 2012) and might also inhibit multiple steps of the pyroptotic pathway (Rathkey et al. 2018). LDC7559 inhibited NET formation, presumably by binding and suppressing GSDMD pore formation (Sollberger et al. 2018). Bay 11-7082 and disulfiram were discovered through drug screening efforts using an in vitro assay for GSDMD pore formation and disulfiram was shown to suppress pyroptosis in vitro and in vivo (Hu et al. 2018).

In characterizing these drug candidates, a reactive cysteine on GSDMD, C191 (C192 in mouse), was identified as the target for covalent modification by necrosulfonamide, disulfiram, and Bay 11-7082. Based on the structure of the homologous GSDMA3 pore (Ruan et al. 2018), C191 likely lies on the tip of a β-hairpin that sits at the membrane interface (Fig. 4B), and modification of C191 by drug candidates could prevent its membrane insertion and/or oligomerization (Liu et al. 2016). Ongoing efforts to target GSDMD will benefit from a more complete knowledge of its membrane insertion mechanism. Conversely, cellular studies will benefit from potent and specific small-molecule inhibitors of GSDMD.

CONCLUSIONS AND FUTURE PERSPECTIVES

During the last several years, a rich body of research has established the critical function of the GSDM family in programmed cell death. Inflammatory caspases are activated upon sensing PAMPs and DAMPs by inflammasomes, resulting in the cleavage of autoinhibited GSDMD and liberation of GSDMD-NT to form membrane pores. In addition to inflammasome activation, *Yersinia* infection and ELANE trigger cleavage of GSDMD. GSDMD-NT pore formation on the plasma membrane facilitates the release of soluble cytosolic contents including mature IL-1 family cytokines and often leads to lytic, pyroptotic cell death. To prevent pyroptosis, host cells and pathogens have evolved mechanisms for repairing GSDMD-perforated membranes and inactivating GSDMD.

GSDME was also shown to be an executioner of pyroptosis following cleavage by active caspase-3 (Rogers et al. 2017; Wang et al. 2017). Despite our increasing knowledge of GSDMD and GSDME, the biological functions remain elusive for most GSDMs. In fact, although conserved in domain organization, many GSDM family members mediate cellular processes beyond pyroptosis. For example, a main function of murine GSDMA3 is to regulate the production of reactive oxygen species by forming pores on mitochondrial membranes (Lin et al. 2015), and GSDMD functions in NETosis (Chen et al. 2018a; Sollberger et al. 2018). Underlying the diverse functions of the GSDM family are

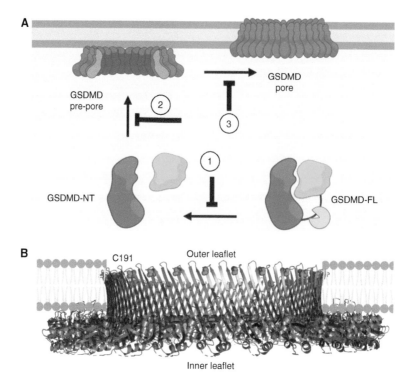

Figure 4. Inhibiting gasdermin D (GSDMD) with small molecules. (*A*) Potential routes of small-molecule intervention include (1) cleavage of GSDMD and release of GSDMD-NT, (2) lipid binding and oligomerization of GSDMD-NT, and (3) membrane insertion of GSDMD-NT. (*B*) Small-molecule drugs disulfiram, necrosulfonamide, and Bay 11-702 target C191 of GSDMD and therefore might disrupt membrane insertion and/or oligomerization of GSDMD-NT. C191, highlighted in red, is located at the tip of a β-hairpin according to a model of the GSDMD pore, generated based on the structure of the GSDMA3 pore (PDB ID: 6CB8). (PyMOL session files are available at osf.io/x7dv8.)

intricate signaling networks that might have been previously underappreciated. GSDM-activating enzymes are not strictly inflammatory caspases, and increasing evidence indicates that cell death pathways are regulated by multiple cross-talking activating enzymes (Vince and Silke 2016; Mascarenhas et al. 2017; Rauch et al. 2017; Taabazuing et al. 2017). A more comprehensive investigation into the mechanisms and regulation of the GSDM family will benefit future understanding of GSDM-mediated diseases and the development of GSDM-targeting therapeutics.

ACKNOWLEDGMENTS

This work was supported by National Institutes of Health (NIH) grants DP1HD087988 and R01AI124491 to H.W. and Albert J. Ryan Fellowships to S.X. and L.R.H. We thank J. Ruan, C. Shen, J. Hu, and A. Brown for discussion and critical reading of the manuscript, and BioRender.com for figure design. The authors apologize for incomplete citations due to space limitations.

REFERENCES

*Reference is also in this collection.

Aglietti RA, Estevez A, Gupta A, Ramirez MG, Liu PS, Kayagaki N, Ciferri C, Dixit VM, Dueber EC. 2016. GsdmD p30 elicited by caspase-11 during pyroptosis forms pores in membranes. *Proc Natl Acad Sci* **113:** 7858–7863. doi:10.1073/pnas.1607769113

Bergsbaken T, Fink SL, Cookson BT. 2009. Pyroptosis: Host cell death and inflammation. *Nat Rev Microbiol* **7:** 99–109. doi:10.1038/nrmicro2070

Broz P, Ruby T, Belhocine K, Bouley DM, Kayagaki N, Dixit VM, Monack DM. 2012. Caspase-11 increases susceptibility to *Salmonella* infection in the absence of caspase-1. *Nature* **490:** 288–291. doi:10.1038/nature11419

Chang M-K, Binder CJ, Miller YI, Subbanagounder G, Silverman GJ, Berliner JA, Witztum JL. 2004. Apoptotic cells with oxidation-specific epitopes are immunogenic and proinflammatory. *J Exp Med* **200:** 1359–1370. doi:10.1084/jem.20031763

Chao KL, Kulakova L, Herzberg O. 2017. Gene polymorphism linked to increased asthma and IBD risk alters gasdermin-B structure, a sulfatide and phosphoinositide binding protein. *Proc Natl Acad Sci* **114:** E1128–E1137. doi:10.1073/pnas.1616783114

Chen KW, Groß CJ, Sotomayor FV, Stacey KJ, Tschopp J, Sweet MJ, Schroder K. 2014. The neutrophil NLRC4 inflammasome selectively promotes IL-1β maturation without pyroptosis during acute *Salmonella* challenge. *Cell Rep* **8:** 570–582. doi:10.1016/j.celrep.2014.06.028

Chen X, He WT, Hu L, Li J, Fang Y, Wang X, Xu X, Wang Z, Huang X, Han J. 2016. Pyroptosis is driven by non-selective gasdermin-D pore and its morphology is different from MLKL channel-mediated necroptosis. *Cell Res* **26:** 1007–1020. doi:10.1038/cr.2016.100

Chen KW, Monteleone M, Boucher D, Sollberger G, Ramnath D, Condon ND, von Pein JB, Broz P, Sweet MJ, Schroder K. 2018a. Noncanonical inflammasome signaling elicits gasdermin D–dependent neutrophil extracellular traps. *Sci Immunol* **3:** eaar6676. doi:10.1126/sciimmunol.aar6676

Chen Q, Shi P, Wang Y, Zou D, Wu X, Wang D, Hu Q, Zou Y, Huang Z, Ren J, et al. 2018b. GSDMB promotes non-canonical pyroptosis by enhancing caspase-4 activity. *J Mol Cell Biol* **11:** 496–508. doi:10.1093/jmcb/mjy056

* Davidson AJ, Wood W. 2019. Phagocyte responses to cell death in flies. *Cold Spring Harb Perspect Biol*. doi:10.1101/cshperspect.a036350

* de Vasconcelos NM, Lamkanfi M. 2019. Recent insights on inflammasomes, gasdermin pores, and pyroptosis. *Cold Spring Harb Perspect Biol*. doi:10.1101/cshperspect.a036392

Ding J, Wang K, Liu W, She Y, Sun Q, Shi J, Sun H, Wang DC, Shao F. 2016. Pore-forming activity and structural autoinhibition of the gasdermin family. *Nature* **535:** 111–116. doi:10.1038/nature18590

DiPeso L, Ji DX, Vance RE, Price JV. 2017. Cell death and cell lysis are separable events during pyroptosis. *Cell Death Discov* **3:** 17070. doi:10.1038/cddiscovery.2017.70

Evavold CL, Ruan J, Tan Y, Xia S, Wu H, Kagan JC. 2018. The pore-forming protein gasdermin D regulates interleukin-1 secretion from living macrophages. *Immunity* **48:** 35–44.e6. doi:10.1016/j.immuni.2017.11.013

Faustin B, Lartigue L, Bruey JM, Luciano F, Sergienko E, Bailly-Maitre B, Volkmann N, Hanein D, Rouiller I, Reed JC. 2007. Reconstituted NALP1 inflammasome reveals two-step mechanism of caspase-1 activation. *Mol Cell* **25:** 713–724. doi:10.1016/j.molcel.2007.01.032

Feng S, Fox D, Man SM. 2018. Mechanisms of gasdermin family members in inflammasome signaling and cell death. *J Mol Biol* **430:** 3068–3080. doi:10.1016/j.jmb.2018.07.002

Fink SL, Cookson BT. 2005. Apoptosis, pyroptosis, and necrosis: Mechanistic description of dead and dying

eukaryotic cells. *Infect Immun* **73:** 1907–1916. doi:10.1128/IAI.73.4.1907-1916.2005

Fink SL, Cookson BT. 2006. Caspase-1-dependent pore formation during pyroptosis leads to osmotic lysis of infected host macrophages. *Cell Microbiol* **8:** 1812–1825. doi:10.1111/j.1462-5822.2006.00751.x

Gaidt MM, Ebert TS, Chauhan D, Schmidt T, Schmid-Burgk JL, Rapino F, Robertson Avril AB, Cooper MA, Graf T, Hornung V. 2016. Human monocytes engage an alternative inflammasome pathway. *Immunity* **44:** 833–846. doi:10.1016/j.immuni.2016.01.012

Galluzzi L, Vitale I, Aaronson SA, Abrams JM, Adam D, Agostinis P, Alnemri ES, Altucci L, Amelio I, Andrews DW, et al. 2018. Molecular mechanisms of cell death: Recommendations of the Nomenclature Committee on Cell Death 2018. *Cell Death Differ* **25:** 486–541. doi:10.1038/s41418-017-0012-4

Geng J, Ito Y, Shi L, Amin P, Chu J, Ouchida AT, Mookhtiar AK, Zhao H, Xu D, Shan B, et al. 2017. Regulation of RIPK1 activation by TAK1-mediated phosphorylation dictates apoptosis and necroptosis. *Nat Commun* **8:** 359. doi:10.1038/s41467-017-00406-w

Gong YN, Guy C, Olauson H, Becker JU, Yang M, Fitzgerald P, Linkermann A, Green DR. 2017. ESCRT-III acts downstream of MLKL to regulate necroptotic cell death and its consequences. *Cell* **169:** 286–300.e16. doi:10.1016/j.cell.2017.03.020

Hagar JA, Powell DA, Aachoui Y, Ernst RK, Miao EA. 2013. Cytoplasmic LPS activates caspase-11: Implications in TLR4-independent endotoxic shock. *Science* **341:** 1250–1253. doi:10.1126/science.1240988

He WT, Wan H, Hu L, Chen P, Wang X, Huang Z, Yang ZH, Zhong CQ, Han J. 2015. Gasdermin D is an executor of pyroptosis and required for interleukin-1β secretion. *Cell Res* **25:** 1285–1298.

Heilig R, Dick MS, Sborgi L, Meunier E, Hiller S, Broz P. 2018. The gasdermin-D pore acts as a conduit for IL-1β secretion in mice. *Eur J Immunol* **48:** 584–592. doi:10.1002/eji.201747404

Hornung V, Ablasser A, Charrel-Dennis M, Bauernfeind F, Horvath G, Caffrey DR, Latz E, Fitzgerald KA. 2009. AIM2 recognizes cytosolic dsDNA and forms a caspase-1-activating inflammasome with ASC. *Nature* **458:** 514–518. doi:10.1038/nature07725

Hu JJ, Liu X, Zhao J, Xia S, Ruan J, Luo X, Kim J, Lieberman J, Wu H. 2018. Identification of pyroptosis inhibitors that target a reactive cysteine in gasdermin D. bioRxiv doi:10.1101/365908.

Jimenez AJ, Maiuri P, Lafaurie-Janvore J, Divoux S, Piel M, Perez F. 2014. ESCRT machinery is required for plasma membrane repair. *Science* **343:** 1247136. doi:10.1126/science.1247136

Jorgensen I, Zhang Y, Krantz BA, Miao EA. 2016. Pyroptosis triggers pore-induced intracellular traps (PITs) that capture bacteria and lead to their clearance by efferocytosis. *J Exp Med* **213:** 2113–2128. doi:10.1084/jem.20151613

Kambara H, Liu F, Zhang X, Liu P, Bajrami B, Teng Y, Zhao L, Zhou S, Yu H, Zhou W, et al. 2018. Gasdermin D exerts anti-inflammatory effects by promoting neutrophil death. *Cell Rep* **22:** 2924–2936. doi:10.1016/j.celrep.2018.02.067

Kayagaki N, Warming S, Lamkanfi M, Vande Walle L, Louie S, Dong J, Newton K, Qu Y, Liu J, Heldens S, et al. 2011.

Non-canonical inflammasome activation targets caspase-11. *Nature* **479:** 117–121. doi:10.1038/nature10558

Kayagaki N, Wong MT, Stowe IB, Ramani SR, Gonzalez LC, Akashi-Takamura S, Miyake K, Zhang J, Lee WP, Muszynski A, et al. 2013. Noncanonical inflammasome activation by intracellular LPS independent of TLR4. *Science* **341:** 1246–1249. doi:10.1126/science.1240248

Kayagaki N, Stowe IB, Lee BL, O'Rourke K, Anderson K, Warming S, Cuellar T, Haley B, Roose-Girma M, Phung QT, et al. 2015. Caspase-11 cleaves gasdermin D for non-canonical inflammasome signalling. *Nature* **526:** 666–671. doi:10.1038/nature15541

Kovacs SB, Miao EA. 2017. Gasdermins: Effectors of pyroptosis. *Trends Cell Biol* **27:** 673–684. doi:10.1016/j.tcb.2017.05.005

Kuang S, Zheng J, Yang H, Li S, Duan S, Shen Y, Ji C, Gan J, Xu XW, Li J. 2017. Structure insight into the basis of GSDMD autoinhibition in cell pyroptosis. *Proc Natl Acad Sci* **114:** 10642–10647. doi:10.1073/pnas.1708194114

Lamkanfi M, Dixit VM. 2012. Inflammasomes and their roles in health and disease. *Annu Rev Cell Dev Biol* **28:** 137–161. doi:10.1146/annurev-cellbio-101011-155745

Lamkanfi M, Dixit VM. 2014. Mechanisms and functions of inflammasomes. *Cell* **157:** 1013–1022. doi:10.1016/j.cell.2014.04.007

Lee BL, Stowe IB, Gupta A, Kornfeld OS, Roose-Girma M, Anderson K, Warming S, Zhang J, Lee WP, Kayagaki N. 2018. Caspase-11 auto-proteolysis is crucial for non-canonical inflammasome activation. *J Exp Med* **215:** 2279–2288. doi:10.1084/jem.20180589

Lei X, Zhang Z, Xiao X, Qi J, He B, Wang J. 2017. Enterovirus 71 inhibits pyroptosis through cleavage of gasdermin D. *J Virol* **91:** e01069. doi:10.1128/JVI.01069-17

Lin PH, Lin HY, Kuo CC, Yang LT. 2015. N-terminal functional domain of gasdermin A3 regulates mitochondrial homeostasis via mitochondrial targeting. *J Biomed Sci* **22:** 44. doi:10.1186/s12929-015-0152-0

Liu X, Zhang Z, Ruan J, Pan Y, Magupalli VG, Wu H, Lieberman J. 2016. Inflammasome-activated gasdermin D causes pyroptosis by forming membrane pores. *Nature* **535:** 153–158. doi:10.1038/nature18629

Liu Z, Wang C, Rathkey JK, Yang J, Dubyak GR, Abbott DW, Xiao TS. 2018. Structures of the gasdermin D C-terminal domains reveal mechanisms of autoinhibition. *Structure* **26:** 778–784.e3. doi:10.1016/j.str.2018.03.002

Liu Z, Wang C, Yang J, Zhou B, Yang R, Ramachandran R, Abbott DW, Xiao TS. 2019. Crystal structures of the full-length murine and human gasdermin D reveal mechanisms of autoinhibition, lipid binding, and oligomerization. *Immunity* **51:** 43–49.e44. doi:10.1016/j.immuni.2019.04.017

Lu A, Magupalli VG, Ruan J, Yin Q, Atianand MK, Vos MR, Schröder GF, Fitzgerald KA, Wu H, Egelman EH. 2014. Unified polymerization mechanism for the assembly of ASC-dependent inflammasomes. *Cell* **156:** 1193–1206. doi:10.1016/j.cell.2014.02.008

Malireddi RKS, Gurung P, Mavuluri J, Dasari TK, Klco JM, Chi H, Kanneganti TD. 2018. TAK1 restricts spontaneous NLRP3 activation and cell death to control myeloid proliferation. *J Exp Med* **215:** 1023–1034. doi:10.1084/jem.20171922

Man SM, Kanneganti TD. 2015. Regulation of inflammasome activation. *Immunol Rev* **265:** 6–21. doi:10.1111/imr.12296

Martinon F, Burns K, Tschopp J. 2002. The inflammasome: A molecular platform triggering activation of inflammatory caspases and processing of proIL-β. *Mol Cell* **10:** 417–426. doi:10.1016/S1097-2765(02)00599-3

Martinon F, Pétrilli V, Mayor A, Tardivel A, Tschopp J. 2006. Gout-associated uric acid crystals activate the NALP3 inflammasome. *Nature* **440:** 237–241. doi:10.1038/nature04516

Mascarenhas DPA, Cerqueira DM, Pereira MSF, Castanheira FVS, Fernandes TD, Manin GZ, Cunha LD, Zamboni DS. 2017. Inhibition of caspase-1 or gasdermin-D enable caspase-8 activation in the Naip5/NLRC4/ASC inflammasome. *PLoS Pathog* **13:** e1006502. doi:10.1371/journal.ppat.1006502

Mayadas TN, Cullere X, Lowell CA. 2014. The multifaceted functions of neutrophils. *Annu Rev Pathol* **9:** 181–218. doi:10.1146/annurev-pathol-020712-164023

McCullough J, Frost A, Sundquist WI. 2018. Structures, functions, and dynamics of ESCRT-III/Vps4 membrane remodeling and fission complexes. *Annu Rev Cell Dev Biol* **34:** 85–109. doi:10.1146/annurev-cellbio-100616-060600

* MacFarlane M. 2019. Mechanisms of caspase activation. *Cold Spring Harb Perspect Biol* doi:10.1101/cshperspect.a036335

* Moldoveanu T, Czabotar PE. 2019. BAX, BAK, and BOK; a coming of age for the BCL-2 family effector proteins. *Cold Spring Harb Perspect Biol* doi:10.1101/cshperspect.a036319

Mulvihill E, Sborgi L, Mari SA, Pfreundschuh M, Hiller S, Müller DJ. 2018. Mechanism of membrane pore formation by human gasdermin-D. *EMBO J* **37:** e98321. doi:10.15252/embj.201798321

Nagata S. 2018. Apoptosis and clearance of apoptotic cells. *Annu Rev Immunol* **36:** 489–517. doi:10.1146/annurev-immunol-042617-053010

Orning P, Weng D, Starheim K, Ratner D, Best Z, Lee B, Brooks A, Xia S, Wu H, Kelliher MA, et al. 2018. Pathogen blockade of TAK1 triggers caspase-8-dependent cleavage of gasdermin D and cell death. *Science* **362:** 1064–1069. doi:10.1126/science.aau2818

Panganiban RA, Sun M, Dahlin A, Park HR, Kan M, Himes BE, Mitchel JA, Iribarren C, Jorgenson E, Randell SH, et al. 2018. A functional splice variant associated with decreased asthma risk abolishes the ability of gasdermin B to induce epithelial cell pyroptosis. *J Allergy Clin Immunol* **142:** 1469–1478.e2. doi:10.1016/j.jaci.2017.11.040

Pétrilli V, Papin S, Dostert C, Mayor A, Martinon F, Tschopp J. 2007. Activation of the NALP3 inflammasome is triggered by low intracellular potassium concentration. *Cell Death Differ* **14:** 1583–1589. doi:10.1038/sj.cdd.4402195

Platnich JM, Chung H, Lau A, Sandall CF, Bondzi-Simpson A, Chen HM, Komada T, Trotman-Grant AC, Brandelli JR, Chun J, et al. 2018. Shiga toxin/lipopolysaccharide activates caspase-4 and gasdermin D to trigger mitochondrial reactive oxygen species upstream of the NLRP3 inflammasome. *Cell Rep* **25:** 1525–1536.e7. doi:10.1016/j.celrep.2018.09.071

Rathkey JK, Zhao J, Liu Z, Chen Y, Yang J, Kondolf HC, Benson BL, Chirieleison SM, Huang AY, Dubyak GR, et al. 2018. Chemical disruption of the pyroptotic pore-forming protein gasdermin D inhibits inflammatory cell death and sepsis. *Sci Immunol* **3:** eaat2738. doi:10.1126/sciimmunol.aat2738

Rauch I, Deets KA, Ji DX, von Moltke J, Tenthorey JL, Lee AY, Philip NH, Ayres JS, Brodsky IE, Gronert K, et al. 2017. NAIP-NLRC4 inflammasomes coordinate intestinal epithelial cell expulsion with eicosanoid and IL-18 release via activation of caspase-1 and -8. *Immunity* **46:** 649–659. doi:10.1016/j.immuni.2017.03.016

Rogers C, Fernandes-Alnemri T, Mayes L, Alnemri D, Cingolani G, Alnemri ES. 2017. Cleavage of DFNA5 by caspase-3 during apoptosis mediates progression to secondary necrotic/pyroptotic cell death. *Nat Commun* **8:** 14128. doi:10.1038/ncomms14128

Ruan J, Xia S, Liu X, Lieberman J, Wu H. 2018. Cryo-EM structure of the gasdermin A3 membrane pore. *Nature* **557:** 62–67. doi:10.1038/s41586-018-0058-6

Rühl S, Shkarina K, Demarco B, Heilig R, Santos JC, Broz P. 2018. ESCRT-dependent membrane repair negatively regulates pyroptosis downstream of GSDMD activation. *Science* **362:** 956–960. doi:10.1126/science.aar7607

Saeki N, Sasaki H. 2012. Gasdermin superfamily: A novel gene family functioning in epithelial cells. In *Endothelium and epithelium: Composition, functions and pathology* (ed. Carrasco J, Mot M), pp. 193–211. Nova Science, New York.

Sarhan J, Liu BC, Muendlein HI, Li P, Nilson R, Tang AY, Rongvaux A, Bunnell SC, Shao F, Green DR, et al. 2018. Caspase-8 induces cleavage of gasdermin D to elicit pyroptosis during *Yersinia* infection. *Proc Natl Acad Sci* **115:** E10888. doi:10.1073/pnas.1809548115

Sborgi L, Rühl S, Mulvihill E, Pipercevic J, Heilig R, Stahlberg H, Farady CJ, Müller DJ, Broz P, Hiller S. 2016. GSDMD membrane pore formation constitutes the mechanism of pyroptotic cell death. *EMBO J* **35:** 1766–1778. doi:10.15252/embj.201694696

Scheffer LL, Sreetama SC, Sharma N, Medikayala S, Brown KJ, Defour A, Jaiswal JK. 2014. Mechanism of Ca^{2+}-triggered ESCRT assembly and regulation of cell membrane repair. *Nat Commun* **5:** 5646. doi:10.1038/ncomms6646

Shi J, Zhao Y, Wang Y, Gao W, Ding J, Li P, Hu L, Shao F. 2014. Inflammatory caspases are innate immune receptors for intracellular LPS. *Nature* **514:** 187–192. doi:10.1038/nature13683

Shi J, Zhao Y, Wang K, Shi X, Wang Y, Huang H, Zhuang Y, Cai T, Wang F, Shao F. 2015. Cleavage of GSDMD by inflammatory caspases determines pyroptotic cell death. *Nature* **526:** 660–665. doi:10.1038/nature15514

Sollberger G, Choidas A, Burn GL, Habenberger P, Di Lucrezia R, Kordes S, Menninger S, Eickhoff J, Nussbaumer P, Klebl B, et al. 2018. Gasdermin D plays a vital role in the generation of neutrophil extracellular traps. *Sci Immunol* **3:** eaar6689. doi:10.1126/sciimmunol.aar6689

Song L, Hobaugh MR, Shustak C, Cheley S, Bayley H, Gouaux JE. 1996. Structure of staphylococcal α-hemolysin, a heptameric transmembrane pore. *Science* **274:** 1859–1865. doi:10.1126/science.274.5294.1859

Sun L, Wang H, Wang Z, He S, Chen S, Liao D, Wang L, Yan J, Liu W, Lei X, et al. 2012. Mixed lineage kinase domain-like protein mediates necrosis signaling downstream of

RIP3 kinase. *Cell* **148:** 213–227. doi:10.1016/j.cell.2011.11.031

Taabazuing CY, Okondo MC, Bachovchin DA. 2017. Pyroptosis and apoptosis pathways engage in bidirectional crosstalk in monocytes and macrophages. *Cell Chem Biol* **24:** 507–514.e4. doi:10.1016/j.chembiol.2017.03.009

Tamura M, Tanaka S, Fujii T, Aoki A, Komiyama H, Ezawa K, Sumiyama K, Sagai T, Shiroishi T. 2007. Members of a novel gene family, *Gsdm*, are expressed exclusively in the epithelium of the skin and gastrointestinal tract in a highly tissue-specific manner. *Genomics* **89:** 618–629. doi:10.1016/j.ygeno.2007.01.003

Tanaka S, Mizushina Y, Kato Y, Tamura M, Shiroishi T. 2013. Functional conservation of *Gsdma* cluster genes specifically duplicated in the mouse genome. *G3 (Bethesda)* **3:** 1843–1850. doi:10.1534/g3.113.007393

Uhlen M, Fagerberg L, Hallstrom BM, Lindskog C, Oksvold P, Mardinoglu A, Sivertsson A, Kampf C, Sjostedt E, Asplund A, et al. 2015. Proteomics. Tissue-based map of the human proteome. *Science* **347:** 1260419. doi:10.1126/science.1260419

van Pee K, Neuhaus A, D'Imprima E, Mills DJ, Kühlbrandt W, Yildiz Ö. 2017. CryoEM structures of membrane pore and prepore complex reveal cytolytic mechanism of Pneumolysin. *eLife* **6:** e23644. doi:10.7554/eLife.23644

Vince JE, Silke J. 2016. The intersection of cell death and inflammasome activation. *Cell Mol Life Sci* **73:** 2349–2367. doi:10.1007/s00018-016-2205-2

Wang Y, Gao W, Shi X, Ding J, Liu W, He H, Wang K, Shao F. 2017. Chemotherapy drugs induce pyroptosis through caspase-3 cleavage of a gasdermin. *Nature* **547:** 99–103. doi:10.1038/nature22393

Wang Y, Yin B, Li D, Wang G, Han X, Sun X. 2018. GSDME mediates caspase-3-dependent pyroptosis in gastric cancer. *Biochem Biophys Res Commun* **495:** 1418–1425. doi:10.1016/j.bbrc.2017.11.156

Wolf AJ, Reyes CN, Liang W, Becker C, Shimada K, Wheeler ML, Cho HC, Popescu NI, Coggeshall KM, Arditi M, et al. 2016. Hexokinase is an innate immune receptor for the detection of bacterial peptidoglycan. *Cell* **166:** 624–636. doi:10.1016/j.cell.2016.05.076

Zanoni I, Tan Y, Di Gioia M, Broggi A, Ruan J, Shi J, Donado CA, Shao F, Wu H, Springstead JR, et al. 2016. An endogenous caspase-11 ligand elicits interleukin-1 release from living dendritic cells. *Science* **352:** 1232–1236. doi:10.1126/science.aaf3036

Zhang Y, Chen X, Gueydan C, Han J. 2018. Plasma membrane changes during programmed cell deaths. *Cell Res* **28:** 9–21. doi:10.1038/cr.2017.133

Zhao Y, Yang J, Shi J, Gong YN, Lu Q, Xu H, Liu L, Shao F. 2011. The NLRC4 inflammasome receptors for bacterial flagellin and type III secretion apparatus. *Nature* **477:** 596–600. doi:10.1038/nature10510

Zhong FL, Mamaï O, Sborgi L, Boussofara L, Hopkins R, Robinson K, Szeverényi I, Takeichi T, Balaji R, Lau A, et al. 2016. Germline NLRP1 mutations cause skin inflammatory and cancer susceptibility syndromes via inflammasome activation. *Cell* **167:** 187–202.e17. doi:10.1016/j.cell.2016.09.001

Zhou R, Yazdi AS, Menu P, Tschopp J. 2011. A role for mitochondria in NLRP3 inflammasome activation. *Nature* **469:** 221–225. doi:10.1038/nature09663

A20 at the Crossroads of Cell Death, Inflammation, and Autoimmunity

Arne Martens[1,2] and Geert van Loo[1,2]

[1]VIB Center for Inflammation Research, 9052 Ghent, Belgium

[2]Department of Biomedical Molecular Biology, Ghent University, 9052 Ghent, Belgium

Correspondence: geert.vanloo@irc.vib-ugent.be

A20 is a potent anti-inflammatory protein, acting by inhibiting nuclear factor κB (NF-κB) signaling and inflammatory gene expression and/or by preventing cell death. Mutations in the *A20/TNFAIP3* gene have been associated with a plethora of inflammatory and autoimmune pathologies in humans and in mice. Although the anti-inflammatory role of A20 is well accepted, fundamental mechanistic questions regarding its mode of action remain unclear. Here, we review new findings that further clarify the molecular and cellular mechanisms by which A20 controls inflammatory signaling and cell death, and discuss new evidence for its involvement in inflammatory and autoimmune disease development.

Inflammation is a protective response to induce repair in conditions of cellular damage and stress. It involves activation of the nuclear factor κB (NF-κB) family of transcription factors leading to the expression of inflammatory cytokines and chemokines to establish an appropriate immune response. Activation of NF-κB also induces the expression of cell-survival genes to protect the cell from dying (Zhang et al. 2017). NF-κB signaling is tightly regulated at multiple levels and strongly depends on reversible modification of signaling proteins, with important roles for phosphorylation and ubiquitination. Ubiquitination is a posttranslational protein modification in which ubiquitin (Ub), a small 76-amino-acid protein, is covalently attached to lysine residues of target proteins by the stepwise activity of an E1 Ub-activating enzyme, E2 Ub-conjugating enzymes, and E3 Ub

protein ligases. Each of the seven lysine residues (K6, K11, K27, K29, K33, K48, and K63) in Ub can themselves be bound to another Ub, leading to the formation of polyubiquitin chains on the target protein. Linear Ub chains (M1), in which Ub is bound via the amino-terminal methionine (M1) residue to another Ub, also happens, and has been shown to be crucially important for NF-κB signaling (Iwai and Tokunaga 2009). Depending on the type of chain, ubiquitination can target the modified protein for proteasomal degradation, or function as a scaffold for proteins that contain a Ub-binding domain (UBD) and mediate downstream signaling. Ubiquitination is reversed by deubiquitinating enzymes (DUBs) that cleave Ub chains from their substrate (Hrdinka and Gyrd-Hansen 2017). Hence, the ubiquitination of proteins, defined by the tight interplay between Ub ligases, Ub-binding proteins, and DUBs, controls NF-κB signaling.

The importance of ubiquitination for the regulation of inflammatory signaling has been studied extensively for the prototype inflammatory pathway induced by the cytokine tumor necrosis factor (TNF) (Verhelst et al. 2011). Binding of TNF to its cognate receptor TNF receptor 1 (TNFR1) mediates NF-κB-dependent gene activation through assembly of a primary membrane-bound signaling complex known as complex I (Ting and Bertrand 2016). This complex is assembled through the recruitment of TNF receptor–associated death domain protein (TRADD), receptor-interacting protein 1 (RIPK1), TNF receptor–associated factor 2 (TRAF2) and/or TRAF5, and cellular inhibitor of apoptosis protein-1 (cIAP1) and cIAP2. cIAP1 and cIAP2 are E3 Ub ligases that on recruitment conjugate RIPK1 and cIAPs themselves with K63-linked polyubiquitin chains, which now serve as a platform for the recruitment of the linear Ub chain assembly complex (LUBAC), consisting of HOIP, HOIL1, and SHARPIN. As a consequence, LUBAC will conjugate several components of the TNFR1 complex with M1-linked chains, followed by the recruitment and activation of the TAB–TAK1 complex and the NEMO–IKK complex via their UBD domains (Haas et al. 2009). This allows TAK1 to phosphorylate and activate the IKK complex, the latter phosphorylating the inhibitor of κB α (IκBα), targeting it for proteasomal degradation. Degradation of IκBα releases NF-κB, which now translocates to the nucleus where it induces the expression of proinflammatory and cell-survival genes (Fig. 1; Zhang et al. 2017). Besides activating NF-κB and inducing inflammatory gene activation and cell survival, TNF can also induce the assembly of a cytosolic death-inducing signaling complex (DISC) composed of TRADD, Fas-associated death domain (FADD), and caspase-8 (complex IIa), or composed of RIPK1, FADD, and caspase-8 (complex IIb, also called the Ripoptosome). In conditions of caspase-8 inhibition, TNF can induce necroptosis by further recruitment of RIPK3 and mixed lineage kinase domain-like pseudokinase (MLKL) assembling the ne-

crosome (Fig. 1; Pasparakis and Vandenabeele 2015; Ting and Bertrand 2016).

Deregulation of inflammatory NF-κB and cell-death signaling has been associated with several (auto)inflammatory diseases and cancer (Kondylis et al. 2017; Lork et al. 2017). Therefore, tight regulation of these pathways is required to avoid chronic inflammation and maintain tissue homeostasis. In this context, numerous (auto)regulatory mechanisms have been described (Renner and Schmitz 2009), and A20, CYLD, and OTULIN have been identified as key players in the negative regulation of NF-κB and cell death in response to TNF (Lork et al. 2017). In this review, we will discuss the molecular mechanisms by which A20 regulates NF-κB signaling and cell death, its own regulation, and our current knowledge on its involvement in inflammatory and autoimmune pathology.

A20 STRUCTURE AND MECHANISM OF NF-κB REGULATION

A20, also known as TNF-α-induced protein 3 (TNFAIP3), was first discovered as a primary response gene that is expressed on stimulation of human endothelial cells with TNF, protecting the cells from TNF-induced cell death (Dixit et al. 1990). Although A20 was initially characterized as an inhibitor of TNF-induced apoptosis, further studies identified A20 as a negative regulator of TNF-induced NF-κB activation, but also of NF-κB signaling downstream from the interleukin 1 receptor (IL-1R), pathogen recognition receptors (PRRs), NOD-like receptors (NLRs), T- and B-cell receptors, and CD40 (Opipari et al. 1990; Catrysse et al. 2014). The basal expression of A20 is low in most cell types, but is rapidly and transiently induced in inflammatory conditions through NF-κB-dependent transcription owing to the presence of two κB elements in the A20 promotor region (Krikos et al. 1992). A20 thus behaves as a prototype negative feedback regulator of NF-κB signaling.

A20 has been characterized as a so-called "Ub-editing" enzyme that inhibits NF-κB signaling by interfering with the ubiquitination status of multiple NF-κB signaling proteins (Wertz et al. 2004). The A20 protein consists

Cite this article as Cold Spring Harb Perspect Biol doi: 10.1101/cshperspect.a036418

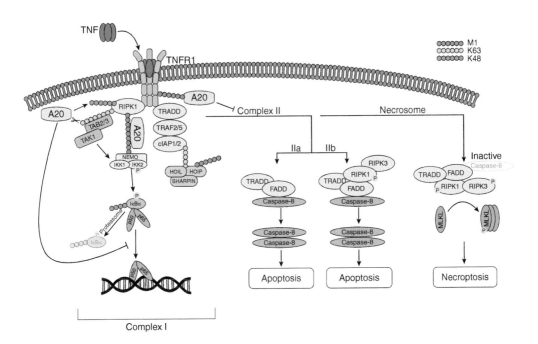

Figure 1. Tumor necrosis factor (TNF)-induced nuclear factor κB (NF-κB) signaling, apoptosis, and necroptosis. Binding of TNF to TNF receptor 1 (TNFR1) induces receptor trimerization that allows the recruitment of TNF receptor-associated death domain protein (TRADD) and receptor-interacting protein 1 (RIPK1). TRADD recruits TRAF2/5 and the E3 ubiquitin (Ub) ligases cIAP1 and cIAP2. cIAP1 and cIAP2 conjugate RIPK1 as well as themselves with K63-linked polyubiquitin chains, which serve as a platform for the recruitment of the LUBAC complex, consisting of HOIP, HOIL1, and SHARPIN. Linear Ub chain assembly complex (LUBAC) conjugates several components of the TNFR1 complex, including NEMO and RIPK1, with linear Ub chains (M1). The TAB2/3–TAK1 complex binds K63-linked chains on RIPK1, whereas the IKK complex is recruited via NEMO binding to M1-linked polyubiquitin. This allows TAK1 to phosphorylate and activate the IKK complex. On activation, IKK2 phosphorylates the inhibitor of κB α (IκBα), targeting IκBα for ubiquitination with K48-linked chains and proteasomal degradation. This releases the p50/p65 NF-κB dimer, which translocates to the nucleus where it induces the expression of NF-κB response genes. A20 is recruited to the TNFR1 signaling complex via M1-linked Ub, which binds to ZnF7. In addition, A20 has been shown to act as a deubiquitinating enzyme (DUB) that removes K63-linked polyubiquitin from different target proteins, including RIPK1 and NEMO. Furthermore, A20 has been shown to target RIPK1 and TNFR1 for proteasomal degradation through its ZnF4 E3 ligase activity. Binding of A20 to M1-linked polyubiquitin prevents downstream signaling by competing with other Ub-binding proteins and by preventing the degradation of M1-linked chains by CYLD (not shown). Loss of M1-linked chains destabilizes complex I and results in the formation of a cytosolic death-inducing signaling complex (DISC), consisting of TRADD, Fas-associated death domain (FADD), and caspase 8 (complex IIa) or RIPK1, FADD, and caspase-8 (complex IIb). Homodimerization and activation of caspase-8 leads to the activation of downstream executioner caspases 3 and 7 (not shown) and apoptotic cell death. If caspase-8 activity is compromised, RIPK1 associates with RIPK3 (the necrosome), which is activated by auto-phosphorylation. The activated necrosome recruits and activates MLKL inducing necroptotic cell death.

of an amino-terminal ovarian tumor (OTU) domain, which has DUB activity, and carboxy-terminal zinc finger (ZnF) domains, with the fourth ZnF domain shown to have Ub E3 ligase activity (Wertz et al. 2004). In addition, the ZnF4 domain was shown to act as a UBD for K63-linked polyubiquitin (Bosanac et al. 2010). More recently, the seventh ZnF (ZnF7) domain of A20 was shown to function as a UBD, but with high binding affinity for M1-linked Ub (Fig. 2; Tokunaga et al. 2012; Verhelst et al. 2012). The Ub editing function of A20 was first shown in

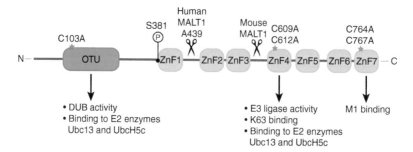

Figure 2. Domain structure of the A20 protein. The amino terminus of A20 contains an ovarian tumor (OTU) domain, which has deubiquitinating enzyme (DUB) activity relying on the catalytic residue Cys103. The carboxy-terminal part of A20 contains seven zinc finger (ZnF) domains. The fourth ZnF domain has K63-linked polyubiquitin-binding affinity and possesses E3 ubiquitin (Ub) ligase activity, whereas the seventh ZnF domain has strong binding affinity for linear (M1) Ub chains. IKK2-mediated phosphorylation of A20 at Ser381 enhances the DUB activity of A20 toward K63-linked polyubiquitin. MALT1 cleaves A20 at Ala439 in human, or between the third and fourth ZnF domains in mouse. Mutations that were introduced to generate OTU (C103A), ZnF4 (C609A, C612A), or ZnF7 (C764A, C767A) domain-specific mutant mice are depicted by red stars. Both the OTU and ZnF4 domains have been shown to bind to E2 enzymes Ubc13 and UbcH5c.

TNFR1 signaling. Using its tandem DUB-E3 ligase activities, A20 was shown to negatively regulate NF-κB signaling by replacing K63-linked chains on RIPK1 with K48-linked chains, which targeted RIPK1 for proteasomal degradation (Wertz et al. 2004). A20 was also shown to deubiquitinate TNFR1 and NEMO after TNF stimulation (Mauro et al. 2006; Wertz et al. 2015). This DUB activity of A20 was further shown to be promoted by IKK2-mediated phosphorylation of A20 at Ser381 (Fig. 2; Mauro et al. 2006; Wertz et al. 2015).

Besides its role in regulating TNF-induced NF-κB signaling, the DUB activity of A20 was shown to regulate TLR4- and NOD2-induced NF-κB activation by removing K63-linked chains from TRAF6 and RIPK2, respectively (Boone et al. 2004; Hitotsumatsu et al. 2008). Furthermore, A20 was shown to negatively regulate T-cell receptor (TCR)-induced NF-κB activation by deubiquitinating MALT1, thereby preventing the interaction between MALT1 and the IKK complex (Mauro et al. 2006; Duwel et al. 2009). Besides functioning as a DUB, A20 can also affect the ubiquitination of signaling mediators indirectly by interfering with the interaction between E2 and E3 enzymes, limiting polyubiquitin formation. In this context, A20 was shown to disrupt the interaction of TRAF6

with the E2 conjugating enzymes Ubc13 and UbcH5c in response to IL-1 and LPS stimulation. Similarly, on TNF stimulation, A20 was shown to prevent the association between Ubc13 and both cIAP1 and TRAF2 (Shembade et al. 2010).

Despite many reports claiming a crucial role of A20 as a DUB regulating NF-κB signaling, the physiological relevance of the deubiquitinase activity of A20 is still unclear. Recently, a number of transgenic mouse lines bearing a point mutation in the catalytic OTU domain (A20-C103A, A20^{OTU} mice) to abrogate A20 DUB activity were generated (Lu et al. 2013; De et al. 2014; Wertz et al. 2015). These mice are grossly normal and do not develop the severe phenotype of A20-deficient mice, which die perinatally because of multiorgan inflammation and cachexia (Lee et al. 2000). However, A20^{OTU} mice were shown to be sensitized to TNF and LPS treatment, and to dextran sodium sulfate (DSS)-induced colitis and experimental autoimmune encephalomyelitis (EAE) (Lu et al. 2013; Wertz et al. 2015). De and colleagues, however, did not observe sensitization of A20^{OTU} mice to LPS-induced pathology, nor did they observe any difference in TNF- or LPS-induced NF-κB signaling or cell death in cells derived from these mice. They concluded that the DUB function of

A20 is dispensable for inflammatory signaling in vivo (De et al. 2014). A20 transgenic mice with a mutation in the E3 ligase activity (A20-C609A/C612A, A20^{ZnF4} mice) have also been generated, but these mice also develop normally without any spontaneous phenotype (Lu et al. 2013; Wertz et al. 2015). Mouse embryo fibroblasts isolated from A20^{OTU} or A20^{ZnF4} mice show increased K63-linked ubiquitination of RIPK1 and TNFR1, resulting in slightly increased mitogen-activated protein kinase (MAPK) and NF-κB signaling on TNF stimulation (Lu et al. 2013; Wertz et al. 2015). Together, these findings indicate that neither the DUB nor E3 ligase functions of A20 critically contribute to its anti-inflammatory function in vivo. Simultaneous inactivation of the OTU and ZnF4 domains may be required to fully abolish its function and phenocopy A20-deficient mice. However, recent in vitro and in vivo evidence indicates that ZnF7 of A20 plays a more prominent role. A20 was shown to impair IKK activation through a non-enzymatic mechanism based on its binding to polyubiquitin via ZnF7 (Skaug et al. 2011). Additional studies have shown that A20 recruitment to TNFR1 and NOD2-associated signaling complexes by binding of ZnF7 to LUBAC-generated M1-linked polyubiquitin is required to inhibit downstream NF-κB signaling (Iwai and Tokunaga 2009; Tokunaga et al. 2012; Verhelst et al. 2012; Draber et al. 2015). Recently, we generated mice bearing a mutation in A20 ZnF7 domain (A20-C764A/C767A, A20^{ZnF7} mice) that abolishes its capacity to bind to M1-linked chains. These mice develop a spontaneous inflammatory arthritis and have reduced bodyweight and splenomegaly, showing an indispensable function for ZnF7 in A20-mediated suppression of inflammation in vivo (Polykratis et al. 2019). The ZnF7 mutation abolished the recruitment of A20 into the TNFR1 signaling complex, and reduced the amount of linear Ub chains within the complex (Draber et al. 2015; Polykratis et al. 2019). These observations indicate that A20 through its ZnF7 stabilizes M1 linkages in the TNFR1 signaling complex, shielding them from degradation by DUBs that are capable of cleaving linear Ub chains in signaling complexes. However, the spontaneous inflammatory phenotype of A20^{ZnF7} mice is mild compared with the severe multiorgan inflammation and postnatal lethality seen in A20 knockout mice (Lee et al. 2000), suggesting that other domains also critically contribute to A20's anti-inflammatory function in vivo. Knockin mice with combined domain mutations may be required to fully inactivate A20 and phenocopy the A20 knockout.

A20 IN THE REGULATION OF CELL DEATH

In addition to its role in the regulation of NF-κB signaling, A20 also acts as a strong inhibitor of cell death in many cell types. Originally, A20 was identified as an inhibitor of TNF-induced apoptosis in endothelial cells (Dixit et al. 1990), but it can also limit apoptosis in thymocytes and fibroblasts (Lee et al. 2000), pancreatic β cells (Liuwantara et al. 2006; Catrysse et al. 2015; Fukaya et al. 2016), hepatocytes (Catrysse et al. 2016), and intestinal epithelial cells (IECs) (Vereecke et al. 2010, 2014; Kattah et al. 2018; Slowicka et al. 2019).

Besides dying from caspase-dependent apoptosis, cells can also undergo caspase-independent but RIPK3- and MLKL-dependent necroptosis (Fig. 1; Pasparakis and Vandenabeele 2015; Ting and Bertrand 2016). A20 has also been proposed as an inhibitor of necroptosis in some cell types. A20-deficient T cells were shown to be more susceptible to anti-CD3/CD28-induced cell death, independent of caspase activation but dependent on RIPK3 and the kinase activity of RIPK1 (Onizawa et al. 2015). In addition, RIPK3 deficiency or inhibition of RIPK1 could considerably delay the early postnatal lethality of A20 knockout mice (Onizawa et al. 2015; Newton et al. 2016). In contrast, MLKL deficiency could not rescue the early lethality of A20 knockout mice, questioning the role of necroptosis in the pathology of these mice (Newton et al. 2016). We recently showed that A20-deficient macrophages can die from RIPK1/RIPK3/MLKL-dependent necroptosis, causing inflammasome activation and arthritis development in myeloid-specific A20-deficient mice (Polykratis et al. 2019). Therefore, inhibition of necroptosis is a critical anti-inflammatory function of A20 in vivo.

Finally, in some cell types, A20 may have proapoptotic functions. A20-deficient B cells and dendritic cells were shown to be protected from Fas-mediated cell death, most likely owing to the up-regulation of NF-κB-dependent expression of antiapoptotic proteins such as Bcl-2 and Bcl-x (Tavares et al. 2010; Kool et al. 2011). A20 was also shown to sensitize smooth muscle cells to apoptosis through a mechanism depending on nitric oxide (NO) production (Patel et al. 2006).

Although it is clear that A20 is a key regulator of cell death, the mechanisms by which it does this are still incompletely understood. On activation of death receptors, A20 was shown to be recruited to the DISC, where it physically interacted with caspase-8. This interaction was suggested to prevent cullin3-mediated ubiquitination of caspase-8, inhibiting caspase-8 activation and apoptosis (Jin et al. 2009). Alternatively, A20 was shown to inhibit TNF-induced apoptosis by preventing the recruitment of RIPK1 and TRADD to TNFR1, thereby inhibiting the recruitment of FADD and caspase-8 (He and Ting 2002). A20 was also shown to inhibit TNF-induced c-jun amino-terminal kinase (JNK) activation and apoptosis by targeting upstream ASK1 for ubiquitination and proteasomal degradation (Won et al. 2010). In the context of its protective function against necroptosis, A20 was shown to restrict RIPK3 ubiquitination, preventing the formation of RIPK1–RIPK3 complexes and necroptosis induction (Onizawa et al. 2015).

Although these latter studies suggest that A20 acts as either a DUB or an E3 ligase, it is becoming more established that the nonenzymatic, polyubiquitin-binding function of A20 plays a prominent role in the regulation of cell death. A reduction of linear Ub chains within TNFR1 complex I was shown to promote cell death by inducing complex II assembly (Ikeda et al. 2011; Peltzer et al. 2014). The ZnF7 domain of A20 stabilizes linear Ub chains in complex I, inhibiting TNFR1-mediated cell death (Draber et al. 2015; Yamaguchi and Yamaguchi 2015). Hence, a mutation in the ZnF7 domain of A20

abolishes the recruitment of A20 and reduces the amount of linear Ub chains in TNFR1 complex I (Tokunaga et al. 2012; Draber et al. 2015; Polykratis et al. 2019). Interestingly, the binding of M1-linked chains by ZnF7 in A20 was shown to protect these chains from degradation by CYLD, thereby stabilizing complex I and preventing the formation of the DISC (Fig. 1; Draber et al. 2015).

MECHANISMS THAT REGULATE A20 ACTIVITY

A20 expression and function is under the control of several regulatory mechanisms, viz. transcriptional, posttranscriptional, and posttranslational. In most cell types, A20 expression levels are low at steady state, but are rapidly up-regulated in inflammatory conditions as a result of NF-κB activation (Krikos et al. 1992; Verstrepen et al. 2010). Different transcription factors work together to fine-tune the expression of A20 and the strength of NF-κB signaling. DREAM (downstream regulatory element antagonist modulator) has been shown to constitutively repress expression of A20 by binding to downstream regulatory elements (DREs) in the A20 promoter, whereas upstream stimulatory factor 1 (USF1) has been shown to bind to the DRE-associated E-box domain in A20 to activate its expression in response to an inflammatory stimulus (Amir-Zilberstein and Dikstein 2008; Tiruppathi et al. 2014). Also, the orphan nuclear estrogen-related receptor α (ERRα), a key metabolic regulator, has been shown to promote expression of A20 in mice (Yuk et al. 2015). ERRα binds to an Esrra consensus motif in the A20 promoter, inducing expression of A20 and suppressing NF-κB signaling in LPS-stimulated macrophages (Yuk et al. 2015). Finally, the histone methyltransferase Ash1l (absent small or homeotic-like) was shown to enhance expression of A20 through H3K4 methylation of the A20 promoter, suppressing inflammatory signaling (Xia et al. 2013).

T lymphocytes constitutively express high levels of A20, which are down-regulated on TCR activation and NF-κB induction (Tewari et al. 1995). Proteasomal degradation of A20, as well as its cleavage by the paracaspase MALT1, contribute to the down-regulation of A20 following TCR stimulation (Coornaert et al. 2008; Duwel et al. 2009). Methylation of the *A20* promotor, preventing expression of A20 and stimulating constitutive NF-κB activation, has been observed in several types of lymphomas (Honma et al. 2009; Chanudet et al. 2010).

Expression of A20 is also regulated by several microRNAs (miRs). In diffuse large B-cell lymphomas, miR-125a and miR-125b were shown to down-regulate expression of A20, leading to constitutive activation of NF-κB, B-cell proliferation, and lymphomagenesis (Kim et al. 2012). miR-125-mediated inhibition of A20 has also been shown in lung infection and in chronic obstructive pulmonary disease (COPD) (Hsu et al. 2017). miR-19, miR-29, miR-221, and miR-let-7 are other miRNAs that have been shown to suppress A20, leading to enhanced NF-κB signaling and/or cell death (Wang et al. 2011; Gantier et al. 2012; Balkhi et al. 2013; Kumar et al. 2015; Langsch et al. 2016; Zhao et al. 2016).

A20 interacts with different partners to perform its regulatory functions. Multiple Ub-binding proteins have been described that bind to A20 and recruit it to its substrates. A20-binding inhibitor of NF-κB (ABIN1) was shown to recruit A20 to polyubiquitinated NEMO to remove Ub chains from NEMO (Heyninck et al. 1999; Mauro et al. 2006; Wagner et al. 2008). Tax1-binding protein 1 (TAX1BP1) negatively regulates NF-κB signaling by recruiting A20 to TRAF6 and RIPK1 (Shembade et al. 2007; Iha et al. 2008). Itch and RING finger protein 11 (RNF11) have been shown to function as subunits of an A20 Ub-editing complex to inhibit NF-κB signaling (Shembade et al. 2008, 2009).

The activity of A20 is also regulated by posttranslational modifications. On receptor activation, A20 is phosphorylated at Ser381 by IKK2, increasing its inhibitory capacity (Hutti et al. 2007). Although A20 hydrolyzes K63-linked Ub chains in vivo via its OTU domain, recombinant A20 expressed in bacteria preferentially cleaves K48-linked chains in vitro (Komander and Barford 2008; Lin et al. 2008). The phosphorylation status of A20 appears to account for

this discrepancy because phosphorylated A20 was shown to efficiently cleave K63-linked chains in vitro (Wertz et al. 2015). In T lymphocytes, A20 can be ubiquitinated by the E3 Ub ligase RNF114, thereby stabilizing A20 and restraining NF-κB responses (Rodriguez et al. 2014). In smooth muscle cells exposed to high glucose levels, A20 can be O-glycosylated and subsequently ubiquitinated and targeted for proteasomal degradation (Shrikhande et al. 2010). Finally, A20 is also regulated by reversible oxidation of the catalytic cysteine residue, modifying its activation state (Kulathu et al. 2013; Lee et al. 2013).

A20 AS A DISEASE SUSCEPTIBILITY GENE

Several single-nucleotide polymorphisms (SNPs) in or near the *A20* gene have been linked to a variety of autoimmune and inflammatory diseases, including systemic lupus erythematosus (SLE), rheumatoid arthritis, psoriasis, type 1 diabetes, Crohn's disease, celiac disease, coronary artery disease in type 2 diabetes, systemic sclerosis, and Sjogren's syndrome (Ma and Malynn 2012; Catrysse et al. 2014). Associations have also been reported for autoimmune hepatitis (de Boer et al. 2014) and primary biliary cirrhosis (Cordell et al. 2015). Most of these disease-associated variants are located in upstream or downstream noncoding regions or in intronic regions of the *A20* gene, possibly affecting the expression of A20 (Graham et al. 2008; Adrianto et al. 2011). Downstream SNPs can influence expression of A20 by affecting the function of cell- and activation-specific enhancers. For example, deletion of a downstream region containing four enhancers was shown to significantly reduce expression of A20, resulting in enhanced inflammatory responses (Sokhi et al. 2018). One of these enhancers harbors the TT > A variant that was linked to SLE susceptibility (Adrianto et al. 2011; Wang et al. 2013a). Deletion of this TT > A enhancer in mice induced spontaneous inflammatory arthritis, thereby establishing the importance of this enhancer in preventing inflammatory pathology and autoimmunity (Sokhi et al. 2018). Besides the many noncoding variants, two SNPs have been identified in exon 3 of *A20* that induce nonsynonymous mutations (rs5029941/A125V and rs2230926/F127C) (Musone et al. 2008; Lodolce et al. 2010). These mutations are suggested to affect the DUB activity of A20, although this has not been evaluated in vivo.

Recently, whole-exome sequencing identified heterozygous loss-of-function mutations in the *A20/TNFAIP3* gene (A20 haploinsufficiency, HA20) of patients with a rare, early-onset autoinflammatory syndrome (Zhou et al. 2016). These nonsense and frameshift mutations are all localized in the DUB or the ZnF4 domains of A20, and the mutant proteins are likely unstable because A20 was not detected in cells from HA20 patients (Zhou et al. 2016). Patients' cells displayed increased NF-κB signaling, NLRP3 inflammasome activity, and increased expression of proinflammatory cytokines (Zhou et al. 2016). Several other cases of germline HA20 have since been identified in patients with early-onset autoimmune disease (Fig. 3; Table 1; Aeschlimann et al. 2018; Duncan et al. 2018). Overall, HA20 patients develop recurrent oral, genital, and/or gastrointestinal ulcers, musculoskeletal and gastrointestinal problems, episodic fever, and recurrent infections. However, disease severity is strongly patient-dependent, ranging from very mild disease to severe multiorgan inflammation, and treatment regimens need to be adjusted to disease severity (Aeschlimann et al. 2018).

Finally, *A20* has been identified as a tumor suppressor gene, because biallelic somatic mutations in *A20* are frequently observed in several B-cell lymphomas, including MALT lymphoma, Hodgkin's lymphoma, diffuse large B-cell lymphoma, and follicular lymphoma. These loss-of-function mutations are associated with constitutive NF-κB signaling and uncontrolled cell proliferation (Compagno et al. 2009; Honma et al. 2009; Kato et al. 2009; Novak et al. 2009; Schmitz et al. 2009; Okosun et al. 2014). Biallelic mutations in *A20* have also been identified in Sézary syndrome, an aggressive variant of cutaneous T-cell lymphoma (Braun et al. 2011). However, A20 has also been described as a tumor promotor, likely connected to its anti-apoptotic functions, and high levels of A20 have been detected in glioma (Guo et al. 2009), glioblastoma (Hjelmeland et al. 2010), and acute lymphoblastic leukemia (Chen et al. 2015). Furthermore, A20 was shown to be up-regulated in human basal-like breast cancers in which it promotes epithelial–mesenchymal transition (EMT) through monoubiquitination and nuclear stabilization of SNAIL1, a transcription factor that drives EMT (Lee et al. 2017).

TISSUE-SPECIFIC FUNCTIONS OF A20

A20 is an important anti-inflammatory protein that acts as a direct inhibitor of NF-κB signaling, or as an inhibitor of proinflammatory cell death. However, inflammatory signaling pathways and

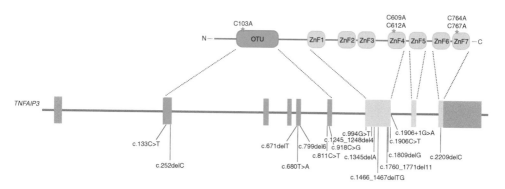

Figure 3. Schematic representation of the *TNFAIP3* gene indicating mutations causing A20 haploinsufficiency (HA20). Exons (1–9) are represented as rectangles. Exons encoding the ovarian tumor (OTU) domain are depicted in green and exons encoding the ZnF domains are depicted in orange; noncoding exons are shown in gray. HA20 mutations depicted in the figure are also listed in Table 1.

Cite this article as *Cold Spring Harb Perspect Biol* doi: 10.1101/cshperspect.a036418

Table 1. Overview of identified HA20 mutations

cDNA alteration	Amino acid alteration	Domain	References
c.680T < A	p.Leu227*	Ovarian tumor (OTU)	Zhou et al. 2016; Aeschlimann et al. 2018
c.671delT	p.Phe224Ser*fs**4*	OTU	
c.811C > T	p.Arg271*	OTU	
c.1809delG	p.Thr604Arg*fs**93*	ZnF4	
c.918C > G	p.Tyr306*	OTU	
c.799delG	p.Pro268Leu*fs**19*	OTU	
c.252delC	p.Trp85Gly*fs**11*	OTU	Shigemura et al. 2016; Ohnishi et al. 2017;
c.133C > T	p.Arg45*	OTU	Takagi et al. 2017; Kadowaki et al. 2018
c.1760_1770del11	p.Ala588Val*fs**80*	Between ZnF3-4	
c.1345delA	p.Asn449Thr*fs**28*	Between ZnF1-2	
c.1906 + 1G > A	p.Phe637Glu*2	Between ZnF4-5	
c.2209delC	p.Gln737Ser*fs**79*	ZnF6	
c.2088 + 5G > C	p.His636Glu*fs**55*	ZnF4	
c.728G > A	p.Cys243Tyr	OTU	
c.1245-1248del4	p.Lys417Ser*fs**4*	Between ZnF1-2	
c.994G > T	p.Glu332*	OTU	Berteau et al. 2019
c.1466_1467delTG	p.Val489Ala*fs**7*	ZnF2	Duncan et al. 2018
c.1906C > T	p.His636*fs**1*	ZnF4	Lawless et al. 2018

cell death are differentially regulated in different cell types; hence, A20 may have cell-specific contributions to prevent inflammation and disease pathogenesis. Because A20 knockout mice die prematurely because of severe multiorgan inflammation (Lee et al. 2000), conditional, lineage-specific knockout strategies are needed to unravel the tissue-specific functions of A20. Several tissue-specific A20-deficient mice have been generated in recent years that clearly show the importance of A20 in maintaining tissue homeostasis by regulating inflammatory responses and cell death (Table 2). In this section, we refer to a number of recent mouse studies that provide new information on the role of A20 in the prevention of cell death and its implications for inflammatory pathology.

Mice lacking A20 in their myeloid cells (A20[Myel-KO]) develop spontaneous polyarthritis with many characteristics of human rheumatoid arthritis, including autoantibodies against type II collagen and rheumatoid arthritis–associated cytokines (Matmati et al. 2011). The arthritis phenotype in A20[Myel-KO] mice was shown to require IL-6 and TLR4-MyD88, but not TNF. Primary macrophages from these mice show sustained NF-κB activation in response to lipo-polysaccharide (LPS), in agreement with a role for A20 as an inhibitor of NF-κB signaling (Matmati et al. 2011). A20-deficient macrophages also express higher levels of STAT1 and STAT1-dependent genes upon stimulation with interferon γ (IFN-γ) or IL-6. Inhibition of JAK-STAT signaling in vivo with the JAK inhibitor tofacitinib was shown to suppress the development of enthesitis in A20[Myel-KO] mice (De Wilde et al. 2017). Interestingly, myeloid A20 deficiency was also shown to promote osteoclastogenesis, suggesting a role for A20 in regulating receptor activator of the NF-κB (RANK)-induced NF-κB signaling. Further studies, preferably using osteoclast-specific A20 targeting, are needed to confirm a direct regulatory role of A20 in regulating osteoclastogenesis and bone formation.

Arthritis development in A20[Myel-KO] mice relies on activation of the Nlrp3 inflammasome and IL-1R signaling (Vande Walle et al. 2014). Thus, A20[Myel-KO] mice crossed into an Nlrp3-, caspase-1/11-, or IL-1R-deficient background no longer develop arthritis. Primary A20-deficient macrophages show enhanced Nlrp3 inflammasome-mediated caspase-1 activation, pyroptosis, and IL-1β secretion by soluble and

Table 2. Phenotypes of tissue-specific A20 knockout and A20 transgenic mice

Cell type	OE/KO	Phenotype	References
Myeloid cells	KO	Spontaneous severe destructive polyarthritis	Matmati et al. 2011; Vande Walle et al. 2014; De Wilde et al. 2017; Polykratis et al. 2019
	KO	Protected from influenza A virus infection	Maelfait et al. 2012
Dendritic cells	KO	Spontaneous colitis, IBD-associated arthritis	Hammer et al. 2011
	KO	Systemic autoimmunity resembling SLE	Kool et al. 2011
	KO	Multiorgan inflammation, hypersensitive to LPS	Xuan et al. 2015
B cells	KO	Development of autoimmune syndrome in older mice (autoreactive immunoglobulins, glomerular immunoglobulin deposits, ↑ germinal center B cells, B-cell resistance to Fas-mediated cell death)	Tavares et al. 2010; Chu et al. 2011; Hovelmeyer et al. 2011
T cells	KO	No spontaneous phenotype, ↓ iNKT cell numbers	Drennan et al. 2016
	KO	No spontaneous phenotype, protected from EAE	Onizawa et al. 2015
	KO	Lymphadenopathy but no detectable pathology, enhanced antitumor activity	Giordano et al. 2014
	KO	↑ Treg cell numbers	Fischer et al. 2017
HSC	KO	Loss of HSC quiescence causing anemia, lymphopenia, and postnatal lethality	Nakagawa et al. 2015
Mast cells	KO	Hypersensitive to allergic airway responses and collagen-induced arthritis	Heger et al. 2014
IEC	KO	No spontaneous pathology, enhanced susceptibility to DSS-induced colitis and to TNF toxicity	Vereecke et al. 2010; Kattah et al. 2018
	OE	Protected from LPS-induced barrier disruption	Kolodziej et al. 2011
	OE	Protected from DSS-induced IEC death and barrier disruption	Rhee et al. 2012
	OE	Hypersensitive to TNF-induced IEC death and systemic inflammation	Garcia-Carbonell et al. 2018
IEC + myeloid cells	KO	Spontaneous severe ileitis and colitis, colorectal cancer	Vereecke et al. 2014
Hepatocytes	KO	Chronic liver inflammation, enhanced sensitivity to TNF and LPS toxicity and to chemically and high fat diet–induced hepatocarcinogenesis	Catrysse et al. 2016
Keratinocytes		Keratinocyte hyperproliferation, ectodermal organ abnormalities, sensitized to experimental psoriasis, and atopic dermatitis	Lippens et al. 2011; Devos et al. 2019
Neurons	KO	No phenotype	McGuire et al. 2013
CNS progenitor cells	KO		
Microglia	KO	↑ Microglia numbers and microglia activation, hypersensitive LPS and EAE	Voet et al. 2018
Astrocytes	KO	Hypersensitive to EAE	Wang et al. 2013b
Airway epithelial cells	KO	Protected from influenza A virus infection, highly sensitive to allergic airway inflammation	Schuijs et al. 2015; Maelfait et al. 2016
Pancreatic β cells	KO	No phenotype	Liuwantara et al. 2006; Catrysse et al. 2015; Fukaya et al. 2016

OE, Overexpression; KO, knockout; IBD, inflammatory bowel disease; SLE, systemic lupus erythematosus; LPS, lipopolysaccharide; iNKT, invariant natural killer T cell; EAE, experimental autoimmune encephalomyelitis; HSC, hematopoietic stem cell; IEC, intestinal epithelial cell; DSS, dextran sodium sulfate; TNF, tumor necrosis factor.

crystalline Nlrp3 stimuli. Myeloid-specific ASC deficiency substantially ameliorated arthritis in A20$^{Myel-KO}$ mice, indicating that cell-intrinsic inflammasome activation in A20-deficient myeloid cells drives pathology (Polykratis et al. 2019). Although A20 was shown to regulate Nlrp3 inflammasome activation in vitro by negatively regulating NF-κB-dependent priming needed for optimal expression of pro-IL-1β and Nlrp3 (Vande Walle et al. 2014), inhibition of IKK/NF-κB signaling in A20-deficient myeloid cells was not sufficient to prevent arthritis development (Polykratis et al. 2019). Rather, regulation of cell death, and particularly of necroptosis, was identified as the key anti-inflammatory function of A20 in preventing the development of arthritis. Indeed, RIPK1/RIPK3/MLKL-dependent macrophage necroptosis was shown to induce inflammasome activation, IL-1β production, and arthritis development in A20$^{Myel-KO}$ mice (Polykratis et al. 2019). These data are supported by real-time single-cell imaging experiments, which show that IL-1β is secreted exclusively by dying A20-deficient macrophages. An earlier study had suggested that inflammasome activation and IL-1β release in A20-deficient macrophages occurs in the absence of cell death (Duong et al. 2015). A20 has also been shown to control microglia activation and neuroinflammation via regulation of Nlrp3 inflammasome activation (Voet et al. 2018). How A20 regulates Nlrp3 activation, and whether this is directly or indirectly preventing microglia cell death, needs further investigation.

An important prosurvival role for A20 has also been shown in IECs. Although IEC-specific A20 knockout mice (A20^{IEC-KO}) do not develop spontaneous disease, they are highly susceptible to DSS-induced colitis and are hypersensitive to a normally sublethal dose of TNF. IECs from A20^{IEC-KO} mice are hypersensitive to apoptosis, leading to loss of intestinal barrier integrity and bacterial infiltration in A20^{IEC-KO} mice (Vereecke et al. 2010). The antiapoptotic property of A20. in IECs was confirmed in transgenic mice with IEC-specific overexpression of A20. These mice are protected from DSS- and LPS-induced IEC apoptosis and loss of barrier func-

tion (Kolodziej et al. 2011; Rhee et al. 2012), but are highly susceptible to TNF-induced IEC apoptosis, intestinal damage, and systemic inflammation. The latter phenotype required RIPK1 and caspase-8, and was mediated by ZnF7 of A20 (Garcia-Carbonell et al. 2018). Thus, A20 binding to linear Ub appears to stabilize the Ripoptosome and potentiate its apoptosis-inducing activity (Garcia-Carbonell et al. 2018). Although A20^{IEC-KO} mice do not develop a spontaneous intestinal phenotype (Vereecke et al. 2010; Kattah et al. 2018), mice lacking A20 in both IECs and myeloid cells develop severe ileitis and colitis, associated with epithelial apoptosis and hyperproliferation, which eventually leads to the development of colon cancer (Vereecke et al. 2014). Combined IEC-specific deletion of A20 and the A20-interacting protein ABIN1 also induces spontaneous intestinal inflammation and severe pathology because of the caspase-8 and RIPK1-dependent death of IECs (Kattah et al. 2018). Similarly, combined deletion of A20 and Atg16L1 in IECs induces spontaneous inflammatory bowel disease (IBD)-like pathology caused by IEC apoptosis (Slowicka et al. 2019). Interestingly, both ABIN-1 and Atg16L1 have been associated with IBD (Hampe et al. 2007; Rioux et al. 2007; Jostins et al. 2012) and have been shown to bind to A20 and to Ub (Wagner et al. 2008; Fujita et al. 2013; Slowicka et al. 2019). Collectively, these data suggest that inflammatory signaling and autophagy cooperatively control intestinal homeostasis by preventing the death of enterocytes that would compromise intestinal barrier integrity.

Finally, A20 was shown to have hepatoprotective activities. Hepatocyte-specific A20-deficient mice (A20^{LPC-KO}) spontaneously develop mild liver inflammation and steatosis, but succumb to a normally sublethal dose of TNF owing to excessive hepatocyte apoptosis. Moreover, chronic liver inflammation and enhanced hepatocyte apoptosis in A20^{LPC-KO} mice increased their susceptibility to chemically and high fat diet–induced hepatocellular carcinoma development (Catrysse et al. 2016). A20 overexpression in liver has been shown to be protective in models of hepatectomy and acute toxic hepatitis, owing to antiapoptotic and anti-inflammatory mechanisms (Arvelo et al. 2002; Ramsey et al. 2009; Damrauer et al. 2011; da Silva et al. 2013).

CONCLUDING REMARKS

A20 was identified nearly 30 years ago (Dixit et al. 1990; Opipari et al. 1990) and has emerged as a critical regulator of inflammatory signaling to preserve tissue immune homeostasis. Although inhibition of NF-κB activation has long been considered the key anti-inflammatory function of A20, it is becoming more evident that its role in preventing cell death is a major mechanism for suppressing inflammation. However, A20 may have different modes of action in different cell types given the phenotypes of different tissue-specific A20 knockout mice. NF-κB inhibitory activities, antiapoptotic functions, as well as antinecroptotic functions, and even proapoptotic functions have been identified in different cell types. Mechanistically, A20 has been proposed to function as a DUB, but the physiological relevance of its function as a DUB has been challenged by the finding that DUB-inactive A20 knockin mice do not develop spontaneous inflammatory disease. Recent evidence suggests that A20 mainly acts as a Ub-binding protein via its ZnF7 domain, which primarily recruits A20 into specific protein complexes to prevent downstream signaling. However, fine-tuning of signaling may require additional A20-specific activities, including A20 DUB activity. Generation of knockin mice with combined domain mutations will be essential to dissect the importance of and redundancies between the different functional domains of A20. Such studies will help to better understand the role of A20 in inflammation and immunity, which might help to develop new therapeutics for the treatment of disease.

ACKNOWLEDGMENTS

A.M. is supported by a grant from the Concerted Research Actions (GOA) of the Ghent University. Research in the G.v.L. laboratory is supported by research grants from the Research Foundation – Flanders (FWO), the Geneeskundige Stichting Koningin Elisabeth (GSKE), the CBC Banque Prize, the Charcot Foundation, the Belgian Foundation against Cancer, and Kom op tegen Kanker.

REFERENCES

Adrianto I, Wen F, Templeton A, Wiley G, King JB, Lessard CJ, Bates JS, Hu Y, Kelly JA, Kaufman KM, et al. 2011. Association of a functional variant downstream of *TNFAIP3* with systemic lupus erythematosus. *Nat Genet* **43:** 253–258. doi:10.1038/ng.766

Aeschlimann FA, Batu ED, Canna SW, Go E, Gül A, Hoffmann P, Leavis HL, Ozen S, Schwartz DM, Stone DL, et al. 2018. A20 haploinsufficiency (HA20): Clinical phenotypes and disease course of patients with a newly recognised NF-κB-mediated autoinflammatory disease. *Ann Rheum Dis* **77:** 728–735. doi:10.1136/annrheumdis-2017-212403

Amir-Zilberstein L, Dikstein R. 2008. Interplay between E-box and NF-κB in regulation of A20 gene by DRB sensitivity-inducing factor (DSIF). *J Biol Chem* **283:** 1317–1323. doi:10.1074/jbc.M706767200

Arvelo MB, Cooper JT, Longo C, Daniel S, Grey ST, Mahiou J, Czismadia E, Abu-Jawdeh G, Ferran C. 2002. A20 protects mice from D-galactosamine/lipopolysaccharide acute toxic lethal hepatitis. *Hepatology* **35:** 535–543. doi:10.1053/jhep.2002.31309

Balkhi MY, Iwenofu OH, Bakkar N, Ladner KJ, Chandler DS, Houghton PJ, London CA, Kraybill W, Perrotti D, Croce CM, et al. 2013. miR-29 acts as a decoy in sarcomas to protect the tumor suppressor A20 mRNA from degradation by HuR. *Sci Signal* **6:** ra63. doi:10.1126/scisignal.2004177

Berteau F, Rouvière B, Nau A, Le Berre R, Sarrabay G, Touitou I, de Moreuil C. 2019. A20 haploinsufficiency (HA20): Clinical phenotypes and disease course of patients with a newly recognised NF-κB-mediated autoinflammatory disease. *Ann Rheum Dis* **78:** e35. doi:10.1136/annrheumdis-2018-213347

Boone DL, Turer EE, Lee EG, Ahmad RC, Wheeler MT, Tsui C, Hurley P, Chien M, Chai S, Hitotsumatsu O, et al. 2004. The ubiquitin-modifying enzyme A20 is required for termination of Toll-like receptor responses. *Nat Immunol* **5:** 1052–1060. doi:10.1038/ni1110

Bosanac I, Wertz IE, Pan B, Yu C, Kusam S, Lam C, Phu L, Phung Q, Maurer B, Arnott D, et al. 2010. Ubiquitin binding to A20 ZnF4 is required for modulation of NF-κB signaling. *Mol Cell* **40:** 548–557. doi:10.1016/j.molcel.2010.10.009

Braun FC, Grabarczyk P, Möbs M, Braun FK, Eberle J, Beyer M, Sterry W, Busse F, Schroder J, Delin M, et al. 2011. Tumor suppressor TNFAIP3 (A20) is frequently deleted in Sézary syndrome. *Leukemia* **25:** 1494–1501. doi:10.1038/leu.2011.101

Catrysse L, Vereecke L, Beyaert R, van Loo G. 2014. A20 in inflammation and autoimmunity. *Trends Immunol* **35:** 22–31. doi:10.1016/j.it.2013.10.005

Catrysse L, Fukaya M, Sze M, Meyerovich K, Beyaert R, Cardozo AK, van Loo G. 2015. A20 deficiency sensitizes

Cite this article as *Cold Spring Harb Perspect Biol* doi: 10.1101/cshperspect.a036418

pancreatic β cells to cytokine-induced apoptosis in vitro but does not influence type 1 diabetes development in vivo. *Cell Death Dis* **6:** e1918. doi:10.1038/cddis.2015.301

Catrysse L, Farhang Ghahremani M, Vereecke L, Youssef SA, McGuire C, Sze M, Weber A, Heikenwalder M, de Bruin A, Beyaert R, et al. 2016. A20 prevents chronic liver inflammation and cancer by protecting hepatocytes from death. *Cell Death Dis* **7:** e2250. doi:10.1038/cddis.2016 .154

Chanudet E, Huang Y, Ichimura K, Dong G, Hamoudi RA, Radford J, Wotherspoon A, Isaacson PG, Ferry J, Du MQ. 2010. A20 is targeted by promoter methylation, deletion and inactivating mutation in MALT lymphoma. *Leukemia* **24:** 483–487. doi:10.1038/leu.2009.234

Chen S, Xing H, Li S, Yu J, Li H, Liu S, Tian Z, Tang K, Rao Q, Wang M, et al. 2015. Up-regulated A20 promotes proliferation, regulates cell cycle progression and induces chemotherapy resistance of acute lymphoblastic leukemia cells. *Leukemia Res* **39:** 976–983. doi:10.1016/j.leukres .2015.06.004

Chu Y, Vahl JC, Kumar D, Heger K, Bertossi A, Wojtowicz E, Soberon V, Schenten D, Mack B, Reutelshofer M, et al. 2011. B cells lacking the tumor suppressor TNFAIP3/A20 display impaired differentiation and hyperactivation and cause inflammation and autoimmunity in aged mice. *Blood* **117:** 2227–2236. doi:10.1182/blood-2010-09-306019

Compagno M, Lim WK, Grunn A, Nandula SV, Brahmachary M, Shen Q, Bertoni F, Ponzoni M, Scandurra M, Califano A, et al. 2009. Mutations of multiple genes cause deregulation of NF-κB in diffuse large B-cell lymphoma. *Nature* **459:** 717–721. doi:10.1038/nature07968

Coornaert B, Baens M, Heyninck K, Bekaert T, Haegman M, Staal J, Sun L, Chen ZJ, Marynen P, Beyaert R. 2008. T cell antigen receptor stimulation induces MALT1 paracaspase–mediated cleavage of the NF-κB inhibitor A20. *Nat Immunol* **9:** 263–271. doi:10.1038/ni1561

Cordell HJ, Han Y, Mells GF, Li Y, Hirschfield GM, Greene CS, Xie G, Juran BD, Zhu D, Qian DC, et al. 2015. International genome-wide meta-analysis identifies new primary biliary cirrhosis risk loci and targetable pathogenic pathways. *Nat Commun* **6:** 8019. doi:10.1038/ ncomms9019

Damrauer SM, Studer P, da Silva CG, Longo CR, Ramsey HE, Csizmadia E, Shrikhande GV, Scali ST, Libermann TA, Bhasin MK, et al. 2011. A20 modulates lipid metabolism and energy production to promote liver regeneration. *PLoS ONE* **6:** e17715. doi:10.1371/journal.pone .0017715

da Silva CG, Studer P, Skroch M, Mahiou J, Minussi DC, Peterson CR, Wilson SW, Patel VI, Ma A, Csizmadia E, et al. 2013. A20 promotes liver regeneration by decreasing SOCS3 expression to enhance IL-6/STAT3 proliferative signals. *Hepatology* **57:** 2014–2025. doi:10.1002/hep .26197

De A, Dainichi T, Rathinam CV, Ghosh S. 2014. The deubiquitinase activity of A20 is dispensable for NF-κB signaling. *EMBO Rep* **15:** 775–783. doi:10.15252/embr .201338305

de Boer YS, van Gerven NM, Zwiers A, Verwer BJ, van Hoek B, van Erpecum KJ, Beuers U, van Buuren HR, Drenth JP, den Ouden JW, et al. 2014. Genome-wide association

study identifies variants associated with autoimmune hepatitis type 1. *Gastroenterology* **147:** 443–452.e5. doi:10.1053/j.gastro.2014.04.022

Devos M, Mogilenko DA, Fleury S, Gilbert B, Becquart C, Quemener S, Dehondt H, Tougaard P, Staels B, Bachert C, et al. 2019. Keratinocyte expression of A20/TNFAIP3 controls skin inflammation associated with atopic dermatitis and psoriasis. *J Invest Dermatol* **139:** 135–145. doi:10 .1016/j.jid.2018.06.191

De Wilde K, Martens A, Lambrecht S, Jacques P, Drennan MB, Debusschere K, Govindarajan S, Coudenys J, Verheugen E, Windels F, et al. 2017. A20 inhibition of STAT1 expression in myeloid cells: A novel endogenous regulatory mechanism preventing development of enthesitis. *Ann Rheum Dis* **76:** 585–592. doi:10.1136/annrheum dis-2016-209454

Dixit VM, Green S, Sarma V, Holzman LB, Wolf FW, O'Rourke K, Ward PA, Prochownik EV, Marks RM. 1990. Tumor necrosis factor-α induction of novel gene products in human endothelial cells including a macrophage-specific chemotaxin. *J Biol Chem* **265:** 2973–2978.

Draber P, Kupka S, Reichert M, Draberova H, Lafont E, de Miguel D, Spilgies L, Surinova S, Taraborrelli L, Hartwig T, et al. 2015. LUBAC-recruited CYLD and A20 regulate gene activation and cell death by exerting opposing effects on linear ubiquitin in signaling complexes. *Cell Rep* **13:** 2258–2272. doi:10.1016/j.celrep.2015.11.009

Drennan MB, Govindarajan S, Verheugen E, Coquet JM, Staal J, McGuire C, Taghon T, Leclercq G, Beyaert R, van Loo G, et al. 2016. NKT sublineage specification and survival requires the ubiquitin-modifying enzyme TNFAIP3/A20. *J Exp Med* **213:** 1973–1981. doi:10 .1084/jem.20151065

Duncan CJA, Dinnigan E, Theobald R, Grainger A, Skelton AJ, Hussain R, Willet JDP, Swan DJ, Coxhead J, Thomas MF, et al. 2018. Early-onset autoimmune disease due to a heterozygous loss-of-function mutation in *TNFAIP3* (A20). *Ann Rheum Dis* **77:** 783–786. doi:10.1136/annr heumdis-2016-210944

Duong BH, Onizawa M, Oses-Prieto JA, Advincula R, Burlingame A, Malynn BA, Ma A. 2015. A20 restricts ubiquitination of pro-interleukin-1β protein complexes and suppresses NLRP3 inflammasome activity. *Immunity* **42:** 55–67. doi:10.1016/j.immuni.2014.12.031

Duwel M, Welteke V, Oeckinghaus A, Baens M, Kloo B, Ferch U, Darnay BG, Ruland J, Marynen P, Krappmann D. 2009. A20 negatively regulates T cell receptor signaling to NF-κB by cleaving Malt1 ubiquitin chains. *J Immunol* **182:** 7718–7728. doi:10.4049/jimmunol.0803313

Fischer JC, Otten V, Kober M, Drees C, Rosenbaum M, Schmickl M, Heidegger S, Beyaert R, van Loo G, Li XC, et al. 2017. A20 restrains thymic regulatory T cell development. *J Immunol* **199:** 2356–2365. doi:10.4049/jimmu nol.1602102

Fujita N, Morita E, Itoh T, Tanaka A, Nakaoka M, Osada Y, Umemoto T, Saitoh T, Nakatogawa H, Kobayashi S, et al. 2013. Recruitment of the autophagic machinery to endosomes during infection is mediated by ubiquitin. *J Cell Biol* **203:** 115–128. doi:10.1083/jcb.201304188

Fukaya M, Brorsson CA, Meyerovich K, Catrysse L, Delaroche D, Vanzela EC, Ortis F, Beyaert R, Nielsen LB, Andersen ML, et al. 2016. A20 inhibits β-cell apoptosis by multiple mechanisms and Predicts residual β-cell function in type 1 diabetes. *Mol Endocrinol* **30:** 48–61. doi:10.1210/me.2015-1176

Gantier MP, Stunden HJ, McCoy CE, Behlke MA, Wang D, Kaparakis-Liaskos M, Sarvestani ST, Yang YH, Xu D, Corr SC, et al. 2012. A miR-19 regulon that controls NF-κB signaling. *Nucleic Acids Res* **40:** 8048–8058. doi:10.1093/nar/gks521

Garcia-Carbonell R, Wong J, Kim JY, Close LA, Boland BS, Wong TL, Harris PA, Ho SB, Das S, Ernst PB, et al. 2018. Elevated A20 promotes TNF-induced and RIPK1-dependent intestinal epithelial cell death. *Proc Natl Acad Sci* **115:** E9192–E9200. doi:10.1073/pnas.1810584115

Giordano M, Roncagalli R, Bourdely P, Chasson L, Buferne M, Yamasaki S, Beyaert R, van Loo G, Auphan-Anezin N, Schmitt-Verhulst AM, et al. 2014. The tumor necrosis factor α-induced protein 3 (TNFAIP3, A20) imposes a brake on antitumor activity of CD8 T cells. *Proc Natl Acad Sci* **111:** 11115–11120. doi:10.1073/pnas.1406259111

Graham RR, Cotsapas C, Davies L, Hackett R, Lessard CJ, Leon JM, Burtt NP, Guiducci C, Parkin M, Gates C, et al. 2008. Genetic variants near TNFAIP3 on 6q23 are associated with systemic lupus erythematosus. *Nat Genet* **40:** 1059–1061. doi:10.1038/ng.200

Guo Q, Dong H, Liu X, Wang C, Liu N, Zhang J, Li B, Cao W, Ding T, Yang Z, et al. 2009. A20 is overexpressed in glioma cells and may serve as a potential therapeutic target. *Exp Opin Ther Targets* **13:** 733–741. doi:10.1517/14728220903045018

Haas TL, Emmerich CH, Gerlach B, Schmukle AC, Cordier SM, Rieser E, Feltham R, Vince J, Warnken U, Wenger T, et al. 2009. Recruitment of the linear ubiquitin chain assembly complex stabilizes the TNF-R1 signaling complex and is required for TNF-mediated gene induction. *Mol Cell* **36:** 831–844. doi:10.1016/j.molcel.2009.10.013

Hammer GE, Turer EE, Taylor KE, Fang CJ, Advincula R, Oshima S, Barrera J, Huang EJ, Hou B, Malynn BA, et al. 2011. Expression of A20 by dendritic cells preserves immune homeostasis and prevents colitis and spondyloarthritis. *Nat Immunol* **12:** 1184–1193. doi:10.1038/ni.2135

Hampe J, Franke A, Rosenstiel P, Till A, Teuber M, Huse K, Albrecht M, Mayr G, De La Vega FM, Briggs J, et al. 2007. A genome-wide association scan of nonsynonymous SNPs identifies a susceptibility variant for Crohn disease in ATG16L1. *Nat Genet* **39:** 207–211. doi:10.1038/ng1954

He KL, Ting AT. 2002. A20 inhibits tumor necrosis factor (TNF) α–induced apoptosis by disrupting recruitment of TRADD and RIP to the TNF receptor 1 complex in Jurkat T cells. *Mol Cell Biol* **22:** 6034–6045. doi:10.1128/MCB.22.17.6034-6045.2002

Heger K, Fierens K, Vahl JC, Aszodi A, Peschke K, Schenten D, Hammad H, Beyaert R, Saur D, van Loo G, et al. 2014. A20-deficient mast cells exacerbate inflammatory responses in vivo. *PLoS Biol* **12:** e1001762. doi:10.1371/journal.pbio.1001762

Heyninck K, Denecker G, De Valck D, Fiers W, Beyaert R. 1999. Inhibition of tumor necrosis factor-induced necrotic cell death by the zinc finger protein A20. *Anticancer Res* **19:** 2863–2868. doi:10.1016/0014-5793(96)00283-9

Hitotsumatsu O, Ahmad RC, Tavares R, Wang M, Philpott D, Turer EE, Lee BL, Shiffin N, Advincula R, Malynn BA, et al. 2008. The ubiquitin-editing enzyme A20 restricts nucleotide-binding oligomerization domain containing 2-triggered signals. *Immunity* **28:** 381–390. doi:10.1016/j.immuni.2008.02.002

Hjelmeland AB, Wu Q, Wickman S, Eyler C, Heddleston J, Shi Q, Lathia JD, Macswords J, Lee J, McLendon RE, et al. 2010. Targeting A20 decreases glioma stem cell survival and tumor growth. *PLoS Biol* **8:** e1000319. doi:10.1371/journal.pbio.1000319

Honma K, Tsuzuki S, Nakagawa M, Tagawa H, Nakamura S, Morishima Y, Seto M. 2009. TNFAIP3/A20 functions as a novel tumor suppressor gene in several subtypes of non-Hodgkin lymphomas. *Blood* **114:** 2467–2475. doi:10.1182/blood-2008-12-194852

Hovelmeyer N, Reissig S, Xuan NT, Adams-Quack P, Lukas D, Nikolaev A, Schlüter D, Waisman A. 2011. A20 deficiency in B cells enhances B-cell proliferation and results in the development of autoantibodies. *Eur J Immunol* **41:** 595–601. doi:10.1002/eji.201041313

Hrdinka M, Gyrd-Hansen M. 2017. The Met1-linked ubiquitin machinery: Emerging themes of (de)regulation. *Mol Cell* **68:** 265–280. doi:10.1016/j.molcel.2017.09.001

Hsu AC, Dua K, Starkey MR, Haw TJ, Nair PM, Nichol K, Zammit N, Grey ST, Baines KJ, Foster PS, et al. 2017. MicroRNA-125a and -b inhibit A20 and MAVS to promote inflammation and impair antiviral response in COPD. *JCI Insight* **2:** e90443. doi:10.1172/jci.insight.90443

Hutti JE, Turk BE, Asara JM, Ma A, Cantley LC, Abbott DW. 2007. IκB kinase β phosphorylates the K63 deubiquitinase A20 to cause feedback inhibition of the NF-κB pathway. *Mol Cell Biol* **27:** 7451–7461. doi:10.1128/MCB.01101-07

Iha H, Peloponese JM, Verstrepen L, Zapart G, Ikeda F, Smith CD, Starost MF, Yedavalli V, Heyninck K, Dikic I, et al. 2008. Inflammatory cardiac valvulitis in TAX1BP1-deficient mice through selective NF-κB activation. *EMBO J* **27:** 629–641. doi:10.1038/emboj.2008.5

Ikeda F, Deribe YL, Skånland SS, Stieglitz B, Grabbe C, Franz-Wachtel M, van Wijk SJ, Goswami P, Nagy V, Terzic J, et al. 2011. SHARPIN forms a linear ubiquitin ligase complex regulating NF-κB activity and apoptosis. *Nature* **471:** 637–641. doi:10.1038/nature09814

Iwai K, Tokunaga F. 2009. Linear polyubiquitination: A new regulator of NF-κB activation. *EMBO Rep* **10:** 706–713. doi:10.1038/embor.2009.144

Jin Z, Li Y, Pitti R, Lawrence D, Pham VC, Lill JR, Ashkenazi A. 2009. Cullin3-based polyubiquitination and p62-dependent aggregation of caspase-8 mediate extrinsic apoptosis signaling. *Cell* **137:** 721–735. doi:10.1016/j.cell.2009.03.015

Jostins L, Ripke S, Weersma RK, Duerr RH, McGovern DP, Hui KY, Lee JC, Schumm LP, Sharma Y, Anderson CA, et al. 2012. Host–microbe interactions have shaped the genetic architecture of inflammatory bowel disease. *Nature* **491:** 119–124. doi:10.1038/nature11582

Kadowaki T, Ohnishi H, Kawamoto N, Hori T, Nishimura K, Kobayashi C, Shigemura T, Ogata S, Inoue Y, Kawai T, et al. 2018. Haploinsufficiency of A20 causes autoinflammatory and autoimmune disorders. *J Allergy Clin Immunol* **141:** 1485–1488.e11. doi:10.1016/j.jaci.2017.10.039

Kato M, Sanada M, Kato I, Sato Y, Takita J, Takeuchi K, Niwa A, Chen Y, Nakazaki K, Nomoto J, et al. 2009. Frequent

 Cite this article as *Cold Spring Harb Perspect Biol* doi: 10.1101/cshperspect.a036418

inactivation of A20 in B-cell lymphomas. *Nature* **459**: 712–716. doi:10.1038/nature07969

Kattah MG, Shao L, Rosli YY, Shimizu H, Whang MI, Advincula R, Achacoso P, Shah S, Duong BH, Onizawa M, et al. 2018. A20 and ABIN-1 synergistically preserve intestinal epithelial cell survival. *J Exp Med* **215**: 1839–1852. doi:10.1084/jem.20180198

Kim SW, Ramasamy K, Bouamar H, Lin AP, Jiang D, Aguiar RC. 2012. MicroRNAs miR-125a and miR-125b constitutively activate the NF-κB pathway by targeting the tumor necrosis factor α-induced protein 3 (TNFAIP3, A20). *Proc Natl Acad Sci* **109**: 7865–7870. doi:10.1073/pnas.1200081109

Kolodziej LE, Lodolce JP, Chang JE, Schneider JR, Grimm WA, Bartulis SJ, Zhu X, Messer JS, Murphy SF, Reddy N, et al. 2011. TNFAIP3 maintains intestinal barrier function and supports epithelial cell tight junctions. *PLoS ONE* **6**: e26352. doi:10.1371/journal.pone.0026352

Komander D, Barford D. 2008. Structure of the A20 OTU domain and mechanistic insights into deubiquitination. *Biochem J* **409**: 77–85. doi:10.1042/BJ20071399

Kondylis V, Kumari S, Vlantis K, Pasparakis M. 2017. The interplay of IKK, NF-κB and RIPK1 signaling in the regulation of cell death, tissue homeostasis and inflammation. *Immunol Rev* **277**: 113–127. doi:10.1111/imr.12550

Kool M, van Loo G, Waelput W, De Prijck S, Muskens F, Sze M, van Praet J, Branco-Madeira F, Janssens S, Reizis B, et al. 2011. The ubiquitin-editing protein A20 prevents dendritic cell activation, recognition of apoptotic cells, and systemic autoimmunity. *Immunity* **35**: 82–96. doi:10.1016/j.immuni.2011.05.013

Krikos A, Laherty CD, Dixit VM. 1992. Transcriptional activation of the tumor necrosis factor α-inducible zinc finger protein, A20, is mediated by kappa B elements. *J Biol Chem* **267**: 17971–17976.

Kulathu Y, Garcia FJ, Mevissen TE, Busch M, Arnaudo N, Carroll KS, Barford D, Komander D. 2013. Regulation of A20 and other OTU deubiquitinases by reversible oxidation. *Nat Commun* **4**: 1569. doi:10.1038/ncomms2567

Kumar M, Sahu SK, Kumar R, Subuddhi A, Maji RK, Jana K, Gupta P, Raffetseder J, Lerm M, Ghosh Z, et al. 2015. MicroRNA let-7 modulates the immune response to *Mycobacterium tuberculosis* infection via control of A20, an inhibitor of the NF-κB pathway. *Cell Host Microbe* **17**: 345–356. doi:10.1016/j.chom.2015.01.007

Langsch S, Baumgartner U, Haemmig S, Schlup C, Schäfer SC, Berezowska S, Rieger G, Dorn P, Tschan MP, Vassella E. 2016. miR-29b mediates NF-κB signaling in KRAS-induced non-small cell lung cancers. *Cancer Res* **76**: 4160–4169. doi:10.1158/0008-5472.CAN-15-2580

Lawless D, Pathak S, Scambler TE, Ouboussad L, Anwar R, Savic S. 2018. A case of adult-onset Still's disease caused by a novel splicing mutation in *TNFAIP3* successfully treated with tocilizumab. *Front Immunol* **9**: 1527. doi:10.3389/fimmu.2018.01527

Lee EG, Boone DL, Chai S, Libby SL, Chien M, Lodolce JP, Ma A. 2000. Failure to regulate TNF-induced NF-κB and cell death responses in A20-deficient mice. *Science* **289**: 2350–2354. doi:10.1126/science.289.5488.2350

Lee JG, Baek K, Soetandyo N, Ye Y. 2013. Reversible inactivation of deubiquitinases by reactive oxygen species in vitro and in cells. *Nat Commun* **4**: 1568. doi:10.1038/ncomms2532

Lee JH, Jung SM, Yang KM, Bae E, Ahn SG, Park JS, Seo D, Kim M, Ha J, Lee J, et al. 2017. A20 promotes metastasis of aggressive basal-like breast cancers through multi-mono-ubiquitylation of Snail1. *Nat Cell Biol* **19**: 1260–1273. doi:10.1038/ncb3609

Lin SC, Chung JY, Lamothe B, Rajashankar K, Lu M, Lo YC, Lam AY, Darnay BG, Wu H. 2008. Molecular basis for the unique deubiquitinating activity of the NF-κB inhibitor A20. *J Mol Biol* **376**: 526–540. doi:10.1016/j.jmb.2007.11.092

Lippens S, Lefebvre S, Gilbert B, Sze M, Devos M, Verhelst K, Vereecke L, McGuire C, Guerin C, Vandenabeele P, et al. 2011. Keratinocyte-specific ablation of the NF-κB regulatory protein A20 (TNFAIP3) reveals a role in the control of epidermal homeostasis. *Cell Death Differ* **18**: 1845–1853. doi:10.1038/cdd.2011.55

Liuwantara D, Elliot M, Smith MW, Yam AO, Walters SN, Marino E, McShea A, Grey ST. 2006. Nuclear factor-κB regulates β-cell death: A critical role for A20 in β-cell protection. *Diabetes* **55**: 2491–2501. doi:10.2337/db06-0142

Lodolce JP, Kolodziej LE, Rhee L, Kariuki SN, Franek BS, McGreal NM, Logsdon MF, Bartulis SJ, Perera MA, Ellis NA, et al. 2010. African-derived genetic polymorphisms in *TNFAIP3* mediate risk for autoimmunity. *J Immunol* **184**: 7001–7009. doi:10.4049/jimmunol.1000324

Lork M, Verhelst K, Beyaert R. 2017. CYLD, A20 and OTULIN deubiquitinases in NF-κB signaling and cell death: So similar, yet so different. *Cell Death Differ* **24**: 1172–1183. doi:10.1038/cdd.2017.46

Lu TT, Onizawa M, Hammer GE, Turer EE, Yin Q, Damko E, Agelidis A, Shifrin N, Advincula R, Barrera J, et al. 2013. Dimerization and ubiquitin mediated recruitment of A20, a complex deubiquitinating enzyme. *Immunity* **38**: 896–905. doi:10.1016/j.immuni.2013.03.008

Ma A, Malynn BA. 2012. A20: linking a complex regulator of ubiquitylation to immunity and human disease. *Nat Rev Immunol* **12**: 774–785. doi:10.1038/nri3313

Maelfait J, Roose K, Bogaert P, Sze M, Saelens X, Pasparakis M, Carpentier I, van Loo G, Beyaert R. 2012. A20 (*Tnfaip3*) deficiency in myeloid cells protects against influenza A virus infection. *PLoS Pathog* **8**: e1002570. doi:10.1371/journal.ppat.1002570

Maelfait J, Roose K, Vereecke L, McGuire C, Sze M, Schuijs MJ, Willart M, Ibanez LI, Hammad H, Lambrecht BN, et al. 2016. A20 deficiency in lung epithelial cells protects against influenza A virus infection. *PLoS Pathog* **12**: e1005410. doi:10.1371/journal.ppat.1005410

Matmati M, Jacques P, Maelfait J, Verheugen E, Kool M, Sze M, Geboes L, Louagie E, McGuire C, Vereecke L, et al. 2011. A20 (*TNFAIP3*) deficiency in myeloid cells triggers erosive polyarthritis resembling rheumatoid arthritis. *Nat Genet* **43**: 908–912. doi:10.1038/ng.874

Mauro C, Pacifico F, Lavorgna A, Mellone S, Iannetti A, Acquaviva R, Formisano S, Vito P, Leonardi A. 2006. ABIN-1 binds to NEMO/IKKγ and co-operates with A20 in inhibiting NF-κB. *J Biol Chem* **281**: 18482–18488. doi:10.1074/jbc.M601502200

McGuire C, Rahman M, Schwaninger M, Beyaert R, van Loo G. 2013. The ubiquitin editing enzyme A20 (TNFAIP3) is upregulated during permanent middle cerebral artery occlusion but does not influence disease outcome. *Cell Death Dis* **4**: e531. doi:10.1038/cddis.2013.55

Musone SL, Taylor KE, Lu TT, Nititham J, Ferreira RC, Ortmann W, Shifrin N, Petri MA, Kamboh MI, Manzi S, et al. 2008. Multiple polymorphisms in the *TNFAIP3* region are independently associated with systemic lupus erythematosus. *Nat Genet* **40**: 1062–1064. doi:10.1038/ng.202

Nakagawa MM, Thummar K, Mandelbaum J, Pasqualucci L, Rathinam CV. 2015. Lack of the ubiquitin-editing enzyme A20 results in loss of hematopoietic stem cell quiescence. *J Exp Med* **212**: 203–216. doi:10.1084/jem.20132544

Newton K, Dugger DL, Maltzman A, Greve JM, Hedehus M, Martin-McNulty B, Carano RA, Cao TC, van Bruggen N, Bernstein L, et al. 2016. RIPK3 deficiency or catalytically inactive RIPK1 provides greater benefit than MLKL deficiency in mouse models of inflammation and tissue injury. *Cell Death Differ* **23**: 1565–1576. doi:10.1038/cdd.2016.46

Novak U, Rinaldi A, Kwee I, Nandula SV, Rancoita PM, Compagno M, Cerri M, Rossi D, Murty VV, Zucca E, et al. 2009. The NF-κB negative regulator *TNFAIP3* (A20) is inactivated by somatic mutations and genomic deletions in marginal zone lymphomas. *Blood* **113**: 4918–4921. doi:10.1182/blood-2008-08-174110

Ohnishi H, Kawamoto N, Seishima M, Ohara O, Fukao T. 2017. A Japanese family case with juvenile onset Behçet's disease caused by *TNFAIP3* mutation. *Allergol Int* **66**: 146–148. doi:10.1016/j.alit.2016.06.006

Okosun J, Bödör C, Wang J, Araf S, Yang CY, Pan C, Boller S, Cittaro D, Bozek M, Iqbal S, et al. 2014. Integrated genomic analysis identifies recurrent mutations and evolution patterns driving the initiation and progression of follicular lymphoma. *Nat Genet* **46**: 176–181. doi:10.1038/ng.2856

Onizawa M, Oshima S, Schulze-Topphoff U, Oses-Prieto JA, Lu T, Tavares R, Prodhomme T, Duong B, Whang MI, Advincula R, et al. 2015. The ubiquitin-modifying enzyme A20 restricts ubiquitination of the kinase RIPK3 and protects cells from necroptosis. *Nat Immunol* **16**: 618–627. doi:10.1038/ni.3172

Opipari AW Jr, Boguski MS, Dixit VM. 1990. The A20 cDNA induced by tumor necrosis factor α encodes a novel type of zinc finger protein. *J Biol Chem* **265**: 14705–14708.

Pasparakis M, Vandenabeele P. 2015. Necroptosis and its role in inflammation. *Nature* **517**: 311–320. doi:10.1038/nature14191

Patel VI, Daniel S, Longo CR, Shrikhande GV, Scali ST, Czismadia E, Groft CM, Shukri T, Motley-Dore C, Ramsey HE, et al. 2006. A20, a modulator of smooth muscle cell proliferation and apoptosis, prevents and induces regression of neointimal hyperplasia. *FASEB J* **20**: 1418–1430. doi:10.1096/fj.05-4981com

Peltzer N, Rieser E, Taraborrelli L, Draber P, Darding M, Pernaute B, Shimizu Y, Sarr A, Draberova H, Montinaro A, et al. 2014. HOIP deficiency causes embryonic lethality by aberrant TNFR1-mediated endothelial cell death. *Cell Rep* **9**: 153–165. doi:10.1016/j.celrep.2014.08.066

Polykratis A, Martens A, Eren RO, Shirasaki Y, Yamagishi M, Yamaguchi Y, Uemura S, Miura M, Holzmann B, Kollias

G, et al. 2019. A20 prevents inflammasome-dependent arthritis by inhibiting macrophage necroptosis through its ZnF7 ubiquitin-binding domain. *Nat Cell Biol* **21**: 731–742. doi:10.1038/s41556-019-0324-3

Ramsey HE, Da Silva CG, Longo CR, Csizmadia E, Studer P, Patel VI, Damrauer SM, Siracuse JJ, Daniel S, Ferran C. 2009. A20 protects mice from lethal liver ischemia/reperfusion injury by increasing peroxisome proliferator-activated receptor-α expression. *Liver Transpl* **15**: 1613–1621. doi:10.1002/lt.21879

Renner F, Schmitz ML. 2009. Autoregulatory feedback loops terminating the NF-κB response. *Trends Biochem Sci* **34**: 128–135. doi:10.1016/j.tibs.2008.12.003

Rhee L, Murphy SF, Kolodziej LE, Grimm WA, Weber CR, Lodolce JP, Chang JE, Bartulis SJ, Messer JS, Schneider JR, et al. 2012. Expression of TNFAIP3 in intestinal epithelial cells protects from DSS- but not TNBS-induced colitis. *Am J Physiol Gastrointest Liver Physiol* **303**: G220–G227. doi:10.1152/ajpgi.00077.2012

Rioux JD, Xavier RJ, Taylor KD, Silverberg MS, Goyette P, Huett A, Green T, Kuballa P, Barmada MM, Datta LW, et al. 2007. Genome-wide association study identifies new susceptibility loci for Crohn disease and implicates autophagy in disease pathogenesis. *Nat Genet* **39**: 596–604. doi:10.1038/ng2032

Rodriguez MS, Egana I, Lopitz-Otsoa F, Aillet F, Lopez-Mato MP, Dorronsoro A, Lobato-Gil S, Sutherland JD, Barrio R, Trigueros C, et al. 2014. The RING ubiquitin E3 RNF114 interacts with A20 and modulates NF-κB activity and T-cell activation. *Cell Death Dis* **5**: e1399. doi:10.1038/cddis.2014.366

Schmitz R, Hansmann ML, Bohle V, Martin-Subero JI, Hartmann S, Mechtersheimer G, Klapper W, Vater I, Giefing M, Gesk S, et al. 2009. *TNFAIP3* (A20) is a tumor suppressor gene in Hodgkin lymphoma and primary mediastinal B cell lymphoma. *J Exp Med* **206**: 981–989. doi:10.1084/jem.20090528

Schuijs MJ, Willart MA, Vergote K, Gras D, Deswarte K, Ege MJ, Madeira FB, Beyaert R, van Loo G, Bracher F, et al. 2015. Farm dust and endotoxin protect against allergy through A20 induction in lung epithelial cells. *Science* **349**: 1106–1110. doi:10.1126/science.aac6623

Shembade N, Harhaj NS, Liebl DJ, Harhaj EW. 2007. Essential role for TAX1BP1 in the termination of TNF-α-, IL-1- and LPS-mediated NF-κB and JNK signaling. *EMBO J* **26**: 3910–3922. doi:10.1038/sj.emboj.7601823

Shembade N, Harhaj NS, Parvatiyar K, Copeland NG, Jenkins NA, Matesic LE, Harhaj EW. 2008. The E3 ligase Itch negatively regulates inflammatory signaling pathways by controlling the function of the ubiquitin-editing enzyme A20. *Nat Immunol* **9**: 254–262. doi:10.1038/ni1563

Shembade N, Parvatiyar K, Harhaj NS, Harhaj EW. 2009. The ubiquitin-editing enzyme A20 requires RNF11 to downregulate NF-κB signalling. *EMBO J* **28**: 513–522. doi:10.1038/emboj.2008.285

Shembade N, Ma A, Harhaj EW. 2010. Inhibition of NF-κB signaling by A20 through disruption of ubiquitin enzyme complexes. *Science* **327**: 1135–1139. doi:10.1126/science.1182364

Shigemura T, Kaneko N, Kobayashi N, Kobayashi K, Takeuchi Y, Nakano N, Masumoto J, Agematsu K. 2016. Novel heterozygous C243Y A20/TNFAIP3 gene mutation is responsible for chronic inflammation in autosomal-dominant Behçet's disease. *RMD Open* **2:** e000223. doi:10.1136/rmdopen-2015-000223

Shrikhande GV, Scali ST, da Silva CG, Damrauer SM, Csizmadia E, Putheti P, Matthey M, Arjoon R, Patel R, Siracuse JJ, et al. 2010. O-glycosylation regulates ubiquitination and degradation of the anti-inflammatory protein A20 to accelerate atherosclerosis in diabetic ApoE-null mice. *PLoS ONE* **5:** e14240. doi:10.1371/journal.pone.0014240

Skaug B, Chen J, Du F, He J, Ma A, Chen ZJ. 2011. Direct, noncatalytic mechanism of IKK inhibition by A20. *Mol Cell* **44:** 559–571. doi:10.1016/j.molcel.2011.09.015

Slowicka K, Serramito-Gómez I, Boada Romero E, Martens A, Sze M, Petta I, Vikkula HK, dR R, Parthoens E, Lippens S, et al. 2019. Physical and functional interaction between A20 and ATG16L1-WD40 domain in the control of intestinal homeostasis. *Nat Commun* **10:** 1834. doi:10.1038/s41467-019-09667-z

Sokhi UK, Liber MP, Frye L, Park S, Kang K, Pannellini T, Zhao B, Norinsky R, Ivashkiv LB, Gong S. 2018. Dissection and function of autoimmunity-associated *TNFAIP3* (A20) gene enhancers in humanized mouse models. *Nat Commun* **9:** 658. doi:10.1038/s41467-018-03081-7

Takagi M, Ogata S, Ueno H, Yoshida K, Yeh T, Hoshino A, Piao J, Yamashita M, Nanya M, Okano T, et al. 2017. Haploinsufficiency of *TNFAIP3* (A20) by germline mutation is involved in autoimmune lymphoproliferative syndrome. *J Allergy Clin Immunol* **139:** 1914–1922. doi:10.1016/j.jaci.2016.09.038

Tavares RM, Turer EE, Liu CL, Advincula R, Scapini P, Rhee L, Barrera J, Lowell CA, Utz PJ, Malynn BA, et al. 2010. The ubiquitin modifying enzyme A20 restricts B cell survival and prevents autoimmunity. *Immunity* **33:** 181–191. doi:10.1016/j.immuni.2010.07.017

Tewari M, Wolf FW, Seldin MF, O'Shea KS, Dixit VM, Turka LA. 1995. Lymphoid expression and regulation of A20, an inhibitor of programmed cell death. *J Immunol* **154:** 1699–1706.

Ting AT, Bertrand MJM. 2016. More to life than NF-κB in TNFR1 signaling. *Trends Immunol* **37:** 535–545. doi:10.1016/j.it.2016.06.002

Tiruppathi C, Soni D, Wang DM, Xue J, Singh V, Thippegowda PB, Cheppudira BP, Mishra RK, Debroy A, Qian Z, et al. 2014. The transcription factor DREAM represses the deubiquitinase A20 and mediates inflammation. *Nat Immunol* **15:** 239–247. doi:10.1038/ni.2823

Tokunaga F, Nishimasu H, Ishitani R, Goto E, Noguchi T, Mio K, Kamei K, Ma A, Iwai K, Nureki O. 2012. Specific recognition of linear polyubiquitin by A20 zinc finger 7 is involved in NF-κB regulation. *EMBO J* **31:** 3856–3870. doi:10.1038/emboj.2012.241

Vande Walle L, Van Opdenbosch N, Jacques P, Fossoul A, Verheugen E, Vogel P, Beyaert R, Elewaut D, Kanneganti TD, van Loo G, et al. 2014. Negative regulation of the NLRP3 inflammasome by A20 protects against arthritis. *Nature* **512:** 69–73. doi:10.1038/nature13322

Vereecke L, Sze M, McGuire C, Rogiers B, Chu Y, Schmidt-Supprian M, Pasparakis M, Beyaert R, van Loo G. 2010. Enterocyte-specific A20 deficiency sensitizes to tumor necrosis factor-induced toxicity and experimental

colitis. *J Exp Med* **207:** 1513–1523. doi:10.1084/jem.20092474

Vereecke L, Vieira-Silva S, Billiet T, van Es JH, McGuire C, Slowicka K, Sze M, van den Born M, De Hertogh G, Clevers H, et al. 2014. A20 controls intestinal homeostasis through cell-specific activities. *Nat Commun* **5:** 5103. doi:10.1038/ncomms6103

Verhelst K, Carpentier I, Beyaert R. 2011. Regulation of TNF-induced NF-κB activation by different cytoplasmic ubiquitination events. *Cytokine Growth Factor Rev* **22:** 277–286. doi:10.1016/j.cytogfr.2011.11.002

Verhelst K, Carpentier I, Kreike M, Meloni L, Verstrepen L, Kensche T, Dikic I, Beyaert R. 2012. A20 inhibits LUBAC-mediated NF-κB activation by binding linear polyubiquitin chains via its zinc finger 7. *EMBO J* **31:** 3845–3855. doi:10.1038/emboj.2012.240

Verstrepen L, Verhelst K, van Loo G, Carpentier I, Ley SC, Beyaert R. 2010. Expression, biological activities and mechanisms of action of A20 (*TNFAIP3*). *Biochem Pharmacol* **80:** 2009–2020. doi:10.1016/j.bcp.2010.06.044

Voet S, McGuire C, Hagemeyer N, Martens A, Schroeder A, Wieghofer P, Daems C, Staszewski O, Vande Walle L, Jordao MJC, et al. 2018. A20 critically controls microglia activation and inhibits inflammasome-dependent neuroinflammation. *Nat Commun* **9:** 2036. doi:10.1038/s41467-018-04376-5

Wagner S, Carpentier I, Rogov V, Kreike M, Ikeda F, Löhr F, Wu CJ, Ashwell JD, Dötsch V, Dikic I, et al. 2008. Ubiquitin binding mediates the NF-κB inhibitory potential of ABIN proteins. *Oncogene* **27:** 3739–3745. doi:10.1038/sj.onc.1211042

Wang CM, Wang Y, Fan CG, Xu FF, Sun WS, Liu YG, Jia JH. 2011. miR-29c targets *TNFAIP3*, inhibits cell proliferation and induces apoptosis in hepatitis B virus-related hepatocellular carcinoma. *Biochem Biophys Res Commun* **411:** 586–592. doi:10.1016/j.bbrc.2011.06.191

Wang S, Wen F, Wiley GB, Kinter MT, Gaffney PM. 2013a. An enhancer element harboring variants associated with systemic lupus erythematosus engages the *TNFAIP3* promoter to influence A20 expression. *PLoS Genet* **9:** e1003750. doi:10.1371/journal.pgen.1003750

Wang X, Deckert M, Xuan NT, Nishanth G, Just S, Waisman A, Naumann M, Schlüter D. 2013b. Astrocytic A20 ameliorates experimental autoimmune encephalomyelitis by inhibiting NF-κB- and STAT1-dependent chemokine production in astrocytes. *Acta Neuropathol* **126:** 711–724. doi:10.1007/s00401-013-1183-9

Wertz IE, O'Rourke KM, Zhou H, Eby M, Aravind L, Seshagiri S, Wu P, Wiesmann C, Baker R, Boone DL, et al. 2004. De-ubiquitination and ubiquitin ligase domains of A20 downregulate NF-κB signalling. *Nature* **430:** 694–699. doi:10.1038/nature02794

Wertz IE, Newton K, Seshasayee D, Kusam S, Lam C, Zhang J, Popovych N, Helgason E, Schoeffler A, Jeet S, et al. 2015. Phosphorylation and linear ubiquitin direct A20 inhibition of inflammation. *Nature* **528:** 370–375. doi:10.1038/nature16165

Won M, Park KA, Byun HS, Sohn KC, Kim YR, Jeon J, Hong JH, Park J, Seok JH, Kim JM, et al. 2010. Novel anti-apoptotic mechanism of A20 through targeting ASK1 to suppress TNF-induced JNK activation. *Cell Death Differ* **17:** 1830–1841. doi:10.1038/cdd.2010.47

Xia M, Liu J, Wu X, Liu S, Li G, Han C, Song L, Li Z, Wang Q, Wang J, et al. 2013. Histone methyltransferase Ash1l suppresses interleukin-6 production and inflammatory autoimmune diseases by inducing the ubiquitin-editing enzyme A20. *Immunity* **39:** 470–481. doi:10.1016/j.immuni.2013.08.016

Xuan NT, Wang X, Nishanth G, Waisman A, Borucki K, Isermann B, Naumann M, Deckert M, Schlüter D. 2015. A20 expression in dendritic cells protects mice from LPS-induced mortality. *Eur J Immunol* **45:** 818–828. doi:10.1002/eji.201444795

Yamaguchi N, Yamaguchi N. 2015. The seventh zinc finger motif of A20 is required for the suppression of TNF-α-induced apoptosis. *FEBS Lett* **589:** 1369–1375. doi:10.1016/j.febslet.2015.04.022

Yuk JM, Kim TS, Kim SY, Lee HM, Han J, Dufour CR, Kim JK, Jin HS, Yang CS, Park KS, et al. 2015. Orphan nuclear receptor ERRα controls macrophage metabolic signaling and A20 Expression to negatively regulate TLR-induced inflammation. *Immunity* **43:** 80–91. doi:10.1016/j.immuni.2015.07.003

Zhang Q, Lenardo MJ, Baltimore D. 2017. 30 years of NF-κB: A blossoming of relevance to human pathobiology. *Cell* **168:** 37–57. doi:10.1016/j.cell.2016.12.012

Zhao D, Zhuang N, Ding Y, Kang Y, Shi L. 2016. MiR-221 activates the NF-κB pathway by targeting A20. *Biochem Biophys Res Commun* **472:** 11–18. doi:10.1016/j.bbrc.2015.11.009

Zhou Q, Wang H, Schwartz DM, Stoffels M, Park YH, Zhang Y, Yang D, Demirkaya E, Takeuchi M, Tsai WL, et al. 2016. Loss-of-function mutations in *TNFAIP3* leading to A20 haploinsufficiency cause an early-onset autoinflammatory disease. *Nat Genet* **48:** 67–73. doi:10.1038/ng.3459

Cite this article as *Cold Spring Harb Perspect Biol* doi: 10.1101/cshperspect.a036418

Regulation of Cell Death and Immunity by XIAP

Philipp J. Jost[1,2,3] and Domagoj Vucic[4]

[1]Medical Department III, School of Medicine, Technical University of Munich, 81675 Munich, Germany

[2]Center for Translational Cancer Research (TranslaTUM), School of Medicine, Technical University of Munich, 81675 Munich, Germany

[3]German Cancer Consortium (DKTK) partner site TUM, DKFZ, 69120 Heidelberg, Germany

[4]Early Discovery Biochemistry Department, Genentech, South San Francisco, California 94080, USA

Correspondence: philipp.jost@tum.de; domagoj@gene.com

X-chromosome-linked inhibitor of apoptosis protein (XIAP) controls cell survival in several regulated cell death pathways and coordinates a range of inflammatory signaling events. Initially identified as a caspase-binding protein, it was considered to be primarily involved in blocking apoptosis from both intrinsic as well as extrinsic triggers. However, XIAP also prevents TNF-mediated, receptor-interacting protein 3 (RIPK3)-dependent cell death, by controlling RIPK1 ubiquitylation and preventing inflammatory cell death. The identification of patients with germline mutations in XIAP (termed XLP-2 syndrome) pointed toward its role in inflammatory signaling. Indeed, XIAP also mediates nucleotide-binding oligomerization domain-containing 2 (NOD2) proinflammatory signaling by promoting RIPK2 ubiquitination within the NOD2 signaling complex leading to NF-κB and MAPK activation and production of inflammatory cytokines and chemokines. Overall, XIAP is a critical regulator of multiple cell death and inflammatory pathways making it an attractive drug target in tumors and inflammatory diseases.

XIAP AND THE INHIBITOR OF APOPTOSIS PROTEIN FAMILY

The evolutionarily conserved family of inhibitor of apoptosis proteins (IAPs) contains structurally related regulators of diverse cellular processes (Varfolomeev and Vucic 2011). IAPs were originally identified in baculoviruses because of their ability to inhibit virus-induced apoptosis and allow viral amplification (Crook et al. 1993; Birnbaum et al. 1994). Subsequently, through functional screens and bioinformatics efforts, IAP genes and proteins were identified in all metazoan organisms including eight human

IAP proteins (Liston et al. 1996; Salvesen and Duckett 2002). Among the human IAP proteins, X-chromosome-linked IAP (XIAP) and cellular IAP1 and 2 (c-IAP1 and c-IAP2) are the most studied, although other IAP proteins (NAIP, ML-IAP, survivin, ILP2, and Apollon) also play important roles in cell survival, cell cycle, inflammation, and overall homeostasis (Fulda and Vucic 2012).

XIAP protein contains three signature baculovirus IAP repeat (BIR) domains, which are conserved 70–80 amino acid zinc–coordinating regions that regulate protein–protein interactions and are instrumental for XIAP function

(Sun et al. 2000; Ndubaku et al. 2009). Many of the BIR-mediated protein–protein interactions use a pocket on the BIR domain that binds to amino-terminal tetrapeptides called IAP-binding motifs (IBMs). The specific BIR sequence of each IAP protein is very important because small variations in the sequence dictate the binding of different interaction partners. The BIR1 domain of XIAP interacts with TAB1 and thus potentially regulates signaling by the TAB-TAK1 complex (Lu et al. 2007). BIR2 and the linker region between BIR1 and BIR2 bind to and inhibit active caspases 3 and 7 (Chai et al. 2001; Huang et al. 2001; Riedl et al. 2001). BIR2 also binds to the kinase receptor-interacting protein 2 (RIPK2), which allows XIAP to ubiquitinate RIPK2 in the nucleotide-binding oligomerization domain-containing 2 (NOD2) signaling pathway (Krieg et al. 2009; Damgaard et al. 2012; Goncharov et al. 2018; Stafford et al. 2018). The BIR3 domain of XIAP binds to and inhibits activation of caspase-9 by blocking its dimerization, which is needed for autocleavage. The BIR3 domain of XIAP also associates with Smac (second mitochondrial activator of caspases)/DIABLO (direct IAP-binding protein with low pI) and other IBM-containing proteins such as serine protease HtrA2, resulting in the antagonism of XIAP antiapoptotic activity (Liu et al. 2000; Sun et al. 2000; Wu et al. 2000).

At the very carboxy terminal of XIAP is a really interesting new gene (RING) domain, which imparts ubiquitin ligase activity (Vaux and Silke 2005), whereas its centrally located ubiquitin-associated (UBA) domain confers binding to monoubiquitin and polyubiquitin chains (Gyrd-Hansen et al. 2008; Blankenship et al. 2009). Ubiquitination involves covalent modification of target proteins with the 76-amino-acid protein ubiquitin, and requires the enzymatic activity of a ubiquitin-activating enzyme (E1), a ubiquitin-conjugating enzyme (E2), and a ubiquitin ligase (E3) (Hershko and Ciechanover 1998). Monoubiquitination occurs when a single ubiquitin molecule is attached to a lysine residue of the substrate protein (Hershko and Ciechanover 1998). However, ubiquitin contains seven lysine residues and a free amino terminus, thus allowing the synthesis of poly-

ubiquitin chains through eight different isopeptide linkages (Pickart and Fushman 2004). XIAP can assemble various polyubiquitin chains on itself and its substrates in collaboration with E2 enzymes of the UbcH5 family (Dynek et al. 2010; Vucic et al. 2011). The most prominent substrates of the E3 activity of XIAP include apoptotic caspases, Smac, and RIPK2 (Morizane et al. 2005; Qin et al. 2016). In many cases, XIAP-mediated ubiquitination leads to substrate degradation (SMAC) (MacFarlane et al. 2002; Galbán and Duckett 2010). However, K63-linked RIPK2 ubiquitination does not affect RIPK2 stability but rather promotes NF-κB and MAPK signaling (Damgaard et al. 2012; Goncharov et al. 2018).

XIAP IN APOPTOSIS

XIAP's ability to bind to and inhibit the caspases that mediate apoptotic cell death was the primary focus of scientists after its initial discovery. Proapoptotic caspases are proteases that dismantle the cell, and therefore must be tightly regulated to prevent uncontrolled cell death or, alternatively, the unwanted survival of damaged or malignant cells. In apoptotic signaling, XIAP binds to caspases 3 and 7 by virtue of a linker region located between BIR domains 1 and 2 (Deveraux et al. 1997; Wilkinson et al. 2004). Indeed, this relatively small linker region is the only part of XIAP that physically interacts with the executioner caspases (Huang et al. 2001). In contrast, XIAP binds to initiator caspase-9 with its BIR3 domain (Deveraux and Reed 1999). XIAP binds to the executioner and initiator caspases with different affinities, but binding serves the same purpose, which is inhibition of apoptosis. Binding of XIAP to caspase-8 is detectable, but unlikely to be relevant biologically because the binding affinity is several magnitudes below that observed for caspases 3 and 7 (Eckelman et al. 2006). XIAP therefore serves as a repressor and regulator of the final steps of apoptotic signaling, acting mostly at the level of activation of executioner caspases.

The family of IAP proteins has several members (Deveraux and Reed 1999; Salvesen and Duckett 2002). Despite their sequence similarities and the presence of BIR domains in other IAP proteins, the ability to directly bind to and inhibit caspases appears restricted to XIAP. It has therefore been referred to as outsider of the family of IAP proteins (Eckelman et al. 2006). XIAP-deficient mice lack overt phenotypes (Harlin et al. 2001; Olayioye et al. 2005), so the field tended to think of XIAP as a protein with relatively low importance in apoptosis signaling. However, it required the right stimulus and a specific condition to understand the relevance of XIAP in apoptosis (Fig. 1). One example is

the activation of FAS (CD95) by FAS ligand (FAS-L), which triggers cell death in most cell types but uses additional signaling pathways depending on the cellular origin. In type I cells, such as lymphocytes, FAS (CD95) stimulation directly activates caspase-8 to trigger executioner caspase processing and cell death. In contrast, in type II cells, such as hepatocytes, caspase-8 also cleaves the proapoptotic BH3 protein BID to amplify the apoptotic signal via the mitochondrial route (Yin et al. 1999; Kaufmann et al. 2007; McKenzie et al. 2008). This mito-

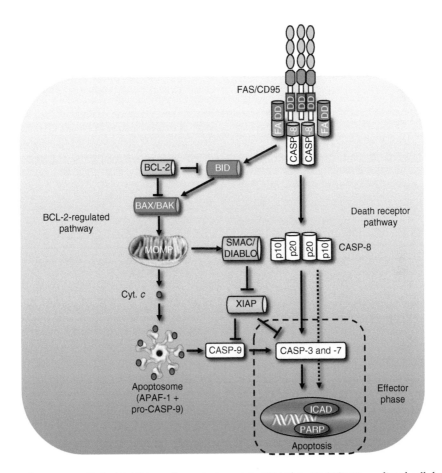

Figure 1. X-chromosome-linked inhibitor of apoptosis protein (XIAP) in FAS/CD95-mediated cell death. The pool of XIAP defines the propensity of cells to undergo apoptotic cell death in response to extrinsic cellular stress such as FAS/CD95 activation. XIAP prevents the activity of processed caspases and thereby increases the threshold for apoptosis activation. XIAP itself is kept in check by proapoptotic factors released from compromised mitochondria such as SMAC/DIABLO that bind and prevent its activity. Cell types with high levels of XIAP such as hepatocytes require amplification of an FAS/CD95-mediated cell death signal via BID-dependent mitochondrial outer membrane potential (MOMP) and subsequent release of SMAC/DIABLO to inhibit XIAP function and thereby allow for apoptosis to occur. (Cyt. *c*) cytochrome *c*.

chondrial amplification step of the apoptotic signal results in the BID-dependent activation of the proapoptotic BCL-2 proteins BAX/BAK eventually resulting in the loss of mitochondrial outer membrane potential (MOMP). The loss of MOMP results in the release of cytochrome c from the intermembrane space promoting formation and activation of a protein complex termed the apoptosome. This protein complex is comprised of procaspase-9, APAF-1, and cytochrome c and on assembly cleaves and activates caspase-9 to induce executioner caspase activation such as caspase-3 and -7. It becomes evident that the fine balance between apoptosis-inducing factors, such as cytochrome c or SMAC/DIABLO, and apoptosis-inhibiting factors such as XIAP defines to a large degree the process of apoptotic cell death execution.

A major difference between type I and type II cells in terms of apoptosis execution is XIAP. Intracellular levels of XIAP are higher in type II cells when compared with type I cells effectively blocking activation of the executioner caspases. BID cleavage in type II cells triggers MOMP, which releases SMAC from mitochondria. SMAC binding to XIAP probably blocks the inhibition of the executioner caspases by XIAP (Jost et al. 2009). Therefore, XIAP fine-tunes sensitivity to FAS-L in different cell types (Varfolomeev et al. 2009).

XIAP IN RIPK3 SIGNALING AND INFLAMMATORY CELL DEATH

After several years of research into the role of XIAP in conventional apoptosis and caspase inhibition, it became clear that XIAP also controls inflammatory forms of cell death. Initial work studied the response of $Xiap^{-/-}$ myeloid cells in inflammation because patients with a germline mutation in $XIAP$ suffer from a hyperinflammatory syndrome termed X-linked lymphoproliferative syndrome type 2 (Rigaud et al. 2006). More detail on this rare but very informative condition is provided in the section entitled, "Genetic Mutations in $XIAP$ Define the XLP-2 Syndrome."

In an effort to understand how XIAP controls inflammation, murine $Xiap^{-/-}$ myeloid

cells were treated with lipopolysaccharide (LPS) as a Toll-like receptor 4 (TLR-4) agonist (Yabal et al. 2014). They released substantial amounts of caspase-1-dependent bioactive IL-1β, whereas similarly treated wild-type cells did not. Moreover, $Xiap^{-/-}$ cells activated a form of programmed and highly inflammatory cell death requiring the kinase RIPK3 (Yabal et al. 2014). Additional genetic experiments revealed that this LPS-induced cell death is mediated by TNF and eventually results in the release of IL-1β (Yabal et al. 2014; Lawlor et al. 2015; Wicki et al. 2016). Indeed, earlier examination of the contribution of individual IAP family members to TNF-mediated cell death found a substantial contribution of XIAP to the suppression of TNF- and RIPK1/RIPK3-mediated cell death in myeloid cells (Wong et al. 2014).

The identification of a role for XIAP in TNF signaling and RIPK3-dependent cell death initially caused some skepticism because, unlike c-IAP1 and c-IAP2, XIAP is not found within the TNF receptor 1 (TNFR1) signaling complex (Bertrand et al. 2008; Varfolomeev et al. 2008; Haas et al. 2009; Vince et al. 2012). Rather, XIAP may be part of a cytosolic protein complex containing RIPK1 and RIPK3 (Yabal et al. 2014). Indeed, the TNF-induced ubiquitylation pattern on cytosolic RIPK1 is controlled by XIAP (Mehrotra et al. 2010). In macrophages or dendritic cells, the genetic ablation of XIAP results in an elevated ubiquitylation of RIPK1 (Yabal et al. 2014). This suggested that XIAP is not the primary ligase in the complex that ubiquitylates RIPK1 but possibly controls the presence or the activation status of an alternative ligase responsible for RIPK1 ubiquitylation that has not been identified to date.

Subsequent work in myeloid cells showed that XIAP represses TLR/MyD88-induced and TNF/TNFR2-dependent proteasomal degradation of TRAF2 and c-IAP1 (Lawlor et al. 2017). Thus, in the absence of XIAP, loss of TRAF2 and c-IAP1 sensitizes the cells to proinflammatory cell death. The mechanistic details of how XIAP prevents the degradation of TRAF2 and cIAP1 still have to be worked out. It is also unclear whether additional functions of XIAP impact on TNF or necroptotic signaling, but it is clear

Cite this article as *Cold Spring Harb Perspect Biol* doi: 10.1101/cshperspect.a036426

that XIAP functions in a distinct manner from c-IAP1 or c-IAP2.

The molecular functions of XIAP that have been identified using $Xiap^{-/-}$ mice largely require the RING domain of XIAP (Schile et al. 2008). Thus, knockin mice expressing carboxy-terminally truncated XIAP deficient for the RING domain that can still bind to activated caspases phenocopy $Xiap^{-/-}$ mice in their exacerbated response to LPS (Yabal et al. 2014). Therefore, XIAP can regulate inflammation independent of it binding to caspases.

XIAP IN INFLAMMATION, NOD SIGNALING

$Xiap^{-/-}$ mice fail to mount an immune response to infection with *Listeria* or to treatment with bacterial peptidoglycans such muramyldipeptide (MDP) (Pedersen et al. 2014). Yet, mutations in *XIAP* are associated with inflammatory diseases such as XLP-2 and very early-onset inflammatory bowel disease (VEO-IBD) (Damgaard et al. 2013; Pedersen et al. 2014). Recognition of invading pathogens and rapid initiation of host defense mechanisms are critical for the maintenance of homeostasis (Chen et al. 2008). Innate immune cells use pathogen-recognition receptors to detect invading pathogens and these include the nucleotide-binding oligomerization domain-containing (NOD)-like receptors (NLRs) (Chen et al. 2008). NOD1 and NOD2 are NLRs that are instrumental for innate immune responses to some bacterial infections (Franchi et al. 2009; Caruso et al. 2014). Mutations in *NOD2* are also associated with inflammatory diseases, including Crohn's disease, early-onset sarcoidosis, and Blau syndrome (Hugot et al. 2001; Ogura et al. 2001; Caso et al. 2015).

Binding of MDP to NOD2 induces recruitment of RIPK2 and XIAP, the latter promoting K63-linked ubiquitination of RIPK2, which then recruits the linear ubiquitin chain assembly complex (LUBAC) (Damgaard et al. 2012). Linear ubiquitination of RIPK2 is required for full activation of the NF-κB and MAPK signaling pathways that drive the expression of many inflammatory cytokines and chemokines (Fig. 2; Damgaard et al. 2012). Several other E3 ligases

reportedly bind to and promote ubiquitination of RIPK2 (Witt and Vucic 2017). Nevertheless, XIAP is essential for efficient NOD2-RIPK2 signaling (Damgaard et al. 2012; Goncharov et al. 2018; Stafford et al. 2018). The related E3 ligases c-IAP1/2 contribute to NOD2-dependent autocrine TNF signaling and amplify cytokine production in vivo (Stafford et al. 2018), but XIAP is more directly involved in signaling by NOD2 (Damgaard et al. 2012). Thus, XIAP-selective antagonists strongly inhibit NOD2-mediated activation of NF-κB and MAPK signaling and subsequent cytokine/chemokine production (Goncharov et al. 2018). XIAP uses its BIR2 domain to engage the kinase domain of RIPK2, and this association allows XIAP to ubiquitinate RIPK2. Mutation of the XIAP-dependent ubiquitination sites in RIPK2 diminishes NOD2 signaling, whereas inactivating the kinase activity of RIPK2 does not (Goncharov et al. 2018). These and other data strongly suggest that ubiquitination of RIPK2 is more important than the catalytic activity of RIPK2 for NOD2 signaling (Goncharov et al. 2018; Hrdinka et al. 2018).

XIAP was also reported to activate NF-κB signaling through its BIR1 domain and interaction with TAB1 (Lu et al. 2007). XIAP BIR1 adopts a dimeric structure in complex with TAB1, and disruption of BIR1 dimerization blocks XIAP BIR1-mediated NF-κB activation (Lu et al. 2007; Cossu et al. 2015). However, the physiological relevance of this BIR1-mediated interaction has not been defined yet.

GENETIC MUTATIONS IN *XIAP* DEFINE THE XLP-2 SYNDROME

Germline mutations in *XIAP* in patients define a genetic susceptibility condition termed X-linked lymphoproliferative syndrome type II (XLP-2) (Rigaud et al. 2006; Latour and Aguilar 2015). The mutations in XIAP are scattered throughout the protein and represent either nonsense mutations, frameshift mutations, or deletions, resulting in substantial irregularities of the protein (Filipovich et al. 2010; Marsh et al. 2010; Pachlopnik Schmid et al. 2011; Yang et al. 2012). The carboxy-terminal RING domain is affected by mutations either caused by nonsense or

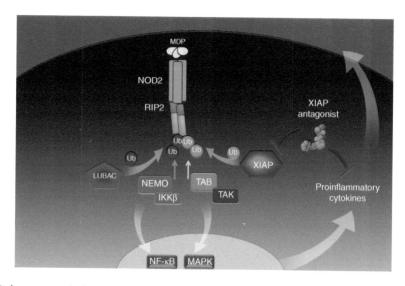

Figure 2. X-chromosome-linked inhibitor of apoptosis protein (XIAP) in inflammation plus XIAP-targeting agents. Muramyldipeptide (MDP)-binding oligomerizes nucleotide-binding oligomerization domain-containing 2 (NOD2) and recruits receptor-interacting protein 2 (RIPK2) and XIAP to promote K63-linked polyubiquitination of RIPK2 (red symbols) and subsequent linear ubiquitin chain assembly complex (LUBAC)-mediated linear polyubiquitination (purple symbols) to activate NF-κB and MAPK signaling and production of inflammatory cytokines. By disrupting RIPK2–XIAP interaction, XIAP-selective antagonists prevent RIPK2 ubiquitination and release of inflammatory cytokines.

frameshift mutations within the body of the encoded gene or, alternatively, by direct localization of the mutations within the RING domain indicating that most mutations interfere with its E3 activity (Damgaard et al. 2013).

XLP-2 boys mostly present at infanthood to early adulthood with symptoms as diverse as cytopenia, fever, hepatosplenomegaly, and elevated levels of acute phase proteins such as ferritin (Latour and Aguilar 2015). The condition clinically may present as hemophagocytic lymphohistiocytosis (HLH) syndrome. XLP-2 onset has often been associated with Epstein–Barr virus (EBV) infections or alternative triggers (Aguilar and Latour 2015). It has therefore been suggested to rename the condition to X-linked HLH disease to encompass the clinical features and the absence of monoclonal lymphoproliferation as suggested in the term XLP-2 (Marsh et al. 2010).

Bone marrow transplantation represents a curative treatment option in severe cases of XLP-2 (Marsh et al. 2013). However, the conditioning regimen and the posttransplant mortal-

ity in XLP-2 patients is high, possibly a result of the exaggerated inflammatory signaling events present during the conditioning treatment phase before allogeneic stem cell transplantation. Indeed, it was recently identified that *Xiap* deficiency in hematopoietic recipient cells drives donor T-cell activation and graft versus host disease (GvHD) in a major mismatch bone marrow transplantation mouse model, thus mimicking allogeneic bone marrow transplantation of XLP-2 patients (Müller et al. 2019). In this setting, the aggravated mortality of allogeneic transplanted mice deficient for XIAP was followed by increased levels of proinflammatory cytokines and donor-derived T lymphocytes (Müller et al. 2019), supporting a critical role of XIAP in preventing aberrant immune activation during allogeneic transplantation. Using XIAP-deficient donor or, alternatively, XIAP-deficient recipient mice indicated that loss of XIAP specifically in the recipient hematopoietic compartment was responsible for aggravation of GvHD after allogeneic stem cell transplantation.

Given the role of EBV infections in triggering XLP-2 and the clinical symptoms associated with this systemic inflammatory syndrome, XLP-2 patients were considered to be affected primarily by hyperinflammation within the hematopoietic system. Therefore, it was a surprise when the group of Sebastian Zeissig identified loss-of-function mutations in *XIAP* as a common occurrence in male patients with early-onset Crohn's disease. This form of inflammatory bowel disease was observed with Mendelian frequency in families carrying mutations in *XIAP* and the penetrance of disease onset was relatively high in affected individuals (Speckmann and Ehl 2014; Zeissig et al. 2015). As the frequency of *XIAP* mutations in children and young adults with IBD-like symptoms or with symptoms associated with HLH was largely unknown, and even today only a fraction of early-onset IBD or HLH patients get screened for *XIAP* mutations, it is important for clinicians to be able to rapidly test for *XIAP* alterations. Rapid detection has become more available by a flow cytometry-based detection assay for the intracellular production of TNF in primary blood-derived monocytic cells from patients suspected to present with XLP-2 syndrome. Simulation of monocytes by a modified form of MDP, termed L18-MDP, the cognate ligand for NOD2, results in TNF production that is detectable within the cells by flow cytometry and aids in the identification of patients incapable of secreting NOD2-dependent TNF (Ammann et al. 2014).

TARGETING XIAP FOR THERAPEUTIC INTERVENTION

Expression of XIAP in several human cancers is associated with poor prognosis (Fulda and Vucic 2012). Besides promoting tumor progression, XIAP has also been implicated in tumor cell mobility, invasion, and metastasis. For example, XIAP, as well as cellular IAPs, and ML-IAP can destabilize kinase C-RAF and consequently modulate MAPK signaling and cell motility (Dogan et al. 2008; Oberoi-Khanuja et al. 2012). In addition, XIAP and survivin cooperate to promote metastasis via the activation of cell motility kinases Src and FAK (Mehrotra et al. 2010). Conversely, antagonism of IAP proteins can block tumor cell migration and invasion (Mehrotra et al. 2010). XIAP and c-IAP1 can also ubiquitinate the Rho GTPase Rac1 and target it for proteasomal degradation (Oberoi et al. 2012). Therefore, regulation of Rac1 stability provides another modality for the modulation of cell migration.

Elevated expression of XIAP in numerous tumor types in combination with its prosurvival functions makes XIAP an attractive target for therapeutic intervention in cancer (Fulda and Vucic 2012). Several clinical trials have targeted XIAP in cancer, either via antisense RNA or by using IAP antagonists (smac mimetics), in the hope of stimulating and/or enhancing tumor cell death (Fulda and Vucic 2012). However, results of these trials were not encouraging and, coupled with a lack of clear predictive biomarkers, IAP antagonists and XIAP antisense reagents have not reached their potential as a cancer treatment option through the activation of cell death.

The recent discovery of the role of XIAP in NOD2 signaling and the development of XIAP-selective antagonists indicate that targeting XIAP in NOD2-mediated inflammatory diseases could be beneficial. IAP antagonists that are selective for XIAP are well positioned for intervention in NOD2-mediated pathologies because, unlike pan-IAP antagonists, they do not activate cell death, c-IAP1/2 autoubiquitination and proteasomal degradation, or NF-κB signaling (Goncharov et al. 2018). In addition, RIPK2 is a dedicated and specific adaptor for the NOD1/NOD2 pathways. Accordingly, XIAP-selective antagonists do not compromise other inflammatory pathways such as TNF signaling (Goncharov et al. 2018). XIAP selectivity is possible because of the uniqueness of the XIAP BIR2 domain that binds to the kinase domain of RIPK2. XIAP-selective antagonists break the direct physical interaction between XIAP and RIPK2. Disruption of XIAP-RIPK2 binding prevents RIPK2 ubiquitination and the assembly of the NOD2 signaling complex, thereby precluding production of inflammatory cytokines (Damgaard et al. 2013; Goncharov et al. 2018).

Previous drug discovery efforts that used pan IAP antagonists have produced reagents with favorable potency and pharmacological properties allowing clinical trials in cancer patients (Fulda and Vucic 2012). Identification of XIAP mutations-associated pathologies urges a cautious approach in targeting XIAP (Aguilar and Latour 2015). Nevertheless, XIAP-selective antagonists have shown no adverse effects and display great efficacy in NOD2-RIPK2 pathway inhibition (Goncharov et al. 2018). Thus, future optimization of XIAP-selective antagonists could yield promising agents for therapeutic intervention in NOD2-mediated diseases such as Crohn's disease, sarcoidosis, and Blau syndrome.

REFERENCES

Aguilar C, Latour S. 2015. X-linked inhibitor of apoptosis protein deficiency: more than an X-linked lymphoproliferative syndrome. *J Clin Immunol* 35: 331–338. doi:10.1007/s10875-015-0141-9

Ammann S, Elling R, Gyrd-Hansen M, Dückers G, Bredius R, Burns SO, Edgar JD, Worth A, Brandau H, Warnatz K, et al. 2014. A new functional assay for the diagnosis of X-linked inhibitor of apoptosis (XIAP) deficiency. *Clin Exp Immunol* 176: 394–400. doi:10.1111/cei.12306

Bertrand MJ, Milutinovic S, Dickson KM, Ho WC, Boudreault A, Durkin J, Gillard JW, Jaquith JB, Morris SJ, Barker PA. 2008. cIAP1 and cIAP2 facilitate cancer cell survival by functioning as E3 ligases that promote RIP1 ubiquitination. *Mol Cell* 30: 689–700. doi:10.1016/j.molcel.2008.05.014

Birnbaum MJ, Clem RJ, Miller LK. 1994. An apoptosis-inhibiting gene from a nuclear polyhedrosis virus encoding a polypeptide with Cys/His sequence motifs. *J Virol* 68: 2521–2528.

Blankenship JW, Varfolomeev E, Goncharov T, Fedorova AV, Kirkpatrick DS, Izrael-Tomasevic A, Phu L, Arnott D, Aghajan M, Zobel K, et al. 2009. Ubiquitin binding modulates IAP antagonist-stimulated proteasomal degradation of c-IAP1 and c-IAP2(1). *Biochem J* 417: 149–165. doi:10.1042/BJ20081885

Caruso R, Warner N, Inohara N, Núñez G. 2014. NOD1 and NOD2: signaling, host defense, and inflammatory disease. *Immunity* 41: 898–908. doi:10.1016/j.immuni.2014.12.010

Caso F, Galozzi P, Costa L, Sfriso P, Cantarini L, Punzi L. 2015. Autoinflammatory granulomatous diseases: from Blau syndrome and early-onset sarcoidosis to NOD2-mediated disease and Crohn's disease. *RMD Open* 1: e000097. doi:10.1136/rmdopen-2015-000097

Chai J, Shiozaki E, Srinivasula SM, Wu Q, Dataa P, Alnemri ES, Shi Y. 2001. Structural basis of caspase-7 inhibition by XIAP. *Cell* 104: 769–780. doi:10.1016/S0092-8674(01)00272-0

Chen G, Shaw MH, Kim YG, Nuñez G. 2008. Nod-like receptors: role in innate immunity and inflammatory disease. *Annu Rev Pathol*. 4: 365–398. doi:10.1146/annurev.pathol.4.110807.092239

Cossu F, Milani M, Grassi S, Malvezzi F, Corti A, Bolognesi M, Mastrangelo E. 2015. NF023 binding to XIAP-BIR1: searching drugs for regulation of the NF-κB pathway. *Proteins* 83: 612–620. doi:10.1002/prot.24766

Crook NE, Clem RJ, Miller LK. 1993. An apoptosis-inhibiting baculovirus gene with a zinc finger-like motif. *J Virol* 67: 2168–2174.

Damgaard RB, Nachbur U, Yabal M, Wong WW, Fiil BK, Kastirr M, Rieser E, Rickard JA, Bankovacki A, Peschel C, et al. 2012. The ubiquitin ligase XIAP recruits LUBAC for NOD2 signaling in inflammation and innate immunity. *Mol Cell* 46: 746–758. doi:10.1016/j.molcel.2012.04.014

Damgaard RB, Fiil BK, Speckmann C, Yabal M, zur Stadt U, Bekker-Jensen S, Jost PJ, Ehl S, Mailand N, Gyrd-Hansen M. 2013. Disease-causing mutations in the XIAP BIR2 domain impair NOD2-dependent immune signalling. *EMBO Mol Med* 5: 1278–1295. doi:10.1002/emmm.201303090

Deveraux QL, Reed JC. 1999. IAP family proteins—suppressors of apoptosis. *Genes Dev* 13: 239–252. doi:10.1101/gad.13.3.239

Deveraux QL, Takahashi R, Salvesen GS, Reed JC. 1997. X-linked IAP is a direct inhibitor of cell-death proteases. *Nature* 388: 300–304. doi:10.1038/40901

Dogan T, Harms GS, Hekman M, Karreman C, Oberoi TK, Alnemri ES, Rapp UR, Rajalingam K. 2008. X-linked and cellular IAPs modulate the stability of C-RAF kinase and cell motility. *Nat Cell Biol* 10: 1447–1455. doi:10.1038/ncb1804

Dynek JN, Goncharov T, Dueber EC, Fedorova AV, Izrael-Tomasevic A, Phu L, Helgason E, Fairbrother WJ, Deshayes K, Kirkpatrick DS, et al. 2010. c-IAP1 and UbcH5 promote K11-linked polyubiquitination of RIP1 in TNF signalling. *EMBO J* 29: 4198–4209. doi:10.1038/emboj.2010.300

Eckelman BP, Salvesen GS, Scott FL. 2006. Human inhibitor of apoptosis proteins: why XIAP is the black sheep of the family. *EMBO Rep* 7: 988–994. doi:10.1038/sj.embor.7400795

Filipovich AH, Zhang K, Snow AL, Marsh RA. 2010. X-linked lymphoproliferative syndromes: brothers or distant cousins? *Blood* 116: 3398–3408. doi:10.1182/blood-2010-03-275909

Franchi L, Warner N, Viani K, Nuñez G. 2009. Function of Nod-like receptors in microbial recognition and host defense. *Immunol Rev* 227: 106–128. doi:10.1111/j.1600-065X.2008.00734.x

Fulda S, Vucic D. 2012. Targeting IAP proteins for therapeutic intervention in cancer. *Nat Rev Drug Discov* 11: 109–124. doi:10.1038/nrd3627

Galbán S, Duckett CS. 2010. XIAP as a ubiquitin ligase in cellular signaling. *Cell Death Differ* 17: 54–60. doi:10.1038/cdd.2009.81

Goncharov T, Hedayati S, Mulvihill MM, Izrael-Tomasevic A, Zobel K, Jeet S, Fedorova AV, Eidenschenk C, deVoss J, Yu K, et al. 2018. Disruption of XIAP-RIP2 association blocks NOD2-mediated inflammatory signaling. *Mol Cell* 69: 551–565.e7. doi:10.1016/j.molcel.2018.01.016

Gyrd-Hansen M, Darding M, Miasari M, Santoro MM, Zender L, Xue W, Tenev T, da Fonseca PC, Zvelebil M, Bujnicki JM, et al. 2008. IAPs contain an evolutionarily conserved ubiquitin-binding domain that regulates NF-κB as well as cell survival and oncogenesis. *Nat Cell Biol* **10:** 1309–1317. doi:10.1038/ncb1789

Haas TL, Emmerich CH, Gerlach B, Schmukle AC, Cordier SM, Rieser E, Feltham R, Vince J, Warnken U, Wenger T, et al. 2009. Recruitment of the linear ubiquitin chain assembly complex stabilizes the TNF-R1 signaling complex and is required for TNF-mediated gene induction. *Mol Cell* **36:** 831–844. doi:10.1016/j.molcel.2009.10.013

Harlin H, Reffey SB, Duckett CS, Lindsten T, Thompson CB. 2001. Characterization of XIAP-deficient mice. *Mol Cell Biol* **21:** 3604–3608. doi:10.1128/MCB.21.10.3604-3608.2001

Hershko A, Ciechanover A. 1998. The ubiquitin system. *Annu Rev Biochem* **67:** 425–479. doi:10.1146/annurev.biochem.67.1.425

Hrdinka M, Schlicher L, Dai B, Pinkas DM, Bufton JC, Picaud S, Ward JA, Rogers C, Suebsuwong C, Nikhar S, et al. 2018. Small molecule inhibitors reveal an indispensable scaffolding role of RIPK2 in NOD2 signaling. *EMBO J* **37:** e99372. doi:10.15252/embj.201899372

Huang Y, Park YC, Rich RL, Segal D, Myszka DG, Wu H. 2001. Structural basis of caspase inhibition by XIAP: differential roles of the linker versus the BIR domain. *Cell* **104:** 781–790.

Hugot JP, Chamaillard M, Zouali H, Lesage S, Cézard JP, Belaiche J, Almer S, Tysk C, O'Morain CA, Gassull M, et al. 2001. Association of NOD2 leucine-rich repeat variants with susceptibility to Crohn's disease. *Nature* **411:** 599–603. doi:10.1038/35079107

Jost PJ, Grabow S, Gray D, McKenzie MD, Nachbur U, Huang DC, Bouillet P, Thomas HE, Borner C, Silke J, et al. 2009. XIAP discriminates between type I and type II FAS-induced apoptosis. *Nature* **460:** 1035–1039. doi:10.1038/nature08229

Kaufmann T, Tai L, Ekert PG, Huang DC, Norris F, Lindemann RK, Johnstone RW, Dixit VM, Strasser A. 2007. The BH3-only protein bid is dispensable for DNA damage- and replicative stress-induced apoptosis or cell-cycle arrest. *Cell* **129:** 423–433. doi:10.1016/j.cell.2007.03.017

Krieg A, Correa RG, Garrison JB, Le Negrate G, Welsh K, Huang Z, Knoefel WT, Reed JC. 2009. XIAP mediates NOD signaling via interaction with RIP2. *Proc Natl Acad Sci* **106:** 14524–14529. doi:10.1073/pnas.0907131106

Latour S, Aguilar C. 2015. XIAP deficiency syndrome in humans. *Semin Cell Dev Biol* **39:** 115–123. doi:10.1016/j.semcdb.2015.01.015

Lawlor KE, Khan N, Mildenhall A, Gerlic M, Croker BA, D'Cruz AA, Hall C, Kaur Spall S, Anderton H, Masters SL, et al. 2015. RIPK3 promotes cell death and NLRP3 inflammasome activation in the absence of MLKL. *Nat Commun* **6:** 6282. doi:10.1038/ncomms7282

Lawlor KE, Feltham R, Yabal M, Conos SA, Chen KW, Ziehe S, Graß C, Zhan Y, Nguyen TA, Hall C, et al. 2017. XIAP loss triggers RIPK3- and caspase-8-driven IL-1β activation and cell death as a consequence of TLR-MyD88-induced cIAP1-TRAF2 degradation. *Cell Rep* **20:** 668–682. doi:10.1016/j.celrep.2017.06.073

Liston P, Roy N, Tamai K, Lefebvre C, Baird S, Cherton-Horvat G, Farahani R, McLean M, Ikeda JE, MacKenzie A, et al. 1996. Suppression of apoptosis in mammalian cells by NAIP and a related family of IAP genes. *Nature* **379:** 349–353. doi:10.1038/379349a0

Liu Z, Sun C, Olejniczak ET, Meadows RP, Betz SF, Oost T, Herrmann J, Wu JC, Fesik SW. 2000. Structural basis for binding of Smac/DIABLO to the XIAP BIR3 domain. *Nature* **408:** 1004–1008. doi:10.1038/35050006

Lu M, Lin SC, Huang Y, Kang YJ, Rich R, Lo YC, Myszka D, Han J, Wu H. 2007. XIAP induces NF-κB activation via the BIR1/TAB1 interaction and BIR1 dimerization. *Mol Cell* **26:** 689–702. doi:10.1016/j.molcel.2007.05.006

MacFarlane M, Merrison W, Bratton SB, Cohen GM. 2002. Proteasome-mediated degradation of Smac during apoptosis: XIAP promotes Smac ubiquitination in vitro. *J Biol Chem* **277:** 36611–36616. doi:10.1074/jbc.M200317200

Marsh RA, Madden L, Kitchen BJ, Mody R, McClimon B, Jordan MB, Bleesing JJ, Zhang K, Filipovich AH. 2010. XIAP deficiency: a unique primary immunodeficiency best classified as X-linked familial hemophagocytic lymphohistiocytosis and not as X-linked lymphoproliferative disease. *Blood* **116:** 1079–1082. doi:10.1182/blood-2010-01-256099

Marsh RA, Rao K, Satwani P, Lehmberg K, Muller I, Li D, Kim MO, Fischer A, Latour S, Sedlacek P, et al. 2013. Allogeneic hematopoietic cell transplantation for XIAP deficiency: an international survey reveals poor outcomes. *Blood* **121:** 877–883. doi:10.1182/blood-2012-06-432500

McKenzie MD, Carrington EM, Kaufmann T, Strasser A, Huang DC, Kay TW, Allison J, Thomas HE. 2008. Pro-apoptotic BH3-only protein Bid is essential for death receptor-induced apoptosis of pancreatic β-cells. *Diabetes* **57:** 1284–1292. doi:10.2337/db07-1692

Mehrotra S, Languino LR, Raskett CM, Mercurio AM, Dohi T, Altieri DC. 2010. IAP regulation of metastasis. *Cancer Cell* **17:** 53–64. doi:10.1016/j.ccr.2009.11.021

Morizane Y, Honda R, Fukami K, Yasuda H. 2005. X-linked inhibitor of apoptosis functions as ubiquitin ligase toward mature caspase-9 and cytosolic Smac/DIABLO. *J Biochem* **137:** 125–132. doi:10.1093/jb/mvi029

Müller N, Fischer JC, Yabal M, Haas T, Poeck H, Jost PJ. 2019. XIAP deficiency in hematopoietic recipient cells drives donor T-cell activation and GvHD in mice. *Eur J Immunol* **49:** 504–507. doi:10.1002/eji.201847818

Ndubaku C, Cohen F, Varfolomeev E, Vucic D. 2009. Targeting inhibitor of apoptosis proteins for therapeutic intervention. *Future Med Chem* **1:** 1509–1525. doi:10.4155/fmc.09.116

Oberoi TK, Dogan T, Hocking JC, Scholz RP, Mooz J, Anderson CL, Karreman C, Meyer zu Heringdorf D, Schmidt G, Ruonala M, et al. 2012. IAPs regulate the plasticity of cell migration by directly targeting Rac1 for degradation. *EMBO J* **31:** 14–28. doi:10.1038/emboj.2011.423

Oberoi-Khanuja TK, Karreman C, Larisch S, Rapp UR, Rajalingam K. 2012. Role of melanoma inhibitor of apoptosis (ML-IAP) protein, a member of the baculoviral IAP repeat (BIR) domain family, in the regulation of C-RAF kinase and cell migration. *J Biol Chem* **287:** 28445–28455. doi:10.1074/jbc.M112.341297

Ogura Y, Bonen DK, Inohara N, Nicolae DL, Chen FF, Ramos R, Britton H, Moran T, Karaliuskas R, Duerr RH, et al. 2001. A frameshift mutation in NOD2 associated with susceptibility to Crohn's disease. *Nature* **411**: 603–606. doi:10.1038/35079114

Olayioye MA, Kaufmann H, Pakusch M, Vaux DL, Lindeman GJ, Visvader JE. 2005. XIAP-deficiency leads to delayed lobuloalveolar development in the mammary gland. *Cell Death Differ* **12**: 87–90. doi:10.1038/sj.cdd.4401524

Pachlopnik Schmid JP, Canioni D, Moshous D, Touzot F, Mahlaoui N, Hauck F, Kanegane H, Lopez-Granados E, Mejstrikova E, Pellier I, et al. 2011. Clinical similarities and differences of patients with X-linked lymphoproliferative syndrome type 1 (XLP-1/SAP deficiency) versus type 2 (XLP-2/XIAP deficiency). *Blood* **117**: 1522–1529. doi:10.1182/blood-2010-07-298372

Pedersen J, LaCasse EC, Seidelin JB, Coskun M, Nielsen OH. 2014. Inhibitors of apoptosis (IAPs) regulate intestinal immunity and inflammatory bowel disease (IBD) inflammation. *Trends Mol Med* **20**: 652–665. doi:10.1016/j.molmed.2014.09.006

Pickart CM, Fushman D. 2004. Polyubiquitin chains: polymeric protein signals. *Curr Opin Chem Biol* **8**: 610–616. doi:10.1016/j.cbpa.2004.09.009

Qin S, Yang C, Zhang B, Li X, Sun X, Li G, Zhang J, Xiao G, Gao X, Huang G, et al. 2016. XIAP inhibits mature Smac-induced apoptosis by degrading it through ubiquitination in NSCLC. *Int J Oncol* **49**: 1289–1296. doi:10.3892/ijo.2016.3634

Riedl SJ, Renatus M, Schwarzenbacher R, Zhou Q, Sun C, Fesik SW, Liddington RC, Salvesen GS. 2001. Structural basis for the inhibition of caspase-3 by XIAP. *Cell* **104**: 791–800. doi:10.1016/S0092-8674(01)00274-4

Rigaud S, Fondanèche MC, Lambert N, Pasquier B, Mateo V, Soulas P, Galicier L, Le Deist F, Rieux-Laucat F, Revy P, et al. 2006. XIAP deficiency in humans causes an X-linked lymphoproliferative syndrome. *Nature* **444**: 110–114. doi:10.1038/nature05257

Salvesen GS, Duckett CS. 2002. IAP proteins: blocking the road to death's door. *Nat Rev Mol Cell Biol* **3**: 401–410. doi:10.1038/nrm830

Schile AJ, Garcia-Fernandez M, Steller H. 2008. Regulation of apoptosis by XIAP ubiquitin-ligase activity. *Genes Dev* **22**: 2256–2266. doi:10.1101/gad.1663108

Speckmann C, Ehl S. 2014. XIAP deficiency is a mendelian cause of late-onset IBD. *Gut* **63**: 1031–1032. doi:10.1136/gutjnl-2013-306474

Stafford CA, Lawlor KE, Heim VJ, Bankovacki A, Bernardini JP, Silke J, Nachbur U. 2018. IAPs regulate distinct innate immune pathways to co-ordinate the response to bacterial peptidoglycans. *Cell Rep* **22**: 1496–1508. doi:10.1016/j.celrep.2018.01.024

Sun C, Cai M, Meadows RP, Xu N, Gunasekera AH, Herrmann J, Wu JC, Fesik SW. 2000. NMR structure and mutagenesis of the third Bir domain of the inhibitor of apoptosis protein XIAP. *J Biol Chem* **275**: 33777–33781. doi:10.1074/jbc.M006226200

Varfolomeev E, Vucic D. 2011. Inhibitor of apoptosis proteins: fascinating biology leads to attractive tumor therapeutic targets. *Future Oncol* **7**: 633–648. doi:10.2217/fon.11.40

Varfolomeev E, Goncharov T, Fedorova AV, Dynek JN, Zobel K, Deshayes K, Fairbrother WJ, Vucic D. 2008. c-IAP1 and c-IAP2 are critical mediators of tumor necrosis factor α (TNF-α)-induced NF-κB activation. *J Biol Chem* **283**: 24295–24299. doi:10.1074/jbc.C800128200

Varfolomeev E, Alicke B, Elliott JM, Zobel K, West K, Wong H, Scheer JM, Ashkenazi A, Gould SE, Fairbrother WJ, et al. 2009. X chromosome-linked inhibitor of apoptosis regulates cell death induction by proapoptotic receptor agonists. *J Biol Chem* **284**: 34553–34560. doi:10.1074/jbc.M109.040139

Vaux DL, Silke J. 2005. IAPs, RINGs and ubiquitylation. *Nat Rev Mol Cell Biol* **6**: 287–297. doi:10.1038/nrm1621

Vince JE, Wong WW, Gentle I, Lawlor KE, Allam R, O'Reilly L, Mason K, Gross O, Ma S, Guarda G, et al. 2012. Inhibitor of apoptosis proteins limit RIP3 kinase-dependent interleukin-1 activation. *Immunity* **36**: 215–227. doi:10.1016/j.immuni.2012.01.012

Vucic D, Dixit VM, Wertz IE. 2011. Ubiquitylation in apoptosis: a post-translational modification at the edge of life and death. *Nat Rev Mol Cell Biol* **12**: 439–452. doi:10.1038/nrm3143

Wicki S, Gurzeler U, Wei-Lynn Wong W, Jost PJ, Bachmann D, Kaufmann T. 2016. Loss of XIAP facilitates switch to TNFα-induced necroptosis in mouse neutrophils. *Cell Death Dis* **7**: e2422. doi:10.1038/cddis.2016.311

Wilkinson JC, Cepero E, Boise LH, Duckett CS. 2004. Upstream regulatory role for XIAP in receptor-mediated apoptosis. *Mol Cell Biol* **24**: 7003–7014. doi:10.1128/MCB.24.16.7003-7014.2004

Witt A, Vucic D. 2017. Diverse ubiquitin linkages regulate RIP kinases-mediated inflammatory and cell death signaling. *Cell Death Differ* **24**: 1160–1171. doi:10.1038/cdd.2017.33

Wong WW, Vince JE, Lalaoui N, Lawlor KE, Chau D, Bankovacki A, Anderton H, Metcalf D, O'Reilly L, Jost PJ, et al. 2014. cIAPs and XIAP regulate myelopoiesis through cytokine production in an RIPK1- and RIPK3-dependent manner. *Blood* **123**: 2562–2572. doi:10.1182/blood-2013-06-510743

Wu G, Chai J, Suber TL, Wu JW, Du C, Wang X, Shi Y. 2000. Structural basis of IAP recognition by Smac/DIABLO. *Nature* **408**: 1008–1012. doi:10.1038/35050012

Yabal M, Müller N, Adler H, Knies N, Groß CJ, Damgaard RB, Kanegane H, Ringelhan M, Kaufmann T, Heikenwälder M, et al. 2014. XIAP restricts TNF- and RIP3-dependent cell death and inflammasome activation. *Cell Rep* **7**: 1796–1808. doi:10.1016/j.celrep.2014.05.008

Yang X, Kanegane H, Nishida N, Imamura T, Hamamoto K, Miyashita R, Imai K, Nonoyama S, Sanayama K, Yamaide A, et al. 2012. Clinical and genetic characteristics of XIAP deficiency in Japan. *J Clin Immunol* **32**: 411–420. doi:10.1007/s10875-011-9638-z

Yin XM, Wang K, Gross A, Zhao Y, Zinkel S, Klocke B, Roth KA, Korsmeyer SJ. 1999. Bid-deficient mice are resistant to Fas-induced hepatocellular apoptosis. *Nature* **400**: 886–891. doi:10.1038/23730

Zeissig Y, Petersen BS, Milutinovic S, Bosse E, Mayr G, Peuker K, Hartwig J, Keller A, Kohl M, Laass MW, et al. 2015. XIAP variants in male Crohn's disease. *Gut* **64**: 66–76. doi:10.1136/gutjnl-2013-306520

Cell Death and Neurodegeneration

Benjamin J. Andreone, Martin Larhammar, and Joseph W. Lewcock

Denali Therapeutics, South San Francisco, California 94080, USA

Correspondence: lewcock@dnli.com

Neurodegenerative disease is characterized by the progressive deterioration of neuronal function caused by the degeneration of synapses, axons, and ultimately the death of nerve cells. An increased understanding of the mechanisms underlying altered cellular homeostasis and neurodegeneration is critical to the development of effective treatments for disease. Here, we review what is known about neuronal cell death and how it relates to our understanding of neurodegenerative disease pathology. First, we discuss prominent molecular signaling pathways that drive neuronal loss, and highlight the upstream cell biology underlying their activation. We then address how neuronal death may occur during disease in response to neuron intrinsic and extrinsic stressors. An improved understanding of the molecular mechanisms underlying neuronal dysfunction and cell death will open up avenues for clinical intervention in a field lacking disease-modifying treatments.

Unlike most cells in the body, which readily turn over to maintain tissue homeostasis, postmitotic neurons harbor minimal regenerative capacity and must survive for the lifetime of an organism to ensure nervous system function. Although neuronal cell death is essential to promote proper nervous system development (Yamaguchi and Miura 2015), neuronal loss in the adult inevitably leads to functional decline and underlies disease progression in neurodegenerative indications ranging from acute insults such as traumatic brain injury (TBI) and stroke, to chronic conditions such as Alzheimer's disease (AD), Parkinson's disease (PD), and amyotrophic lateral sclerosis (ALS). These and other chronic neurodegenerative diseases affect millions of individuals worldwide, with no treatments available that are able to slow the rate of decline.

Human neuropathological studies using postmortem AD, PD, and ALS tissue have identified stereotyped patterns of neuronal degeneration that develop sequentially at different brain and spinal cord regions and match the clinical severity of disease (Braak and Braak 1991; Braak et al. 2003; Brettschneider et al. 2015). This loss of neurons is often accompanied by additional pathologies, including impaired axonal transport and synaptic function, mitochondrial and lysosomal dysfunction, oxidative stress, microglial activation, and protein aggregates that can be visualized in diseased neurons. A combination of aging, genetic risk, and environmental factors is thought to disrupt neuronal homeostasis and contribute to these pathologies—the "initiators" of neuronal degeneration. The subsequent engagement of known cell death pathways in neurons—the "executioners" of cell death—has been described across multiple neurodegenerative diseases, suggesting that common molecular signaling events may at least in part underlie disease-associated functional decline.

Although our understanding of the genetics and pathology of neurodegenerative disease has improved rapidly in recent years, the biology of neurodegeneration is complex, and there remains an unsatisfying disconnect between the initiators that can be observed as a component of disease pathology and the subsequent activation of neuronal cell death executioners. Determining which cell death pathways are relevant during disease progression and the mechanisms by which they are activated remains a key question in the field, and unraveling these mechanisms has the potential to drive the development of a new generation of targeted therapeutics that modify neurodegenerative disease progression.

PATHWAYS REGULATING NEURONAL CELL DEATH

Much of what we know about the executioners of cell death in neurons comes from studies of acute neuronal injury or neuronal development. Here, we discuss several of these pathways and highlight what is known about their role in neuronal degeneration.

Evidence for Apoptosis, Necroptosis, and Other Death Pathways in Neurons

Apoptosis is a form of programmed cell death characterized by DNA fragmentation, degradation of cytoskeletal and nuclear proteins, and eventual phagocytosis by immune cells. It is executed by the caspase family of proteins, which make up a proteolytic cascade that ultimately results in cell death (Elmore 2007). Neuronal apoptosis occurs during development of the central nervous system to remove excess neurons, and can be observed in adult neurons following acute insults ranging from excitotoxicity to mechanical injury (Yuan and Yankner 2000). Apoptosis can be initiated downstream from death receptor activation via caspase-8 (extrinsic pathway), but more commonly occurs following

mitochondrial damage via caspase-9 (intrinsic pathway) (Fig. 1A). These pathways converge on caspase-3, which is required for neuronal apoptosis (Kuida et al. 1996). Caspase-7, another executioner caspase important for apoptosis in other cell types, may also contribute but appears to play a less essential role in neurons (Zhang et al. 2000; Slee et al. 2001). Consistent with these findings, mice lacking caspase-3 expression display excess neurons at birth and reduced apoptosis in a number of contexts (Kuida et al. 1996; Leonard et al. 2002).

Mitochondrial damage-mediated apoptosis in neurons requires the proapoptotic Bcl-2 family protein Bax, which forms channels in the mitochondrial membrane that promote the release of cytochrome c from the mitochondria and subsequent activation of caspase-9 (Kroemer et al. 2007). This pathway functions to regulate neuronal apoptosis even in scenarios in which the primary insult does not involve mitochondrial damage, such as trophic factor withdrawal (Kristiansen and Ham 2014), suggesting that mitochondrial permeabilization broadly contributes to neuronal apoptosis. Neurons lacking Bax expression are strongly protected from degeneration and do not display caspase activation even after prolonged insult (Deckwerth et al. 1996; Vila et al. 2001; Libby et al. 2005).

Necroptosis is a more recently discovered form of programmed cell death that is caspase-independent and harbors cellular characteristics of necrosis (Vanden Berghe et al. 2014; Zhang et al. 2017). Various stimuli can induce necroptosis, including tumor necrosis factor (TNF), interferon, and Toll-like receptor (TLR) signaling, as well as viral infection. Downstream from TNF receptor 1 (TNFR1), a complex containing receptor-interacting kinase proteins 1 and 3 (RIPK1/3) phosphorylates mixed lineage kinase domain-like protein (MLKL) to induce its oligomerization (Fig. 1B). Other pathways can lead to MLKL phosphorylation in a RIPK1-independent fashion (Silke et al. 2015). Phosphorylated MLKL subsequently forms pores in the plasma membrane, leading to neuronal death and release of proinflammatory factors from the cell. Several reports indicate that RIPK signaling contributes to neuronal loss in various in vivo mod-

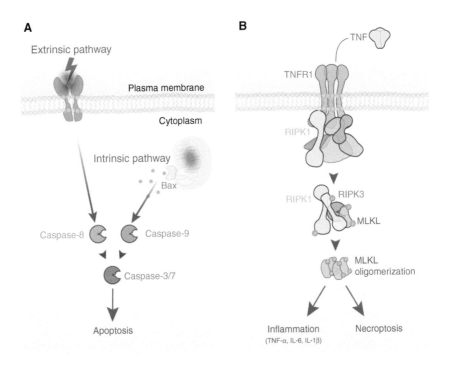

Figure 1. Apoptosis and necroptosis are executioners of neuronal cell death. (*A*) Neuronal apoptosis is executed by the caspase family of proteases. In the extrinsic pathway, death receptor engagement by secreted or membrane-bound death ligands leads to caspase-8 activation. In the intrinsic pathway, mitochondrial damage activates caspase-9 through back-mediated release of mitochondrial proteins. The intrinsic and extrinsic pathways converge on caspase-3 (and to a lesser extent in neurons, caspase-7), ultimately resulting in cell death. (*B*) Neuronal necroptosis is induced by several stimuli, including TNF signaling. Downstream from receptor activation, RIPK3 phosphorylates MLKL, leading to plasma membrane permeabilization and cell death. In addition to necroptosis, RIPK activity also mediates a cell-autonomous inflammation response that exacerbates neurodegeneration. TNF, Tumor necrosis factor; TNFR1, TNF receptor 1; RIPK, receptor-interacting kinase protein; MLKL, mixed lineage kinase domain-like protein; IL, interleukin.

els (Ito et al. 2016; Caccamo et al. 2017; Yang et al. 2017; Zhang et al. 2017; Iannielli et al. 2018). In some studies, RIPK1 has been shown to act directly in neurons to regulate degeneration (Re et al. 2014; Iannielli et al. 2018; Arrazola et al. 2019), whereas others suggest that RIPK1 regulates necroptosis in other central nervous system cell types (Ofengeim et al. 2015; Ito et al. 2016). Interestingly, RIPK1 has also been shown to mediate a cell-autonomous proinflammatory response, rather than cell death, in peripheral immune cell populations and microglia (Najjar et al. 2016; Ofengeim et al. 2017; Yuan et al. 2019). Taken together, these results suggest that RIPK1 function may contribute both directly and indirectly to neuronal degeneration.

In addition to apoptosis and necroptosis, several other cell death pathways have been observed in neurons, including ferroptosis (iron-dependent necrosis), autolysis (lysosomal cell death), and autophagy (death by "self-eating"), among others. Complicating the picture of neuronal cell death is the observation that there is significant cross talk between several of these pathways, a topic that has been reviewed previously (Fricker et al. 2018).

The Neuronal Stress Response: A Master Regulator of Cell Death

Cell death in neurons often occurs following a stress response via signals such as the c-Jun amino-terminal kinase (JNK) pathway. The JNK signaling pathway has been extensively characterized in neurons, and is required for a variety of functions, including mediating the response to neuronal injury (Borsello and Forloni 2007; Tedeschi and Bradke 2013; Yarza et al. 2015; Siu et al. 2018). The mixed lineage dual leucine zipper kinase (DLK) is an essential upstream regulator of stress-dependent JNK signaling in neurons (Holzman et al. 1994; Ghosh et al. 2011), and is activated in response to neurotrophic growth factors, neuronal and axonal injury, oxidative stress, and misfolded proteins, among other factors (Tedeschi and Bradke 2013; Simon and Watkins 2018; Siu et al. 2018). DLK-dependent phosphorylation of JNK results in its translocation to the nucleus and phosphorylation of transcription factors including c-Jun, which induce a transcriptional stress response culminating in apoptosis in the central nervous system (Yang et al. 1997; Behrens et al. 1999; Fernandes et al. 2012). DLK appears to act as a central node that regulates neuronal degeneration during development, following neuronal injury, and in chronic neurodegenerative disease. Neurons lacking DLK signaling fail to mount a normal transcriptional injury response, display attenuated caspase activation, and are strongly protected from degeneration following insult (Pozniak et al. 2013; Watkins et al. 2013; Patel et al. 2015; Le Pichon et al. 2017; Siu et al. 2018).

DLK activity can also impact protein kinase R-like endoplasmic reticulum kinase (PERK)-dependent phosphorylation of the translation eukaryotic initiation factor-2α (eIF2α) in neurons following acute injury (Larhammar et al. 2017). PERK signaling is a component of the unfolded protein response (UPR) as outlined below, and DLK-dependent activation of this pathway has been shown to contribute to neuronal degeneration. These results suggest that DLK signaling acts broadly downstream from neuronal insults through impacting multiple stress response pathways. Recent reviews provide more information regarding how cellular context and other signaling pathways impact the disparate outcomes downstream from DLK signaling (Tedeschi and Bradke 2013; Simon and Watkins 2018; Siu et al. 2018).

Axon Degeneration: Programmed Destruction of Neuronal Processes

Axon degeneration, characterized by the elimination of axonal processes and loss of neuronal connectivity, is an early clinical feature of most neurodegenerative disorders, ranging from acute insults to chronic neurodegenerative disease (Burke and O'Malley 2013; Johnson et al. 2013; Benarroch 2015; Cashman and Höke 2015; Tagliaferro and Burke 2016; Kim et al. 2018). Recent data have shown that axon degeneration is an active process, dependent on both local axonal signaling events and transcriptional regulation within the neuronal cell body. Furthermore, it can occur in the absence of, or in conjunction with, neuronal apoptosis and is thought to contribute to functional decline in many disease indications (Simon et al. 2016; Simon and Watkins 2018).

Multiple pathways have been shown to contribute to axon degeneration, with the mechanism being dependent on the context examined (Fig. 2; Gerdts et al. 2016). In many experimental paradigms, including trophic factor withdrawal and addition of chemotherapeutic compounds known to cause neuropathy, the DLK/JNK signaling pathway acts as a central regulator of axon degeneration (Miller et al. 2009). DLK-dependent degeneration of axons appears to be governed by similar pathways to those that regulate neuronal cell death and requires the activation of caspase signaling (Ghosh et al. 2011).

In Wallerian degeneration paradigms, in which the axon is lesioned and disconnected from the cell body, sterile α and TIR motif-containing protein 1 (SARM1) is necessary and sufficient for axon degeneration, and axons of SARM1 knockout mice are protected in several in vivo neurodegeneration models (Osterloh et al. 2012; Gerdts et al. 2013, 2016; Henninger et al. 2016; Turkiew et al. 2017; Ziogas and Koliatsos 2018). Upon axonal damage or stress, SARM1, which harbors intrinsic NADase activity, becomes activated and hydrolyses NAD, resulting in reduction of axonal ATP levels and ultimately degradation of neurofilaments and axonal fragmentation via the calpain family of proteases (Gerdts et al. 2013, 2016). Axonal in-

Cite this article as *Cold Spring Harb Perspect Biol* doi: 10.1101/cshperspect.a036434

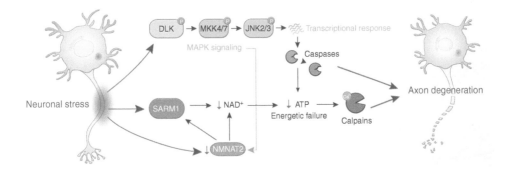

Figure 2. Neuronal signaling pathways regulating axon degeneration. Axon degeneration is an active process involving several signaling pathways engaged by diverse neuronal stressors, both extracellular (e.g., axonal injury, toxins/chemotherapeutic agents) and intracellular (e.g., impaired protein homeostasis and mitochondrial dysfunction). Axonal damage results in both depletion of the protective axonal maintenance factor NMNAT2 and activation of SARM1, leading to a reduction of NAD$^+$ levels, depletion of axonal ATP, and activation of proteolytic calpains. Neuronal stress also results in phosphorylation of DLK, a master kinase of the neuronal stress response that regulates both axon degeneration and neuronal apoptosis. DLK activation leads to phosphorylation of downstream kinases MKK4/7 and JNK2/3, and ultimately induces a prodegenerative caspase signaling cascade that activates calpains. There is cross talk between the SARM1 and DLK pathways; DLK signaling promotes NMNAT2 turnover, thereby enhancing SARM1-mediated axon degeneration. SARM1, sterile α and TIR motif-containing protein 1; NMNAT2, nicotinamide nucleotide adenylyltransferase 2; JNK, c-Jun amino-terminal kinase; DLK, dual leucine zipper kinase; MAPK, mitogen-activated protein kinase.

jury also decreases the levels of the NAD biosynthetic enzyme NMNAT2, which normally acts in part to inhibit SARM1 activity (Gilley et al. 2015). This reduction in NMNAT2 both decreases NAD levels and increases SARM1 signaling, thereby exacerbating axon degeneration. Furthermore, the DLK signaling pathway has been found to promote degradation of NMNAT2, thus promoting SARM1-mediated axon degeneration, although the functional contribution of DLK signaling to Wallerian degeneration appears relatively minor (Xiong et al. 2012; Summers et al. 2016; Walker et al. 2017). Interestingly, both trophic factor withdrawal and Wallerian degeneration pathways culminate in the activation of the calpains. Accordingly, inhibition of calpains either with small molecules or via overexpression of the calpain inhibitor calpastatin is sufficient to protect axons from degeneration (Yang et al. 2013). Additional signaling mechanisms that contribute to axon degeneration have been reviewed elsewhere (Rishal and Fainzilber 2014; Simon and Watkins 2018).

MECHANISMS OF NEURONAL CELL DEATH IN DISEASE

It is thought that several diverse stressors can initiate intracellular events in neurons that lead to neuronal cell death in chronic neurodegenerative disease, often in parallel. These initiators include toxic aggregation of misfolded proteins, mitochondrial dysfunction, neuroinflammation, defects in the endolysosomal system, and oxidative stress (Dawson and Dawson 2017; Hetz and Saxena 2017). Here, we review how several prevalent initiators impact neuronal death in disease, with an emphasis on the pathologies observed in AD and PD.

Protein Homeostasis

A common pathological trait of neurodegenerative disorders, including AD and PD, is the aggregation of toxic misfolded proteins that are associated with synaptic dysfunction and degeneration of neuronal populations. In AD, this consists of extracellular β-amyloid (Aβ) plaques and intracellular neurofibrillary tangles of hyperphosphorylated tau, while PD pathology is characterized by intracellular inclusions of α-synuclein (α-syn), termed Lewy bodies (Ross and Poirier 2004; Smith and Mallucci 2016; Hetz and Saxena 2017). Although the precise mechanisms that trigger the conversion from normal to misfolded protein are unresolved, Aβ, tau, and α-syn comprise a portion of the "metastable proteome," a set of proteins whose cellular concentrations are high relative to their intrinsic solubility and are therefore biased to misfolding (Ciryam et al. 2013, 2016; Kundra et al. 2017). This bias increases with aging, the greatest risk factor for neurodegenerative disease. The biochemical properties

of misfolded proteins facilitate the recruitment of normal proteins into growing toxic oligomers, accelerating the process of aggregation (Fig. 3A; Jarrett and Lansbury 1993; Meisl et al. 2017). Furthermore, in preclinical models, misfolded tau and α-syn have been shown to propagate from cell to cell (Emmanouilidou et al. 2010; Kfoury et al. 2012; Brettschneider et al. 2015; Soto and Pritzkow 2018). Initially controversial, cumulative evidence suggests that aggregates spread via neuronal synapses—similar to prions—allowing misfolded protein pathology to spread throughout the brain in a connectome-dependent fashion that matches disease progression (Clavaguera et al. 2009; Desplats et al. 2009; Luk et al. 2012; Iba et al. 2013; Guo et al. 2016).

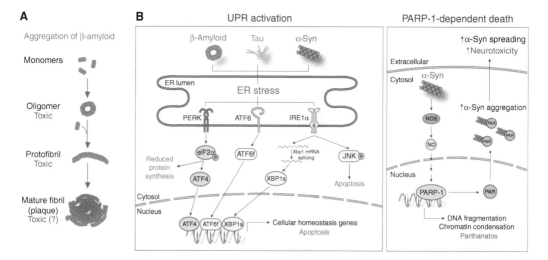

Figure 3. Impaired protein homeostasis is an initiator of neuronal cell death. (A) Aggregation of misfolded proteins, including β-amyloid, is a common pathological trait of chronic neurodegeneration disease. Monomeric proteins misfold and aggregate into toxic oligomers that recruit normal proteins into growing protofibrils, eventually forming mature plaques. Recent evidence suggests that oligomers are particularly toxic in disease, a topic of ongoing debate (Verma et al. 2015). (B, *left*) Protein aggregates cause endoplasmic reticulum (ER) stress and activate the UPR, inducing PERK, ATF6, and IRE1α signaling cascades. PERK activation leads to the phosphorylation of eIF2α and subsequent reduction of protein synthesis and expression of the transcription factor ATF4. ATF6 and IRE1α signaling lead to the expression of the transcription factors ATF6f and XBP1s, respectively. These transcription factors induce cellular homeostasis gene expression, but can also drive expression of proapoptotic genes under prolonged ER stress. JNK signaling is also activated downstream from IRE1α, promoting apoptosis. (B, *right*) In PD, α-syn fibrils active NOS, leading to PARP-1 activation and death of dopaminergic neurons via parthanatos. PAR produced by PARP-1 activation binds α-syn, increasing its aggregation, cell-to-cell spreading, and subsequent neurotoxicity. UPR, Unfolded protein response; PERK, PKR-like ER kinase; ATF4/6, activating transcription factor 4/6; IRE1α, inositol-requiring protein 1α; eIF2α, eukaryotic translation initiation factor 2α; XBP1, X-box binding protein 1; JNK, c-Jun amino-terminal kinase; NOS, nitric oxide synthase; NO, nitric oxide; PARP-1, poly(adenosine 5′-diphosphate-ribose) polymerase-1.

Cite this article as *Cold Spring Harb Perspect Biol* doi: 10.1101/cshperspect.a036434

The presence of pathology characterized by protein aggregates has led to the widely held hypothesis that protein misfolding plays a role in neurodegenerative cell death. But how do misfolded proteins kill neurons? It is possible that the formation of aggregates sequesters proteins from their endogenous localization and precludes them from performing their physiological functions required for neuronal survival. In the case of α-syn, which is normally found at the presynapse and promotes SNARE-complex assembly, evidence exists that this function may be disrupted in disease (Burre et al. 2010; Lashuel et al. 2013). Formation of α-syn aggregates in cultured neurons leads to decreased expression of other SNARE proteins, impairments in synaptic connectivity and transmission, and eventually neuronal death, arguing that the endogenous localization and function of α-syn is critical for synaptic signaling and neuronal survival (Volpicelli-Daley et al. 2011). However, no similar data supporting a sequestration hypothesis exists for tau. Indeed, reduction of endogenous tau levels in a human β-amyloid precursor protein AD mouse model both protects against excitotoxicity and abrogates behavioral deficits in these mice (Roberson et al. 2007), arguing that absence of normal tau function is not likely to underlie the neuronal dysfunction and ultimately cell death observed in AD.

A large body of evidence suggests that toxic gain-of-function, rather than loss of endogenous function, is the primary driver of neuronal cell death following aggregation of most disease-associated proteins. Under physiological conditions, proteins are folded into their native form in the endoplasmic reticulum (ER) with the help of cytosolic and resident ER chaperones, whereas misfolded proteins are targeted and trafficked to the autophagosome, proteasome, and lysosome for degradation (Hetz and Saxena 2017). This process of protein homeostasis, or "proteostasis," governs protein production and is critical for overall neuronal health. The presence of misfolded proteins results in the activation of multiple stress response pathways including the UPR, which attempts to restore proteostasis through the up-regulation of ER chaperones, reduction of protein synthesis, and regulation of protein secretion and degradation. The UPR is mediated by three distinct branches that initiate unique signaling cascades upon ER stress: PERK, activating transcription factor 6 (ATF6), and inositol-requiring protein 1α (IRE1α) (Fig. 3B; Walter and Ron 2011; Pavitt and Ron 2012). The PERK pathway leads to activation of eIF2α, resulting in the global shutdown of translation and expression of the transcription factor ATF4, which up-regulates genes that allow cells to respond to insult. However, during severe or sustained ER stress that is thought to occur in chronic neurodegenerative disease, the UPR pathways can shift toward proapoptotic signaling (Smith and Mallucci 2016; Hetz and Saxena 2017). For example, ATF4 induces expression of proapoptotic genes under severe stress conditions, including the Bcl-2 family members Puma and Bim (Galehdar et al. 2010; Li et al. 2014). Additionally, other arms of the UPR can also initiate proapoptotic signaling, such as the activation of JNK signaling triggered by IRE1α (Urano et al. 2000).

Several lines of evidence suggest that misfolded proteins promote the induction of ER stress and the UPR in chronic neurodegenerative disease. For example, in PD, misfolded α-syn accumulates in the ER lumen, where it interacts with and decreases the function of ER chaperones (Bellucci et al. 2011). In addition, α-syn causes ER stress by inhibiting the normal trafficking of proteins from the ER to the Golgi apparatus, causing molecular traffic jams that inhibit protein maturation (Credle et al. 2015). In AD, both Aβ and tau aggregates lead to ER stress through distinct mechanisms. Aβ oligomers that are internalized by neurons accumulate in the ER lumen and aberrantly open calcium channels, resulting in toxic increases in intracellular calcium levels (Demuro et al. 2010; Shtifman et al. 2010). Tau aggregates inhibit the ability of neurons to eliminate misfolded proteins by associating with and inhibiting molecular components of ER-associated degradation (ERAD) (Abisambra et al. 2013; Meier et al. 2015). Taken together, it is reasonable to conclude that protein misfolding in neurons results in prolonged ER stress through multiple mechanisms and may ultimately contribute to the induction of apoptotic cell death.

ER stress and the UPR are important mediators of neurotoxicity downstream from protein misfolding, yet the specific mechanisms leading to cell death downstream from protein aggregation are still poorly defined. A recent study shed light on this question by showing that α-syn aggregates activate the poly(adenosine 5′-diphosphate-ribose) polymerase-1 (PARP-1)-dependent cell death pathway known as parthanatos in vitro and in vivo (Fig. 3B) (Kam et al. 2018). Mechanistically, pathologic α-syn activates nitric oxide synthase, resulting in the activation of the PARP-1 enzyme by nitric oxide. PARP-1 activation subsequently leads to large-scale DNA fragmentation, chromatin condensation, and ultimately cell death (Fatokun et al. 2014). Furthermore, PAR produced through PARP-1 activation binds to α-syn and promotes its aggregation, causing a positive feedback loop that exacerbates neuronal parthanatos. Although genetic and pharmacological inhibition of PARP-1 prevented neuronal cell death in vivo, blockers of necroptosis and autophagy had little effect, suggesting that parthanatos is a major driver of α-syn toxicity in the disease models used. Interestingly, caspase inhibitors had a partial protective effect on cell death in the same study, indicating that apoptosis also plays a role in neuronal death in PD.

Mitochondrial Dysfunction

Another initiator thought to contribute to cytotoxicity in disease is impaired mitochondrial function. Increasing evidence suggests that mitochondrial dysfunction is a cause of neurodegeneration, rather than a consequence, and may contribute to disease pathology and neuronal loss in both AD and PD (Devi et al. 2008; Chinta et al. 2010; Swerdlow et al. 2010; Dawson and Dawson 2017). Early studies showing that apoptosis is induced in neurons following treatment with the mitochondrial toxins 1-methyl-4-phenyl-1,2,3,6-tertahydropyridine (MPTP) and rotenone provided the first evidence that mitochondrial damage may be a particularly relevant stressor in this cell type (Vila and Przedborski

2003). These discoveries were complemented by human genetic data showing that loss-of-function mutations in PTEN-induced kinase 1 (PINK1) and Parkin, two proteins required for mitochondrial quality control, are associated with increased PD risk (Pickrell and Youle 2015). Studies using cultured neurons from PINK1 and Parkin knockout mice suggested that these proteins normally mediate the removal of damaged mitochondria from neuronal axons via mitophagy, a specialized form of autophagy (Ashrafi et al. 2014). PINK1 and Parkin also attenuate apoptosis under stress conditions by limiting Bax recruitment to the mitochondrial membrane (Johnson et al. 2012). Interestingly, PINK1 and Parkin knockout mouse models do not display obvious loss of dopaminergic neurons in vivo (Kitada et al. 2009), indicating that rodents may compensate for their loss through yet-unknown mechanisms. However, a recent study showed that these knockout mice display an activation of the innate immune system on mitochondrial stress (Sliter et al. 2018), suggesting that mitophagy regulated by PINK1 and Parkin may function to suppress neuroinflammation, another initiator of neuronal death discussed below.

There is a body of data linking mitochondrial damage to neuronal apoptosis, although a recent study showed that mitochondrial dysfunction can also result in necroptosis in dopaminergic neurons (Iannielli et al. 2018). Induced pluripotent stem cell (iPSC)-derived neurons from PD patients with mutations in optic atrophy type 1 (OPA1), a resident mitochondria gene involved in mitochondrial fusion and respiratory efficiency (Patten et al. 2014), showed severe energetic deficits, mitochondrial fragmentation, and oxidative stress. Interestingly, pharmacological inhibition of necroptosis via necrostatin-1 both improved the health of these iPSC-neurons and significantly protected against dopaminergic neuron death by MPTP in vivo, suggesting that necroptosis may play a role in neuronal cell death caused by mitochondrial damage. However, as necrostatin-1 also blocks RIPK1-dependent apoptosis in certain contexts (Vandenabeele et al. 2013), a role for apoptosis in the paradigms tested cannot be completely excluded.

Cite this article as *Cold Spring Harb Perspect Biol* doi: 10.1101/cshperspect.a036434

Neuroinflammation

It has become increasingly clear that disease-related neurodegeneration is a multifaceted process involving multiple cell types in the brain. Emerging evidence implicates microglia, the resident innate immune cells of the brain, as major players promoting pathology in neurodegenerative disease, particularly in AD (Heneka et al. 2015; Hansen et al. 2018). Indeed, a large percentage of AD risk genes are highly (and in many cases specifically) expressed in microglia (Srinivasan et al. 2016; Zhang et al. 2016). Based on current understanding, microglia contribute to neuronal cell death through multiple mechanisms, including release of cytotoxic inflammatory cytokines and direct phagocytosis of damaged synapses and neurons.

Increased levels of proinflammatory cytokines secreted by microglia have been reported in disease models and brains of patients with AD and PD, suggesting that these factors may contribute to neuronal cell death (Griffin et al. 1989; Mogi et al. 1994; Zhou et al. 2016). In line with these observations, Aβ and α-syn aggregates have been shown to activate the NLRP3 inflammasome, a danger sensor of the innate immune system that drives production of the proinflammatory cytokines interleukin (IL)-1β and IL-18 (Codolo et al. 2013; Heneka et al. 2013; Voet et al. 2019). These cytokines in turn bind to their cognate receptors on neurons, initiating cytotoxic events such as aberrant calcium influx and activation of the JNK signaling pathway (Curran et al. 2003; Viviani et al. 2003; Allan et al. 2005). Similar to what was observed with α-syn mediated activation of parthanatos in neurons (Kam et al. 2018), NLRP3 inflammasome activation in microglia has been shown to propagate Aβ aggregation and spreading, creating a positive feedback loop that may exacerbate neuronal cell death (Venegas et al. 2017). Activated microglia can also produce additional cytotoxic species, including TNF-α, which has been shown to promote neuronal cell death through the potentiation of neural excitotoxicity, and the induction of neuronal apoptosis via signaling through neuronally expressed death receptors

(Guadagno et al. 2013; Neniskyte et al. 2014; Olmos and Lladó 2014).

Another potential mechanism by which microglia contribute to neuronal cell death is the aberrant co-option of microglial phagocytic capacity. During normal development, microglia contribute to neural circuit refinement through the pruning of synapses (Schafer et al. 2012). This synaptic pruning by microglia relies on the classical complement pathway, a component of the innate immune system normally used to clear pathogens. Histological evidence shows that the complement pathway becomes reactivated in AD microglia, which may render them once again capable of engulfing synapses (Zanjani et al. 2005). Consistent with this idea, genetic ablation or pharmacological inhibition of molecular components of complement protects against neuron loss in several mouse models, arguing that microglia contribute to neuronal cell death via synapse phagocytosis (Fonseca et al. 2004; Hong et al. 2016; Shi et al. 2017; Dejanovic et al. 2018). Additional work showed that microglia are capable of directly phagocytosing neurons after Aβ stimulation, a process termed phagoptosis (Neniskyte et al. 2011). Phagoptosis occurs without the activation of apoptosis or necroptosis in the neurons being engulfed, suggesting that it is a primary executioner of cell death and not secondary to the activation of other death pathways.

Selective Vulnerability

During neurodegenerative disease progression, specific brain regions and neuronal subsets within those regions are more likely to experience cytotoxic events, a phenomenon termed "selective vulnerability" (Fu et al. 2018). In AD, the first neurons lost in disease include pyramidal neurons in the entorhinal cortex, followed by neuronal populations in the hippocampus, and eventually the neocortex. In PD, clinical motor phenotypes are primarily caused by a loss of dopaminergic neurons in the substantia nigra pars compacta (SNpc) region of the striatum, with neuronal loss in other regions such as the pedunculopontine nucleus and locus coeruleus also being reported. Why one set of neurons succumbs to disease pathology while neighboring cells remain healthy remains a perplexing question in the field, particularly in light of the fact that the proteins susceptible to aggregation, and other factors leading to neurodegeneration, are present throughout the brain.

Studies comparing vulnerable and spared neurons in disease have shed light onto intrinsic physiological differences between these populations that may underlie selective vulnerability (Greene et al. 2005; Liang et al. 2008; Freer et al. 2016). Expression profiling in mouse and human has shown that vulnerable neuronal populations in AD and PD express reduced levels of genes involved in both protein homeostasis and calcium buffering compared with spared populations, likely predisposing them to higher levels of protein misfolding and ER stress (Hof et al. 1993; Chung et al. 2005; Wang and Mattson 2014; Freer et al. 2016). The SNpc dopaminergic neurons impacted in PD also have abundant deletions in mitochondrial DNA compared with other dopaminergic neuron populations, which may lead to increased mitochondrial dysfunction (Bender et al. 2006).

In AD, excitatory neurons appear more vulnerable to cell death than inhibitory neurons, with cholinergic neurons of the basal forebrain being particularly susceptible (Davies and Maloney 1976). Although early studies uncovered an association between deficits in excitatory neurotransmission and AD progression (Francis et al. 1999), a mechanistic understanding of how cell death pathways are more readily activated in excitatory neurons remains elusive. A recent study analyzing human single-cell RNA sequencing databases uncovered a tau protein homeostasis signature present in inhibitory but not excitatory neurons, and identified inhibitory neuron-enriched Bcl-2-associated athanogene 3 (BAG3) as a master regulator of tau accumulation (Fu et al. 2019). The diminished expression of BAG3 in excitatory neurons provides one of the first molecular links between selective vulnerability and toxic tau aggregation, an initiator of cell death. The identification of additional links will be facilitated by the expanding reper-

toire of single-cell sequencing techniques, as well as the advent of technologies that enable the differentiation of human iPSCs into various neuronal populations. Indeed, iPSCs from AD patients that have been differentiated into excitatory and inhibitory neuronal populations show differential susceptibility to tau and Aβ pathology (Muratore et al. 2017). Future studies have the opportunity to determine how distinct neuronal populations differentially respond to protein misfolding, mitochondrial dysfunction, and neuroinflammatory insults, and investigate how these responses lead to activation of distinct cell death pathways.

LINKING THE INITIATORS AND EXECUTIONERS OF NEURODEGENERATIVE DISEASE

Although several examples have been discussed in this review, there is relatively little data linking the initiator pathologies that have been observed in neurodegenerative disease to the executioner pathways known to mediate neuronal cell death. This stems in part from the fact that it is difficult to capture all aspects of a chronic neurodegenerative disease, such as AD or PD, in a single model. For instance, a number of the more prominent animal models for these disorders present with protein aggregate pathology (i.e., Aβ plaques or α-syn aggregates), but do not show appreciable cell death (Wirths and Bayer 2010).

To determine the relevance of executioner pathways in chronic neurodegenerative disease, researchers have looked for evidence of their activity in brains of human AD and PD patients. In the case of necroptosis, recent studies have found evidence of pathway activity in AD and other neurodegenerative indications (Ofengeim et al. 2015; Caccamo et al. 2017), although direct evidence of necrotic cell death has been more difficult to visualize. Similar studies have been conducted to identify apoptotic neurons with mixed results (Shimohama 2000; Venderova and Park 2012). Neurons undergoing apoptosis in the brains of AD and PD patients are relatively rare, although it is unclear whether this indicates that apoptosis is not a major

Cite this article as *Cold Spring Harb Perspect Biol* doi: 10.1101/cshperspect.a036434

driver of cell death or if this transient event is simply difficult to visualize in postmortem samples. In contrast, the activation of upstream stress response pathways is more readily visible in the brains of patients with chronic neurodegenerative disease. For instance, the presence UPR activation markers, particularly phosphorylated PERK and eIF2α, has been reported in the brains of AD, PD, and ALS patients (Chang et al. 2002; Hoozemans et al. 2007, 2009; Atkin et al. 2008; Cornejo and Hetz 2013; Mercado et al. 2016). It is also well established that phosphorylated JNK levels are elevated in disease-affected regions of postmortem AD brains (Shoji et al. 2000; Zhu et al. 2001; Le Pichon et al. 2017). Given the functional relationship between these pathways and cell death observed in preclinical models, it is logical to conclude that similar mechanisms may underlie neuronal loss in disease. Further unraveling the interplay between the initiators and executioners of neuronal cell death will be an important step toward understanding neurodegenerative disease genesis and progression (Fig. 4). Additional studies and new models designed to specifically investigate the interactions between these pathways should be a key future direction of the field.

POTENTIAL FOR THERAPEUTICS TARGETING NEURONAL CELL DEATH

Neurodegenerative disease is one of the largest unmet medical needs of our time, with few to no disease-modifying treatments available to patients. The majority of therapeutic strategies that have been explored thus far have focused on disrupting disease-associated pathologies, or initiators of neuronal cell death, based on the framework defined in this review. Although many of these strategies are still under development, results from those that target Aβ plaques suggest that disrupting this pathology is not sufficient to slow the course of degeneration, at least when dosed in patients with preexisting pathology (Lane et al. 2012; Castellani et al. 2019). Alternative strategies targeting tau or α-syn may have greater potential to impact disease, as these pathologies tend to be more correlated with functional decline (Arriagada et al. 1992; Brier et al. 2016; Bejanin et al. 2017; Marks et al. 2017; Kametani and Hasegawa 2018). Ongoing studies will shed light on their therapeutic potential.

As mechanistic insight into the role of executioners in chronic neurodegenerative disease continues to emerge, opportunities to develop

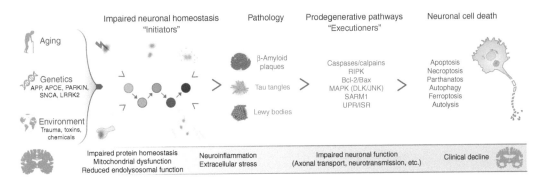

Figure 4. A model linking the initiators and executioners of neurodegenerative disease. A combination of aging, genetics, and environment contribute to the development of chronic neurodegenerative disease. These factors eventually lead to impaired neuronal homeostasis in response to neuron intrinsic and extrinsic disease initiators, including impaired protein homeostasis, mitochondrial dysfunction, neuroinflammation, among others. Over time, this altered homeostasis can manifest as pathologies commonly observed in disease, such as β-amyloid plaques and tau tangles in Alzheimer's disease (AD) or Lewy bodies in Parkinson's disease (PD). Neurons respond to these assaults by activating context-dependent, prodegenerative executioner pathways, precipitating impaired function and ultimately cell death. Disease-specific, stereotyped patterns of degeneration correlate with the clinical severity of neurodegenerative disease.

therapeutics that engage disease-relevant pathways may also arise. Two examples of this approach are molecules targeting DLK and RIPK1 that have recently advanced into clinical studies. Small molecule inhibitors for DLK have been developed that are capable of attenuating JNK signaling and downstream neuronal degeneration in several preclinical models including axonal injury, MPTP, and genetic models of chronic neurodegenerative disease (Patel et al. 2015; Larhammar et al. 2017; Le Pichon et al. 2017). Genentech/Roche has advanced the first DLK inhibitor into clinical trials for ALS (NCT02655614), although results have not yet been reported. In the case of RIPK1, preclinical data from genetic models and mice treated with RIPK1 inhibitors, including necrostatin-1, suggest a potential role for RIPK1 signaling in a number of disease contexts, including AD, ALS, and multiple sclerosis (Re et al. 2014; Ofengeim et al. 2015, 2017; Yang et al. 2017; Arrazola et al. 2019). Denali Therapeutics has recently advanced a CNS penetrant RIPK1 inhibitor into clinical trials, which is being tested in patients with AD and ALS (NCT03757325, NCT03757351). Although both of these programs are in the early stages of clinical study, results from these programs may provide insight into the potential of targeting these pathways for disease treatment.

CONCLUDING REMARKS

Despite having been studied for more than a century (Beard 1896), there remain significant gaps in our understanding of neuronal cell death, particularly as it pertains to chronic neurodegenerative disease. The advent of new technologies, such as iPSC-derived cells from disease patients, as well as an improved understanding of the molecular mechanisms by which vulnerable neuronal populations degenerate in disease will enable a deeper insight into the functional interplay between the initiators and executioners of neuronal cell death described here. Ultimately, these advances may translate into disease-modifying therapies that target context-specific cell death pathways, leading to an improvement in global health.

ACKNOWLEDGMENTS

We apologize to our colleagues whose research we could not cite or discuss owing to space limitations.

REFERENCES

Abisambra JF, Jinwal UK, Blair LJ, O'Leary JC III, Li Q, Brady S, Wang L, Guidi CE, Zhang B, Nordhues BA, et al. 2013. Tau accumulation activates the unfolded protein response by impairing endoplasmic reticulum-associated degradation. *J Neurosci* **33**: 9498–9507. doi:10.1523/jneurosci.5397-12.2013

Allan SM, Tyrrell PJ, Rothwell NJ. 2005. Interleukin-1 and neuronal injury. *Nat Rev Immunol* **5**: 629–640. doi:10.1038/nri1664

Arrazola MS, Saquel C, Catalan RJ, Barrientos SA, Hernandez DE, Catenaccio A, Court FA. 2019. Axonal degeneration is mediated by necroptosis activation. *J Neurosci* **39**: 3832–3844. doi:10.1523/jneurosci.0881-18.2019

Arriagada PV, Growdon JH, Hedley-Whyte ET, Hyman BT. 1992. Neurofibrillary tangles but not senile plaques parallel duration and severity of Alzheimer's disease. *Neurology* **42**: 631–639. doi:10.1212/WNL.42.3.631

Ashrafi G, Schlehe JS, LaVoie MJ, Schwarz TL. 2014. Mitophagy of damaged mitochondria occurs locally in distal neuronal axons and requires PINK1 and Parkin. *J Cell Biol* **206**: 655–670. doi:10.1083/jcb.201401070

Atkin JD, Farg MA, Walker AK, McLean C, Tomas D, Horne MK. 2008. Endoplasmic reticulum stress and induction of the unfolded protein response in human sporadic amyotrophic lateral sclerosis. *Neurobiol Dis* **30**: 400–407. doi:10.1016/j.nbd.2008.02.009

Beard J. 1896. The history of a transient nervous apparatus in certain *Ichthyopsida*. An account of the development and degeneration of ganglion cells and nerve fibers. *Zool Jahrb* **9**: 319–426.

Behrens A, Sibilia M, Wagner EF. 1999. Amino-terminal phosphorylation of c-Jun regulates stress-induced apoptosis and cellular proliferation. *Nat Genet* **21**: 326–329. doi:10.1038/6854

Bejanin A, Schonhaut DR, La Joie R, Kramer JH, Baker SL, Sosa N, Ayakta N, Cantwell A, Janabi M, Lauriola M, et al. 2017. Tau pathology and neurodegeneration contribute to cognitive impairment in Alzheimer's disease. *Brain* **140**: 3286–3300. doi:10.1093/brain/awx243

Bellucci A, Navarria L, Zaltieri M, Falarti E, Bodei S, Sigala S, Battistin L, Spillantini M, Missale C, Spano P. 2011. Induction of the unfolded protein response by α-synuclein in experimental models of Parkinson's disease. *J Neurochem* **116**: 588–605. doi:10.1111/j.1471-4159.2010.07143.x

Benarroch EE. 2015. Acquired axonal degeneration and regeneration: Recent insights and clinical correlations. *Neurology* **84**: 2076–2085. doi:10.1212/WNL.0000000000001601

Bender A, Krishnan KJ, Morris CM, Taylor GA, Reeve AK, Perry RH, Jaros E, Hersheson JS, Betts J, Klopstock T, et al. 2006. High levels of mitochondrial DNA deletions in substantia nigra neurons in aging and Parkinson disease. *Nat Genet* **38**: 515–517. doi:10.1038/ng1769

Borsello T, Forloni G. 2007. JNK signalling: A possible target to prevent neurodegeneration. *Curr Pharm Des* **13**: 1875–1886. doi:10.2174/138161207780858384

Braak H, Braak E. 1991. Neuropathological stageing of Alzheimer-related changes. *Acta Neuropathol* **82**: 239–259. doi:10.1007/BF00308809

Braak H, Del Tredici K, Rüb U, de Vos RA, Jansen Steur EN, Braak E. 2003. Staging of brain pathology related to sporadic Parkinson's disease. *Neurobiol Aging* **24**: 197–211. doi:10.1016/S0197-4580(02)00065-9

Brettschneider J, Del Tredici K, Lee VM, Trojanowski JQ. 2015. Spreading of pathology in neurodegenerative diseases: A focus on human studies. *Nat Rev Neurosci* **16**: 109–120. doi:10.1038/nrn3887

Brier MR, Gordon B, Friedrichsen K, McCarthy J, Stern A, Christensen J, Owen C, Aldea P, Su Y, Hassenstab J, et al. 2016. Tau and Aβ imaging, CSF measures, and cognition in Alzheimer's disease. *Sci Transl Med* **8**: 338ra66. doi:10.1126/scitranslmed.aaf2362

Burke RE, O'Malley K. 2013. Axon degeneration in Parkinson's disease. *Exp Neurol* **246**: 72–83. doi:10.1016/j.expneurol.2012.01.011

Burre J, Sharma M, Tsetsenis T, Buchman V, Etherton MR, Sudhof TC. 2010. α-synuclein promotes SNARE-complex assembly in vivo and in vitro. *Science* **329**: 1663–1667. doi:10.1126/science.1195227

Caccamo A, Branca C, Piras IS, Ferreira E, Huentelman MJ, Liang WS, Readhead B, Dudley JT, Spangenberg EE, Green KN, et al. 2017. Necroptosis activation in Alzheimer's disease. *Nat Neurosci* **20**: 1236–1246. doi:10.1038/nn.4608

Cashman CR, Höke A. 2015. Mechanisms of distal axonal degeneration in peripheral neuropathies. *Neurosci Lett* **596**: 33–50. doi:10.1016/j.neulet.2015.01.048

Castellani RJ, Plascencia-Villa G, Perry G. 2019. The amyloid cascade and Alzheimer's disease therapeutics: Theory versus observation. *Lab Invest* **99**: 958–970. doi:10.1038/s41374-019-0231-z

Chang RC, Suen KC, Ma CH, Elyaman W, Ng HK, Hugon J. 2002. Involvement of double-stranded RNA-dependent protein kinase and phosphorylation of eukaryotic initiation factor-2α in neuronal degeneration. *J Neurochem* **83**: 1215–1225. doi:10.1046/j.1471-4159.2002.01237.x

Chinta SJ, Mallajosyula JK, Rane A, Andersen JK. 2010. Mitochondrial α-synuclein accumulation impairs complex I function in dopaminergic neurons and results in increased mitophagy in vivo. *Neurosci Lett* **486**: 235–239. doi:10.1016/j.neulet.2010.09.061

Chung YH, Joo KM, Nam RH, Cho MH, Kim DJ, Lee WB, Cha CI. 2005. Decreased expression of calretinin in the cerebral cortex and hippocampus of SOD1G93A transgenic mice. *Brain Res* **1035**: 105–109. doi:10.1016/j.brainres.2004.12.022

Ciryam P, Tartaglia GG, Morimoto RI, Dobson CM, Vendruscolo M. 2013. Widespread aggregation and neurodegenerative diseases are associated with supersaturated proteins. *Cell Rep* **5**: 781–790. doi:10.1016/j.celrep.2013.09.043

Ciryam P, Kundra R, Freer R, Morimoto RI, Dobson CM, Vendruscolo M. 2016. A transcriptional signature of Alzheimer's disease is associated with a metastable subproteome at risk for aggregation. *Proc Natl Acad Sci* **113**: 4753–4758. doi:10.1073/pnas.1516604113

Clavaguera F, Bolmont T, Crowther RA, Abramowski D, Frank S, Probst A, Fraser G, Stalder AK, Beibel M, Staufenbiel M, et al. 2009. Transmission and spreading of tauopathy in transgenic mouse brain. *Nat Cell Biol* **11**: 909–913. doi:10.1038/ncb1901

Codolo G, Plotegher N, Pozzobon T, Brucale M, Tessari I, Bubacco L, de Bernard M. 2013. Triggering of inflammasome by aggregated α-synuclein, an inflammatory response in synucleinopathies. *PLoS ONE* **8**: e55375. doi:10.1371/journal.pone.0055375

Cornejo VH, Hetz C. 2013. The unfolded protein response in Alzheimer's disease. *Semin Immunopathol* **35**: 277–292. doi:10.1007/s00281-013-0373-9

Credle JJ, Forcelli PA, Delannoy M, Oaks AW, Permaul E, Berry DL, Duka V, Wills J, Sidhu A. 2015. α-Synuclein-mediated inhibition of ATF6 processing into COPII vesicles disrupts UPR signaling in Parkinson's disease. *Neurobiol Dis* **76**: 112–125. doi:10.1016/j.nbd.2015.02.005

Curran BP, Murray HJ, O'Connor JJ. 2003. A role for c-Jun N-terminal kinase in the inhibition of long-term potentiation by interleukin-1β and long-term depression in the rat dentate gyrus in vitro. *Neuroscience* **118**: 347–357. doi:10.1016/S0306-4522(02)00941-7

Davies P, Maloney AJ. 1976. Selective loss of central cholinergic neurons in Alzheimer's disease. *Lancet* **308**: 1403. doi:10.1016/S0140-6736(76)91936-X

Dawson TM, Dawson VL. 2017. Mitochondrial mechanisms of neuronal cell death: Potential therapeutics. *Annu Rev Pharmacol Toxicol* **57**: 437–454. doi:10.1146/annurev-pharmtox-010716-105001

Deckwerth TL, Elliott JL, Knudson CM, Johnson EM Jr, Snider WD, Korsmeyer SJ. 1996. BAX is required for neuronal death after trophic factor deprivation and during development. *Neuron* **17**: 401–411. doi:10.1016/S0896-6273(00)80173-7

Dejanovic B, Huntley MA, De Mazière A, Meilandt WJ, Wu T, Srinivasan K, Jiang Z, Gandham V, Friedman BA, Ngu H, et al. 2018. Changes in the synaptic proteome in tauopathy and rescue of tau-induced synapse loss by C1q antibodies. *Neuron* **100**: 1322–1336.e7. doi:10.1016/j.neuron.2018.10.014

Demuro A, Parker I, Stutzmann GE. 2010. Calcium signaling and amyloid toxicity in Alzheimer disease. *J Biol Chem* **285**: 12463–12468. doi:10.1074/jbc.R109.080895

Desplats P, Lee HJ, Bae EJ, Patrick C, Rockenstein E, Crews L, Spencer B, Masliah E, Lee SJ. 2009. Inclusion formation and neuronal cell death through neuron-to-neuron transmission of α-synuclein. *Proc Natl Acad Sci* **106**: 13010–13015. doi:10.1073/pnas.0903691106

Devi L, Raghavendran V, Prabhu BM, Avadhani NG, Anandatheerthavarada HK. 2008. Mitochondrial import and accumulation of α-synuclein impair complex I in human dopaminergic neuronal cultures and Parkinson disease brain. *J Biol Chem* **283**: 9089–9100. doi:10.1074/jbc.M710012200

Elmore S. 2007. Apoptosis: A review of programmed cell death. *Toxicol Pathol* **35:** 495–516. doi:10.1080/0192 6230701320337

Emmanouilidou E, Melachroinou K, Roumeliotis T, Garbis SD, Ntzouni M, Margaritis LH, Stefanis L, Vekrellis K. 2010. Cell-produced α-synuclein is secreted in a calcium-dependent manner by exosomes and impacts neuronal survival. *J Neurosci* **30:** 6838–6851. doi:10.1523/jneuro sci.5699-09.2010

Fatokun AA, Dawson VL, Dawson TM. 2014. Parthanatos: Mitochondrial-linked mechanisms and therapeutic opportunities. *Br J Pharmacol* **171:** 2000–2016. doi:10 .1111/bph.12416

Fernandes KA, Harder JM, Fornarola LB, Freeman RS, Clark AF, Pang IH, John SW, Libby RT. 2012. JNK2 and JNK3 are major regulators of axonal injury-induced retinal ganglion cell death. *Neurobiol Dis* **46:** 393–401. doi:10.1016/j .nbd.2012.02.003

Fonseca MI, Zhou J, Botto M, Tenner AJ. 2004. Absence of C1q leads to less neuropathology in transgenic mouse models of Alzheimer's disease. *J Neurosci* **24:** 6457–6465. doi:10.1523/jneurosci.0901-04.2004

Francis PT, Palmer AM, Snape M, Wilcock GK. 1999. The cholinergic hypothesis of Alzheimer's disease: A review of progress. *J Neurol Neurosurg Psychiatry* **66:** 137–147. doi:10.1136/jnnp.66.2.137

Freer R, Sormanni P, Vecchi G, Ciryam P, Dobson CM, Vendruscolo M. 2016. A protein homeostasis signature in healthy brains recapitulates tissue vulnerability to Alzheimer's disease. *Sci Adv* **2:** e1600947. doi:10.1126/sciadv .1600947

Fricker M, Tolkovsky AM, Borutaite V, Coleman M, Brown GC. 2018. Neuronal cell death. *Physiol Rev* **98:** 813–880. doi:10.1152/physrev.00011.2017

Fu H, Hardy J, Duff KE. 2018. Selective vulnerability in neurodegenerative diseases. *Nat Neurosci* **21:** 1350–1358. doi:10.1038/s41593-018-0221-2

Fu H, Possenti A, Freer R, Nakano Y, Hernandez Villegas NC, Tang M, Cauhy PVM, Lassus BA, Chen S, Fowler SL, et al. 2019. A tau homeostasis signature is linked with the cellular and regional vulnerability of excitatory neurons to tau pathology. *Nat Neurosci* **22:** 47–56. doi:10.1038/ s41593-018-0298-7

Galehdar Z, Swan P, Fuerth B, Callaghan SM, Park DS, Cregan SP. 2010. Neuronal apoptosis induced by endoplasmic reticulum stress is regulated by ATF4-CHOP-mediated induction of the Bcl-2 homology 3-only member PUMA. *J Neurosci* **30:** 16938–16948. doi:10.1523/jneuro sci.1598-10.2010

Gerdts J, Summers DW, Sasaki Y, DiAntonio A, Milbrandt J. 2013. Sarm1-mediated axon degeneration requires both SAM and TIR interactions. *J Neurosci* **33:** 13569–13580. doi:10.1523/jneurosci.1197-13.2013

Gerdts J, Summers DW, Milbrandt J, DiAntonio A. 2016. Axon self-destruction: New links among SARM1, MAPKs, and NAD$^+$ metabolism. *Neuron* **89:** 449–460. doi:10.1016/j.neuron.2015.12.023

Ghosh AS, Wang B, Pozniak CD, Chen M, Watts RJ, Lewcock JW. 2011. DLK induces developmental neuronal degeneration via selective regulation of proapoptotic JNK activity. *J Cell Biol* **194:** 751–764. doi:10.1083/jcb .201103153

Gilley J, Orsomando G, Nascimento-Ferreira I, Coleman MP. 2015. Absence of SARM1 rescues development and survival of NMNAT2-deficient axons. *Cell Rep* **10:** 1974–1981. doi:10.1016/j.celrep.2015.02.060

Greene JG, Dingledine R, Greenamyre JT. 2005. Gene expression profiling of rat midbrain dopamine neurons: Implications for selective vulnerability in parkinsonism. *Neurobiol Dis* **18:** 19–31. doi:10.1016/j.nbd.2004.10.003

Griffin WS, Stanley LC, Ling C, White L, MacLeod V, Perrot LJ, White CL III, Araoz C. 1989. Brain interleukin 1 and S-100 immunoreactivity are elevated in Down syndrome and Alzheimer disease. *Proc Natl Acad Sci* **86:** 7611–7615. doi:10.1073/pnas.86.19.7611

Guadagno J, Xu X, Karajgikar M, Brown A, Cregan SP. 2013. Microglia-derived TNFα induces apoptosis in neural precursor cells via transcriptional activation of the Bcl-2 family member Puma. *Cell Death Dis* **4:** e538. doi:10.1038/ cddis.2013.59

Guo JL, Narasimhan S, Changolkar L, He Z, Stieber A, Zhang B, Gathagan RJ, Iba M, McBride JD, Trojanowski JQ, et al. 2016. Unique pathological tau conformers from Alzheimer's brains transmit tau pathology in nontransgenic mice. *J Exp Med* **213:** 2635–2654. doi:10.1084/jem .20160833

Hansen DV, Hanson JE, Sheng M. 2018. Microglia in Alzheimer's disease. *J Cell Biol* **217:** 459–472. doi:10.1083/jcb .201709069

Heneka MT, Kummer MP, Stutz A, Delekate A, Schwartz S, Vieira-Saecker A, Griep A, Axt D, Remus A, Tzeng TC, et al. 2013. NLRP3 is activated in Alzheimer's disease and contributes to pathology in APP/PS1 mice. *Nature* **493:** 674–678. doi:10.1038/nature11729

Heneka MT, Golenbock DT, Latz E. 2015. Innate immunity in Alzheimer's disease. *Nat Immunol* **16:** 229–236. doi:10 .1038/ni.3102

Henninger N, Bouley J, Sikoglu EM, An J, Moore CM, King JA, Bowser R, Freeman MR, Brown RH Jr. 2016. Attenuated traumatic axonal injury and improved functional outcome after traumatic brain injury in mice lacking *Sarm1*. *Brain* **139:** 1094–1105. doi:10.1093/brain/aw w001

Hetz C, Saxena S. 2017. ER stress and the unfolded protein response in neurodegeneration. *Nat Rev Neurol* **13:** 477–491. doi:10.1038/nrneurol.2017.99

Hof PR, Nimchinsky EA, Celio MR, Bouras C, Morrison JH. 1993. Calretinin-immunoreactive neocortical interneurons are unaffected in Alzheimer's disease. *Neurosci Lett* **152:** 145–149. doi:10.1016/0304-3940(93)90504-E

Holzman LB, Merritt SE, Fan G. 1994. Identification, molecular cloning, and characterization of dual leucine zipper bearing kinase. A novel serine/threonine protein kinase that defines a second subfamily of mixed lineage kinases. *J Biol Chem* **269:** 30808–30817.

Hong S, Beja-Glasser VF, Nfonoyim BM, Frouin A, Li S, Ramakrishnan S, Merry KM, Shi Q, Rosenthal A, Barres BA, et al. 2016. Complement and microglia mediate early synapse loss in Alzheimer mouse models. *Science* **352:** 712–716. doi:10.1126/science.aad8373

Hoozemans JJ, van Haastert ES, Eikelenboom P, de Vos RA, Rozemuller JM, Scheper W. 2007. Activation of the unfolded protein response in Parkinson's disease. *Biochem*

Biophys Res Commun **354:** 707–711. doi:10.1016/j.bbrc
.2007.01.043

Hoozemans JJ, van Haastert ES, Nijholt DA, Rozemuller AJ, Eikelenboom P, Scheper W. 2009. The unfolded protein response is activated in pretangle neurons in Alzheimer's disease hippocampus. *Am J Pathol* **174:** 1241–1251. doi:10.2353/ajpath.2009.080814

Iannielli A, Bido S, Folladori L, Segnali A, Cancellieri C, Maresca A, Massimino L, Rubio A, Morabito G, Caporali L, et al. 2018. Pharmacological inhibition of necroptosis protects from dopaminergic neuronal cell death in Parkinson's disease models. *Cell Rep* **22:** 2066–2079. doi:10.1016/j.celrep.2018.01.089

Iba M, Guo JL, McBride JD, Zhang B, Trojanowski JQ, Lee VM. 2013. Synthetic tau fibrils mediate transmission of neurofibrillary tangles in a transgenic mouse model of Alzheimer's-like tauopathy. *J Neurosci* **33:** 1024–1037. doi:10.1523/jneurosci.2642-12.2013

Ito Y, Ofengeim D, Najafov A, Das S, Saberi S, Li Y, Hitomi J, Zhu H, Chen H, Mayo L, et al. 2016. RIPK1 mediates axonal degeneration by promoting inflammation and necroptosis in ALS. *Science* **353:** 603–608. doi:10.1126/science.aaf6803

Jarrett JT, Lansbury PT Jr. 1993. Seeding "one-dimensional crystallization" of amyloid: A pathogenic mechanism in Alzheimer's disease and scrapie? *Cell* **73:** 1055–1058. doi:10.1016/0092-8674(93)90635-4

Johnson BN, Berger AK, Cortese GP, Lavoie MJ. 2012. The ubiquitin E3 ligase parkin regulates the proapoptotic function of Bax. *Proc Natl Acad Sci* **109:** 6283–6288. doi:10.1073/pnas.1113248109

Johnson VE, Stewart W, Smith DH. 2013. Axonal pathology in traumatic brain injury. *Exp Neurol* **246:** 35–43. doi:10.1016/j.expneurol.2012.01.013

Kam TI, Mao X, Park H, Chou SC, Karuppagounder SS, Umanah GE, Yun SP, Brahmachari S, Panicker N, Chen R, et al. 2018. Poly(ADP-ribose) drives pathologic α-synuclein neurodegeneration in Parkinson's disease. *Science* **362:** eaat8407. doi:10.1126/science.aat8407

Kametani F, Hasegawa M. 2018. Reconsideration of amyloid hypothesis and tau hypothesis in Alzheimer's disease. *Front Neurosci* **12:** 25. doi:10.3389/fnins.2018.00025

Kfoury N, Holmes BB, Jiang H, Holtzman DM, Diamond MI. 2012. Trans-cellular propagation of Tau aggregation by fibrillar species. *J Biol Chem* **287:** 19440–19451. doi:10.1074/jbc.M112.346072

Kim M, Ahn JS, Park W, Hong SK, Jeon SR, Roh SW, Lee S. 2018. Diffuse axonal injury (DAI) in moderate to severe head injured patients: Pure DAI vs. non-pure DAI. *Clin Neurol Neurosurg* **171:** 116–123. doi:10.1016/j.clineuro.2018.06.011

Kitada T, Tong Y, Gautier CA, Shen J. 2009. Absence of nigral degeneration in aged parkin/DJ-1/PINK1 triple knockout mice. *J Neurochem* **111:** 696–702. doi:10.1111/j.1471-4159.2009.06350.x

Kristiansen M, Ham J. 2014. Programmed cell death during neuronal development: The sympathetic neuron model. *Cell Death Differ* **21:** 1025–1035. doi:10.1038/cdd.2014.47

Kroemer G, Galluzzi L, Brenner C. 2007. Mitochondrial membrane permeabilization in cell death. *Physiol Rev* **87:** 99–163. doi:10.1152/physrev.00013.2006

Kuida K, Zheng TS, Na S, Kuan C, Yang D, Karasuyama H, Rakic P, Flavell RA. 1996. Decreased apoptosis in the brain and premature lethality in CPP32-deficient mice. *Nature* **384:** 368–372. doi:10.1038/384368a0

Kundra R, Ciryam P, Morimoto RI, Dobson CM, Vendruscolo M. 2017. Protein homeostasis of a metastable subproteome associated with Alzheimer's disease. *Proc Natl Acad Sci* **114:** E5703–E5711. doi:10.1073/pnas.1618417114

Lane RF, Shineman DW, Steele JW, Lee LB, Fillit HM. 2012. Beyond amyloid: The future of therapeutics for Alzheimer's disease. *Adv Pharmacol* **64:** 213–271. doi:10.1016/B978-0-12-394816-8.00007-6

Larhammar M, Huntwork-Rodriguez S, Jiang Z, Solanoy H, Sengupta Ghosh A, Wang B, Kaminker JS, Huang K, Eastham-Anderson J, Siu M, et al. 2017. Dual leucine zipper kinase-dependent PERK activation contributes to neuronal degeneration following insult. *eLife* **6:** e20725. doi:10.7554/eLife.20725

Lashuel HA, Overk CR, Oueslati A, Masliah E. 2013. The many faces of α-synuclein: From structure and toxicity to therapeutic target. *Nat Rev Neurosci* **14:** 38–48. doi:10.1038/nrn3406

Leonard JR, Klocke BJ, D'Sa C, Flavell RA, Roth KA. 2002. Strain-dependent neurodevelopmental abnormalities in caspase-3-deficient mice. *J Neuropathol Exp Neurol* **61:** 673–677. doi:10.1093/jnen/61.8.673

Le Pichon CE, Meilandt WJ, Dominguez S, Solanoy H, Lin H, Ngu H, Gogineni A, Sengupta Ghosh A, Jiang Z, Lee SH, et al. 2017. Loss of dual leucine zipper kinase signaling is protective in animal models of neurodegenerative disease. *Sci Transl Med* **9:** eaag0394. doi:10.1126/scitranslmed.aag0394

Li Y, Guo Y, Tang J, Jiang J, Chen Z. 2014. New insights into the roles of CHOP-induced apoptosis in ER stress. *Acta Biochim Biophys Sin (Shanghai)* **46:** 629–640. doi:10.1093/abbs/gmu048

Liang WS, Dunckley T, Beach TG, Grover A, Mastroeni D, Ramsey K, Caselli RJ, Kukull WA, McKeel D, Morris JC, et al. 2008. Altered neuronal gene expression in brain regions differentially affected by Alzheimer's disease: A reference data set. *Physiol Genomics* **33:** 240–256. doi:10.1152/physiolgenomics.00242.2007

Libby RT, Li Y, Savinova OV, Barter J, Smith RS, Nickells RW, John SW. 2005. Susceptibility to neurodegeneration in a glaucoma is modified by Bax gene dosage. *PLoS Genet* **1:** 17–26. doi:10.1371/journal.pgen.0010004

Luk KC, Kehm V, Carroll J, Zhang B, O'Brien P, Trojanowski JQ, Lee VM. 2012. Pathological α-synuclein transmission initiates Parkinson-like neurodegeneration in nontransgenic mice. *Science* **338:** 949–953. doi:10.1126/science.1227157

Marks SM, Lockhart SN, Baker SL, Jagust WJ. 2017. Tau and β-amyloid are associated with medial temporal lobe structure, function, and memory encoding in normal aging. *J Neurosci* **37:** 3192–3201. doi:10.1523/jneurosci.3769-16.2017

Meier S, Bell M, Lyons DN, Ingram A, Chen J, Gensel JC, Zhu H, Nelson PT, Abisambra JF. 2015. Identification of novel tau interactions with endoplasmic reticulum proteins in Alzheimer's disease brain. *J Alzheimers Dis* **48:** 687–702. doi:10.3233/JAD-150298

Meisl G, Rajah L, Cohen SAI, Pfammatter M, Šarić A, Hell-strand E, Buell AK, Aguzzi A, Linse S, Vendruscolo M, et al. 2017. Scaling behaviour and rate-determining steps in filamentous self-assembly. *Chem Sci* **8:** 7087–7097. doi:10.1039/C7SC01965C

Mercado G, Castillo V, Soto P, Sidhu A. 2016. ER stress and Parkinson's disease: Pathological inputs that converge into the secretory pathway. *Brain Res* **1648:** 626–632. doi:10.1016/j.brainres.2016.04.042

Miller BR, Press C, Daniels RW, Sasaki Y, Milbrandt J, Di-Antonio A. 2009. A dual leucine kinase-dependent axon self-destruction program promotes Wallerian degeneration. *Nat Neurosci* **12:** 387–389. doi:10.1038/nn.2290

Mogi M, Harada M, Kondo T, Riederer P, Inagaki H, Mi-nami M, Nagatsu T. 1994. Interleukin-1β, interleukin-6, epidermal growth factor and transforming growth factor-α are elevated in the brain from parkinsonian patients. *Neurosci Lett* **180:** 147–150. doi:10.1016/0304-3940(94)90508-8

Muratore CR, Zhou C, Liao M, Fernandez MA, Taylor WM, Lagomarsino VN, Pearse RV II, Rice HC, Negri JM, He A, et al. 2017. Cell-type dependent Alzheimer's disease phenotypes: Probing the biology of selective neuronal vulnerability. *Stem Cell Rep* **9:** 1868–1884. doi:10.1016/j.stemcr.2017.10.015

Najjar M, Saleh D, Zelic M, Nogusa S, Shah S, Tai A, Finger JN, Polykratis A, Gough PJ, Bertin J, et al. 2016. RIPK1 and RIPK3 kinases promote cell-death-independent inflammation by Toll-like receptor 4. *Immunity* **45:** 46–59. doi:10.1016/j.immuni.2016.06.007

Neniskyte U, Neher JJ, Brown GC. 2011. Neuronal death induced by nanomolar amyloid β is mediated by primary phagocytosis of neurons by microglia. *J Biol Chem* **286:** 39904–39913. doi:10.1074/jbc.M111.267583

Neniskyte U, Vilalta A, Brown GC. 2014. Tumour necrosis factor α-induced neuronal loss is mediated by microglial phagocytosis. *FEBS Lett* **588:** 2952–2956. doi:10.1016/j.febslet.2014.05.046

Ofengeim D, Ito Y, Najafov A, Zhang Y, Shan B, DeWitt JP, Ye J, Zhang X, Chang A, Vakifahmetoglu-Norberg H, et al. 2015. Activation of necroptosis in multiple sclerosis. *Cell Rep* **10:** 1836–1849. doi:10.1016/j.celrep.2015.02.051

Ofengeim D, Mazzitelli S, Ito Y, DeWitt JP, Mifflin L, Zou C, Das S, Adiconis X, Chen H, Zhu H, et al. 2017. RIPK1 mediates a disease-associated microglial response in Alzheimer's disease. *Proc Natl Acad Sci* **114:** E8788–E8797. doi:10.1073/pnas.1714175114

Olmos G, Lladó J. 2014. Tumor necrosis factor α: A link between neuroinflammation and excitotoxicity. *Mediators Inflamm* **2014:** 861231. doi:10.1155/2014/861231

Osterloh JM, Yang J, Rooney TM, Fox AN, Adalbert R, Powell EH, Sheehan AE, Avery MA, Hackett R, Logan MA, et al. 2012. dSarm/Sarm1 is required for activation of an injury-induced axon death pathway. *Science* **337:** 481–484. doi:10.1126/science.1223899

Patel S, Cohen F, Dean BJ, De La Torre K, Deshmukh G, Estrada AA, Ghosh AS, Gibbons P, Gustafson A, Huestis MP, et al. 2015. Discovery of dual leucine zipper kinase (DLK, MAP3K12) inhibitors with activity in neurodegeneration models. *J Med Chem* **58:** 401–418. doi:10.1021/jm5013984

Patten DA, Wong J, Khacho M, Soubannier V, Mailloux RJ, Pilon-Larose K, MacLaurin JG, Park DS, McBride HM, Trinkle-Mulcahy L, et al. 2014. OPA1-dependent cristae modulation is essential for cellular adaptation to metabolic demand. *EMBO J* **33:** 2676–2691. doi:10.15252/embj.201488349

Pavitt GD, Ron D. 2012. New insights into translational regulation in the endoplasmic reticulum unfolded protein response. *Cold Spring Harb Perspect Biol* **4:** a012278. doi:10.1101/cshperspect.a012278

Pickrell AM, Youle RJ. 2015. The roles of PINK1, parkin, and mitochondrial fidelity in Parkinson's disease. *Neuron* **85:** 257–273. doi:10.1016/j.neuron.2014.12.007

Pozniak CD, Sengupta Ghosh A, Gogineni A, Hanson JE, Lee SH, Larson JL, Solanoy H, Bustos D, Li H, Ngu H, et al. 2013. Dual leucine zipper kinase is required for excitotoxicity-induced neuronal degeneration. *J Exp Med* **210:** 2553–2567. doi:10.1084/jem.20122832

Re DB, Le Verche V, Yu C, Amoroso MW, Politi KA, Phani S, Ikiz B, Hoffmann L, Koolen M, Nagata T, et al. 2014. Necroptosis drives motor neuron death in models of both sporadic and familial ALS. *Neuron* **81:** 1001–1008. doi:10.1016/j.neuron.2014.01.011

Rishal I, Fainzilber M. 2014. Axon–soma communication in neuronal injury. *Nat Rev Neurosci* **15:** 32–42. doi:10.1038/nrn3609

Roberson ED, Scearce-Levie K, Palop JJ, Yan F, Cheng IH, Wu T, Gerstein H, Yu GQ, Mucke L. 2007. Reducing endogenous tau ameliorates amyloid β-induced deficits in an Alzheimer's disease mouse model. *Science* **316:** 750–754. doi:10.1126/science.1141736

Ross CA, Poirier MA. 2004. Protein aggregation and neuro-degenerative disease. *Nat Med* **10**(Suppl)**:** S10–S17. doi:10.1038/nm1066

Schafer DP, Lehrman EK, Kautzman AG, Koyama R, Mardinly AR, Yamasaki R, Ransohoff RM, Greenberg ME, Barres BA, Stevens B. 2012. Microglia sculpt postnatal neural circuits in an activity and complement-dependent manner. *Neuron* **74:** 691–705. doi:10.1016/j.neuron.2012.03.026

Shi Q, Chowdhury S, Ma R, Le KX, Hong S, Caldarone BJ, Stevens B, Lemere CA. 2017. Complement C3 deficiency protects against neurodegeneration in aged plaque-rich APP/PS1 mice. *Sci Transl Med* **9:** eaaf6295. doi:10.1126/scitranslmed.aaf6295

Shimohama S. 2000. Apoptosis in Alzheimer's disease—An update. *Apoptosis* **5:** 9–16. doi:10.1023/A:1009625323388

Shoji M, Iwakami N, Takeuchi S, Waragai M, Suzuki M, Kanazawa I, Lippa CF, Ono S, Okazawa H. 2000. JNK activation is associated with intracellular β-amyloid accumulation. *Brain Res Mol Brain Res* **85:** 221–233. doi:10.1016/S0169-328X(00)00245-X

Shtifman A, Ward CW, Laver DR, Bannister ML, Lopez JR, Kitazawa M, LaFerla FM, Ikemoto N, Querfurth HW. 2010. Amyloid-β protein impairs Ca^{2+} release and contractility in skeletal muscle. *Neurobiol Aging* **31:** 2080–2090. doi:10.1016/j.neurobiolaging.2008.11.003

Silke J, Rickard JA, Gerlic M. 2015. The diverse role of RIP kinases in necroptosis and inflammation. *Nat Immunol* **16:** 689–697. doi:10.1038/ni.3206

Simon DJ, Watkins TA. 2018. Therapeutic opportunities and pitfalls in the treatment of axon degeneration. *Curr Opin*

Neurol **31:** 693–701. doi:10.1097/WCO.000000000000
0621

Simon DJ, Pitts J, Hertz NT, Yang J, Yamagishi Y, Olsen O, Tešić Mark M, Molina H, Tessier-Lavigne M. 2016. Axon degeneration gated by retrograde activation of somatic pro-apoptotic signaling. *Cell* **164:** 1031–1045. doi:10.1016/j.cell.2016.01.032

Siu M, Sengupta Ghosh A, Lewcock JW. 2018. Dual leucine zipper kinase inhibitors for the treatment of neurodegeneration. *J Med Chem* **61:** 8078–8087. doi:10.1021/acs.jmedchem.8b00370

Slee EA, Adrain C, Martin SJ. 2001. Executioner caspase-3, -6, and -7 perform distinct, non-redundant roles during the demolition phase of apoptosis. *J Biol Chem* **276:** 7320–7326. doi:10.1074/jbc.M008363200

Sliter DA, Martinez J, Hao L, Chen X, Sun N, Fischer TD, Burman JL, Li Y, Zhang Z, Narendra DP, et al. 2018. Parkin and PINK1 mitigate STING-induced inflammation. *Nature* **561:** 258–262. doi:10.1038/s41586-018-0448-9

Smith HL, Mallucci GR. 2016. The unfolded protein response: Mechanisms and therapy of neurodegeneration. *Brain* **139:** 2113–2121. doi:10.1093/brain/aww101

Soto C, Pritzkow S. 2018. Protein misfolding, aggregation, and conformational strains in neurodegenerative diseases. *Nat Neurosci* **21:** 1332–1340. doi:10.1038/s41593-018-0235-9

Srinivasan K, Friedman BA, Larson JL, Lauffer BE, Goldstein LD, Appling LL, Borneo J, Poon C, Ho T, Cai F, et al. 2016. Untangling the brain's neuroinflammatory and neurodegenerative transcriptional responses. *Nat Commun* **7:** 11295. doi:10.1038/ncomms11295

Summers DW, Gibson DA, DiAntonio A, Milbrandt J. 2016. SARM1-specific motifs in the TIR domain enable NAD$^+$ loss and regulate injury-induced SARM1 activation. *Proc Natl Acad Sci* **113:** E6271–E6280. doi:10.1073/pnas.1601506113

Swerdlow RH, Burns JM, Khan SM. 2010. The Alzheimer's disease mitochondrial cascade hypothesis. *J Alzheimers Dis* **20:** S265–S279. doi:10.3233/JAD-2010-100339

Tagliaferro P, Burke RE. 2016. Retrograde axonal degeneration in Parkinson disease. *J Parkinsons Dis* **6:** 1–15. doi:10.3233/JPD-150769

Tedeschi A, Bradke F. 2013. The DLK signalling pathway—A double-edged sword in neural development and regeneration. *EMBO Rep* **14:** 605–614. doi:10.1038/embor.2013.64

Turkiew E, Falconer D, Reed N, Höke A. 2017. Deletion of Sarm1 gene is neuroprotective in two models of peripheral neuropathy. *J Peripher Nerv Syst* **22:** 162–171. doi:10.1111/jns.12219

Urano F, Wang X, Bertolotti A, Zhang Y, Chung P, Harding HP, Ron D. 2000. Coupling of stress in the ER to activation of JNK protein kinases by transmembrane protein kinase IRE1. *Science* **287:** 664–666. doi:10.1126/science.287.5453.664

Vandenabeele P, Grootjans S, Callewaert N, Takahashi N. 2013. Necrostatin-1 blocks both RIPK1 and IDO: Consequences for the study of cell death in experimental disease models. *Cell Death Differ* **20:** 185–187. doi:10.1038/cdd.2012.151

Vanden Berghe T, Linkermann A, Jouan-Lanhouet S, Walczak H, Vandenabeele P. 2014. Regulated necrosis: The expanding network of non-apoptotic cell death pathways. *Nat Rev Mol Cell Biol* **15:** 135–147. doi:10.1038/nrm3737

Venderova K, Park DS. 2012. Programmed cell death in Parkinson's disease. *Cold Spring Harb Perspect Med* **2:** a009365. doi:10.1101/cshperspect.a009365

Venegas C, Kumar S, Franklin BS, Dierkes T, Brinkschulte R, Tejera D, Vieira-Saecker A, Schwartz S, Santarelli F, Kummer MP, et al. 2017. Microglia-derived ASC specks cross-seed amyloid-β in Alzheimer's disease. *Nature* **552:** 355–361. doi:10.1038/nature25158

Verma M, Vats A, Taneja V. 2015. Toxic species in amyloid disorders: Oligomers or mature fibrils. *Ann Indian Acad Neurol* **18:** 138–145. doi:10.4103/0972-2327.150606

Vila M, Przedborski S. 2003. Targeting programmed cell death in neurodegenerative diseases. *Nat Rev Neurosci* **4:** 365–375. doi:10.1038/nrn1100

Vila M, Jackson-Lewis V, Vukosavic S, Djaldetti R, Liberatore G, Offen D, Korsmeyer SJ, Przedborski S. 2001. Bax ablation prevents dopaminergic neurodegeneration in the 1-methyl-4-phenyl-1,2,3,6-tetrahydropyridine mouse model of Parkinson's disease. *Proc Natl Acad Sci* **98:** 2837–2842. doi:10.1073/pnas.051633998

Viviani B, Bartesaghi S, Gardoni F, Vezzani A, Behrens MM, Bartfai T, Binaglia M, Corsini E, Di Luca M, Galli CL, et al. 2003. Interleukin-1β enhances NMDA receptor-mediated intracellular calcium increase through activation of the Src family of kinases. *J Neurosci* **23:** 8692–8700. doi:10.1523/jneurosci.23-25-08692.2003

Voet S, Srinivasan S, Lamkanfi M, van Loo G. 2019. Inflammasomes in neuroinflammatory and neurodegenerative diseases. *EMBO Mol Med* **11:** e10248. doi:10.15252/emmm.201810248

Volpicelli-Daley LA, Luk KC, Patel TP, Tanik SA, Riddle DM, Stieber A, Meaney DF, Trojanowski JQ, Lee VM. 2011. Exogenous α-synuclein fibrils induce Lewy body pathology leading to synaptic dysfunction and neuron death. *Neuron* **72:** 57–71. doi:10.1016/j.neuron.2011.08.033

Walker LJ, Summers DW, Sasaki Y, Brace EJ, Milbrandt J, DiAntonio A. 2017. MAPK signaling promotes axonal degeneration by speeding the turnover of the axonal maintenance factor NMNAT2. *eLife* **6:** e22540. doi:10.7554/eLife.22540

Walter P, Ron D. 2011. The unfolded protein response: From stress pathway to homeostatic regulation. *Science* **334:** 1081–1086. doi:10.1126/science.1209038

Wang Y, Mattson MP. 2014. L-type Ca^{2+} currents at CA1 synapses, but not CA3 or dentate granule neuron synapses, are increased in 3xTgAD mice in an age-dependent manner. *Neurobiol Aging* **35:** 88–95. doi:10.1016/j.neurobiolaging.2013.07.007

Watkins TA, Wang B, Huntwork-Rodriguez S, Yang J, Jiang Z, Eastham-Anderson J, Modrusan Z, Kaminker JS, Tessier-Lavigne M, Lewcock JW. 2013. DLK initiates a transcriptional program that couples apoptotic and regenerative responses to axonal injury. *Proc Natl Acad Sci* **110:** 4039–4044. doi:10.1073/pnas.1211074110

Wirths O, Bayer TA. 2010. Neuron loss in transgenic mouse models of Alzheimer's disease. *Int J Alzheimers Dis* **2010:** 723782. doi:10.4061/2010/723782

Xiong X, Hao Y, Sun K, Li J, Li X, Mishra B, Soppina P, Wu C, Hume RI, Collins CA. 2012. The Highwire ubiquitin ligase promotes axonal degeneration by tuning levels of Nmnat protein. *PLoS Biol* **10:** e1001440. doi:10.1371/journal.pbio.1001440

Yamaguchi Y, Miura M. 2015. Programmed cell death in neurodevelopment. *Dev Cell* **32:** 478–490. doi:10.1016/j.devcel.2015.01.019

Yang DD, Kuan CY, Whitmarsh AJ, Rincón M, Zheng TS, Davis RJ, Rakic P, Flavell RA. 1997. Absence of excitotoxicity-induced apoptosis in the hippocampus of mice lacking the Jnk3 gene. *Nature* **389:** 865–870. doi:10.1038/39899

Yang J, Weimer RM, Kallop D, Olsen O, Wu Z, Renier N, Uryu K, Tessier-Lavigne M. 2013. Regulation of axon degeneration after injury and in development by the endogenous calpain inhibitor calpastatin. *Neuron* **80:** 1175–1189. doi:10.1016/j.neuron.2013.08.034

Yang R, Hu K, Chen J, Zhu S, Li L, Lu H, Li P, Dong R. 2017. Necrostatin-1 protects hippocampal neurons against ischemia/reperfusion injury via the RIP3/DAXX signaling pathway in rats. *Neurosci Lett* **651:** 207–215. doi:10.1016/j.neulet.2017.05.016

Yarza R, Vela S, Solas M, Ramirez MJ. 2015. c-Jun N-terminal kinase (JNK) signaling as a therapeutic target for Alzheimer's disease. *Front Pharmacol* **6:** 321. doi:10.3389/fphar.2015.00321

Yuan J, Yankner BA. 2000. Apoptosis in the nervous system. *Nature* **407:** 802–809. doi:10.1038/35037739

Yuan J, Amin P, Ofengeim D. 2019. Necroptosis and RIPK1-mediated neuroinflammation in CNS diseases. *Nat Rev Neurosci* **20:** 19–33. doi:10.1038/s41583-018-0093-1

Zanjani H, Finch CE, Kemper C, Atkinson J, McKeel D, Morris JC, Price JL. 2005. Complement activation in very early Alzheimer disease. *Alzheimer Dis Assoc Disord* **19:** 55–66. doi:10.1097/01.wad.0000165506.60370.94

Zhang Y, Goodyer C, LeBlanc A. 2000. Selective and protracted apoptosis in human primary neurons microinjected with active caspase-3, -6, -7, and -8. *J Neurosci* **20:** 8384–8389. doi:10.1523/jneurosci.20-22-08384.2000

Zhang Y, Sloan SA, Clarke LE, Caneda C, Plaza CA, Blumenthal PD, Vogel H, Steinberg GK, Edwards MS, Li G, et al. 2016. Purification and characterization of progenitor and mature human astrocytes reveals transcriptional and functional differences with mouse. *Neuron* **89:** 37–53. doi:10.1016/j.neuron.2015.11.013

Zhang S, Tang MB, Luo HY, Shi CH, Xu YM. 2017. Necroptosis in neurodegenerative diseases: A potential therapeutic target. *Cell Death Dis* **8:** e2905. doi:10.1038/cddis.2017.286

Zhou Y, Lu M, Du RH, Qiao C, Jiang CY, Zhang KZ, Ding JH, Hu G. 2016. MicroRNA-7 targets Nod-like receptor protein 3 inflammasome to modulate neuroinflammation in the pathogenesis of Parkinson's disease. *Mol Neurodegener* **11:** 28. doi:10.1186/s13024-016-0094-3

Zhu X, Castellani RJ, Takeda A, Nunomura A, Atwood CS, Perry G, Smith MA. 2001. Differential activation of neuronal ERK, JNK/SAPK and p38 in Alzheimer disease: The "two hit" hypothesis. *Mech Ageing Dev* **123:** 39–46. doi:10.1016/S0047-6374(01)00342-6

Ziogas NK, Koliatsos VE. 2018. Primary traumatic axonopathy in mice subjected to impact acceleration: A reappraisal of pathology and mechanisms with high-resolution anatomical methods. *J Neurosci* **38:** 4031–4047. doi:10.1523/jneurosci.2343-17.2018

Cite this article as *Cold Spring Harb Perspect Biol* doi: 10.1101/cshperspect.a036434

Programmed Cell Death in the Evolutionary Race against Bacterial Virulence Factors

Carolyn A. Lacey and Edward A. Miao

Department of Microbiology and Immunology, Center for Gastrointestinal Biology and Disease, and Lineberger Comprehensive Cancer Center, University of North Carolina at Chapel Hill, Chapel Hill, North Carolina 27599, USA

Correspondence: edmiao1@gmail.com

Innate immune sensors can recognize when host cells are irrevocably compromised by pathogens, and in response can trigger programmed cell death (pyroptosis, apoptosis, and necroptosis). Innate sensors can directly bind microbial ligands; for example, NAIP/NLRC4 detects flagellin/rod/needle, whereas caspase-11 detects lipopolysaccharide. Other sensors are guards that monitor normal function of cellular proteins; for instance, pyrin monitors Rho GTPases, whereas caspase-8 and receptor-interacting protein kinase (RIPK)3 guards RIPK1 transcriptional signaling. Some proteins that need to be guarded can be duplicated as decoy domains, as seen in the integrated decoy domains within NLRP1 that watch for microbial attack. Here, we discuss the evolutionary battle between pathogens and host innate immune sensors/guards, illustrated by the Red Queen hypothesis. We discuss in depth four pathogens, and how they either fail in this evolutionary race (*Chromobacterium violaceum*, *Burkholderia thailandensis*), or how the evolutionary race generates increasingly complex virulence factors and host innate immune signaling pathways (*Yersinia* species, and enteropathogenic *Escherichia coli* [EPEC]).

The innate immune system can combat intracellular bacteria by inducing programmed cell death, which eliminates the infected cell niche. Killing host cells can also be useful in the innate immune response in cases in which a cell has been irrevocably reprogrammed to act in the benefit of the pathogen. Programmed cell death can be initiated by a variety of interconnected pathways, resulting in pyroptosis, apoptosis, or necroptosis. There are notable recent advances in our understanding of how programmed cell death either restricts bacterial pathogens, or fails to do so as certain pathogens have evolved to evade this defense. The evolutionary race between pathogen virulence and host defense is vital in determining whether a bacterial pathogen can cause disease, or is readily cleared.

Direct sensors, guards, and decoys initiate specific forms of programmed cell death, including pyroptosis, necroptosis, and apoptosis. Pyroptosis and necroptosis result in membrane rupture that releases soluble cytosolic contents to the extracellular space, whereas apoptosis converts a cell into apoptotic bodies that retain cellular contents within membranes. Cell death may lead to enhanced clearance of the pathogen; however, inappropriate or excessive cell death can be extremely detrimental to the host. These tightly regulated cell death mechanisms are often triggered by the activation of specific caspase proteases. Pyroptosis results from activation of caspase-1 or murine caspase-11/human caspase-4, -5 (Jorgensen et al. 2017). Apoptosis ensues when apoptotic initiators (caspase-8, -9) activate apoptotic effectors (caspase-3, -6, -7) (Taylor et al. 2008). Lastly, necroptosis occurs when RIPK3 phosphorylates the pseudokinase MLKL (Dondelinger et al. 2016a; Vanden Berghe et al. 2016). Many of these signaling pathways use death fold family domains to drive homotypic interactions, including death domains (DDs), death effector domains (DEDs), caspase activation and recruitment domains (CARDs), and pyrin domains (PYDs) (Weber and Vincenz 2001).

RED QUEEN HYPOTHESIS

The evolutionary biologist Leigh Van Valen proposed that each species must constantly evolve to avoid extinction in the face of competitors who are also constantly evolving. To illustrate this race, Van Valen drew upon the imagery in Lewis Carol's book *Through the Looking Glass*, in which the protagonist, Alice, engages in a footrace with the Red Queen. Alice soon finds that they have been running, but have stayed in the same place. The Red Queen informs Alice "Now, here, you see, it takes all the running you can do, to keep in the same place." Van Valen's Red Queen hypothesis proposes that organisms must constantly evolve to maintain their place in an environment where competitors are also constantly evolving (Van Valen 1973).

The Red Queen hypothesis also applies to host–pathogen interactions; as hosts evolve defenses against infection, pathogens must evolve virulence factors to overcome those defenses. This constant evolutionary race by both competitors ensures that neither the host nor pathogen go extinct. In the perpetual drive to outrun each other, hosts evolve increasingly complex defense mechanisms while, simultaneously, pathogens evolve equally complex virulence factors. Typically, pathogens are able to keep up with the evolutionary challenge in the race with the

innate immune system. However, most pathogens lose the race against the adaptive immune system, which eventually resolves the infection, but not before the pathogen transmits to a new host. Thus, the host (which we herein personify as Alice) constantly evolves new immunologic defenses. Meanwhile, pathogens (which we herein personify as Red Queens), must constantly evolve novel virulence factors.

INNATE IMMUNE SENSORS: DIRECT SENSORS, GUARDS, AND DECOYS

Programmed cell death can be initiated by a variety of sensors in the innate immune system. Notable among these are the nucleotide-binding domain, leucine-rich repeat (NLR) superfamily of cytosolic sensors, which cause cell death in eukaryotes ranging from plants to animals. Jones, Vance, and Dangl proposed that NLRs fall into three categories: direct sensors, guards, and decoys (Jones et al. 2016). These concepts not only apply to NLRs, but are also broadly applicable to many pathways in the innate immune system (Fig. 1).

"Direct sensors" bind microbial ligands. For example, TLR4 and its coreceptor MD2 directly bind lipopolysaccharide (LPS) in the extracellular space, triggering a transcriptional response. In the cytosol, caspase-11 is the direct sensor for LPS, triggering pyroptotic cell death. Similarly, extracellular flagellin detection by TLR5 drives a transcriptional response, whereas cytosolic flagellin detection by NAIP/NLRC4 triggers pyroptosis.

"Guards" monitor a cellular protein(s) for evidence of attack by virulence factors, and activate innate immune signaling in response. Guards can directly bind to the guarded protein, but we propose they may also monitor the enzymatic activity of the guarded protein. If the guarded protein is functioning normally, guards do not respond. However, if a guarded protein is attacked by a pathogen virulence factor, then the guard responds by activating and inducing a new response, often programmed cell death. Note that guards do not prevent the guarded protein from being attacked; rather they sound the alarm when such an attack occurs. More

Cite this article as *Cold Spring Harb Perspect Biol* doi: 10.1101/cshperspect.a036459

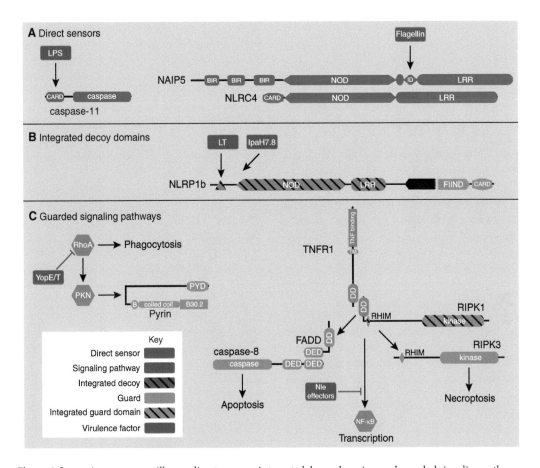

Figure 1. Innate immune surveillance: direct sensors, integrated decoy domains, and guarded signaling pathways. (*A*) Direct sensors include both caspase-11 and NAIP5/NLRC4. Caspase-11 activates when it directly binds to bacterial lipopolysaccharide (LPS) via its caspase activation and recruitment domain (CARD). NAIP5 directly binds bacterial flagellin; the NAIP5-flagellin complex then oligomerizes with NLRC4 molecules to form a caspase-1 activating inflammasome. (*B*) Integrated decoy domains are similar to domains of other signaling pathways, acting as lures for bacterial virulence factors. Once the decoy domain is attacked, this triggers signaling by the innate immune guard domains within the same protein. For example, the anthrax lethal toxin (LT) intends to target host mitogen-activated protein kinase (MAPK) but also cleaves the amino-terminal integrated decoy domain of NLRP1. IpaH7.8 is an E3 ubiquitinase that presumably intends to target another host protein, but also targets the integrated decoy domains of NLRP1 for degradation. Either of these attack events cause the NLRP1 signaling guard domains (the FIIND-CARD fragment) to form an active inflammasome. (*C*) Guards can watch a protein or a specific signaling pathway. Pyrin monitors RhoA signaling to protein kinase (PKN). When the *Yersinia* effectors YopE and T attack RhoA, pyrin detects the loss of PKN activation and in response forms an active inflammasome. Integrated guard domains can also be added within normal signaling pathways; for example, receptor-interacting protein kinase (RIPK)1 has two guard functions that survey the tumor necrosis factor (TNF) transcriptional response, one of which uses an integrated guard domain. When bacterial virulence factors, such as Nle effectors, attack the TNF transcriptional response, the first guard pathway is triggered by exposed death domains (DDs) within RIPK1, which are normally occupied by interacting with the TNF receptor signaling complex. When these DDs are abnormally exposed, they are detected by the DD containing guard adaptor Fas-associated death domain protein (FADD), which in turn signals to the guard caspase-8 to initiate apoptosis. Additionally, RIPK1 also encodes a RHIM and a kinase domain that can be considered integrated guard domains. These domains activate RIPK3, but only when both the transcriptional response and the apoptosis guard pathways are inhibited.

generally applied, the guard concept encompasses sensors that monitor whether signaling pathways are functioning normally. In this context, pyrin is a guard for Rho GTPases, and caspase-8/RIPK3 are guards for the tumor necrosis factor (TNF) signaling pathway. Extending the guard concept further, a guard function may be added to the protein that needs to be guarded within the same open reading frame (an "integrated guard"). Integrated guard domains are added to normal signaling proteins and are not strictly required for their normal function. The kinase domain of RIPK1 may be an example of an integrated guard domain whereby RIPK1 monitors its own scaffolding function.

"Decoys" are duplicates of the protein that needs to be monitored—the guard watches the decoy. Decoys act like lures, tempting virulence factors to attack them, triggering cell death in response. The decoy and guard can also be combined into a single protein; "integrated decoy domains" are decoys that are encoded in the same open reading frame as the guard domains. NLRP1 may be an example of an integrated decoy that monitors for attack on other NLR proteins.

PYROPTOSIS

Caspase-1 is activated by a variety of inflammasome sensors that act as direct sensors (NAIP/NLRC4, AIM2), guards (pyrin), or decoys (NLRP1), or whose mode of sensing remains to be elucidated (NLRP3). Inflammasomes can signal directly to caspase-1 if they contain a CARD (NLRC4, NLRP1), or indirectly via the ASC adaptor if they signal via a PYD (AIM2, pyrin, NLRP3). In contrast, caspase-11 is a combined direct sensor and protease that activates itself when it binds to LPS.

Either activated caspase-1 or caspase-11 can independently cleave and activate gasdermin D (Kayagaki et al. 2015; Shi et al. 2015) (for a gasdermin review, see Kovacs and Miao 2017). Cleavage of gasdermin D separates the amino-terminal pore-forming domain from the carboxy-terminal regulatory domain. Twenty-seven gasdermin pore-forming domains oligomerize to form an 18-nm pore that is large enough to allow the egress of all small molecules

as well as small proteins (Ruan et al. 2018). In addition to gasdermin D, caspase-1 (but not caspase-11[Ramirez et al. 2018]) also cleaves IL-1β (4.5 nm) and IL-18 (5.0 nm), which will easily pass through the gasdermin pore. Sodium, and thereby water, enter the cell, increasing the cytosolic turgor pressure until the membrane ruptures. We define this rupture event as pyroptosis.

Pyroptosis is often described as lytic programmed cell death, which often conjures an image of a fully dispersed cell. Indeed, the membrane rupture is large enough to allow all soluble cytosolic contents to immediately escape from the cell. However, the torn plasma membrane remains otherwise intact, and retains the organelles, cytoskeleton, nucleus, and intracellular bacteria. These bacteria remain viable, but trapped within the remains of the pyroptotic cell. We termed the corpse of the pyroptotic cell as a pore-induced intracellular trap (PIT); as apoptosis converts cells into apoptotic bodies, pyroptosis converts cells into PITs. PITs simultaneously attract neutrophils, which efferocytose both the PIT and its trapped bacteria (Jorgensen et al. 2016a,b). The neutrophil then generates reactive oxygen species and kills the detained pathogen (Miao et al. 2010a).

NAIP–NLRC4 Defending against *Chromobacterium violaceum*

Perhaps the best understood inflammasome is NAIP/NLRC4, which detects the activity of bacterial type III secretion systems (T3SS). T3SS are syringe-like mechanisms that inject effector proteins into the cytosol of host cells. These effector proteins reprogram the host cell to the benefit of the pathogen. However, T3SS also aberrantly translocates flagellin, rod, and needle proteins into the host cytosol. Flagellin binds to mouse NAIP5 or NAIP6, T3SS rod proteins bind to NAIP2, whereas the T3SS needle binds to NAIP1 (Vance 2015). The structural basis for detection and activation have recently been delineated (Hu et al. 2015; Zhang et al. 2015; Tenthorey et al. 2017; Yang et al. 2018). For example, a single flagellin protein molecule binds to a single NAIP5, triggering a conformational change that exposes a polymerization interface on

NAIP5. This interface recruits NLRC4, causing a similar conformational change that recruits another NLRC4, and so on until 10 NLRC4s have been oligomerized. This creates a hub-like structure composed of one NAIP and 10 NLRC4 proteins that is termed an inflammasome. The CARD domains of the 10 NLRC4 proteins are clustered in the middle of the inflammasome hub, and initiate caspase-1 polymerization. Thus, it is possible that a single molecule of flagellin, rod, or needle protein can result in pyroptosis. This seems to be an incredibly decisive system to destroy cells that have been compromised by T3SS injection.

One would expect that the existence of NLRC4 would make it very difficult for bacterial pathogens to use a T3SS. Yet, approximately half of all Gram-negative bacterial pathogens use T3SS. The Red Queen's race may have selected for pathogens that uniformly evade or inhibit NLRC4. For example, flagellin repression strategies are common and at least one case of an NLRC4-evasive T3SS has been described (Miao et al. 2010a,b). Remarkably, engineering bacteria to reverse flagellin repression generates stains that are eliminated by NLRC4 with exquisite sensitivity.

At least one bacterium, *C. violaceum,* appears to have no ability to evade NLRC4. *C. violaceum* is a ubiquitous environmental pathogen that encodes a T3SS, but whose natural host is unknown (Batista and da Silva Neto 2017). It infects people with significant immunologic defects, primarily those with chronic granulomatous disease (Batista and da Silva Neto 2017). Inflammasome responses clear *C. violaceum* infection in mouse models. Pyroptosis is required for splenic clearance, in which macrophages are most likely infected. In contrast, in the liver, inflammasome-driven IL-18 primes a natural killer (NK) cell response in which perforin-mediated cytotoxicity clears the hepatocyte niche, presumably by triggering apoptosis. Just as normal people are resistant to *C. violaceum* infection, wild-type (WT) mice resist high dose challenge (1,000,000 CFUs). However, $Nlrc4^{-/-}$ mice succumb to low-dose challenge (100 CFUs) (Maltez et al. 2015). We estimate that this represents a >50,000-fold

change in the effective 100% lethal dose (LD_{100}; Table 1) when comparing WT to inflammasome-deficient mice (Maltez and Miao 2016). The remarkable change in the effective lethal dose is a phenotype that is nearly unheard of in the inflammasome literature, equaled only by *B. thailandensis* (discussed in the next section).

Caspase-11 Defending against *Burkholderia thailandensis*

Caspase-11 is the most direct pathway to programmed cell death, in that it requires just two proteins: caspase-11 and gasdermin D. Caspase-11 itself serves both as a sensor and as a catalytic protease. The CARD domain of caspase-11 directly binds to LPS in the cytosol, which triggers CARD-oligomerization and activation of the protease (Hagar et al. 2013; Kayagaki et al. 2013; Shi et al. 2014). Active caspase-11 then cleaves gasdermin D and induces pyroptosis (Kayagaki et al. 2015; Shi et al. 2015).

By sensing cytosolic LPS, caspase-11 discriminates cytosol-invasive bacteria from vacuolar or extracellular bacteria (Aachoui et al. 2013). Caspase-11 efficiently detects *B. thailandensis*, a cytosol invasive bacterium, in vivo and is incredibly effective at clearing the bacterium. This defense pathway is so effective that a systemic challenge of 20,000,000 CFUs are cleared within 1 day. In contrast, $Casp11^{-/-}$ mice succumb following challenge with as few as 100 CFUs (Aachoui et al. 2015). This is the strongest in vivo effect for caspase-11 against any infectious challenge (Maltez and Miao 2016). We estimate a 1,000,000-fold change in the effective LD_{100} (Table 1). It is remarkable that the strongest phenotypes for caspase-1 and caspase-11 in defense against infection arise from two ubiquitous environmental soil microbes, which only infect humans under extraordinary settings of host compromise such as patients with chronic granulomatous disease or after near drowning (Macher et al. 1982; Glass et al. 2006). This suggests that caspase-1/11 provide near permanent evolutionary victories over ubiquitous environmental pathogens, which are running the Red Queen's race against other hosts that lack caspase-1/11. (This hypothesis is de-

Table 1. Inflammasome and necroptosis survival studies

Pathogen	Mutation	Dose	Route	Time to death (median days)		% Survival		Δ lethal dose[a]	Micro-biota ctrl	Background	KO used in paper to get results shown in Time and Survival columns (and other KO with similar results)	References
				WT	KO	WT	KO					
BACTERIA versus INFLAMMASOMES												
Acinetobacter baumanii		5×10^8	in	1	2	88%	25%	>2	ns	C57BL/6	$Casp11^{-/-}$	Wang et al. 2017a
Bacillus anthracis Ames 35 —vegetative		10^5	sc	∞	4	100%	0%	>5	ns	C57BL/6NTac	"WT" are Nlrp1$^{S/S}$, KO are Nlrp1$^{R/R}$	Moayeri et al. 2010
—vegetative		10^5	iv	∞	4	100%	0%	>5	ns	C57BL/6NTac	"WT" are Nlrp1$^{S/S}$, KO are Nlrp1$^{R/R}$	Moayeri et al. 2010
—spores		10^7	sc	∞	4.5	100%	20%	>2	ns	C57BL/6NTac	"WT" are Nlrp1$^{S/S}$, KO are Nlrp1$^{R/R}$	Moayeri et al. 2010
—spores		2×10^7	sc	∞	1	100%	0%	>5	L	BALB/c	All Nlrp1S/S; $Casp1/11^{+/+}$ or $Casp1/11^{-/-}$	Moayeri et al. 2010
Bacillus anthracis Ames— spores		4×10^2	ip	3	3	50%	25%	>1	L	C57BL/6	"WT" are Nlrp1b^{129Tg}, KO are no Tg	Terra et al. 2010
Bacillus anthracis Sterne— spores		2.5×10^7	ip	∞	4	100%	0%	>5	L	C57BL/6	"WT" are Nlrp1b^{129Tg}, KO are no Tg	Terra et al. 2010
Bacillus anthracis Sterne— spores		10^6	ip	∞	3	100%	0%	>5	ns	C3H/HeJCr	$Casp1/11^{-/-}$	Kang et al. 2008
Burkholderia cepacia		10^6	ip	∞	∞	100%	100%	1	N	C57BL/6	$Casp1/11^{-/-}$	Maltez et al. 2015
Burkholderia cenocepacia		2×10^8	ip	∞	4	100%	17%	>2	ns	C57BL/6	$Mefv^{-/-}$	Aubert et al. 2016
Burkholderia pseudomallei		10^2	in	4	4	65%	0%	>5	ns	C57BL/6	$Casp1/11^{-/-}$ ($Nlrc4^{-/-}$, $Asc^{-/-}$)	Ceballos-Olvera et al. 2011
		25	in	∞	4	100%	0%	>5	ns	C57BL/6	$Casp1/11^{-/-}$ ($Nlrc4^{-/-}$, $Asc^{-/-}$)	Ceballos-Olvera et al. 2011
		10^2	in	4	5	65%	20%	>1	ns	C57BL/6	$Nlrp3^{-/-}$	Ceballos-Olvera et al. 2011

Cite this article as *Cold Spring Harb Perspect Biol* doi: 10.1101/cshperspect.a036459

Organism	Dose	Route							Strain	Genotype	Reference
Burkholderia thailandensis (unpassaged)	10^2	in	∞	2–3	100%	0%	>5	ns	C57BL/6	*Casp1/11*$^{-/-}$	Breitbach et al. 2009
	10^2	in	∞	3.5	100%	0%	>5	N	C57BL/6	*Casp1/11*$^{-/-}$ (*Casp11*$^{-/-}$)	Aachoui et al. 2013
	2×10^7	ip	∞	16	100%	13%	>20	N	C57BL/6	*Casp1/11*$^{-/-}$	Aachoui et al. 2013, 2015
	2×10^6	ip	∞	14	100%	20%	>20	N	C57BL/6	*Casp1/11*$^{-/-}$	Aachoui et al. 2013
	2×10^7	ip	∞	∞	100%	100%	1	N	C57BL/6	*Nlrc4*$^{-/-}$	Aachoui et al. 2013
	2×10^6	ip	∞	∞	100%	100%	1	N	C57BL/6	*Nlrc4*$^{-/-}$ *Asc*$^{-/-}$	Aachoui et al. 2013
	10^4	in	∞	5	100%	0%	>5	N	C57BL/6	*Casp1/11*$^{-/-}$	Aachoui et al. 2013
	10^4	in	∞	∞	100%	100%	1	N	C57BL/6	*Nlrc4*$^{-/-}$ *Asc*$^{-/-}$ (*Nlrc4*$^{-/-}$, *Asc*$^{-/-}$, *Nlrp3*$^{-/-}$)	Aachoui et al. 2013
Burkholderia thailandensis (one mouse passage)	2×10^7	ip	∞	1	100%	0%	>1,000,000	N	C57BL/6	*Casp1/11*$^{-/-}$ (*Casp11*$^{-/-}$)	Aachoui et al. 2013
	2×10^7	ip	∞	1	100%	0%	>1,000,000	N	C57BL/6	*Casp1/11*$^{-/-}$ (*Casp11*$^{-/-}$)	Aachoui et al. 2015; Maltez et al. 2015
	10^6	ip	∞	2	100%	0%	>1,000,000	N	C57BL/6	*Casp1/11*$^{-/-}$	Aachoui et al. 2015
	10^5	ip	∞	2	100%	0%	>1,000,000	N	C57BL/6	*Casp1/11*$^{-/-}$	Aachoui et al. 2015
	10^4	ip	∞	2	100%	0%	>1,000,000	N	C57BL/6	*Casp1/11*$^{-/-}$ (*Casp11*$^{-/-}$)	Aachoui et al. 2015; Maltez et al. 2015
	10^3	ip	∞	3	100%	0%	>1,000,000	N	C57BL/6	*Casp1/11*$^{-/-}$	Aachoui et al. 2015
	10^2	ip	∞	3	100%	0%	>1,000,000	N	C57BL/6	*Casp1/11*$^{-/-}$ (*Casp11*$^{-/-}$)	Aachoui et al. 2015
	2×10^7	ip	∞	2	100%	0%	>5	N	C57BL/6	*Nlrc4*$^{-/-}$ *Asc*$^{-/-}$ (*Nlrc4*$^{-/-}$ *Nlrp3*$^{-/-}$)	Aachoui et al. 2015
	10^4	ip	∞	∞	100%	100%	>5	N	C57BL/6	*Nlrc4*$^{-/-}$ *Asc*$^{-/-}$	Aachoui et al. 2015
	10^2	ip	∞	∞	100%	100%	>5	N	C57BL/6	*Nlrc4*$^{-/-}$ *Asc*$^{-/-}$	Aachoui et al. 2015
	2×10^7	ip	∞	∞	100%	100%	1	N	C57BL/6	*Nlrc4*$^{-/-}$ (*Nlrp3*$^{-/-}$, *Nlrc3*$^{-/-}$, *Nlrc5*$^{-/-}$, *Nlrp12*$^{-/-}$)	Aachoui et al. 2015
Chromobacterium violaceum	10^6	ip	∞	n.d.	100%	n.d.	>50,000	N	C57BL/6	*Casp1/11*$^{-/-}$ (*Nlrc4*$^{-/-}$, *Nlrc4*$^{-/-}$ *Asc*$^{-/-}$)	Maltez et al. 2015

Continued

Table 1. Continued

Pathogen	Mutation	Dose	Route	Time to death (median days)		% Survival		Δ lethal dose[a]	Micro-biota ctrl	Background	KO used in paper to get results shown in Time and Survival columns (and other KO with similar results)	References
				WT	KO	WT	KO					
		10^4	ip	∞	3	100%	0%	>50,000	N	C57BL/6	$Casp1/11^{-/-}$ ($Nlrc4^{-/-}$, $Nlrc4^{-/-} Asc^{-/-}$)	Maltez et al. 2015
		10^2	ip	∞	4	100%	0%	>50,000	N	C57BL/6	$Casp1/11^{-/-}$ ($Nlrc4^{-/-}$, $Nlrc4^{-/-} Asc^{-/-}$)	Maltez et al. 2015
Ehrlichia (*Ixodes ovatus ehrlichia*)		$\sim10^3$	ip	9	9	100%	100%	1	ns	C57BL/6	$Casp1/11^{-/-}$ ($Nlrp3^{-/-}$)	Yang et al. 2015
Escherichia coli O18:K1:H7		Range	ip	ns	ns	ns	ns	4	L	CeH/HeJ	$Casp1/11^{-/-}$	Joshi et al. 2002
Francisella tularensis subsp. *novicida*		1.5×10^5	sc	4	3	65%	0%	>2	L	C57BL/6	$Aim2^{-/-}$	Fernandes-Alnemri et al. 2010
		1.5×10^5	sc	6	4	30%	0%	"	N, (L)	C57BL/6	$Casp1/11^{-/-}$ ($Asc^{-/-}$)	Mariathasan et al. 2005
		1.5×10^5	sc	5	4	22%	0%	<1	L	C57BL/6	$Nlrc4^{-/-}$	Mariathasan et al. 2005
		5×10^3	sc	3	2.5	75%	0%	"	ns	ns	$Casp1/11^{-/-}$	Meunier et al. 2015
		7.5×10^4	sc	4	4	65%	0%	"	ns	ns	$Casp1/11^{-/-}$ ($Aim2^{-/-}$)	Man et al. 2015
Francisella turlarensis subsp. *Holartica*	LVS	2×10^4	in	7	7	0%	0%	1	ns	C57BL/6	$Nlrp3^{-/-}$	Duffy et al. 2016
Francisella philomiragia		10^6	ip	∞	∞	100%	100%	1	N	C57BL/6	$Casp1/11^{-/-}$	Maltez et al. 2015
Klebsiella pneumoniae		7.4×10^4	it	2.1	1.9	15%	0%	>1	ns	C57BL/6	$Nlrp3^{-/-}$ ($Asc^{-/-}$)	Willingham et al. 2009
		7.4×10^4	it	2	2.5	0%	0%	1	ns	C57BL/6	$Nlrc4^{-/-}$	Willingham et al. 2009
		10^3	in	5	5	75%	40%	>1	ns	C57BL/6	$Nlrc4^{-/-}$	Cai et al. 2012
		10^4	in	4	6	50%	15%	>1	ns	C57BL/6	$Nlrc4^{-/-}$	Cai et al. 2012

Cite this article as *Cold Spring Harb Perspect Biol* doi: 10.1101/cshperspect.a036459

Organism		Route	Dose							Strain	Genotype	Reference
Listeria monocytogenes		iv	10^6	5	3–4	35%–65%	0%		>2	ns	C57BL/6[J]	$Casp1/11^{-/-}$	Tsuji et al. 2004
		iv	10^5	5	6	0%	50%		<2	N	C57BL/6	$Casp11^{-/-}$ ($Nlrp6^{-/-}$)	Hara et al. 2018
Mycobacterium tuberculosis		in	250–350	200	148	0%	0%		1	ns	C57BL/6	$Asc^{-/-}$	Mcelvania-Tekippe et al. 2010
		in	250–350	170	170	0%	0%		1	ns	C57BL/6	$Casp1/11^{-/-}$ ($Nlrp3^{-/-}$, $Nlrc4^{-/-}$)	Mcelvania-Tekippe et al. 2010
		in	50–100	200	110	90%	0%		>2	ns	C57BL/6	$Casp1/11^{-/-}$ ($Asc^{-/-}$)	Mayer-Barber et al. 2010
Pseudomonas aeruginosa		it	10^6	∞	6	0%	0%		>5	L	C57BL/6	$Aim2^{-/-}$	Saiga et al. 2012
		in	2×10^7	1.5	1.7	20%	65%		<1	ns	C57BL/6	$Nlrc4^{-/-}$	Faure et al. 2014
	$\Delta popBD$	in	10^8	2.5	1.5	44%	0%		>1	ns	C57BL/6	$Nlrc4^{-/-}$	Faure et al. 2014
		it	7×10^5	3	8	90%	100%		<1	ns	C57BL/6	$Nlrc4^{-/-}$	Tolle et al. 2015
		in	3×10^7	∞	4	100%	0%		>5	ns	C57BL/6	$Nlrc4^{-/-}$	Iannitti et al. 2016
		in	3×10^7	∞	∞	100%	100%		1	ns	C57BL/6	$Nlrp3^{-/-}$	Iannitti et al. 2016
		in	3.5×10^5	2.0	1.8	12%	4%		>1	ns	C57BL/6	$Casp1/11^{-/-}$	Hughes et al. 2018
		in	3.5×10^5	2.0	2.9	12%	46%		<1	ns	C57BL/6	$Asc^{-/-}$	Hughes et al. 2018
Salmonella enterica sv. Typhimurium		ip	100	4.5	5.5	0%	0%		1	ns	C57BL/6	$Casp1/11^{-/-}$	Monack et al. 2000
		po	10^5	9	6	0%	0%		1	L	C57BL/6	$Casp1/11^{-/-}$ ($Asc^{-/-}$)	Lara-Tejero et al. 2006
		po	10^5	9	9	0%	0%		1	L	C57BL/6	$Nlrp3^{-/-}$	Lara-Tejero et al. 2006
		po	10^6	19	11	30%	20%		>1	L	$Nramp1^{+/+}$	$Casp1/11^{-/-}$	Lara-Tejero et al. 2006
		po	10^8	8	5.5	0%	0%		1	ns	C57BL/6	$Casp1/11^{-/-}$	Raupach et al. 2006
		po	2×10^{10}	∞	20	100%	22%		>2	ns	129SvJ	$Casp1/11^{-/-}$	Raupach et al. 2006
		ip	10^8	3	8	0%	0%		1	ns	C57BL/6	$Casp1/11^{-/-}$	Raupach et al. 2006
		po+s	$1\text{–}5 \times 10^5$	6	6	0%	0%		1	ns	C57BL/6	$Nlrc4^{-/-}$	Franchi et al. 2012
		ip	10^5	4	5	0%	0%		1	ns	C57BL/6	$Nlrc4^{-/-}$	Franchi et al. 2012

Continued

Table 1. Continued

Pathogen	Mutation	Dose	Route	Time to death (median days) WT	KO	% Survival WT	KO	Δ lethal dose[a]	Microbiota ctrl	Background	KO used in paper to get results shown in Time and Survival columns (and other KO with similar results)	References
		$1\text{–}5\times10^5$	po	8	7	0%	0%	1	ns	BALB/c	$Nlrc4^{-/-}$	Franchi et al. 2012
		$1\text{–}5\times10^5$	po+s	6	4	0%	0%	1	ns	BALB/c	$Nlrc4^{-/-}$	Franchi et al. 2012
		10^5	ip	4	5	0%	0%	1	ns	BALB/c	$Nlrc4^{-/-}$	Franchi et al. 2012
		10^3	ip	6	5	0%	0%	1	N	C57BL/6	$Nlrc4^{-/-}$ ($Naip2^{-/-}$, $Naip5^{-/-}$)	Zhao et al. 2016
Salmonella enterica sv. Typhimurium	$FliC^{ON}$	10^2	ip	∞	7	100%	0%	>5	N	C57BL/6	$Nlrc4^{-/-}$	Miao et al. 2010a
	$FliC^{ON}$	10^3	ip	∞	6	100%	0%	>5	ns	C57BL/6	$Nlrc4^{-/-}$ ($Naip5^{-/-}$)	Zhao et al. 2016
	$FliC^{ON}$	10^3	ip	∞	∞	100%	100%	1	ns	C57BL/6	$Naip2^{-/-}$	Zhao et al. 2016
	$PrgJ^{ON}$	10^3	ip	∞	5	100%	0%	>5	ns	C57BL/6	$Nlrc4^{-/-}$ ($Naip2^{-/-}$)	Zhao et al. 2016
	$PrgJ^{ON}$	10^3	ip	∞	∞	100%	100%	1	ns	C57BL/6	$Naip5^{-/-}$	Zhao et al. 2016
Shigella flexneri		2×10^8	in	0.8	1.9	75%	20%	>2	ns	C57BL/6	$Casp1/11^{-/-}$	Sansonetti et al. 2000
Staphylococcus aureus		10^4	ic	0.8	1	60%	25%	>1	ns	C57BL/6	$Casp1/11^{-/-}$ ($Asc^{-/-}$, $Aim2^{-/-}$)	Hanamsagar et al. 2014
		10^4	ic	5	5	95%	70%	>1	ns	C57BL/6	$Nlrp3^{-/-}$	Hanamsagar et al. 2014
Streptococcus agalactiae (group B)		10^8	ro	2	1	84%	40%	>1	ns	C57BL/6	$Casp1/11^{-/-}$	Kitur et al. 2016
		2×10^8	iv	0.5	1.5	0%	60%	>2	N	C57BL/6	$Casp11^{-/-}$ ($Nlrp6^{-/-}$)	Hara et al. 2018
		10^5	ip	∞	1	100%	40%	>2	ns	C57BL/6	$Casp1/11^{-/-}$ ($Asc^{-/-}$, $Nlrp3^{-/-}$)	Costa et al. 2012
Streptococcus pneumoniae		10^5	in	3	2.5	87%	90%	1	ns	C57BL/6	$Casp1/11^{-/-}$	Albiger et al. 2007
Vibrio vulnificus		10^3	in	6	5	90%	50%	>1	ns	C57BL/6	$Nlrp3^{-/-}$, $Aim2^{-/-}$	Rodriguez et al. 2019
		1.5×10^4	ip	∞	1	100%	60%	>1	ns	C57BL/6	$Casp1/11^{-/-}$ ($Asc^{-/-}$, $Nlrp3^{-/-}$)	Toma et al. 2010

Cite this article as *Cold Spring Harb Perspect Biol* doi: 10.1101/cshperspect.a036459

Continued

		Dose	Route	n	n	%	%	%	Fold	N	Strain	Genotype	Reference
Yersinia pestis		10^4	in	3	3	0%	0%	0%	1	N	C57BL/6	$Casp1/11^{-/-}$	Sivaraman et al. 2015
Yersinia pestis	ΔyopJM	2×10^2	sc	7	6	72%	25%		>1	ns	C57BL/6	$Casp1/11^{-/-}$	Ratner et al. 2016
Yersinia pestis		10^1	sc	7	5	0%	0%		1	ns	C57BL/6	$Nlrp12^{-/-}$	Vladimer et al. 2012
Yersinia pestis	pYtbLpxL	5×10^2	sc	8	8	100%	20%		>2	ns	C57BL/6	$Nlrp12^{-/-}$ $(Nlrp3^{-/-})$	Vladimer et al. 2012
Yersinia pseudotuberculosis		10^3	ip	6	4	0%	0%		1	ns	C57BL/6	$Casp1/11^{-/-}$	LaRock and Cookson 2012
Yersinia pseudotuberculosis	ΔyopM	10^3	ip	8	6	100%	0%		>5	ns	C57BL/6	$Casp1/11^{-/-}$	LaRock and Cookson 2012
Yersinia pseudotuberculosis		2×10^3	iv	7	7	0%	0%		1	ns	C57BL/6	$Mefv^{-/-}$	Chung et al. 2016
Yersinia pseudotuberculosis	ΔyopM	2×10^3	iv	13	7	89%	0%		>2	ns	C57BL/6	$Mefv^{-/-}$	Chung et al. 2016
Yersinia pseudotuberculosis	Comp. mutant	1×10^9	po	5	5	0%	0%		1	ns	C57BL/6	$Casp1/11^{-/-}$	Zheng et al. 2012
VIRUSES versus INFLAMMASOMES													
Encephalomyocarditis virus		2xLD50	ip	5	5	10%	15%		>1	N	C57BL/6	$Casp1/11^{-/-}$	Rajan et al. 2011
Influenza A virus		6×10^4	in	8	7	65%	40%		>1	ns	C57BL/6	$Casp1/11^{-/-}$ $(Asc^{-/-},$ $Nlrp3^{-/-})$	Allen et al. 2009
		6×10^4	in	8	7	68%	71%		1	ns	C57BL/6	$Nlrc4^{-/-}$	Allen et al. 2009
		8×10^3	in	11	10	65%	35%		>1	ns	C57BL/6	$Casp1/11^{-/-},$ $(Nlrp3^{-/-})$	Thomas et al. 2009
		8×10^3	in	12	9	83%	85		1	ns	C57BL/6	$Nlrc4^{-/-}$	Thomas et al. 2009
		10^1	in	8	8	100%	0%		>5	ns	C57BL/6	$Casp1/11^{-/-}$ $(Asc^{-/-})$	Ichinohe et al. 2009
		10^1	in	8	8	100%	100%		1	ns	C57BL/6	$Nlrp3^{-/-}$	Ichinohe et al. 2009
		1.25×10^2	in	8	8	100%	100%		1	ns	C57BL/6	$Nlrp3^{-/-}$ $(Aim2^{-/-})$	Rodriguez et al. 2019
Influenza A virus H7N9		5×10^4	in	9	9	18%	64%		<1	ns	C57BL/6	$Nlrp3^{-/-}$ $(Nlrp3^{-/-},$ $Asc^{-/-})$	Ren et al. 2017
Herpes simplex virus-1		1.5×10^6	scr	11	12.5	71%	73%		1	ns	C57BL/6	$Casp1/11^{-/-}$	Milora et al. 2014
Murine hepatitis virus A59		5×10^3	ic	10	10	90%	35%		>2	ns	C57BL/6	$Casp1/11^{-/-}$	Zalinger et al. 2017
Murine hepatitis virus-3		10^2	ip	4	5	0%	28%		<1	ns	C57BL/6	$Casp1/11^{-/-}$ $(Nlrp3^{-/-})$	Guo et al. 2015

Table 1. *Continued*

Pathogen	Mutation	Dose	Route	Time to death (median days) WT	KO	% Survival WT	KO	Δ lethal dose[a]	Micro-biota ctrl	Background	KO used in paper to get results shown in Time and Survival columns (and other KO with similar results)	References
Murine norovirus		10^2	po	6	7	0%	0%	1		C57BL/6 $Stat1^{-/-}$	$Nlrp3^{-/-}$ ($Gsdmd^{-/-}$)	Dubois et al. 2019
Vaccinia		3×10^6	iv	6	7	0%	0%	1	ns	C57BL/6	$Casp1^{-/-}$ ($Asc^{-/-}$)	Wang et al. 2017b
Vesicular stomatitis virus		2×10^5	in	7	7	40%	20%	>1	N	C57BL/6	$Casp1/11^{-/-}$	Rajan et al. 2011
		2×10^7	iv	5	5	0%	0%	1	ns	C57BL/6	$Casp1^{-/-}$ ($Asc^{-/-}$)	Wang et al. 2017b
West Nile virus		10^2	sc	11	11	81%	50%	>1	ns	C57BL/6	$Casp1/11^{-/-}$ ($Nlrp3^{-/-}$)	Ramos et al. 2012
		10^2	sc	11	12	75%	75%	1	ns	C57BL/6	$Nlrc4^{-/-}$	Ramos et al. 2012
		10^2	sc	11	10	42%	11%	>1	ns	C57BL/6	$Asc^{-/-}$	Kumar et al. 2013
		10^1	sc	11	11	89%	58%	>1	ns	C57BL/6	$Asc^{-/-}$	Kumar et al. 2013
FUNGI versus INFLAMMASOMES												
Aspergillus fumigatus		2×10^7	in	∞	∞	100%	100%	1	ns	C57BL/6	$Aim2^{-/-} Nlrp3^{-/-}$	Man et al. 2017
		2×10^7	in	∞	∞	100%	100%	1	ns	C57BL/6	$Nlrp3^{-/-}$, ($Nlrc4^{-/-}$)	Iannitti et al. 2016
Aspergillus fumigatus (imm. suppressed)		10^5	in	6	4	70%	0%	>2	ns	C57BL/6	$Casp1/11^{-/-}$ ($Aim2^{-/-}$ $Nlrp3^{-/-}$, $Asc^{-/-}$)	Man et al. 2017
		10^5	in	6	5	70%	50%	>1	ns	C57BL/6	$Nlrp3^{-/-}$	Man et al. 2017
		10^5	in	6	5.5	70%	67%	>1	ns	C57BL/6	$Aim2^{-/-}$	Man et al. 2017
		5×10^5	in	5	4	43%	0%	>1	ns	C57BL/6	$Casp1^{-/-}$ ($Casp1/11^{-/-}$)	Karki et al. 2015
		5×10^5	in	5	5	43%	0%	>1	ns	C57BL/6	$Casp11^{-/-}$	Karki et al. 2015
Candida albicans		10^5	iv	9	5	40%	0%	>1	ns	C57BL/6	$Nlrp3^{-/-}$	Gross et al. 2009
		2×10^5	iv	18	17	83%	50%	>1	ns	C57BL/6	$Casp1/11^{-/-}$ ($Asc^{-/-}$)	Van De Veerdonk et al. 2011
		5×10^6	ts	N/A	3	100%	60%	>1	ns	C57BL/6	$Nlrc4^{-/-}$	Tomalka et al. 2011
		ns	ts	∞	5	97%	60%	>1	ns	C57BL/6	$Casp1/11^{-/-}$ ($Asc^{-/-}$)	Hise et al. 2009
Paracoccidioides brasiliensis		2×10^6	iv	98	93	73%	0%	>2	ns	C57BL/6	$Casp1/11^{-/-}$	Ketelut-Carneiro et al. 2015

Cite this article as *Cold Spring Harb Perspect Biol* doi: 10.1101/cshperspect.a036459

	Dose	Route					Fold		Strain	Genotype	Reference
	2×10^6	iv	98	93	73%	30%	>1	ns	C57BL/6	$Asc^{-/-}$	Ketelut-Carneiro et al. 2015
	10^6	it	85	50	54%	0%	>2	ns	C57BL/6	$Casp1/11^{-/-}$	Feriotti et al. 2017
	10^6	it	85	120	54%	0%	>2	ns	C57BL/6	$Asc^{-/-}$ ($Nlrp3^{-/-}$)	Feriotti et al. 2017

PARASITES versus INFLAMMASOMES

	Dose	Route					Fold		Strain	Genotype	Reference
Plasmodium berghei sporozoites	10	iv	9	12	45%	73%	<1	ns	C57BL/6	$Nlrp3^{-/-}$	Dostert et al. 2009
Plasmodium berghei sporozoites	10^4	iv	6.5	6.5	0%	0%	1	ns	C57BL/6	$Casp1/11^{-/-}$	Kordes et al. 2011
Plasmodium berghei iRBCs	10^4	iv	6.5	6.5	0%	0%	1	ns	C57BL/6	$Casp1/11^{-/-}$	Kordes et al. 2011
Plasmodium berghei iRBCs	10^6	ip	6	6	0%	0%	1	ns	C57BL/6	$Casp1/11^{-/-}$ ($Asc^{-/-}$)	Reimer et al. 2010
	10^6	ip	6	8	0%	0%	1	ns	C57BL/6	$Nlrp3^{-/-}$	Reimer et al. 2010
Plasmodium chabaudi adami	5×10^4	ip	11	12	0%	0%	1	ns	C57BL/6	$Nlrp3^{-/-}$	Shio et al. 2009
Toxoplasma gondii tachyzoites	10^4	ip	10	9	75%	10%	>2	ns	C57BL/6	$Casp1/11^{-/-}$ ($Asc^{-/-}$, $Nlrp3^{-/-}$, $Nlrp1abc^{-/-}$) $Asc^{-/-}$	Gorfu et al. 2014
	10^3	ip	13	13	50%	30%	>1	ns	C57BL/6	$Asc^{-/-}$	Coutermarsh-Ott et al. 2016
	10^3	ip	13	∞	50%	100%	<2	ns	C57BL/6	$Casp11^{-/-}$	Coutermarsh-Ott et al. 2016
Trypanosoma cruzi trypomastigotes	10^3	ip	28	22	90%	0%	>2	ns	C57BL/6	$Asc^{-/-}$	Silva et al. 2013
	10^3	ip	28	28	90%	17%	>2	ns	C57BL/6	$Casp1/11^{-/-}$	Silva et al. 2013
	10^3	ip	28	28	90%	67%	>1	ns	C57BL/6	$Nlrp3^{-/-}$	Silva et al. 2013
	10^3	sc	22	28	85%	92%	1	ns	C57BL/6	$Casp1/11^{-/-}$ ($Nlrp3^{-/-}$)	Gonçalves et al. 2013
	10^3	ip	∞	26	100%	50%	>2	ns	C57BL/6	$Casp1/11^{-/-}$	Paroli et al. 2018
	10^3	ip	∞	∞	100%	100%	1	ns	C57BL/6	$Nlrp3^{-/-}$	Paroli et al. 2018

Table 1. *Continued*

Pathogen	Mutation	Dose	Route	Time to death (median days)		% Survival		Δ lethal dose[a]	Microbiota ctrl	Background	KO used in paper to get results shown in Time and Survival columns (and other KO with similar results)	References
				WT	KO	WT	KO					
BACTERIA versus NECROPTOSIS												
Salmonella enterica sv.		10^2	iv	11	10	0%	0%	1	ns	C57BL/6	*Ripk3*$^{-/-}$	Robinson et al. 2012
Typhimurium		10^8	po	10	7	0%	0%	1	ns	C57BL/6	*Mlkl*$^{-/-}$	Yu et al. 2018
Staphylococcus aureus		10^8	ro	2	4	84%	88%	1	ns	C57BL/6	*Ripk3*$^{-/-}$	Kitur et al. 2016
		10^8	ro	2	3	84%	44%	>1	ns	C57BL/6	*Mlkl*$^{-/-}$	Kitur et al. 2016
Yersinia pestis	*pEcLpxL*	5×10^2	sc	∞	11	100%	83%	>1	ns	C57BL/6	*Ripk3*$^{-/-}$ (BMT)	Weng et al. 2014
		2×10^5	fp	5	7	36%	70%	<1	ns	C57BL/6	*Ripk1*$^{D138N/D138N}$	Arifuzzaman et al. 2018
Yersinia pseudotuberculosis		9×10^7	po	>6	>6	100%	100%	1	ns	C57BL/6	*Ripk3*$^{-/-}$ (BMT)	Philip et al. 2014
VIRUSES versus NECROPTOSIS												
Influenza A		13 HAU	in	11	13	78%	62%	1	ns	C57BL/6	*Ripk3*$^{-/-}$	Rodrigue-Gervais et al. 2014
		4×10^3	in	11	11	71%	38%	>1	ns	C57BL/6	*Ripk3*$^{-/-}$	Nogusa et al. 2016
		4×10^3	in	10	11	74%	81%	1	ns	C57BL/6	*Mlkl*$^{-/-}$	Nogusa et al. 2016
		10^3	in	8.5	8	67%	100%	<1	ns	C57BL/6	*Zbp1*$^{-/-}$	Kuriakose et al. 2016
		10^3	in	∞	10	100%	22%	>2	L	C57BL/6	*Zbp1*$^{-/-}$	Thapa et al. 2016
Herpes simplex virus-1		2×10^7	ip	∞	6	100%	58%	>2	ns	ns	*Ripk3*$^{-/-}$	Wang et al. 2014
		10^7	iv	∞	5	100%	50%	>2	ns	ns	*Ripk3*$^{-/-}$	Huang et al. 2015

Cite this article as *Cold Spring Harb Perspect Biol* doi: 10.1101/cshperspect.a036459

				3.5	3.5								
Herpes simplex virus-1	ΔICP6	10^7	iv	3.5	3.5	38%	56%	1		ns	ns	$Ripk3^{-/-}$	Huang et al. 2015
Vaccinia		2×10^6	ip	9	7	86%	0%	>2		ns	C57BL/6	$Ripk3^{-/-}$	Cho et al. 2009
West Nile virus		10^2	sc	11	11	56%	0%	>2		ns	C57BL/6	$Ripk3^{-/-}$	Daniels et al. 2017
		10^2	sc	11	12	40%	46%	1		ns	C57BL/6	$Mlkl^{-/-}$	Daniels et al. 2017
		10^2	sc	11	11	56%	0%	>2		ns	C57BL/6	$Ripk1^{KD/KD}$	Daniels et al. 2017
		10^1	ic	9	7	0%	0%	0		ns	C57BL/6	$Ripk3^{-/-}$	Daniels et al. 2017
Zika virus		10^3	ic	7	6.5	73%	0%	>2		ns	C57BL/6N	$Ripk3^{-/-}$	Daniels et al. 2019
		10^3	ic	7.5	7	60%	0%	>2		ns	C57BL/6J	$Ripk1^{KD/KD}$	Daniels et al. 2019
		10^3	ic	7.5	7.5	60%	70%	1		ns	C57BL/6J	$Mlkl^{-/-}$	Daniels et al. 2019
		10^3	ic	6.5	7	82%	7%	>2		L	C57BL/6	$Ripk3^{fl/fl}$ CamKIIα-Cre$^+$	Daniels et al. 2019
		10^3	ic	7.5	8	80%	20%	>2		ns	C57BL/6J	$Zbp1^{-/-}$	Daniels et al. 2019

The investigators intend to continue to update this table and thus would appreciate notification of omissions or errors. This table includes experiments performed with mice defective for necroptosis genes, but does not include apoptosis knockout (KO) mice such as $Casp8^{-/-}$ mice.

BMT, KO mice instead are wild-type (WT) mice that received knockout bone marrow; iv, intravenous; sc, subcutaneous; scr, scratch to flank skin; po, per oral; po+s, per oral pretreated with streptomycin; in, intranasal; ro, retroorbital intravenous; it, intratracheal; ic, intracranial; fp, footpad; ts, tongue scratch; L, littermate; CO; cohoused; N; not littermate and not cohoused; ns, not stated.

[a]Estimated change in lethality between WT and KO mice. A difference in survival percentages of (1) 0% change was estimated to be onefold increase in infectious dose (or changes that were not statistically significant); (2) 1%–49% was estimated to be greater than onefold increase in the infectious dose; (3) 50%–99% was estimated to be greater than twofold; and (4) >100% was estimated to be greater than fivefold. Where multiple doses were examined, the high dose was divided by the low dose and then multiplied by the aforementioned estimator. As an example, in Ceballos-Olvera et al. (2011), WT mice had 100% lethality at 200 CFU but many survived at 100 CFU and $Casp1^{-/-}Casp11^{-/-}$ mice succumbed to 25 CFU; 200/25 = 8, and so the change in lethal dose is listed as greater than eightfold since $Casp1^{-/-}Casp11^{-/-}$ mice may succumb to even lower doses. Note that many infectious models have only examined one dose, so the change in lethal dose may turn out to be much larger than listed in this table once additional doses are tested. A final caveat is the all these manuscripts use mice on the C57BL/6 background, which carries MX1 and Nramp1 mutations that cause susceptibility to viral and bacterial infection, respectively; it is difficult to detect increases in the lethal dose when the lethal dose is already very low in WT C57BL/6 mice.

scribed in more depth in Maltez and Miao 2016 and Box 1 of Jorgensen et al. 2017.)

Some pathogens alter their LPS structure to evade detection by caspase-11 (Hagar et al. 2013; Kayagaki et al. 2013; Paciello et al. 2013; Yang et al. 2019). Additionally, some bacteria avoid recognition by caspase-11 by remaining within the vacuolar space (Aachoui et al. 2013). There are several examples of caspase-11 responding to pathogens that are typically considered to be vacuolar. In these cases, there must be either vacuolar leakage or rupture to introduce LPS into the cytosol. In such cases, caspase-11 can reduce vacuolar pathogen burdens (Lacey et al. 2018). Nevertheless, it is likely that caspase-11 evolved to combat cytosol-invasive pathogens.

LPS is an incredibly abundant ligand. Thus, LPS sensing risks detection of LPS that enters the cytosol aberrantly, as in endotoxic shock models (Hagar et al. 2013; Kayagaki et al. 2013). This risk is partially ameliorated by tight regulation of caspase-11; caspase-11 cannot be activated in the absence of interferon (IFN)-γ or type I IFN signaling (Broz et al. 2012; Rathinam et al. 2012; Aachoui et al. 2015).

The Evolutionary Race between *Yersinia* and the Pyrin Inflammasome

One of the most notable examples of the never-ending race between host and pathogen are pathogenic *Yersinia* spp. The causative agent of plague (*Yersinia pestis*) and the enteric pathogens (*Yersinia pseudotuberculosis* and *Yersinia enterocolitica*) all encode a T3SS enabling them to infect humans and other mammals. These T3SS inject effectors called *Yersinia* outer proteins (Yops) into the cytosol of host cells (Bliska et al. 2013). Given that *Yersinia* expresses a T3SS, it should be detected by the NLRC4 inflammasome. Additionally, the NLRP3 inflammasome may detect the YopB and YopD translocon proteins. However, YopK reportedly restricts both these detection events although through an unclear mechanism (Brodsky et al. 2010; Zwack et al. 2015). A more explicit example of *Yersinia* species running the Red Queen's race has recently been shown in *Yersinia*'s ability to evade the pyrin inflammasome.

During bacterial infection, neutrophils are the first immune cells recruited to the site of infection. Neutrophils phagocytose bacteria by activating Rho GTPases, such as RhoA, Rac1, and CDC42, which polymerize actin driving phagocytosis (Mao and Finnemann 2015). The primary virulence strategy of *Yersinia* spp. is to inhibit phagocytosis and replicate extracellularly (Ke et al. 2013). Once neutrophils arrive, they are the predominant cell type targeted by the *Yersinia* T3SS (Pechous et al. 2013). *Yersinia* prevents actin polymerization in part by using YopE and YopT. YopE facilitates GTPase hydrolysis, keeping RhoA in the inactive GDP-bound state, whereas YopT is a cysteine protease that cleaves the carboxyl terminus of Rho GTPases, releasing them from the membrane (Black and Bliska 2000; Shao et al. 2003). Thus, both YopE and YopT disable Rho GTPase activity to prevent phagocytosis (Grosdent et al. 2002). At this point, the pathogen is winning the evolutionary race (Fig. 2).

In an effort to combat antiphagocytic effectors, the host evolved a functional guard called pyrin (encoded by *Mefv*). Pyrin essentially monitors for normal biochemical function of the Rho GTPases; when the Rho GTPases are perturbed, pyrin activates caspase-1 (Xu et al. 2014). This guard function is accomplished not by direct interaction between pyrin and Rho GTPases, but via monitoring Rho effector protein kinases N1 and N2 (PKN1 and PKN2; also called PRK1/PRK2). Although the exact biochemical mechanism remains unclear, current literature suggests that RhoA activates PKN1/PKN2, activating their normal effector functions (Thumkeo et al. 2013) and also permitting PKN1/PKN2 to phosphorylate pyrin (Gao et al. 2016; Park et al. 2016). Thus, when RhoA is present and is capable of inducing phagocytosis, PKN1/PKN2 phosphorylate pyrin and repress inflammasome assembly. However, if RhoA is degraded or enzymatically modified by YopT and/or YopE, PKN1/PKN2 no longer phosphorylate pyrin. By an unclear mechanism, this lack of phosphorylation results in pyrin activation (Chung et al. 2016). Exactly how inflammasome activation at this point helps the host remains to be elucidated. Induction of pyroptosis should delete phagocytes that have

Cite this article as *Cold Spring Harb Perspect Biol* doi: 10.1101/cshperspect.a036459

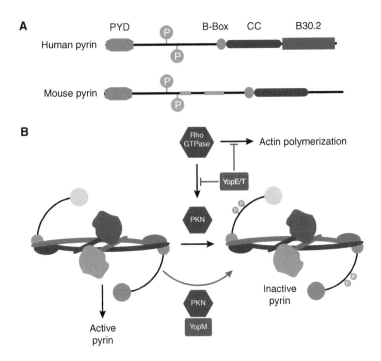

Figure 2. The pyrin inflammasome guards Rho GTPases. Pyrin was the first gene described to encode the pyrin domain (PYD) that is also found in many other inflammasomes. PYD signals through the PYD-CARD adaptor protein ASC, and thereby to caspase-1. In addition to the PYD, pyrin is a member of the tripartite motif (TRIM) family, and thus contains a B-Box, a coiled coil (CC), and a B30.2 domain (Kawai and Akira 2011; Weinert et al. 2015). (*A*) Domain structure of human and mouse pyrin. Slim red boxes indicate regions that are present in the human but absent in the mouse, or vice versa. (*B*) Rho GTPases are activated in response to signals for immune cell motility and/or phagocytosis. As part of their effector functions, Rho GTPases activate actin polymerization as well as effector protein kinases (including PKN1 and PKN2). Pyrin (shown in dimeric form) guards these Rho GTPase signaling pathways by mechanisms that are only partially understood. Successful Rho GTPase signaling will activate PKN1/PKN2, which phosphorylate and thereby inactivate pyrin. Thus, pyrin acts as a checkpoint to verify Rho GTPase function. If phosphorylation fails, pyrin becomes an active inflammasome. *Yersinia* encodes YopE and YopT that prevent phagocytosis by attacking Rho GTPases, but at the cost of preventing pyrin phosphorylation. Another *Yersinia* effector, YopM, reattaches PKN1/PKN2 to pyrin, driving pyrin phosphorylation. Thus, *Yersinia* successfully blocks phagocytosis with YopE/YopT while simultaneously defusing the pyrin guard with YopM.

been debilitated by the Yops. Additionally, the capase-1-driven release of IL-1β should recruit new neutrophils to the site of infection. In total, the pyrin inflammasome prevents the pathogen from creating a favorable environment for extracellular bacterial replication. Pyrin is also important for detection of a variety of pathogens that produce toxins or T3SS effectors that perturb Rho GTPases, indicating that pyrin is a general guard for actin cytoskeletal function (Xu et al. 2014; Aubert et al. 2016).

The evolutionary race was not finished; *Yersinia* spp. evolved an additional virulence factor in an effort to overthrow the host. *Yersinia* developed the effector YopM, which recruits PKN1/PKN2 to pyrin, driving pyrin phosphorylation even in the absence of RhoA activity (Mcdonald et al. 2003; Chung et al. 2016). Thus, WT mice are susceptible to WT *Yersinia* infection, but resistant to *yopM* mutants. *Yersinia yopM* mutants, on the other hand, are virulent in mice deficient in pyrin or caspase-1 (LaRock and Cookson 2012; Chung et al. 2016) (also see Table 1).

The basic function of YopM, to inhibit pyrin, is conserved among *Yersinia* spp.

Nevertheless, the Red Queen's race appears to be ongoing as YopM is polymorphic between *Yersinia* isolates and species (Chung et al. 2016). Similarly, there are several key differences between mouse and human pyrin, suggesting some YopM variants may work effectively against the pyrin of one host, but fail against others (Fig. 2). This extreme evolutionary pressure may push the host into a precarious position. Autosomal recessive mutations in pyrin result in the most common autoinflammatory disease worldwide, familial Mediterranean fever (FMF) (Özen 2018). These mutant pyrin proteins are not phosphorylated to the same degree by PKN1/PKN2 (Park et al. 2016). Thus, pyrin is more easily activated, resulting in high IL-1β levels and reoccurring inflammation and fever, but only in the homozygous state (Park et al. 2016). FMF is highly prevalent in parts of the Mediterranean region, thus populations that are native to the area may have evolved this gain of function if it confers a selective advantage in the heterozygous state to a pathogen. Given the impact of plague throughout history, it is tempting to speculate that *Y. pestis* provided the selective pressure to induce expansion of pyrin mutations.

NLRP1—One NLR to Guard Them All

Although it was the first inflammasome identified (Martinon et al. 2002), how the NLRP1 inflammasome detects pathogens has only recently been elucidated. NLRP1 in humans and mice contains a unique domain structure among NLRs (Fig. 3). Like NLRC4, NLRP1 has a CARD that directly activates caspase-1, yet it is located on the carboxyl terminus rather than the amino terminus. NLRP1 is also unique in that it includes a carboxy-terminal function-to-find domain (FIIND) directly upstream of its CARD. The FIIND undergoes constitutive autoproteolysis, but remains noncovalently associated with the rest of NLRP1. This noncovalent association is required for activation of NLRP1 (D'Osualdo et al. 2011; Finger et al. 2012; Frew et al. 2012). Furthermore, the minimal active component of NLRP1 is actually the cleaved FIIND-CARD fragment alone; the NOD and

LRR are surprisingly dispensable. Activation of this cleaved-FIIND-CARD inflammasome occurs only when the other domains of NLRP1 are degraded (Xu et al. 2018; Chui et al. 2019; Sandstrom et al. 2019). This amino-terminal degradation is thought to be induced, at least in part, by pathogen-mediated mechanisms (Frew et al. 2012).

NLRP1 was first recognized as a sensor of anthrax lethal toxin (LT), a metalloprotease produced by *Bacillus anthracis* (Boyden and Dietrich 2006). LT cleaves MAP kinase kinases (MAPKKs) to prevent innate immune signaling (Turk 2007). The host response to this in certain mouse strains seems to be to integrate a decoy domain into NLRP1b, such that now LT cleaves NLRP1b in addition to its primary target. This cleavage event occurs at the amino terminus and activates NLRP1 (Levinsohn et al. 2012). Proteolysis of amino-terminal residues induces a process known as the N-end rule (Lucas and Ciulli 2017), wherein amino acid residues at the new amino terminus are modified by cellular E3 ubiquitin ligases and targeted for degradation by the proteasome. Thus, when LT cleaves NLRP1b, the amino-terminal portion of NLRP1 is ubiquitinated and degraded by the proteasome. The cleaved-FIIND-CARD disassociates from the rest of NLRP1 during this process because of its noncovalent association (Squires et al. 2007; Wickliffe et al. 2008; Xu et al. 2018; Chui et al. 2019; Sandstrom et al. 2019), and is then free to oligomerize and form potent inflammasomes (Xu et al. 2018; Chui et al. 2019; Sandstrom et al. 2019).

FIIND or similar domains are found in other innate immune genes, including *CARD8* (present in humans but absent in mice) and *PIDD*, suggesting that activation by proteolysis may be useful for regulating other pathways (Tinel et al. 2007; D'Osualdo et al. 2011). Given this mechanism of NLRP1 activation, there needs to be a way to safely degrade old NLRP1, CARD8, and PIDD during cell homeostasis without activating caspase-1. In this regard, inhibitors of serine dipeptidases, Dpp, induce activation of NLRP1 and/or CARD8 without proteolytic cleavage, but the proteasome is still required (Okondo et al. 2017; 2018; Johnson et al. 2018; Zhong

Cite this article as *Cold Spring Harb Perspect Biol* doi: 10.1101/cshperspect.a036459

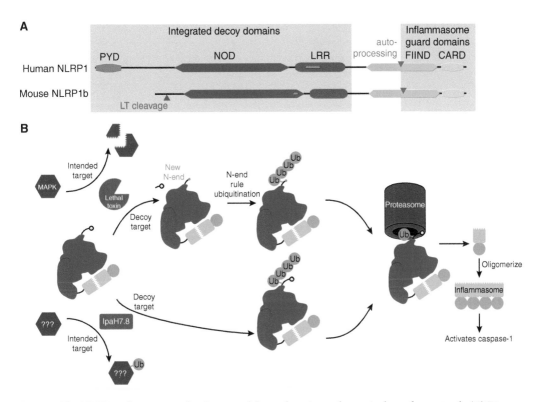

Figure 3. The NLRP1 inflammasome has integrated decoy domains to detect virulence factor attack. (*A*) Diagram of the human NLRP1 protein, and one of the NLRP1 proteins in mice (NLRP1b). Domains are indicated by colored boxes. Slim red boxes indicate regions that are present in the human, but absent in the mouse protein, or vice versa. Red triangle indicates the lethal toxin cleavage site in mouse NLRP1b. Purple triangle indicates the autoprocessing site within the FIIND domain. (*B*) Schematic of murine NLRP1b pathways. In one mode of activation (*top*), anthrax lethal toxin intends to cleave MAP kinase kinases; however, a decoy sequence in NLRP1b is also cleaved by lethal toxin. This exposes a new amino terminus (New N-end) in NLRP1b, which is detected by the N-end rule ubiquitinases that attach a ubiquitin to a nearby lysine residue. In a second mode (*bottom*) a bacterial effector such as IpaH7.8 intends to attack and ubiquitinate a different cellular protein, but also inadvertently ubiquitinates the decoy domains of NLRP1b. Ubiquitinated NLRP1b is then degraded by the proteasome, but when the precleaved FIIND domain approaches the proteasome, the carboxy-terminal FIIND fragment and attached CARD domain dissociate and are therefore not degraded. This dissociation results from the FIIND-CARD domains not being covalently attached to the rest of the protein. The liberated FIIND-CARD then oligomerizes to form an inflammasome, clustering the CARD domains that activate caspase-1. The FIIND-CARD inflammasome is different from typical inflammasomes in which the NOD domain drives oligomerization and clustering of an amino-terminal CARD domain. (From Lacey and Miao 2019; adapted, with permission, from the authors.)

et al. 2018). Thus, Dpp may participate in controlled NLRP1 and/or CARD8 degradation during normal homeostasis, thereby reducing the risk of NLRP1 autoactivation.

Nonproteolytic pathogen-mediated mechanisms may also target NLRP1 for degradation in the proteasome. *Shigella flexneri* has a T3SS that secretes the E3 ubiquitin ligase, IpaH7.8 (Rohde et al. 2007; Singer et al. 2008; Zhu et al. 2008),

which is detected by the NLRP1b inflammasome (Sandstrom et al. 2019). NLRP1 is ubiquitinated by IpaH7.8, which targets it for proteasomal degradation, and consequently releases the FIIND-CARD fragment that activates caspase-1. Ubiquitination and degradation of inhibitory proteins is also used in normal signaling pathways; for example, the NEMO/IKKα/IKKβ complex ubiquitinates IκB to drive its degradation, releasing nuclear factor (NF)-κB to translocate to the nucleus. In contrast, we speculate that NLRP1 acts as a decoy to detect ubiquitin attack on other NLRs. Thus, the NOD and LRR domains of NLRP1 would be defined as integrated decoy domains (Fig. 1). This idea would be supported if future research discovers that IpaH7.8 actually evolved to target a different NLR for degradation. The host may use these accessory integrated decoy domains to trick pathogens into targeting NLRP1 for degradation, creating a tripwire that activates the inflammasome.

The number of *NLRP1* genes and the domain structure of NLRP1 varies not only between species, but also within a species (Boyden and Dietrich 2006; D'Osualdo et al. 2011; Lilue et al. 2018). Unlike mouse macrophages, human macrophages exposed to IpaH7.8 do not activate caspase-1 (Muehlbauer et al. 2007). Thus, *Shigella* is winning the Red Queen's race against humans and losing against mice (Sharma et al. 2017). The versatility of NLRP1 is illustrated by the integration of an amino-terminal PYD in human NLRP1 (Moayeri et al. 2012), which perhaps evolved as an extra integrated decoy domain to lure virulence factors that attack PYDs. Pathogens that activate human NLRP1 have not yet been discovered, perhaps because NLRP1 eradicates them before disease development. If NLRs are the subject of attack by virulence factors, then NLRP1 could be considered the one NLR to guard them all.

Guarding Transcriptional Signaling with Apoptosis or Necroptosis

TNF is an important proinflammatory cytokine that is often targeted by bacterial virulence factors. Thus, it makes sense that TNF signaling is carefully guarded. TNF receptor signaling can have one of three outcomes: gene transcription, apoptosis, or necroptosis. The primary goal of TNF signaling is likely an NF-κB transcriptional response. When this transcriptional response is inhibited, guard functions detect the defective signaling pathway and, in response, trigger either apoptosis, pyroptosis, or necroptosis. These guard functions are likely the result of the ongoing evolutionary battle between host and

pathogen, dating back to the dawn of primitive multicellular organisms (Quistad et al. 2014).

In the Red Queen's race, a hypothetical pathogen would attempt to inhibit the TNF transcriptional response with a virulence factor to dampen the host immune response. One key signaling point within the TNF signaling pathway is RIPK1, because it gets modified with polyubiquitin chains that recruit the TAK1 and IKK complexes needed to activate NF-κB (Dondelinger et al. 2016a). To counter this attack, the host has evolved guards for TNF transcriptional signaling mediated by the RIPK1 axis. Once the TNF guards detect that a virulence factor has intercepted the TNF to NF-κB pathway, the "interpretation" is that the cell has been irrevocably compromised. Therefore, the conservative response, erring on the side of assuming the worst, is to kill the cell.

The first pathway that evolved to guard RIPK1 was a branch to apoptosis (Lamkanfi et al. 2002; Dondelinger et al. 2016b). The DD of RIPK1 normally recruits it to the DD of TNF receptor 1 or the DD in the adaptor TRADD (Dondelinger et al. 2016a). When RIPK1 modification is perturbed, its DD becomes exposed and is detected by the guard protein Fas-associated death domain (FADD) protein (the caspase-8 adaptor composed of a DD and a DED). FADD binds to RIPK1 via DD–DD interactions, and then recruits caspase-8 to initiate apoptosis. Thus, FADD and caspase-8 are guards for RIPK1 (Fig. 1). This is mimicked in vitro through the addition of IAP antagonists, inhibitors of TAK1, or inhibitors of IKK, which inhibit the TNF transcriptional response and induce apoptosis (Dondelinger et al. 2016a). Although enacting apoptosis and killing a cell is a dramatic response, it allows the host to remove cells that have been irrevocably compromised. Thus, the host prevents the pathogen from hijacking cells and the pathogen is denied the ability to create its preferred environment.

Not only does caspase-8 guard RIPK1, but RIPK1 can conversely guard caspase-8. Many pathogens have evolved virulence factors to inhibit TNF transcriptional signaling, while simultaneously inhibiting caspase-8-mediated apoptosis (Kaiser et al. 2013). In response to

Cite this article as *Cold Spring Harb Perspect Biol* doi: 10.1101/cshperspect.a036459

this double attack, the host appears to have evolved another guard pathway attached to RIPK1, monitoring for abnormalities in caspase-8. Now, instead of RIPK1 triggering apoptosis, it instead induces a completely different form of programmed cell death termed necroptosis (Dondelinger et al. 2016a). Evolution selected for the addition of a RHIM domain and a kinase domain to RIPK1. The RIPK1 kinase domain autophosphorylates RIPK1 to promote RHIM–RHIM interactions that recruit RIPK3. This induces RIPK3 oligomerization and autophosphorylation. Then RIPK3 phosphorylates the pseudokinase MLKL, leading to cell lysis. In this pathway, the RHIM and kinase domains of RIPK1 can be considered integrated guard domains (Fig. 1) that signal to the guard protein RIPK3.

In summary, for a pathogen to prevent TNF (or the similarly guarded TLR3 and TLR4)-induced activation of NF-κB while also maintaining cell viability, the pathogen must run through multiple steps of evolution. The pathogen must simultaneously achieve its primary goal of blocking NF-κB, secondarily it must also block apoptosis, and finally necroptosis. This network of pathways and guards probably makes it incredibly difficult for pathogens to readily add NF-κB-inhibiting virulence factors to their repertoire.

EPEC TRIPLE ATTACKS TRANSCRIPTION, APOPTOSIS, AND NECROPTOSIS

Enteropathogenic *E. coli* (EPEC) is a human-specific pathogen that causes diarrhea and lives in close association with host cells (Fig. 4). It uses the locus of enterocyte effacement (LEE) T3SS to reprogram intestinal epithelial cells (IECs), permitting extracellular adherence of EPEC in the intestinal lumen. The translocated effectors cause the IEC microvilli to efface, and then induce the formation of a dense actin network, creating a pedestal on which EPEC closely adheres. The bacterium likely evolved this strategy to gain first access to oxygen and nutrients that diffuse across the IEC, gaining a replication advantage over luminal commensals (Lopez et al. 2016). EPEC must reprogram the IEC,

while also preventing the IEC from noticing it is compromised. Despite the fact that the EPEC rod protein is detected by the NAIP/NLRC4 inflammasome, NLRC4 is inefficient at detecting the live bacteria (Miao et al. 2010b). Detection by NLRC4 in IECs would otherwise trigger immediate extrusion (Rauch et al. 2017), which would be devastating to the virulence strategy of EPEC. We therefore speculate that EPEC has an undiscovered strategy to evade NLRC4.

There is mounting evidence that EPEC inhibits the transcriptional responses of the IEC to which it is attached, effectively shutting down the TNF signaling pathway (Pearson and Hartland 2014). EPEC attacks TAB2 and TAB3, which are adaptors for TAK1 downstream from RIPK1. EPEC accomplishes this by injecting the T3SS effector NleE, which methylates cysteines in the zinc finger domains of TAB2 and TAB3, thus inhibiting NF-κB responses (Zhang et al. 2011). EPEC also has two metalloproteases, NleC and NleD, which degrade key proteins in the transcriptional response. NleC degrades NF-κB by cleaving its p65 (RelA) subunit (Yen et al. 2010; Baruch et al. 2011; Mühlen et al. 2011; Pearson et al. 2011). Meanwhile, NleD degrades JNK and p38, which are activators of the AP1 transcription factor (Baruch et al. 2011). Another effector, NleL, also attacks JNK, but by ubiquitinating it, hence blocking JNK phosphorylation and activation (Sheng et al. 2017). Finally, NleH1 and NleH2 inhibit NF-κB-driven gene expression in part by NleH1, preventing IKKβ phosphorylation of RPS3, a specifier subunit of certain NF-κB complexes (Gao et al. 2009; Royan et al. 2010; Wan et al. 2011). The fact that EPEC translocates a plethora of T3SS effectors that redundantly attack NF-κB and AP1 indicate that preventing transcriptional responses is very important to its virulence strategy. However, attacks on these transcriptional signaling pathways theoretically should be detected by their guards.

EPEC must inhibit the caspase-8 apoptosis guard pathway to maintain its adherent niche on the IEC. The T3SS effector NleB attacks DD-containing proteins by adding an *N*-acetylglucosamine (called GlcNAcylation), with a preference for attacking the caspase-8 adaptor FADD

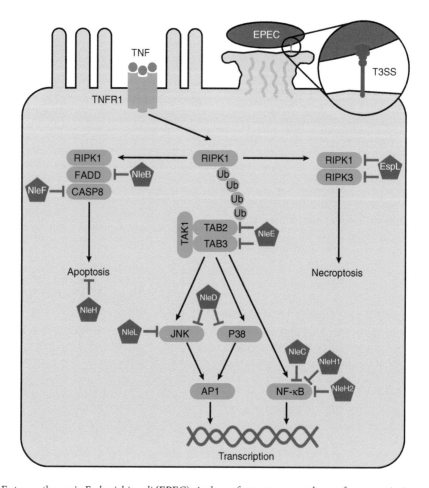

Figure 4. Enteropathogenic *Escherichia coli* (EPEC) virulence factors target pathways for transcription, apoptosis, and necroptosis. Tumor necrosis factor (TNF) receptor signaling can trigger one of three responses. The cell will first try to initiate transcriptional signaling pathways. RIPK1 acts as a scaffold for polyubiquitin chains, on which assemble TAB2 and TAB3 in a complex with TAK1. Downstream from TAK1, JNK and p38 signal to the transcription factor AP1, and other complexes signal to nuclear factor (NF)-κB. Transcription can be blocked by EPEC effectors: NleE inhibits TAB2 and TAB3, NleD obstructs JNK and p38, and NleC and NleH1/2 block NF-κB. When transcription pathways are hindered, guards of RIPK1 initiate apoptosis. RIPK1 death domain (DD) becomes exposed allowing homotypic interactions with the DD of the adaptor FADD, which in turn recruits caspase-8 to initiate apoptosis. EPEC has additional virulence strategies to inhibit the apoptotic guard pathway: NleB and NleF attack FADD and caspase-8, respectively. Finally, the host cell has a third guard pathway in which RIPK1 uses its integrated guard RHIM and kinase domains to activate the guard, RIPK3. RIPK3 signaling then induces necroptosis of the host cell. Similar to transcription and apoptosis, EPEC has a virulence factor, EspL, to inhibit signaling to necroptosis.

(Li et al. 2013; Pearson et al. 2013; Scott et al. 2017). Thus, NleB prevents guard signaling from RIPK1 to caspase-8 and prevents apoptosis. Another effector, NleF, also attacks caspase-8 (as well as some other caspases) by direct binding to inhibit its catalytic activity (Blasche et al. 2013). In addition to attack on caspase-8

apoptotic guard function, EPEC also attacks other aspects of apoptosis—NleH binds Bax inhibitor-1 (BI-1) to prevent cell-intrinsic initiation of apoptosis (Hemrajani et al. 2010).

Lastly, EPEC also attacks the necroptotic guard pathway. EspL is a cysteine protease that cleaves the RHIM domains in RIPK1, RIPK3, and other RHIM-containing proteins, thereby preventing oligomerization and activation of RIPK3 (Pearson et al. 2017). This should prevent the IEC necroptosis guard pathway from activating in response to simultaneous attacks upon transcriptional signaling and apoptosis. The fact that EPEC delivers multiple effectors to prevent apoptosis and necroptosis illustrates that maintaining the viability of the IEC is important to the virulence strategy of EPEC, allowing it to replicate in its adherent niche.

The compendium of EPEC effectors that attack NF-κB signaling and simultaneously attack the two guard pathways to apoptosis and necroptosis are an excellent illustration of the Red Queen's race between pathogen and host. We expect that other bacterial pathogens that attack NF-κB signaling will similarly need to simultaneously block apoptosis and necroptosis. This should be particularly important for pathogens such as EPEC that replicate in intimate contact with the host cell it has reprogramed. Similarly, intracellular pathogens like *Salmonella* replicate inside a single host cell, and thus must also keep that host cell alive. *S.* Typhimurium attacks NF-κB signaling, which should trigger the apoptotic and necroptotic guard pathways. However, *S.* Typhimurium encodes SseK proteins that are similar to the EPEC apoptosis inhibitor NleB. Although homologs of the EPEC EspL necroptosis inhibitor are not present in the commonly used strains of *S.* Typhimurium, we predict that *S.* Typhimurium inhibits necroptosis by using undiscovered effectors. *Yersinia* species also attack transcriptional responses. For example, YopJ attacks TAK1, but at the cost of also triggering the apoptotic guard functions (Paquette et al. 2012). In contrast to EPEC and *S.* Typhimurium, the host cell does not need to remain viable for *Yersinia* to replicate. Thus, it remains to be determined whether *Yersinia* actually benefits from the apoptotic guard pathway triggered by its inhibition of the transcriptional response. Given that all pathogens are running the Red Queen's race against the host, it seems likely that many other pathogens will have virulence strategies that have equal complexity to those illustrated by the EPEC T3SS effectors.

CONCLUDING REMARKS

Evolution is limited by organism replication rates, thus it is easy to assume multicellular organisms are at a disadvantage as many have life span that are significantly longer than bacteria, which allows them less time to mutate and evolve between generations. The solution to being faced with a pathogen that step-for-step keeps up with the continuing evolution of the innate immune system was to generate an arm of the immune system that can evolve at a rate faster than bacterial evolution. The adaptive immune system accomplishes this feat—B cells and T cells evolve new antibody and T-cell receptors within one week. Ultimately, almost all infections in which pathogens are running the Red Queen's race against innate immunity are cleared by the adaptive immune system.

REFERENCES

Aachoui Y, Leaf IA, Hagar JA, Fontana MF, Campos CG, Zak DE, Tan MH, Cotter PA, Vance RE, Aderem A, et al. 2013. Caspase-11 protects against bacteria that escape the vacuole. *Science* **339:** 975–978. doi:10.1126/science.1230751

Aachoui Y, Kajiwara Y, Leaf IA, Mao D, Ting JPY, Coers J, Aderem A, Buxbaum JD, Miao EA. 2015. Canonical inflammasomes drive IFN-γ to prime caspase-11 in defense against a cytosol-invasive bacterium. *Cell Host Microbe* **18:** 320–332. doi:10.1016/j.chom.2015.07.016

Albiger B, Dahlberg S, Sandgren A, Wartha F, Beiter K, Katsuragi H, Akira S, Normark S, Henriques-Normark B. 2007. Toll-like receptor 9 acts at an early stage in host defence against pneumococcal infection. *Cell Microbiol* **9:** 633–644. doi:10.1111/j.1462-5822.2006.00814.x

Allen IC. 2009. The NLRP3 inflammasome mediates in vivo innate immunity to influenza A virus through recognition of viral RNA. *Immunity* **30:** 556–565. doi:10.1016/j.immuni.2009.02.005

Arifuzzaman M, Ang WXG, Choi HW, Nilles ML, St John AL, Abraham SN. 2018. Necroptosis of infiltrated macrophages drives *Yersinia pestis* dispersal within buboes. *JCI Insight* **3:** 122188. doi:10.1172/jci.insight.122188

Aubert DF, Xu H, Yang J, Shi X, Gao W, Li L, Bisaro F, Chen S, Valvano MA, Shao F. 2016. A *Burkholderia* type VI effector deamidates Rho GTPases to activate the pyrin inflammasome and trigger inflammation. *Cell Host Microbe* **19:** 664–674. doi:10.1016/j.chom.2016.04.004

Baruch K, Gur-Arie L, Nadler C, Koby S, Yerushalmi G, Ben-Neriah Y, Yogev O, Shaulian E, Guttman C, Zarivach R, et al. 2011. Metalloprotease type III effectors that specifically cleave JNK and NF-κB. *EMBO J* **30:** 221–231. doi:10.1038/emboj.2010.297

Batista JH, da Silva Neto JF. 2017. *Chromobacterium violaceum* pathogenicity: updates and insights from genome sequencing of novel *Chromobacterium* species. *Front Microbiol* **8**: 2213. doi:10.3389/fmicb.2017.02213

Black DS, Bliska JB. 2000. The RhoGAP activity of the *Yersinia pseudotuberculosis* cytotoxin YopE is required for antiphagocytic function and virulence. *Mol Microbiol* **37**: 515–527. doi:10.1046/j.1365-2958.2000.02021.x

Blasche S, Mörtl M, Steuber H, Siszler G, Nisa S, Schwarz F, Lavrik I, Gronewold TMA, Maskos K, Donnenberg MS, et al. 2013. The *E. coli* effector protein NleF is a caspase inhibitor. *PLoS ONE* **8**: e58937. doi:10.1371/journal.pone.0058937

Bliska JB, Wang X, Viboud GI, Brodsky IE. 2013. Modulation of innate immune responses by *Yersinia* type III secretion system translocators and effectors. *Cell Microbiol* **15**: 1622–1631. doi:10.1111/cmi.12164

Boyden ED, Dietrich WF. 2006. Nalp1b controls mouse macrophage susceptibility to anthrax lethal toxin. *Nat Genet* **38**: 240–244. doi:10.1038/ng1724

Breitbach K, Sun GW, Köhler J, Eske K, Wongprompitak P, Tan G, Liu Y, Gan YH, Steinmetz I. 2009. Caspase-1 mediates resistance in murine melioidosis. **77**: 1589–1595.

Brodsky IE, Palm NW, Sadanand S, Ryndak MB, Sutterwala FS, Flavell RA, Bliska JB, Medzhitov R. 2010. A *Yersinia* effector protein promotes virulence by preventing inflammasome recognition of the type III secretion system. *Cell Host Microbe* **7**: 376–387. doi:10.1016/j.chom.2010.04.009

Broz P, Ruby T, Belhocine K, Bouley DM, Kayagaki N, Dixit VM, Monack DM. 2012. Caspase-11 increases susceptibility to *Salmonella* infection in the absence of caspase-1. *Nature* **490**: 288–291. doi:10.1038/nature11419

Cai S, Batra S, Wakamatsu N, Pacher P, Jeyaseelan S. 2012. NLRC4 inflammasome-mediated production of IL-1β modulates mucosal immunity in the lung against Gram-negative bacterial infection. *J Immunol* **188**: 5623–5635. doi:10.4049/jimmunol.1200195

Ceballos-Olvera I, Sahoo M, Miller MA, Del Barrio L, Re F. 2011. Inflammasome-dependent pyroptosis and IL-18 protect against *Burkholderia pseudomallei* lung infection while IL-1β is deleterious. *PLoS Pathog* **7**: e1002452. doi:10.1371/journal.ppat.1002452

Cho YS, Challa S, Moquin D, Genga R, Ray TD, Guildford M, Chan FK. 2009. Phosphorylation-driven assembly of the RIP1-RIP3 complex regulates programmed necrosis and virus-induced inflammation. *Cell* **137**: 1112–1123. doi:10.1016/j.cell.2009.05.037

Chui AJ, Okondo MC, Rao SD, Gai K, Griswold AR, Johnson DC, Ball DP, Taabazuing CY, Orth EL, Vittimberga BA, et al. 2019. N-terminal degradation activates the NLRP1B inflammasome. *Science* **364**: 82–85. doi:10.1126/science.aau1208

Chung LK, Park YH, Zheng Y, Brodsky IE, Hearing P, Kastner DL, Chae JJ, Bliska JB. 2016. The *Yersinia* virulence factor YopM hijacks host kinases to inhibit type III effector-triggered activation of the pyrin inflammasome. *Cell Host Microbe* **20**: 296–306. doi:10.1016/j.chom.2016.07.018

Costa A, Gupta R, Signorino G, Malara A, Cardile F, Biondo C, Midiri A, Galbo R, Trieu-Cuot P, Papasergi S, et al. 2012. Activation of the NLRP3 inflammasome by group B

Streptococci. *J Immunol* **188**: 1953–1960. doi:10.4049/jimmunol.1102543

Coutermarsh-Ott SL, Doran JT, Campbell C, Williams TM, Lindsay DS, Allen IC. 2016. Caspase-11 modulates inflammation and attenuates *Toxoplasma gondii* pathogenesis. *Mediators Inflamm* **2016**: 9848263. doi:10.1155/2016/9848263

Daniels BP, Snyder AG, Olsen TM, Orozco S, Oguin TH III, Tait SWG, Martinez J, Gale M Jr, Loo YM, Oberst A. 2017. RIPK3 restricts viral pathogenesis via cell death-independent neuroinflammation. *Cell* **169**: 301–313.e11. doi:10.1016/j.cell.2017.03.011

Daniels BP, Kofman SB, Smith JR, Norris GT, Snyder AG, Kolb JP, Gao X, Locasale JW, Martinez J, Gale M Jr, et al. 2019. The nucleotide sensor ZBP1 and kinase RIPK3 induce the enzyme IRG1 to promote an antiviral metabolic state in neurons. *Immunity* **50**: 64–76.e4. doi:10.1016/j.immuni.2018.11.017

Dondelinger Y, Darding M, Bertrand MJM, Walczak H. 2016a. Poly-ubiquitination in TNFR1-mediated necroptosis. *Cell Mol Life Sci* **73**: 2165–2176. doi:10.1007/s00018-016-2191-4

Dondelinger Y, Hulpiau P, Saeys Y, Bertrand MJM, Vandenabeele P. 2016b. An evolutionary perspective on the necroptotic pathway. *Trends Cell Biol* **26**: 721–732. doi:10.1016/j.tcb.2016.06.004

Dostert C, Guarda G, Romero JF, Menu P, Gross O, Tardivel A, Suva ML, Stehle JC, Kopf M, Stamenkovic I, et al. 2009. Malarial hemozoin is a Nalp3 inflammasome activating danger signal. *PLoS ONE* **4**: e6510. doi:10.1371/journal.pone.0006510

D'Osualdo A, Weichenberger CX, Wagner RN, Godzik A, Wooley J, Reed JC. 2011. CARD8 and NLRP1 undergo autoproteolytic processing through a ZU5-like domain. *PLoS ONE* **6**: e27396. doi:10.1371/journal.pone.0027396

Dubois H, Sorgeloos F, Sarvestani ST, Martens L, Saeys Y, Mackenzie JM, Lamkanfi M, van Loo G, Goodfellow I, Wullaert A. 2019. Nlrp3 inflammasome activation and Gasdermin D-driven pyroptosis are immunopathogenic upon gastrointestinal norovirus infection. *PLoS Pathog* **15**: e1007709. doi:10.1371/journal.ppat.1007709

Duffy EB, Periasamy S, Hunt D, Drake JR, Harton JA. 2016. FcγR mediates TLR2- and Syk-dependent NLRP3 inflammasome activation by inactivated *Francisella tularensis* LVS immune complexes. *J Leukoc Biol* **100**: 1335–1347. doi:10.1189/jlb.2A1215-555RR

Faure E, Mear JB, Faure K, Normand S, Couturier-Maillard A, Grandjean T, Balloy V, Ryffel B, Dessein R, Chignard M, et al. 2014. *Pseudomonas aeruginosa* type-3 secretion system host defense by exploiting the NLRC4-coupled inflammasome. *Am J Respir Crit Care Med* **189**: 799–811. doi:10.1164/rccm.201307-1358OC

Feriotti C, de Araújo EF, Loures FV, da Costa TA, Galdino NAL, Zamboni DS, Calich VLG. 2017. NOD-like receptor P3 inflammasome controls protective Th1/Th17 immunity against pulmonary paracoccidioidomycosis. *Front Immunol* **8**: 786. doi:10.3389/fimmu.2017.00786

Fernandes-Alnemri T, Yu JW, Juliana C, Solorzano L, Kang S, Wu J, Datta P, McCormick M, Huang L, McDermott E, et al. 2010. The AIM2 inflammasome is critical for innate immunity to *Francisella tularensis*. *Nat Immunol* **11**: 385–393. doi:10.1038/ni.1859

Cite this article as *Cold Spring Harb Perspect Biol* doi: 10.1101/cshperspect.a036459

Finger JN, Lich JD, Dare LC, Cook MN, Brown KK, Durai-swami C, Bertin J, Bertin JJ, Gough PJ. 2012. Autolytic proteolysis within the function to find domain (FIIND) is required for NLRP1 inflammasome activity. *J Biol Chem* **287:** 25030–25037. doi:10.1074/jbc.M112.378323

Franchi L, Kamada N, Nakamura Y, Burberry A, Kuffa P, Suzuki S, Shaw MH, Kim YG, Núñez G. 2012. NLRC4-driven production of IL-1β discriminates between path-ogenic and commensal bacteria and promotes host intes-tinal defense. *Nat Immunol* **13:** 449–456. doi:10.1038/ni .2263

Frew BC, Joag VR, Mogridge J. 2012. Proteolytic processing of nlrp1b is required for inflammasome activity. *PLoS Pathog* **8:** e1002659. doi:10.1371/journal.ppat.1002659

Gao X, Wan F, Mateo K, Callegari E, Wang D, Deng W, Puente J, Li F, Chaussee MS, Finlay BB, et al. 2009. Bac-terial effector binding to ribosomal protein S3 subverts NF-κB function. *PLoS Pathog* **5:** e1000708. doi:10.1371/ journal.ppat.1000708

Gao W, Yang J, Liu W, Wang Y, Shao F. 2016. Site-specific phosphorylation and microtubule dynamics control pyrin inflammasome activation. *Proc Natl Acad Sci* **113:** E4857–E4866. doi:10.1073/pnas.1601700113

Glass MB, Gee JE, Steigerwalt AG, Cavuoti D, Barton T, Hardy RD, Godoy D, Spratt BG, Clark TA, Wilkins PP. 2006. Pneumonia and septicemia caused by *Burkholderia thailandensis* in the United States. *J Clin Microbiol* **44:** 4601–4604. doi:10.1128/JCM.01585-06

Gonçalves VM, Matteucci KC, Buzzo CL, Miollo BH, Fer-rante D, Torrecilhas AC, Rodrigues MM, Alvarez JM, Bortoluci KR. 2013. NLRP3 controls *Trypanosoma cruzi* infection through a caspase-1-dependent IL-1R-indepen-dent NO production. *PLoS Negl Trop Dis* **7:** e2469. doi:10 .1371/journal.pntd.0002469

Gorfu G, Cirelli KM, Melo MB, Mayer-Barber K, Crown D, Koller BH, Masters S, Sher A, Leppla SH, Moayeri M, et al. 2014. Dual role for inflammasome sensors NLRP1 and NLRP3 in murine resistance to *Toxoplasma gondii*. *mBio* **5:** e01117. doi:10.1128/mBio.01117-13

Grosdent N, Maridonneau-Parini I, Sory MP, Cornelis GR. 2002. Role of Yops and adhesins in resistance of *Yersinia enterocolitica* to phagocytosis. *Infect Immun* **70:** 4165–4176. doi:10.1128/IAI.70.8.4165-4176.2002

Gross O, Poeck H, Bscheider M, Dostert C, Hannesschläger N, Endres S, Hartmann G, Tardivel A, Schweighoffer E, Tybulewicz V, et al. 2009. Syk kinase signalling couples to the Nlrp3 inflammasome for anti-fungal host defence. *Nature* **459:** 433–436. doi:10.1038/nature07965

Guo S, Yang C, Diao B, Huang X, Jin M, Chen L, Yan W, Ning Q, Zheng L, Wu Y, et al. 2015. The NLRP3 inflam-masome and IL-1β accelerate immunologically mediated pathology in experimental viral fulminant hepatitis. *PLoS Pathog* **11:** e1005155. doi:10.1371/journal.ppat.1005155

Hagar JA, Powell DA, Aachoui Y, Ernst RK, Miao EA. 2013. Cytoplasmic LPS activates caspase-11: implications in TLR4-independent endotoxic shock. *Science* **341:** 1250–1253. doi:10.1126/science.1240988

Hanamsagar R, Aldrich A, Kielian T. 2014. Critical role for the AIM2 inflammasome during acute CNS bacterial infection. *J Neurochem* **129:** 704–711. doi:10.1111/jnc .12669

Hara H, Seregin SS, Yang D, Fukase K, Chamaillard M, Alnemri ES, Inohara N, Chen GY, Núñez G. 2018. The NLRP6 inflammasome recognizes lipoteichoic acid and regulates Gram-positive pathogen infection. *Cell* **175:** 1651–1664.e14. doi:10.1016/j.cell.2018.09.047

Hemrajani C, Berger CN, Robinson KS, Marches O, Mous-nier A, Frankel G. 2010. NleH effectors interact with Bax inhibitor-1 to block apoptosis during enteropathogenic *Escherichia coli* infection. *Proc Natl Acad Sci* **107:** 3129–3134. doi:10.1073/pnas.0911609106

Hise AG, Tomalka J, Ganesan S, Patel K, Hall BA, Brown GD, Fitzgerald KA. 2009. An essential role for the NLRP3 inflammasome in host defense against the human fungal pathogen *Candida albicans*. *Cell Host Microbe* **5:** 487–497. doi:10.1016/j.chom.2009.05.002

Hu Z, Zhou Q, Zhang C, Fan S, Cheng W, Zhao Y, Shao F, Wang HW, Sui SF, Chai J. 2015. Structural and biochem-ical basis for induced self-propagation of NLRC4. *Science* **350:** 399–404. doi:10.1126/science.aac5489

Huang Z, Wu SQ, Liang Y, Zhou X, Chen W, Li L, Wu J, Zhuang Q, Chen C, Li J, et al. 2015. RIP1/RIP3 binding to HSV-1 ICP6 initiates necroptosis to restrict virus propa-gation in mice. *Cell Host Microbe* **17:** 229–242. doi:10 .1016/j.chom.2015.01.002

Hughes AJ, Knoten CA, Morris AR, Hauser AR. 2018. ASC acts in a caspase-1-independent manner to worsen acute pneumonia caused by *Pseudomonas aeruginosa*. *J Med Microbiol* **67:** 1168–1180. doi:10.1099/jmm.0.000782

Iannitti RG, Napolioni V, Oikonomou V, De Luca A, Galosi C, Pariano M, Massi-Benedetti C, Borghi M, Puccetti P, Lucidi V, et al. 2016. IL-1 receptor antagonist ameliorates inflammasome-dependent inflammation in murine and human cystic fibrosis. *Nat Commun* **7:** 10791. doi:10 .1038/ncomms10791

Ichinohe T, Lee HK, Ogura Y, Flavell R, Iwasaki A. 2009. Inflammasome recognition of influenza virus is essential for adaptive immune responses. *J Exp Med* **206:** 79–87. doi:10.1084/jem.20081667

Johnson DC, Taabazuing CY, Okondo MC, Chui AJ, Rao SD, Brown FC, Reed C, Peguero E, de Stanchina E, Kentsis A, et al. 2018. DPP8/DPP9 inhibitor-induced pyroptosis for treatment of acute myeloid leukemia. *Nat Med* **24:** 1151–1156. doi:10.1038/s41591-018-0082-y

Jones JDG, Vance RE, Dangl JL. 2016. Intracellular innate immune surveillance devices in plants and animals. *Sci-ence* **354:** aaf6395–aaf6395. doi:10.1126/science.aaf6395

Jorgensen I, Lopez JP, Laufer SA, Miao EA. 2016a. IL-1β, IL-18, and eicosanoids promote neutrophil recruitment to pore-induced intracellular traps following pyroptosis. *Eur J Immunol* **46:** 2761–2766. doi:10.1002/eji.201646647

Jorgensen I, Zhang Y, Krantz BA, Miao EA. 2016b. Pyrop-tosis triggers pore-induced intracellular traps (PITs) that capture bacteria and lead to their clearance by efferocy-tosis. *J Exp Med* **213:** 2113–2128. doi:10.1084/jem .20151613

Jorgensen I, Rayamajhi M, Miao EA. 2017. Programmed cell death as a defence against infection. *Nat Rev Immunol* **17:** 151–164. doi:10.1038/nri.2016.147

Joshi VD, Kalvakolanu DV, Hebel JR, Hasday JD, Cross AS. 2002. Role of caspase 1 in murine antibacterial host de-fenses and lethal endotoxemia. *Infect Immun* **70:** 6896–6903. doi:10.1128/IAI.70.12.6896-6903.2002

Kaiser WJ, Upton JW, Mocarski ES. 2013. Viral modulation of programmed necrosis. *Curr Opin Virol* **3**: 296–306. doi:10.1016/j.coviro.2013.05.019

Kang TJ, Basu S, Zhang L, Thomas KE, Vogel SN, Baillie L, Cross AS. 2008. *Bacillus anthracis* spores and lethal toxin induce IL-1β via functionally distinct signaling pathways. *Eur.J Immunol* **38**: 1574–1584. doi:10.1002/eji.200838141

Karki R, Man SM, Malireddi RKS, Gurung P, Vogel P, Lamkanfi M, Kanneganti TD. 2015. Concerted activation of the AIM2 and NLRP3 inflammasomes orchestrates host protection against *Aspergillus* infection. *Cell Host Microbe* **17**: 357–368. doi:10.1016/j.chom.2015.01.006

Kawai T, Akira S. 2011. Regulation of innate immune signalling pathways by the tripartite motif (TRIM) family proteins. *EMBO Mol Med* **3**: 513–527. doi:10.1002/emmm.201100160

Kayagaki N, Wong MT, Stowe IB, Ramani SR, Gonzalez LC, Akashi-Takamura S, Miyake K, Zhang J, Lee WP, Muszynski A, et al. 2013. Noncanonical inflammasome activation by intracellular LPS independent of TLR4. *Science* **341**: 1246–1249. doi:10.1126/science.1240248

Kayagaki N, Stowe IB, Lee BL, O'Rourke K, Anderson K, Warming S, Cuellar T, Haley B, Roose-Girma M, Phung QT, et al. 2015. Caspase-11 cleaves gasdermin D for noncanonical inflammasome signalling. *Nature* **526**: 666–671. doi:10.1038/nature15541

Ke Y, Chen Z, Yang R. 2013. *Yersinia pestis*: Mechanisms of entry into and resistance to the host cell. *Front Cell Infect Microbiol* **3**: 106. doi:10.3389/fcimb.2013.00106

Ketelut-Carneiro N, Silva GK, Rocha FA, Milanezi CM, Cavalcanti-Neto FF, Zamboni DS, Silva JS. 2015. IL-18 triggered by the Nlrp3 inflammasome induces host innate resistance in a pulmonary model of fungal infection. *J Immunol* **194**: 4507–4517. doi:10.4049/jimmunol.1402321

Kitur K, Wachtel S, Brown A, Wickersham M, Paulino F, Peñaloza HF, Soong G, Bueno S, Parker D, Prince A. 2016. Necroptosis promotes *Staphylococcus aureus* clearance by inhibiting excessive inflammatory signaling. *Cell Rep* **16**: 2219–2230. doi:10.1016/j.celrep.2016.07.039

Kordes M, Matuschewski K, Hafalla JCR. 2011. Caspase-1 activation of interleukin-1β (IL-1β) and IL-18 is dispensable for induction of experimental cerebral malaria. *Infect Immun* **79**: 3633–3641. doi:10.1128/IAI.05459-11

Kovacs SB, Miao EA. 2017. Gasdermins: Effectors of pyroptosis. *Trends Cell Biol* **27**: 673–684. doi:10.1016/j.tcb.2017.05.005

Kumar M, Roe K, Orillo B, Muruve DA, Nerurkar VR, Gale M Jr, Verma S. 2013. Inflammasome adaptor protein apoptosis-associated speck-like protein containing CARD (ASC) is critical for the immune response and survival in West Nile virus encephalitis. *J Virol* **87**: 3655–3667. doi:10.1128/JVI.02667-12

Kuriakose T, Man SM, Malireddi RK, Karki R, Kesavardhana S, Place DE, Neale G, Vogel P, Kanneganti TD. 2016. ZBP1/DAI is an innate sensor of influenza virus triggering the NLRP3 inflammasome and programmed cell death pathways. *Sci Immunol* **1**: aag2045. doi:10.1126/sciimmunol.aag2045

Lacey CA, Miao EA. 2019. NLRP1 - One NLR to guard them all. *EMBO J* **38**: e102494. doi:10.15252/embj.2019102494

Lacey CA, Mitchell WJ, Dadelahi AS, Skyberg JA. 2018. Caspase-1 and caspase-11 mediate pyroptosis, inflammation, and control of *Brucella* joint infection. *Infect Immun* **86**: e00361-18. doi:10.1128/IAI.00361-18

Lamkanfi M, Declercq W, Kalai M, Saelens X, Vandenabeele P. 2002. Alice in caspase land. A phylogenetic analysis of caspases from worm to man. *Cell Death Differ* **9**: 358–361. doi:10.1038/sj.cdd.4400989

Lara-Tejero M, Sutterwala FS, Ogura Y, Grant EP, Bertin J, Coyle AJ, Flavell RA, Galán JE. 2006. Role of the caspase-1 inflammasome in *Salmonella typhimurium* pathogenesis. *J Exp Med* **203**: 1407–1412. doi:10.1084/jem.20060206

LaRock CN, Cookson BT. 2012. The *Yersinia* virulence effector YopM binds caspase-1 to arrest inflammasome assembly and processing. *Cell Host Microbe* **12**: 799–805. doi:10.1016/j.chom.2012.10.020

Levinsohn JL, Newman ZL, Hellmich KA, Fattah R, Getz MA, Liu S, Sastalla I, Leppla SH, Moayeri M. 2012. Anthrax lethal factor cleavage of nlrp1 is required for activation of the inflammasome. *PLoS Pathog* **8**: e1002638. doi:10.1371/journal.ppat.1002638

Li S, Zhang L, Yao Q, Li L, Dong N, Rong J, Gao W, Ding X, Sun L, Chen X, et al. 2013. Pathogen blocks host death receptor signalling by arginine GlcNAcylation of death domains. *Nature* **501**: 242–246. doi:10.1038/nature12436

Lilue J, Doran AG, Fiddes IT, Abrudan M, Armstrong J, Bennett R, Chow W, Collins J, Collins S, Czechanski A, et al. 2018. Sixteen diverse laboratory mouse reference genomes define strain-specific haplotypes and novel functional loci. *Nat Genet* **50**: 1574–1583.

Lopez CA, Miller BM, Rivera-Chávez F, Velazquez EM, Byndloss MX, Chávez-Arroyo A, Lokken KL, Tsolis RM, Winter SE, Bäumler AJ. 2016. Virulence factors enhance *Citrobacter rodentium* expansion through aerobic respiration. *Science* **353**: 1249–1253. doi:10.1126/science.aag3042

Lucas X, Ciulli A. 2017. Recognition of substrate degrons by E3 ubiquitin ligases and modulation by small-molecule mimicry strategies. *Curr Opin Struct Biol* **44**: 101–110. doi:10.1016/j.sbi.2016.12.015

Macher AM, Casale TB, Fauci AS. 1982. Chronic granulomatous disease of childhood and *Chromobacterium violaceum* infections in the southeastern United States. *Ann Intern Med* **97**: 51–55. doi:10.7326/0003-4819-97-1-51

Maltez VI, Miao EA. 2016. Reassessing the evolutionary importance of inflammasomes. *J Immunol* **196**: 956–962. doi:10.4049/jimmunol.1502060

Maltez VI, Tubbs AL, Cook KD, Aachoui Y, Falcone EL, Holland SM, Whitmire JK, Miao EA. 2015. Inflammasomes coordinate pyroptosis and natural killer cell cytotoxicity to clear infection by a ubiquitous environmental bacterium. *Immunity* **43**: 987–997. doi:10.1016/j.immuni.2015.10.010

Man SM, Karki R, Malireddi RK, Neale G, Vogel P, Yamamoto M, Lamkanfi M, Kanneganti TD. 2015. The transcription factor IRF1 and guanylate-binding proteins target activation of the AIM2 inflammasome by *Francisella* infection. *Nat Immunol* **16**: 467–475. doi:10.1038/ni.3118

Man SM, Karki R, Briard B, Burton A, Gingras S, Pelletier S, Kanneganti TD. 2017. Differential roles of caspase-1 and

Cite this article as *Cold Spring Harb Perspect Biol* doi: 10.1101/cshperspect.a036459

caspase-11 in infection and inflammation. *Sci Rep* **7:** 45126. doi:10.1038/srep45126

Mao Y, Finnemann SC. 2015. Regulation of phagocytosis by Rho GTPases. *Small GTPases* **6:** 89–99. doi:10.4161/21541248.2014.989785

Mariathasan S, Weiss DS, Dixit VM, Monack DM. 2005. Innate immunity against *Francisella tularensis* is dependent on the ASC/caspase-1 axis. *J Exp Med* **202:** 1043–1049. doi:10.1084/jem.20050977

Martinon F, Burns K, Tschopp J. 2002. The inflammasome: A molecular platform triggering activation of inflammatory caspases and processing of pro-IL-β. *Mol Cell* **10:** 417–426. doi:10.1016/S1097-2765(02)00599-3

Mayer-Barber KD, Barber DL, Shenderov K, White SD, Wilson MS, Cheever A, Kugler D, Hieny S, Caspar P, Núñez G, et al. 2010. Caspase-1 independent IL-1β production is critical for host resistance to mycobacterium tuberculosis and does not require TLR signaling in vivo. *J Immunol* **184:** 3326–3330. doi:10.4049/jimmunol.0904189

Mcdonald C, Vacratsis PO, Bliska JB, Dixon JE. 2003. The *Yersinia* virulence factor YopM forms a novel protein complex with two cellular kinases. *J Biol Chem* **278:** 18514–18523. doi:10.1074/jbc.M301226200

Mcelvania-Tekippe E, Allen IC, Hulseberg PD, Sullivan JT, McCann JR, Sandor M, Braunstein M, Ting JP. 2010. Granuloma formation and host defense in chronic *Mycobacterium tuberculosis* infection requires PYCARD/ASC but not NLRP3 or caspase-1. *PLoS ONE* **5:** e12320. doi:10.1371/journal.pone.0012320

Meunier E, Wallet P, Dreier RF, Costanzo S, Anton L, Rühl S, Dussurgey S, Dick MS, Kistner A, Rigard M, et al. 2015. Guanylate-binding proteins promote activation of the AIM2 inflammasome during infection with *Francisella novicida*. *Nat Immunol* **16:** 476–484. doi:10.1038/ni.3119

Miao EA, Leaf IA, Treuting PM, Mao DP, Dors M, Sarkar A, Warren SE, Wewers MD, Aderem A. 2010a. Caspase-1-induced pyroptosis is an innate immune effector mechanism against intracellular bacteria. *Nat Immunol* **11:** 1136–1142. doi:10.1038/ni.1960

Miao EA, Mao DP, Yudkovsky N, Bonneau R, Lorang CG, Warren SE, Leaf IA, Aderem A. 2010b. Innate immune detection of the type III secretion apparatus through the NLRC4 inflammasome. *Proc Natl Acad Sci* **107:** 3076–3080. doi:10.1073/pnas.0913087107

Milora KA, Miller SL, Sanmiguel JC, Jensen LE. 2014. Interleukin-1α released from HSV-1-infected keratinocytes acts as a functional alarmin in the skin. *Nat Commun* **5:** 5230. doi:10.1038/ncomms6230

Moayeri M, Crown D, Newman ZL, Okugawa S, Eckhaus M, Cataisson C, Liu S, Sastalla I, Leppla SH. 2010. Inflammasome sensor Nlrp1b-dependent resistance to anthrax is mediated by caspase-1, IL-1 signaling and neutrophil recruitment. *PLoS Pathog* **6:** e1001222. doi:10.1371/journal.ppat.1001222

Moayeri M, Sastalla I, Leppla SH. 2012. Anthrax and the inflammasome. *Microbes Infect* **14:** 392–400. doi:10.1016/j.micinf.2011.12.005

Monack DM, Hersh D, Ghori N, Bouley D, Zychlinsky A, Falkow S. 2000. *Salmonella* exploits caspase-1 to colonize Peyer's patches in a murine typhoid model. *J Exp Med* **192:** 249–258. doi:10.1084/jem.192.2.249

Muehlbauer SM, Evering TH, Bonuccelli G, Squires RC, Ashton AW, Porcelli SA, Lisanti MP, Brojatsch J. 2007. Anthrax lethal toxin kills macrophages in a strain-specific manner by apoptosis or caspase-1-mediated necrosis. *Cell Cycle* **6:** 758–766. doi:10.4161/cc.6.6.3991

Mühlen S, Ruchaud-Sparagano M-H, Kenny B. 2011. Proteasome-independent degradation of canonical NF-κB complex components by the NleC protein of pathogenic *Escherichia coli*. *J Biol Chem* **286:** 5100–5107. doi:10.1074/jbc.M110.172254

Nogusa S, Thapa RJ, Dillon CP, Liedmann S, Oguin TH III, Ingram JP, Rodriguez DA, Kosoff R, Sharma S, Sturm O, et al. 2016. RIPK3 activates parallel pathways of MLKL-driven necroptosis and FADD-mediated apoptosis to protect against influenza A virus. *Cell Host Microbe* **20:** 13–24. doi:10.1016/j.chom.2016.05.011

Okondo MC, Johnson DC, Sridharan R, Go EB, Chui AJ, Wang MS, Poplawski SE, Wu W, Liu Y, Lai JH, et al. 2017. DPP8 and DPP9 inhibition induces pro-caspase-1-dependent monocyte and macrophage pyroptosis. *Nat Chem Biol* **13:** 46–53. doi:10.1038/nchembio.2229

Okondo MC, Rao SD, Taabazuing CY, Chui AJ, Poplawski SE, Johnson DC, Bachovchin DA. 2018. Inhibition of Dpp8/9 activates the Nlrp1b inflammasome. *Cell Chem Biol* **25:** 262–267.e5. doi:10.1016/j.chembiol.2017.12.013

Özen S. 2018. Update on the epidemiology and disease outcome of familial Mediterranean fever. *Best Pract Res Clin Rheumatol* **32:** 254–260. doi:10.1016/j.berh.2018.09.003

Paciello I, Silipo A, Lembo-Fazio L, Curcurù L, Zumsteg A, Noël G, Ciancarella V, Sturiale L, Molinaro A, Bernardini ML. 2013. Intracellular *Shigella* remodels its LPS to dampen the innate immune recognition and evade inflammasome activation. *Proc Natl Acad Sci* **110:** E4345–E4354. doi:10.1073/pnas.1303641110

Paquette N, Conlon J, Sweet C, Rus F, Wilson L, Pereira A, Rosadini CV, Goutagny N, Weber ANR, Lane WS, et al. 2012. Serine/threonine acetylation of TGFβ-activated kinase (TAK1) by *Yersinia pestis* YopJ inhibits innate immune signaling. *Proc Natl Acad Sci* **109:** 12710–12715. doi:10.1073/pnas.1008203109

Park YH, Wood G, Kastner DL, Chae JJ. 2016. Pyrin inflammasome activation and RhoA signaling in the autoinflammatory diseases FMF and HIDS. *Nat Immunol* **17:** 914–921. doi:10.1038/ni.3457

Paroli AF, Gonzalez PV, Díaz-Luján C, Onofrio LI, Arocena A, Cano RC, Carrera-Silva EA, Gea S. 2018. NLRP3 Inflammasome and caspase-1/11 pathway orchestrate different outcomes in the host protection against *Trypanosoma cruzi* acute infection. *Front Immunol* **9:** 913. doi:10.3389/fimmu.2018.00913

Pearson JS, Hartland EL. 2014. The inflammatory response during enterohemorrhagic *Escherichia coli* infection. *Microbiol Spectr* **2:** EHEC-0012–2013. doi:10.1128/microbiolspec.EHEC-0012-2013

Pearson JS, Riedmaier P, Marches O, Frankel G, Hartland EL. 2011. A type III effector protease NleC from enteropathogenic *Escherichia coli* targets NF-κB for degradation. *Mol Microbiol* **80:** 219–230. doi:10.1111/j.1365-2958.2011.07568.x

Pearson JS, Giogha C, Ong SY, Kennedy CL, Kelly M, Robinson KS, Lung TWF, Mansell A, Riedmaier P, Oates CVL, et al. 2013. A type III effector antagonizes death receptor signalling during bacterial gut infection. *Nature* **501**: 247–251. doi:10.1038/nature12524

Pearson JS, Giogha C, Mühlen S, Nachbur U, Pham CLL, Zhang Y, Hildebrand JM, Oates CV, Lung TWF, Ingle D, et al. 2017. EspL is a bacterial cysteine protease effector that cleaves RHIM proteins to block necroptosis and inflammation. *Nat Microbiol* **2**: 16258. doi:10.1038/nmicro biol.2016.258

Pechous RD, Sivaraman V, Price PA, Stasulli NM, Goldman WE. 2013. Early host cell targets of *Yersinia pestis* during primary pneumonic plague. *PLoS Pathog* **9**: e1003679. doi:10.1371/journal.ppat.1003679

Philip NH, Dillon CP, Snyder AG, Fitzgerald P, Wynosky-Dolfi MA, Zwack EE, Hu B, Fitzgerald L, Mauldin EA, Copenhaver AM, et al. 2014. Caspase-8 mediates caspase-1 processing and innate immune defense in response to bacterial blockade of NF-κB and MAPK signaling. *Proc Natl Acad Sci* **111**: 7385–7390. doi:10.1073/pnas.140 3252111

Quistad SD, Stotland A, Barott KL, Smurthwaite CA, Hilton BJ, Grasis JA, Wolkowicz R, Rohwer FL. 2014. Evolution of TNF-induced apoptosis reveals 550 My of functional conservation. *Proc Natl Acad Sci* **111**: 9567–9572. doi:10 .1073/pnas.1405912111

Rajan JV, Rodriguez D, Miao EA, Aderem A. 2011. The NLRP3 inflammasome detects encephalomyocarditis virus and vesicular stomatitis virus infection. *J Virol* **85**: 4167–4172. doi:10.1128/JVI.01687-10

Ramirez MLG, Poreba M, Snipas SJ, Groborz K, Drag M, Salvesen GS. 2018. Extensive peptide and natural protein substrate screens reveal that mouse caspase-11 has much narrower substrate specificity than caspase-1. *J Biol Chem* **293**: 7058–7067. doi:10.1074/jbc.RA117.001329

Ramos HJ, Lanteri MC, Blahnik G, Negash A, Suthar MS, Brassil MM, Sodhi K, Treuting PM, Busch MP, Norris PJ, et al. 2012. IL-1β signaling promotes CNS-intrinsic immune control of West Nile virus infection. *PLoS Pathog* **8**: e1003039. doi:10.1371/journal.ppat.1003039

Rathinam VAK, Vanaja SK, Waggoner L, Sokolovska A, Becker C, Stuart LM, Leong JM, Fitzgerald KA. 2012. TRIF licenses caspase-11-dependent NLRP3 inflammasome activation by Gram-negative bacteria. *Cell* **150**: 606–619. doi:10.1016/j.cell.2012.07.007

Ratner D, Orning MP, Starheim KK, Marty-Roix R, Proulx MK, Goguen JD, Lien E. 2016. Manipulation of interleukin-1β and interleukin-18 production by *Yersinia pestis* effectors YopJ and YopM and redundant impact on virulence. *J Biol Chem* **291**: 9894–9905. doi:10.1074/jbc .M115.697698

Rauch I, Deets KA, Ji DX, Moltke Von J, Tenthorey JL, Lee AY, Philip NH, Ayres JS, Brodsky IE, Gronert K, et al. 2017. NAIP-NLRC4 inflammasomes coordinate intestinal epithelial cell expulsion with eicosanoid and IL-18 release via activation of caspase-1 and -8. *Immunity* **46**: 649–659. doi:10.1016/j.immuni.2017.03.016

Raupach B, Peuschel SK, Monack DM, Zychlinsky A. 2006. Caspase-1-mediated activation of interleukin-1β (IL-1β) and IL-18 contributes to innate immune defenses against *Salmonella enterica* serovar Typhimurium infection. *Infect Immun* **74**: 4922–4926. doi:10.1128/IAI.00417-06

Reimer T, Shaw MH, Franchi L, Coban C, Ishii KJ, Akira S, Horii T, Rodriguez A, Núñez G. 2010. Experimental ce-

rebral malaria progresses independently of the Nlrp3 inflammasome. *Eur J Immunol* **40**: 764–769. doi:10.1002/eji .200939996

Ren R, Wu S, Cai J, Yang Y, Ren X, Feng Y, Chen L, Qin B, Xu C, Yang H, et al. 2017. The H7N9 influenza A virus infection results in lethal inflammation in the mammalian host via the NLRP3-caspase-1 inflammasome. *Sci Rep* **7**: 7625. doi:10.1038/s41598-017-07384-5

Robinson N, McComb S, Mulligan R, Dudani R, Krishnan L, Sad S. 2012. Type I interferon induces necroptosis in macrophages during infection with *Salmonella enterica* serovar Typhimurium. *Nat Immunol* **13**: 954–962. doi:10 .1038/ni.2397

Rodrigue-Gervais IG, Labbé K, Dagenais M, Dupaul-Chicoine J, Champagne C, Morizot A, Skeldon A, Brincks EL, Vidal SM, Griffith TS, et al. 2014. Cellular inhibitor of apoptosis protein cIAP2 protects against pulmonary tissue necrosis during influenza virus infection to promote host survival. *Cell Host Microbe* **15**: 23–35. doi:10.1016/j .chom.2013.12.003

Rodriguez AE, Bogart C, Gilbert CM, McCullers JA, Smith AM, Kanneganti TD, Lupfer CR. 2019. Enhanced IL-1β production is mediated by a TLR2-MYD88-NLRP3 signaling axis during coinfection with influenza A virus and *Streptococcus pneumoniae*. *PLoS ONE* **14**: e0212236. doi:10.1371/journal.pone.0212236

Rohde JR, Breitkreutz A, Chenal A, Sansonetti PJ, Parsot C. 2007. Type III secretion effectors of the IpaH family are E3 ubiquitin ligases. *Cell Host Microbe* **1**: 77–83. doi:10 .1016/j.chom.2007.02.002

Royan SV, Jones RM, Koutsouris A, Roxas JL, Falzari K, Weflen AW, Kim A, Bellmeyer A, Turner JR, Neish AS, et al. 2010. Enteropathogenic *E. coli* non-LEE encoded effectors NleH1 and NleH2 attenuate NF-κB activation. *Mol Microbiol* **78**: 1232–1245. doi:10.1111/j.1365-2958 .2010.07400.x

Ruan J, Xia S, Liu X, Lieberman J, Wu H. 2018. Cryo-EM structure of the gasdermin A3 membrane pore. *Nature* **557**: 62–67. doi:10.1038/s41586-018-0058-6

Saiga H, Kitada S, Shimada Y, Kamiyama N, Okuyama M, Makino M, Yamamoto M, Takeda K. 2012. Critical role of AIM2 in *Mycobacterium tuberculosis* infection. *Int Immunol* **24**: 637–644. doi:10.1093/intimm/dxs062

Sandstrom A, Mitchell PS, Goers L, Mu EW, Lesser CF, Vance RE. 2019. Functional degradation: a mechanism of NLRP1 inflammasome activation by diverse pathogen enzymes. *Science* **364**: eaau1330. doi:10.1126/science .aau1330

Sansonetti PJ, Phalipon A, Arondel J, Thirumalai K, Banerjee S, Akira S, Takeda K, Zychlinsky A. 2000. Caspase-1 activation of IL-1β and IL-18 are essential for *Shigella flexneri*–induced inflammation. *Immunity* **12**: 581–590. doi:10.1016/S1074-7613(00)80209-5

Scott NE, Giogha C, Pollock GL, Kennedy CL, Webb AI, Williamson NA, Pearson JS, Hartland EL. 2017. The bacterial arginine glycosyltransferase effector NleB preferentially modifies Fas-associated death domain protein (FADD). *J Biol Chem* **292**: 17337–17350. doi:10.1074/ jbc.M117.805036

Shao F, Vacratsis PO, Bao Z, Bowers KE, Fierke CA, Dixon JE. 2003. Biochemical characterization of the *Yersinia* YopT protease: cleavage site and recognition elements

in Rho GTPases. *Proc Natl Acad Sci* **100**: 904–909. doi:10
.1073/pnas.252770599

Sharma D, Yagnik B, Baksi R, Desai N, Padh H, Desai P.
2017. Shigellosis murine model established by intraperitoneal and intranasal route of administration: a comparative comprehension overview. *Microbes Infect* **19**: 47–54.
doi:10.1016/j.micinf.2016.09.002

Sheng X, You Q, Zhu H, Chang Z, Li Q, Wang H, Wang C,
Wang H, Hui L, Du C, et al. 2017. Bacterial effector NleL
promotes enterohemorrhagic *E. coli*-induced attaching
and effacing lesions by ubiquitylating and inactivating
JNK. *PLoS Pathog* **13**: e1006534. doi:10.1371/journal
.ppat.1006534

Shi J, Zhao Y, Wang Y, Gao W, Ding J, Li P, Hu L, Shao F.
2014. Inflammatory caspases are innate immune receptors for intracellular LPS. *Nature* **514**: 187–192. doi:10
.1038/nature13683

Shi J, Zhao Y, Wang K, Shi X, Wang Y, Huang H, Zhuang Y,
Cai T, Wang F, Shao F. 2015. Cleavage of GSDMD by
inflammatory caspases determines pyroptotic cell death.
Nature **526**: 660–665. doi:10.1038/nature15514

Shio MT, Eisenbarth SC, Savaria M, Vinet AF, Bellemare MJ,
Harder KW, Sutterwala FS, Bohle DS, Descoteaux A,
Flavell RA, et al. 2009. Malarial hemozoin activates the
NLRP3 inflammasome through Lyn and Syk kinases.
PLoS Pathog **5**: e1000559. doi:10.1371/journal.ppat
.1000559

Silva GK, Costa RS, Silveira TN, Caetano BC, Horta CV,
Gutierrez FR, Guedes PM, Andrade WA, De Niz M, Gaz-
zinelli RT, et al. 2013. Apoptosis-associated speck-like
protein containing a caspase recruitment domain inflammasomes mediate IL-1β response and host resistance to
Trypanosoma cruzi infection. *J Immunol* **191**: 3373–3383.
doi:10.4049/jimmunol.1203293

Singer AU, Rohde JR, Lam R, Skarina T, Kagan O, Dileo R,
Chirgadze NY, Cuff ME, Joachimiak A, Tyers M, et al.
2008. Structure of the *Shigella* T3SS effector IpaH defines
a new class of E3 ubiquitin ligases. *Nat Struct Mol Biol* **15**:
1293–1301. doi:10.1038/nsmb.1511

Sivaraman V, Pechous RD, Stasulli NM, Miao EA, Goldman
WE. 2015. *Yersinia pestis* activates both IL-1β and IL-1
receptor antagonist to modulate lung inflammation during pneumonic plague. *PLoS Pathog* **11**: e1004688. doi:10
.1371/journal.ppat.1004688

Squires RC, Muehlbauer SM, Brojatsch J. 2007. Proteasomes
control caspase-1 activation in anthrax lethal toxin-me-
diated cell killing. *J Biol Chem* **282**: 34260–34267. doi:10
.1074/jbc.M705687200

Taylor RC, Cullen SP, Martin SJ. 2008. Apoptosis: Controlled demolition at the cellular level. *Nat Rev Mol Cell
Biol* **9**: 231–241. doi:10.1038/nrm2312

Tenthorey JL, Haloupek N, López-Blanco JR, Grob P, Adam-
son E, Hartenian E, Lind NA, Bourgeois NM, Chacón P,
Nogales E, et al. 2017. The structural basis of flagellin
detection by NAIP5: a strategy to limit pathogen immune
evasion. *Science* **358**: 888–893. doi:10.1126/science.aao
1140

Terra JK, Cote CK, France B, Jenkins AL, Bozue JA, Welkos
SL, LeVine SM, Bradley KA. 2010. Cutting edge: resistance to *Bacillus anthracis* infection mediated by a lethal
toxin sensitive allele of *Nalp1b/Nlrp1b*. *J Immunol* **184**:
17–20. doi:10.4049/jimmunol.0903114

Thapa RJ, Ingram JP, Ragan KB, Nogusa S, Boyd DF, Benitez
AA, Sridharan H, Kosoff R, Shubina M, Landsteiner VJ, et
al. 2016. DAI senses influenza A virus genomic RNA and
activates RIPK3-dependent cell death. *Cell Host Microbe*
20: 674–681. doi:10.1016/j.chom.2016.09.014

Thomas PG, Dash P, Aldridge JR, Ellebedy AH, Reynolds C,
Funk AJ, Martin WJ, Lamkanfi M, Webby RJ, Boyd KL, et
al. 2009. The intracellular sensor NLRP3 mediates key
innate and healing responses to influenza A virus via
the regulation of caspase-1. *Immunity* **30**: 566–575.
doi:10.1016/j.immuni.2009.02.006

Thumkeo D, Watanabe S, Narumiya S. 2013. Physiological
roles of Rho and Rho effectors in mammals. *Eur J Cell Biol*
92: 303–315. doi:10.1016/j.ejcb.2013.09.002

Tinel A, Janssens S, Lippens S, Cuenin S, Logette E, Jaccard
B, Quadroni M, Tschopp J. 2007. Autoproteolysis of
PIDD marks the bifurcation between pro-death caspase-
2 and pro-survival NF-κB pathway. *EMBO J* **26**: 197–208.
doi:10.1038/sj.emboj.7601473

Tolle L, Yu FS, Kovach MA, Ballinger MN, Newstead MW,
Zeng X, Nunez G, Standiford TJ. 2015. Redundant and
cooperative interactions between TLR5 and NLRC4 in
protective lung mucosal immunity against *Pseudomonas
aeruginosa*. *J Innate Immun* **7**: 177–186. doi:10.1159/
000367790

Toma C, Higa N, Koizumi Y, Nakasone N, Ogura Y, McCoy
AJ, Franchi L, Uematsu S, Sagara J, Taniguchi S, et al.
2010. Pathogenic *Vibrio* activate NLRP3 inflammasome
via cytotoxins and TLR/nucleotide-binding oligomerization domain-mediated NF-κB signaling. *J Immunol* **184**:
5287–5297. doi:10.4049/jimmunol.0903536

Tomalka J, Ganesan S, Azodi E, Patel K, Majmudar P, Hall
BA, Fitzgerald KA, Hise AG. 2011. A novel role for the
NLRC4 inflammasome in mucosal defenses against the
fungal pathogen *Candida albicans*. *PLoS Pathog* **7**:
e1002379. doi:10.1371/journal.ppat.1002379

Tsuji NM, Tsutsui H, Seki E, Kuida K, Okamura H, Nakanishi K, Flavell RA. 2004. Roles of caspase-1 in *Listeria*
infection in mice. *Int Immunol* **16**: 335–343. doi:10.1093/
intimm/dxh041

Turk BE. 2007. Manipulation of host signalling pathways by
anthrax toxins. *Biochem J* **402**: 405–417. doi:10.1042/
BJ20061891

Vance RE. 2015. The NAIP/NLRC4 inflammasomes. *Curr
Opin Immunol* **32**: 84–89. doi:10.1016/j.coi.2015.01.010

Vanden Berghe T, Hassannia B, Vandenabeele P. 2016. An
outline of necrosome triggers. *Cell Mol Life Sci* **73**: 2137–
2152. doi:10.1007/s00018-016-2189-y

Van De Veerdonk FL, Joosten LA, Shaw PJ, Smeekens SP,
Malireddi RK, van der Meer JW, Kullberg BJ, Netea MG,
Kanneganti TD. 2011. The inflammasome drives protective Th1 and Th17 cellular responses in disseminated
candidiasis. *Eur J Immunol* **41**: 2260–2268. doi:10.1002/
eji.201041226

Van Valen L. 1973. A new evolutionary law. *Evol Theory* **1**:
1–30.

Vladimer GI, Weng D, Paquette SW, Vanaja SK, Rathinam
VA, Aune MH, Conlon JE, Burbage JJ, Proulx MK, Liu Q,
et al. 2012. The NLRP12 inflammasome recognizes *Yersinia pestis*. *Immunity* **37**: 96–107. doi:10.1016/j.immuni
.2012.07.006

Wan F, Weaver A, Gao X, Bern M, Hardwidge PR, Lenardo MJ. 2011. IKKβ phosphorylation regulates RPS3 nuclear translocation and NF-κB function during infection with *Escherichia coli* strain O157:H7. *Nat Immunol* **12:** 335–343. doi:10.1038/ni.2007

Wang X, Li Y, Liu S, Yu X, Li L, Shi C, He W, Li J, Xu L, Hu Z, et al. 2014. Direct activation of RIP3/MLKL-dependent necrosis by herpes simplex virus 1 (HSV-1) protein ICP6 triggers host antiviral defense. *Proc Natl Acad Sci* **111:** 15438–15443. doi:10.1073/pnas.1412767111

Wang W, Shao Y, Li S, Xin N, Ma T, Zhao C, Song M. 2017a. Caspase-11 plays a protective role in pulmonary *Acinetobacter baumannii* infection. *Infect Immun* **85:** e00350. doi:10.1128/IAI.00350-17

Wang Y, Ning X, Gao P, Wu S, Sha M, Lv M, Zhou X, Gao J, Fang R, Meng G, et al. 2017b. Inflammasome activation triggers caspase-1-mediated cleavage of cGAS to regulate responses to DNA virus infection. *Immunity* **46:** 393–404. doi:10.1016/j.immuni.2017.02.011

Weber CH, Vincenz C. 2001. The death domain superfamily: a tale of two interfaces? *Trends Biochem Sci* **26:** 475–481. doi:10.1016/S0968-0004(01)01905-3

Weinert C, Morger D, Djekic A, Grütter MG, Mittl PRE. 2015. Crystal structure of TRIM20 C-terminal coiled-coil/B30.2 fragment: implications for the recognition of higher order oligomers. *Sci Rep* **5:** 10819. doi:10.1038/srep10819

Weng D, Marty-Roix R, Ganesan S, Proulx MK, Vladimer GI, Kaiser WJ, Mocarski ES, Pouliot K, Chan FK, Kelliher MA, et al. 2014. Caspase-8 and RIP kinases regulate bacteria-induced innate immune responses and cell death. *Proc Natl Acad Sci* **111:** 7391–7396. doi:10.1073/pnas.1403477111

Wickliffe KE, Leppla SH, Moayeri M. 2008. Killing of macrophages by anthrax lethal toxin: involvement of the N-end rule pathway. *Cell Microbiol* **10:** 1352–1362. doi:10.1111/j.1462-5822.2008.01131.x

Willingham SB, Allen IC, Bergstralh DT, Brickey WJ, Huang MT, Taxman DJ, Duncan JA, Ting JP. 2009. NLRP3 (NALP3, Cryopyrin) facilitates in vivo caspase-1 activation, necrosis, and HMGB1 release via inflammasome-dependent and -independent pathways. *J Immunol* **183:** 2008–2015. doi:10.4049/jimmunol.0900138

Xu H, Yang J, Gao W, Li L, Li P, Zhang L, Gong Y-N, Peng X, Xi JJ, Chen S, et al. 2014. Innate immune sensing of bacterial modifications of Rho GTPases by the pyrin inflammasome. *Nature* **513:** 237–241. doi:10.1038/nature13449

Xu H, Shi J, Gao H, Liu Y, Yang Z, Shao F, Dong N. 2019. The N-end rule ubiquitin ligase UBR2 mediates NLRP1B inflammasome activation by anthrax lethal toxin. *EMBO J* **38:** e101996. doi:10.15252/embj.2019101996

Yang Q, Stevenson HL, Scott MJ, Ismail N. 2015. Type I interferon contributes to noncanonical inflammasome activation, mediates immunopathology, and impairs protective immunity during fatal infection with lipopolysaccharide-negative Ehrlichiae. *Am J Pathol* **185:** 446–461. doi:10.1016/j.ajpath.2014.10.005

Yang X, Yang F, Wang W, Lin G, Hu Z, Han Z, Qi Y, Zhang L, Wang J, Sui SF, et al. 2018. Structural basis for specific flagellin recognition by the NLR protein NAIP5. *Cell Res* **28:** 35–47. doi:10.1038/cr.2017.148

Yang C, Briones M, Chiou J, Lei L, Patton MJ, Ma L, McClarty G, Caldwell HD. 2019. *Chlamydia trachomatis* lipopolysaccharide evades the canonical and noncanonical inflammatory pathways to subvert innate immunity. *mBio* **10:** e00595.

Yen H, Ooka T, Iguchi A, Hayashi T, Sugimoto N, Tobe T. 2010. NleC, a type III secretion protease, compromises NF-κB activation by targeting p65/RelA. *PLoS Pathog* **6:** e1001231. doi:10.1371/journal.ppat.1001231

Yu SX, Chen W, Liu ZZ, Zhou FH, Yan SQ, Hu GQ, Qin XX, Zhang J, Ma K, Du CT, et al. 2018. Non-hematopoietic MLKL protects against *Salmonella* mucosal infection by enhancing inflammasome activation. *Front Immunol* **9:** 119. doi:10.3389/fimmu.2018.00119

Zalinger ZB, Elliott R, Weiss SR. 2017. Role of the inflammasome-related cytokines Il-1 and Il-18 during infection with murine coronavirus. *J Neurovirol* **23:** 845–854. doi:10.1007/s13365-017-0574-4

Zhang L, Ding X, Cui J, Xu H, Chen J, Gong YN, Hu L, Zhou Y, Ge J, Lu Q, et al. 2011. Cysteine methylation disrupts ubiquitin-chain sensing in NF-κB activation. *Nature* **481:** 204–208. doi:10.1038/nature10690

Zhang L, Chen S, Ruan J, Wu J, Tong AB, Yin Q, Li Y, David L, Lu A, Wang WL, et al. 2015. Cryo-EM structure of the activated NAIP2-NLRC4 inflammasome reveals nucleated polymerization. *Science* **350:** 404–409. doi:10.1126/science.aac5789

Zhao Y, Shi J, Shi X, Wang Y, Wang F, Shao F. 2016. Genetic functions of the NAIP family of inflammasome receptors for bacterial ligands in mice. *J Exp Med* **213:** 647–656. doi:10.1084/jem.20160006

Zheng Y, Lilo S, Mena P, Bliska JB. 2012. YopJ-induced caspase-1 activation in *Yersinia*-infected macrophages: independent of apoptosis, linked to necrosis, dispensable for innate host defense. *PLoS ONE* **7:** e36019. doi:10.1371/journal.pone.0036019

Zhong FL, Robinson K, Teo DET, Tan KY, Lim C, Harapas CR, Yu C-H, Xie WH, Sobota RM, Au VB, et al. 2018. Human DPP9 represses NLRP1 inflammasome and protects against autoinflammatory diseases via both peptidase activity and FIIND domain binding. *J Biol Chem* **293:** 18864–18878. doi:10.1074/jbc.RA118.004350

Zhu Y, Li H, Hu L, Wang J, Zhou Y, Pang Z, Liu L, Shao F. 2008. Structure of a *Shigella* effector reveals a new class of ubiquitin ligases. *Nat Struct Mol Biol* **15:** 1302–1308. doi:10.1038/nsmb.1517

Zwack EE, Snyder AG, Wynosky-Dolfi MA, Ruthel G, Philip NH, Marketon MM, Francis MS, Bliska JB, Brodsky IE. 2015. Inflammasome activation in response to the *Yersinia* type III secretion system requires hyperinjection of translocon proteins YopB and YopD. *mBio* **6:** e02095. doi:10.1128/mBio.02095-14

Cell-Cycle Cross Talk with Caspases and Their Substrates

Patrick Connolly,[1,4] Irmina Garcia-Carpio,[1,4] and Andreas Villunger[1,2,3,4]

[1]Division of Developmental Immunology, Biocenter, Medical University of Innsbruck, Innsbruck 6020, Austria

[2]Ludwig Boltzmann Institute for Rare and Undiagnosed Diseases, Vienna 1090, Austria

[3]CeMM Research Center for Molecular Medicine of the Austrian Academy of Sciences, Vienna 1090, Austria

Correspondence: andreas.villunger@i-med.ac.at

Caspases play central roles in mediating both cell death and inflammation. It has more recently become evident that caspases also drive other biological processes. Most prominently, caspases have been shown to be involved in differentiation. Several stem and progenitor cell types rely on caspases to initiate and execute their differentiation processes. These range from neural and glial cells, to skeletal myoblasts and osteoblasts, and several cell types of the hematopoietic system. Beyond differentiation, caspases have also been shown to play roles in other "noncanonical" processes, including cell proliferation, arrest, and senescence, thereby contributing to the mechanisms that regulate tissue homeostasis at multiple levels. Remarkably, caspases directly influence the course of the cell cycle in both a positive and negative manner. Caspases both cleave elements of the cell-cycle machinery and are themselves substrates of cell-cycle kinases. Here we aim to summarize the breadth of interactions between caspases and cell-cycle regulators. We also highlight recent developments in this area.

Caspases are a class of aspartic acid–specific endopeptidases that are best known for their roles in cell death and inflammation. Activation of caspases typically occurs as the result of complex intracellular signaling cascades, with multiple "checks and balances" to prevent their inappropriate activation. Cytoplasmic adaptor molecules recruit apical caspases into high molecular weight complexes that facilitate self-processing for autoactivation (Fig. 1). Once activated, apical caspase-8 or -9 processes and activates downstream effector caspases by proteolysis to induce apoptotic cell death. Others have roles in directing inflammatory responses, for example, caspase-1 or mouse caspase-11 (caspase-4 and -5 in humans) are activated by different types of inflammasomes for cytokine processing and executing a lytic type of cell death called pyroptosis (Bratton and Salvesen 2010; Julien and Wells 2017; Van Opdenbosch and Lamkanfi 2019).

Along with their traditional roles in apoptosis, caspases have more recently been shown to be crucial mediators of a wider range of biological processes. The most well studied of these has been cellular differentiation. Caspases have been shown to be essential to the differentiation of

---

[4]All authors contributed equally to this work.

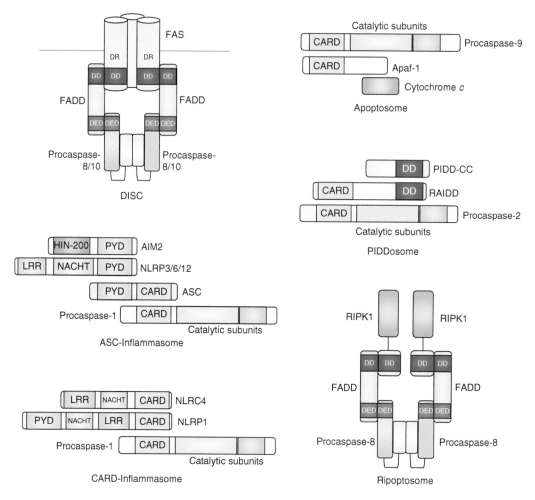

Figure 1. Schematic representations of different high-molecular weight caspase activation complexes. The inflammasome is represented in both ASC-dependent and -independent configurations.

neural and glial cells, skeletal myoblasts and osteoblasts, as well as several cell types of the hematopoietic system, including erythrocytes (for review, see Crawford and Wells 2011; Connolly et al. 2014). In these systems, caspases are prevented from unrestrained runaway activation, which would normally lead to apoptotic cell death. Caspases cleave a limited set of substrates to affect the morphological and transcriptional changes that are characteristic of differentiation. Indeed, it has been argued that apoptosis itself is an extreme form of terminal differentiation, in analogy to the cornification of keratinocytes that causes their death (Fernando and Megeney 2007).

It may seem paradoxical to speak of activation of apoptotic initiator and effector caspases in the absence of cell death, but there are several well-studied mechanisms through which this can occur. One such mechanism is minority mitochondrial outer membrane permeabilization (miMOMP) (Ichim et al. 2015), a process in which only a fraction of a cell's mitochondria undergo permeabilization. This produces a low level of caspase activation, which is insufficient to trigger apoptosis. It has been shown that this process is used to trigger cytokine secretion in epithelial cells in response to viral, bacterial, and protozoal infection (Brokatzky et al. 2019). Moreover, it has even been reported that cells can survive and recover even from full apoptotic caspase activity. In a process termed "anastasis," cells can reverse activation of caspases if a proapoptotic stimulus is removed, even at late stages of apoptosis, when morphological transformation of the cell has already occurred (Tang et al. 2012). This suggests that there exist mechanisms that can suppress "full-blown" caspase activation, apparently through a transcriptional mechanism that rapidly induces expression of antiapoptotic factors.

Another process in which caspases have shown "noncanonical" roles is in cell-cycle control and proliferation. This is admittedly a less well-studied area; however, a number of in vitro and in vivo studies have laid out a case for a potential role for caspases in regulation of the cell cycle. During both apoptosis and many forms of differentiation, caspases effect a terminal withdrawal from the cell cycle into a postmitotic state. In both cases, this is mediated both directly by proteolytic degradation of cell-cycle components, such as p21, p27, CDK11, or the retinoblastoma protein (Jänicke et al. 1996; Beyaert et al. 1997; Levkau et al. 1998; Podmirseg et al. 2016) and indirectly through cleavage of other targets, such as CHK1 or PAK2 that control proliferation rates in response to stress (Rudel and Bokoch 1997; Matsuura et al. 2008; Larsen et al. 2010). However, this is not the fullest extent to which caspases influence the cell cycle. There is also evidence for roles for caspases both in promoting normal cell-cycle progression, and in the response to perturbations in the cell cycle, in which caspases help mediate cell fate decisions. Along this line it is worth mentioning that when cells slip out of extended mitotic arrest, caspase-9 and -7 become activated (Orth et al. 2012). This leads to cleavage of the inhibitor of caspase-activated DNase, iCAD, releasing the endonuclease CAD, which causes limited DNA damage, as well as a p53 response. Hain et al. (2016) report a similar mechanism, tracing the caspase-induced DNA damage to telomeres, driving a DNA-PK-dependent cell-cycle arrest.

In this review, we summarize the evidence for interactions between caspases that act as initiators or effectors of apoptosis and the cell-cycle machinery. These interactions can be divided into two groups: modification of caspases by cell-cycle proteins, chiefly kinases such as CDK1, and (selective) proteolytic processing of cell-cycle proteins or their regulators by caspases. This area of research has undergone a reemergence in the recent past, particularly with respect to in vivo studies. It should be noted up front that the data outlined below often rely on measurement of cell proliferation in response to stimulation. Much of this work was completed before the discovery of necroptotic cell death as a response to caspase-8 inhibition (Holler et al. 2000; Degterev et al. 2005). As such, it is possible that necroptosis induction also contributes to the observed reduction in cell proliferation. Clearly, these phenomena should be reappraised in light of our current knowledge on the role of caspase-8 as an inhibitor of necroptosis. Finally, we highlight some of the most recent discoveries related to caspase-cell-cycle cross talk in the final section and advocate to reinvestigate the role of caspase substrates controlling cell-cycle progression in nonapoptotic settings.

LINKING CASPASES TO CELL-CYCLE REENTRY AND CELL-CYCLE PROGRESSION

Entry into the Cell Cycle from Quiescence

Many cell types, particularly stem and progenitor cells, remain in a quiescent but metabolically active state, termed G_0. Upon tissue damage, infection or extensive cell loss, stem and progenitor cells reenter the cell cycle. There are several older examples in which caspase activity has been proposed to influence this transition into the cell cycle, particularly in the hematopoietic system.

An interesting example of this was shown in B cells where costimulation through CD40 and CD180 induced the cleavage and activation of both caspase-6 and -8 in a time-dependent manner, peaking at 12 and 24 h poststimulation, respectively (Olson et al. 2003). Chemical inhibition of either caspase greatly reduced the rate of proliferation triggered by either CD40/CD180 or LPS stimulation. Mechanistically, it was proposed that caspase-6-mediated cleavage of the transcriptional repressor SATB1 derepresses the transcription of cyclin D1/2 and CDK4, facilitating exit from G_0 (Fig. 2). A caveat here is that the observed failure to proliferate could also be partly the result of non-cell-cycle effects, such as induction of necroptosis upon prolonged caspase-8 inhibition. Similarly, Watanabe et al. showed that upon stimulation of wild-type B cells with the CpG PAMP, activating TLR9, caspase-6 undergoes proteolytic cleavage to its active form, and that chemical inhibition of caspase-6 inhibited the resulting B-cell proliferation (Watanabe et al. 2008). Although the selectivity of pharmacological caspase-inhibitors is limited and caspase-8 may have been inhibited in parallel, *Casp6* knockout mice showed a greatly increased fraction of splenic B cells in G_1 phase. Knockout B cells differentiated more readily into plasma cells, and secreted IgG and IgM levels were higher. Hyperphosphorylation of the retinoblastoma protein was also observed in LPS-stimulated B cells lacking caspase-6 (Watanabe et al. 2008). Hence, caspase-6 might

be a negative regulator of B-cell proliferation, and its absence appears to promote both proliferation and differentiation of this lineage. It remains to be clarified how caspase-6 may be activated in these settings, if not by its upstream sibling, caspase-3 that itself needs to be processed by caspase-9 (Slee et al. 1999). It deserves to be mentioned here that caspase-6 has been shown to cleave RIPK1 (van Raam et al. 2013). It is therefore possible that loss of caspase-6 may simply allow for greater RIPK1-mediated NF-κB activation upon stimulation, which would promote downstream differentiation signals.

In earlier work, Olson and colleagues demonstrated that unstimulated B cells show basal activation of caspase-3, as well as cleaved PARP1, potentially reflecting low level apoptosis in situ, that both disappear upon stimulation in tissue culture (Olson et al. 2003). This phenomenon coincides with another report demonstrating caspase-3 to be a negative regulator of B-cell proliferation (Woo et al. 2003), similar to caspase-6 (Watanabe et al. 2008). Caspase-3-defi-

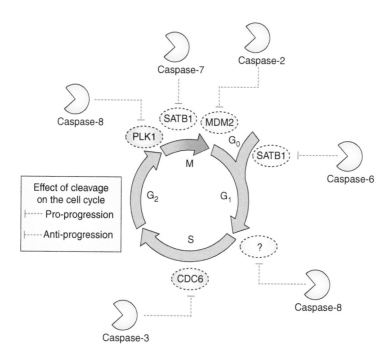

Figure 2. Substrates of caspases reported to be cleaved at different phases of the cell cycle in nonapoptotic contexts. Cleavage events are represented as either promoting or inhibiting cell-cycle progression.

Cite this article as *Cold Spring Harb Perspect Biol* doi: 10.1101/cshperspect.a036475

cient mice show splenomegaly and lymphade-nopathy, showing both higher B-cell numbers and increased proliferation. Isolated B cells were also hyperresponsive to mitogenic stimuli, showing increased CDK2/4 activity and higher PCNA levels. Strangely, this hyperproliferation phenotype is mediated through the cell-cycle inhibitor CDKN1A/(p21), as *Casp3* knockout mice show higher levels of p21, and codeletion of both *Casp3* and *p21* abolished the phenotype. Although this seems paradoxical, there is evidence that, under certain circumstances, p21 can actually promote the formation of active cyclin B/CDK1 complexes (LaBaer et al. 1997), suggesting that a caspase-processed form of p21 may be less potent in doing so. Interestingly, caspase-3 knockout mice also show an increased number of T cells, sufficient to maintain a normal B:T ratio, suggesting that T-cell activation or proliferation may also be negatively regulated by caspase-3 (Woo et al. 2003). This would suggest that processing of p21 by caspase-3 is needed for proper cell-cycle control in lymphocytes. One may also speculate that in the absence of caspase-3, homeostatic apoptosis may be inefficient and mitochondrial DNA release (McArthur et al. 2018; Riley et al. 2018) may cause inflammation-driven expansion of lymphocytes. Hence, conditional deletion of effector caspases in leukocytes may reveal interesting new insights about their noncanonical roles in signaling.

Transition into S Phase

The decision to transition from G_1 into S phase is governed largely by CDK4/6-mediated phosphorylation of Rb, removing its inhibition of E2F family transcription factors. It was suggested early on that caspase-8 is a regulator of S-phase entry both in vitro and in vivo.

It has been shown that CD3-mediated stimulation of resting human T cells causes rapid cleavage of caspase-8 (Kennedy et al. 1999). No concomitant caspase-3 activation was noted, suggesting this phenomenon is distinct from any stimulus-induced apoptosis that can drive caspase-8 processing as a secondary consequence. Treatment with the caspase-8 inhibitor

IETD-fmk prevented the induction of DNA synthesis and interleukin (IL)-2 secretion. Similarly, Arechiga et al. (2007) showed that caspase-8 knockout T cells are defective in S-phase entry upon CD3/CD28 costimulation. In another study, it was shown that caspase-inhibited T cells remain as a G_0/G_1 population upon CD3 stimulation, and fail to undergo CDK1/2 or cyclin A up-regulation or phosphorylation of Rb (Arechiga et al. 2007). Addition of exogenous IL-2 was able to restore T-cell proliferation under conditions of caspase-8 inhibition (Falk et al. 2004). Again, a role for caspase-8 in suppressing necroptosis upon T-cell stimulation needs to be considered here. How IL-2 signaling would counteract necrosome formation and cell death remains to be investigated. An attractive alternative explanation, however, would be that efficient T-cell activation-induced proliferation depends also on the formation of a heterodimer of caspase-8 with the paracaspase MALT1, that, similar to interaction with cFLIP$_L$, alters substrate specificity (Oberst et al. 2011) and thereby facilitates BCL10-driven NF-κB activation and IL-2 transcription (Kawadler et al. 2008).

Beyond T cells, others reported similar findings in primary hepatocytes. Stimulation of primary rat hepatocytes with epidermal growth factor (EGF) stimulates a robust proliferation response (Gilot et al. 2005). During this time, caspase-8 cleavage is observed. In the absence of caspase-8, this proliferation response is blunted. The influence of caspase-8 upon S-phase entry is thought to be mediated through a FAS/FADD/cFLIP$_L$ complex and leads to increased S6 kinase activity through cyclin E/CDK2 activation. Salmena et al. (2003) suggest that caspase-8 stimulates proliferation through proteolysis of an unidentified cyclin E/CDK2 inhibitor. Both p21 and p27, the key inhibitors of cyclin E/CDK2 complexes, would qualify as they are known to be caspase substrates (Table 1).

G_2/Mitosis

Progression through mitosis is regulated by a series of checkpoints to monitor completeness of DNA replication, and ensure chromosomal integrity and proper spindle attachment. Cas-

Table 1. Cell-cycle regulators that are caspase substrates

Gene name	Protein name	Species	Cleavage sequence	References
ATP2B4	PMCA4	H	DEID (1080)	Pászty et al. 2002
BUB1	BUB1	H	Undefined	Baek et al. 2005
BUB1B	BUB1 beta/BUBR1	H	DTCD (610), DVCD (579)	Kim et al. 2005
CABLES2	CDK5 and ABL1 enzyme substrate 2	H	ISLDGRPP	Mahrus et al. 2008
CCNA1	Cyclin A1/A2	X	DEPD (90)	Stack and Newport 1997
CCNE1	Cyclin E	H	LDVD (275)	Mazumder et al. 2002
CCNT2	Cyclin T2	H	DVRDHYIA	Mahrus et al. 2008
CDC27	Cdc27	H	Undefined	Zhou et al. 1998
CDC42	Cdc42	H	DLRD (121)	Tu and Cerione 2001
CDC5L	Cell division cycle 5–like protein	H	HESDFSGV	Mahrus et al. 2008
CDC6	Cdc6	H	LVRD (99), SEVD (422)	Pelizon et al. 2002
CDKN1A	Cyclin-dependent kinase inhibitor 1	H	DHVD (112)	Gervais et al. 1998
CDKN1B	p27Kip1	H	DPSD (139), ESQD (108)	Levkau et al. 1998
CENPF	Centromere protein F	H	Undefined	Dix et al. 2008
CHAF1A	Chromatin assembly factor 1 sub A	H	EQPDSLVD	Mahrus et al. 2008
CKAP5	Cytoskeleton-associated protein 5	H	IEND (1495)	Dix et al. 2008
EIF4EBP1	eIF4E-BP1	H	VLGD (25)	Bushell et al. 2000
EIF4G1	eIF4G 1	H	DLLD (492), DRLD (1136)	Marissen et al. 2000
EIF4G2	eIF4G 2	H	DETD (790)	Marissen et al. 2000
KIF11	Kinesin-like protein KIF23	H	Undefined	Dix et al. 2008
MAX	Max	H	IEVE (10), SAFD (135)	Krippner-Heidenreich et al. 2001
MCM2	MCM2	H	Undefined	Schwab et al. 1998
MCM3	MCM3	H	Undefined	Schwab et al. 1998
MCM4	MCM4	H	Undefined	Schwab et al. 1998
MCM5	DNA RLF MCM5	H	FYSDSFGG	Mahrus et al. 2008
MCM6	DNA RLF MCM6	H	SGVDGYET	Mahrus et al. 2008
MDM2	MDM2	H	DVPD (359)	Erhardt et al. 1997
MDM4	MDMX	M	DVPD (361)	Gentiletti et al. 2002
MKI67	Antigen KI-67	H	THTDKVPG	Mahrus et al. 2008
PCM1	Pericentriolar material 1 protein	H	EDGDGAGA	Mahrus et al. 2008
POLA1	DNA polymerase α catalytic subunit	H	Undefined	Dix et al. 2008
POLE	DNA polymerase ε	H	DQLD (216), DMED (1214)	Liu and Linn 2000
RB1	Rb	H	DEAD (886)	Jänicke et al. 1996
TMEM30A	Cell-cycle control protein 50A	H	DEVDGGPP	Mahrus et al. 2008
TP53	p53	H	Undefined	Sayan et al. 2006
WEE1	Wee1	H	Undefined	Zhou et al. 1998

(H) human, (X) *Xenopus*, (M) mouse.

pases have been found to regulate several of these processes. One of the major regulators of the spindle assembly checkpoint (SAC) is the kinase BUB1b/BUBR1, an integral part of the mitotic checkpoint complex (MCC). Upon extended mitotic arrest, induced, for example, by microtubule targeting agents (MTAs) such as paclitaxel, caspases cleave BUBR1 at two separate sites (Baek et al. 2005; Kim et al. 2005). This cleavage removes the mitotic "brake," allowing cells to exit from the forced mitotic arrest. Transfection with an uncleavable BUBR1 mutant greatly prolongs the duration of mitotic arrest, and leads to a polyploid or aneuploid phenotype. This raises the question of whether caspase activity noted in cells during normal mitosis (Hashimoto et al. 2008) exerts regulatory roles that contribute to faithful chromosome segregation and genomic stability (see below).

In addition, data from high content transcriptomics suggest that caspase-3 is the only caspase enriched during mitosis (Hsu et al. 2006). Notably, pretreatment of hepatoma cell lines with the caspase inhibitor z-DEVD-FMK

Cite this article as *Cold Spring Harb Perspect Biol* doi: 10.1101/cshperspect.a036475

prevented the efficient induction of mitotic arrest by the MTA nocodazole, suggesting a putative role for caspase-3 in mitotic entry. This study is contradicted, however, by another report (Lee et al. 2011), which showed that inhibition or depletion of multiple caspases, either alone or in combination, affected neither the duration nor magnitude of the mitotic checkpoint in MTA-treated cells. Unexpectedly, the authors noted that caspase inhibition increased the proportion of cells that died in MTA-induced mitotic arrest. This mechanism has only been demonstrated under the scenario of extended mitotic arrest. It remains to be seen whether caspase-mediated proteolysis plays a role in normal SAC disassembly and which type of cell death ensues upon chronic SAC activation when caspases are inhibited in hepatocellular carcinoma cells. It is worth mentioning that other cell types, including HeLa cervical carcinoma, A549 lung adenocarcinoma or colon carcinoma cell lines (Gascoigne and Taylor 2008; Haschka et al. 2015) mainly respond with mitotic slippage when caspases are blocked during mitotic arrest, in favor of the competing network hypothesis (see below).

Hashimoto et al. (2008) showed by immunofluorescence colocalization an accumulation of the active form of caspase-3 during mitosis in HepG2 cells. The authors also showed a partial activation of caspase-7, -8, and -9, although this may be accounted for by cells that fail to complete mitosis and undergo cell death. Knockdown of caspase-7 also appears to induce a delay or arrest in mitosis progression, which leads to cell death in a subset of HepG2 cells. This effect was reversed by transfection with an shRNA-resistant caspase-7. Caspase-7 cleaves the chromosomal passenger proteins CENP-C and INCENP, although in this context cleavage induced their mislocalization and the displacement of Aurora B from the centromeres. Although the effect on mitotic progression overall was not investigated, one would expect that inhibition of the normal function of Aurora B would lead to defects in chromosome condensation, spindle orientation, and cytokinesis, favoring cell death induction (Kelly and Funabiki 2009).

CONNECTING CELL-CYCLE DEFECTS AND APOPTOSIS

When cells experience severe genotoxic stress, they initiate apoptosis from within a certain phase of the cell cycle. This is often preceded by a cell-cycle arrest response (after DNA damage–induced activation of the ATM/CHK2/p53 axis), arresting cells mostly in G_1 phase, or replication stress–associated activation of ATR/CHK1 in S-phase and after experiencing subsequent DNA damage at the G_2/M boundary (Burgess and Misteli 2015). Errors in spindle to kinetochore attachments trigger activation of the SAC in M phase, preventing activation of the APC/C E3–ligase complex needed for mitotic exit (Etemad and Kops 2016).

During the last decades, significant resources have been directed to develop therapeutic strategies that interfere with mitotic events to eliminate rapidly proliferating cancer cells. A wide variety of cancers are commonly treated with cytotoxic drugs, such as taxanes and vinka alkaloids, which interfere with microtubules and disturb the mitotic spindle dynamics (Chan et al. 2012; Mitchison 2012). Despite the considerable clinical efficacy of antimitotic drugs, toxicity and drug-resistance phenotypes have posed a challenge for the application of these therapies. The effect of antimitotic drugs in nondividing cells such as neurons, causing neuropathies, has become a pivotal issue in the field and a crucial factor affecting patient quality of life (Hagiwara and Sunada 2004; McGrogan et al. 2008; Ho and Mackey 2014). To address these problems, extensive efforts have been made to develop a second generation of antimitotic agents targeting kinases that also play a role in spindle formation, like PLK1 (Polo-like kinase1) or Aurora kinases. However, the efficacy of these new compounds seems to be less promising than the classic microtubule poisons (Komlodi-Pasztor et al. 2012; Gutteridge et al. 2016).

It is still unclear whether the success of antimitotic drugs to reduce cancer cells relies on their ability to promote cell death during prolonged mitotic arrest or not (Haschka et al. 2018). This idea is challenged by recent data indicating that division failures caused by antimitotic drugs can lead to micronuclei formation and an inflammatory response that seems to be key for the clinical efficacy of these drugs (Shi and Mitchison 2017). Of note, taxol concentrations used to treat primary breast tumors do not produce mitotic arrest of these cells ex vivo. Instead, the cells advance through mitosis and exhibit multipolar cell divisions that result in chromosome missegregation prior to cell death (Zasadil et al. 2014). Thus, it remains unclear how antimitotic drugs promote tumor clearance. However, in experimental settings, cells arrested in mitosis typically have one of two fates: they die during mitosis through the activation of mechanisms that resemble caspase-dependent apoptosis, or they exit mitosis, often without cell division, in a process known as "slippage" or "mitotic checkpoint adaptation." Cells that undergo slippage can arrest in the next G_1 phase, reenter the cell cycle, or die in this new, often tetraploid, G_1 state (Rieder and Maiato 2004). Continued proliferation of genomically unstable cells that failed mitosis can lead to aneuploidy or chromotrypsis, both drivers of tumorigenesis and the development of drug resistance (Santaguida and Amon 2015). To explain the different fates of cells after extended mitotic arrest, Gascoigne and Taylor introduced the "competing network" hypothesis (Gascoigne and Taylor 2008; Topham and Taylor 2013). In this model, cell fate during mitotic arrest is regulated by two distinct signaling networks and the threshold that is crossed first decides cell fate (Figs. 3 and 4).

Activation of the SAC is required for effective cell death induction during mitosis. Drugs that interfere with microtubules or that cause mitotic spindle defects, such as reduced tension at kinetochores, activate the SAC and cause prometaphase arrest. The SAC inhibits the anaphase-promoting complex APC/C, preventing cyclin-dependent kinase 1 (CDK1) inactivation and mitotic exit, and then the apoptotic threshold is crossed (Lara-Gonzalez et al. 2012). Different kinases such as PLK1, BUB1/BUBR1, and Aurora B play critical roles in SAC signaling, including activation and amplification of the MCC, the main effector of SAC. MCC assembly inhibits APC/C activation and prevents proteo-

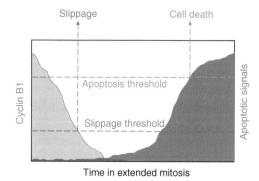

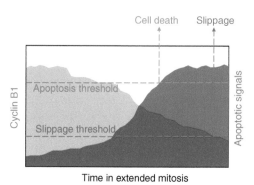

Figure 3. The competing networks model of mitotic cell fate. During extended mitosis, cyclin B levels are gradually depleted by the APC/C E3 ligase complex that controls mitotic exit, while proapoptotic signaling gradually accumulates due to changes in the BCL2 network. Depending on which threshold is reached first, cells may undergo slippage (*top*) or cell death (*bottom*).

lytic degradation of the separase inhibitor, securin, and the CDK1 activator, cyclin B1. Despite continued SAC activation, cyclin B1 levels are slowly degraded during prolonged mitosis by the APC/C in a noncanonical fashion, making it impossible for the cell to maintain high CDK1 activity. The gradual decline of cyclin B1 defines the period of time to mitotic exit (Brito and Rieder 2006).

Several lines of evidence suggest that prolonged SAC activation triggers the intrinsic mitochondrial apoptosis pathway (Topham and Taylor 2013). Interactions between antiapoptotic proteins (for example, BCL2 and MCL1) and proapoptotic proteins (for example, BID, NOXA, and BIM) control activation of the apoptotic effectors BAX and BAK. For an exten-

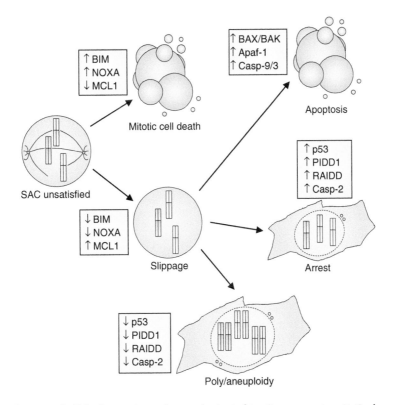

Figure 4. Contributions of cell death proteins to abnormal mitotic fates. In response to mitotic abnormalities such as unattached chromosomes, cells enter into a prolonged mitotic arrest. This arrest is resolved through either mitotic cell death or slippage. Cells that have undergone slippage can die through apoptotic cell death, undergo arrest and senescence, or continue to cycle in a polyploid or aneuploidy state.

sive review of BCL2 family proteins modified during mitosis and their role in mitotic cell death, see Haschka et al. (2018). CDK1 plays a pivotal role during mitotic arrest. During an unperturbed cell division, CDK1 phosphorylates caspase-8 and -9 and inactivates the caspase cascade (Allan and Clarke 2007; Matthess et al. 2010, 2014). However, during prolonged mitosis, CDK1 also phosphorylates antiapoptotic proteins such as BCL2 or BCLX to reduce their antiapoptotic potential. Together with MCL1 degradation by noncanonical APC/C activity, this primes cells for apoptosis (Eichhorn et al. 2013; Allan et al. 2018). If apoptotic signals reach the death threshold first, MOMP is activated, culminating in activation of caspase-9, -3, and -7 and cell death.

Diverse studies have shown a direct correlation between caspase activation and cell death during mitotic arrest. Nevertheless, an open question here is whether these two networks are truly independent (Fava and Villunger 2014). Caspase activity is reported to be dispensable for mitotic slippage and inhibition of caspases in mitosis does not alter the degradation kinetics of endogenous cyclin B1 (Huang et al. 2010; Lee et al. 2011). However, the expression levels of some caspases as well as their activation are reportedly regulated in a cell-cycle, phase-specific manner, including in mitosis. Thus, there may be a tighter connection between both pathways (Hsu et al. 2006; Hashimoto et al. 2008). A correlation between antimitotic drug sensitivity and XIAP levels has also been established (Shi et al. 2008). In addition, preventing slippage by depleting CDC20 by siRNA triggers cell death (Huang et al. 2009). Allan and Clarke (2007) showed that caspase-9 is phosphorylated and inactivated by cyclin B/CDK1 in mitosis, proposing that gradual degradation of cyclin B1 results in gradual activation of caspase-9. Also, Kim et al. (2005) proposed that the apoptosis pathway directly promotes slippage by caspase-dependent proteolysis of the SAC protein BUBR1. Moreover, depletion of BAX, BAK and the mitochondrial fission protein DRP1 from U2OS cells delayed mitotic adaptation, which strongly indicates cross talk between mediators of mitotic apoptosis and SAC adaptation (Díaz-Martínez et al. 2014). More recently, it has been reported that a mitotic delay promotes DNA damage that depends on subapoptotic activation of the classical caspase-9/-3/-7 pathway. Cleavage of iCAD liberates the caspase-activated DNase responsible for the DNA damage (Hain et al. 2016). This CAD activity may drive mutagenesis and genomic instability in cells surviving limited caspase activation after minority MOMP, as suggested by others (Ichim et al. 2015).

Altogether, it seems clear that prolonged mitotic arrest promotes caspase activation downstream of BCL2 family-regulated MOMP. What is less clear at present is whether caspases are activated at low levels (e.g., by miMOMP) (Ichim et al. 2015) to help override the SAC and avoid death in mitosis, or whether lack of caspase activation under such conditions favors alternative cell death pathways that are usually more inflammatory, particularly ferroptosis and necroptosis. We propose that caspase inhibition would be a potential strategy to induce inflammatory cell death upon antimitotic drug therapy, and this might promote antitumor immunity.

CELL-CYCLE REGULATORS AS CASPASE SUBSTRATES

It has long been noted that cell-cycle proteins are subjected to caspase-mediated proteolysis during apoptosis (Table 1). For example, the CDK inhibitors p21 and p27 (Levkau et al. 1998), as well as WEE1 kinase and the APC/C subunit CDC27 (Zhou et al. 1998) are cleaved. Paradoxically, cleavage of the latter causes futile activation of cell-cycle-promoting cyclin complexes

involving CDK2 and CDK1 while caspase-processed p27 can no longer promote cell migration and invasion (Podmirseg et al. 2016).

There are several additional examples of caspase-mediated cleavage of cell-cycle proteins inducing their activation and/or translocation. Rb is cleaved by caspase-3 and -7 (Fattman et al. 2001). This cleaved form binds to cyclin D3 and represses transcriptional activity of E2F1 (Jänicke et al. 1996). Cleavage of the checkpoint kinase CHK1 by caspase-3 and -7 produces a truncated protein with hyperactive kinase activity (Matsuura et al. 2008; Okita et al. 2013). Interestingly, ectopic expression of this truncated CHK1 produces chromatin condensation and disruptions of nuclear architecture, reminiscent of changes seen during apoptosis. Precisely how this phenotype is induced remains unclear.

The cell-cycle driver CDC25A also undergoes caspase cleavage during apoptosis (Mazars et al. 2009). The resulting carboxy-terminal fragment is catalytically active and localizes to the nucleus, where it tightly associates with and dephosphorylates CDK2. Expression of the cleavage product is sufficient to induce apoptosis owing to excessive cyclin B/CDK1 activity (Chou et al. 2010), perhaps through CDK1 phosphorylation of BCL2 family members (Zhou et al. 2014).

Cleavage of the DNA replication regulator CDC6 by caspase-3 during apoptosis generates an amino-terminal fragment that lacks the nuclear export sequence while retaining the ATPase domain. Consequently, it is unable to undergo cytoplasmic translocation upon phosphorylation by CDK2 (Yim et al. 2003; Schories et al. 2004). Expression of this truncated form of CDC6 blocks MCM2 loading onto chromatin, promoting an ATM/ATR response, p53 accumulation, and BAX-mediated apoptosis (Yim et al. 2006).

A key question is whether all these cleavage events are relevant to apoptotic cell death. Arguably not, because caspases are activated during apoptosis at the point of no return (i.e., MOMP, has been passed already). It has been noted in other contexts involving caspase activity, such as cell differentiation, that cleavage of caspase substrates can produce entities with novel activities and functions. This phenomenon is not unique to the execution of apoptosis. For example, caspase-3 cleaves and activates the kinase MST1 during myogenic differentiation (Fernando et al. 2002). Hence it is possible that the apoptotic cleavage events described above may support other functions in nonapoptotic contexts.

CASPASES AS SUBSTRATES OF CELL-CYCLE REGULATORS

Modulation of caspase function by phosphorylation has been noted repeatedly. For example, the oncogenic nonreceptor tyrosine kinases SRC and c-ABL phosphorylate and inactivate caspase-8 and -9, respectively (Cursi et al. 2006; Serrano et al. 2017). Phosphorylation of caspase-3 and -7 by CK2 and PAK2, respectively, also appears to be inhibitory. Phosphorylation most likely affects dimer stability and/or substrate accessibility (Duncan et al. 2011; Li et al. 2011). Cyclin B/CDK1 has been shown to phosphorylate caspase-9 on Thr125, which is located between the CARD prodomain and large catalytic subunit. This site is modified by ERK in S phase, and during normal mitosis and mitotic arrest, when ERK activity is usually low. This suggests that caspase-9 is modified by different kinases depending on cell-cycle stage. Expression of caspase-9(T125A) was shown to accelerate cell death in response to nocodazole treatment but otherwise had no effect on mitotic progression (Allan and Clarke 2007). How phosphorylation of caspase-9 Thr125 affects recruitment into or autoactivation in the apoptosome remains to be addressed. Similarly, cyclin B/CDK1 is reported to phosphorylate caspase-2 at Ser340 during mitosis and in mitotically arrested cells, while in interphase this modification is removed by PP1 phosphatase (Andersen et al. 2009; Nutt et al. 2009). Here, the phospho site is located between the large and small subunits of caspase-2 and it may interfere with autoprocessing, which is needed for full activation. Caspase-2(S340A) sensitizes U2OS cells to death in mitosis in response to nocodazole (Andersen et al. 2009). Notably, caspase-8

was also shown to be phosphorylated in mitosis as a mechanism to limit the sensitivity of mitotic cells to FAS or TNFR ligation (Powley et al. 2016).

Collectively, these studies suggest that the apoptosis effector machinery is inactivated to avoid precocious activation of the caspase cascade during normal mitosis. This mechanism may also act as a fail-safe against partial activation of the apoptotic cascade when mitotic timing errors occur in response to spindle poisons (Orth et al. 2012) and during normal mitosis (e.g., because of chromosome alignment defects or lagging chromosomes) (Soto et al. 2019). It is interesting that phosphates are found in or near the linker region of all human apical caspases, including those controlling inflammation and pyroptosis, but are not in effector caspases (Andersen et al. 2009). Lytic cell death and inflammatory cytokine release might be regulated by proline-directed kinases, other than CDKs, because innate immune cells that are most susceptible to pyroptosis are usually terminally differentiated. However, homeostatic proliferation of tissue resident macrophages has been noted and epithelial cells also express caspase-1 and can undergo lytic cell death (Van Opdenbosch and Lamkanfi 2019).

RECENT DEVELOPMENTS LINKING CASPASES TO GENOMIC STABILITY AND (STERILE) INFLAMMATION

Although apoptotic caspases can modulate cell-cycle progression and cell fate by processing a set of substrates directly involved in cell-cycle control, they can also affect cell fate indirectly by promoting apoptosis and thereby repressing inflammatory responses. For example, hematopoietic stem cells (HSCs) deficient in caspase-9 exhaust prematurely but clear viral infections more effectively. Caspase-9 maintains stem cell dormancy by reducing interferon (IFN)-mediated stem cell mobilization, which is required to satisfy blood cell demand during viral infection. Specifically, activation of caspase-9-dependent apoptosis by BAK and BAX not only kills the cell, it also prevents mitochondrial DNA released into the cytosol by BAX and BAK from stimulating IFN production by the cGAS/ STING/IRF3 pathway (Rongvaux et al. 2014; White et al. 2014; McArthur et al. 2018; Riley et al. 2018). Caspase-3 and -7, the key downstream effectors of caspase-9, have been reported to cull IFN signaling under normal conditions by cleaving cGAS, IRF3, as well as an alternative viral sensor that promotes IFN production called MAVS (Ning et al. 2019).

At the moment, it remains unclear under which physiological or pathological conditions caspase activation is compromised to the degree that mitochondrial DNA actively drives IFN signaling, or, potentially even MLKL-dependent necrotic cell death via activation of the DNA sensor DAI/ZBP1. Considering that many viruses encode proteins that inhibit caspases, viral infections seem a plausible scenario. Cells and tissues that silence expression of the caspase-9 adaptor APAF1 may also be affected. Of note, the cGAS/STING/IRF3 signaling axis also drives inflammation in response to chromosome missegregation and the subsequent formation of micronuclei. Here, breakdown of the brittle nuclear envelope formed around chromosomes that were not included into the main nucleus exposes nuclear dsDNA that is recognized by cGAS (Harding et al. 2017; Mackenzie et al. 2017). Whether caspases suppress inflammatory signaling in cells presenting with micronuclei remains to be investigated.

For a long time, the role of caspase-2, the caspase most closely related to the prototypical *Caenorhabditis elegans* caspase Ced-3, remained mysterious (Kumar 2009; Fava et al. 2012). Most studies pointed to a role for caspase-2 in DNA damage–induced cell death, or potentially death-receptor-mediated apoptosis. Caspase-2, similar to caspase-8, can cleave the BCL2 family protein BID into tBID, which can then trigger activation of BAX and BAK (Upton et al. 2008; Fava et al. 2012). In the context of DNA damage–induced apoptosis, it was considered an effector of p53, because a component of the PIDDosome complex that activates caspase-2, PIDD1, is a transcriptional target of p53 (Lin et al. 2000; Tinel and Tschopp 2004). However, this issue remained controversial because multiple cell types derived from animals lacking

PIDD1, caspase-2, or another PIDDosome component called RAIDD/CRADD exhibited normal p53-dependent apoptosis. Moreover, in contrast to animals with impaired p53 responses upon DNA damage (e.g., *Puma/Bbc3*, *Atm, or Chk2*-deficient mice), absence of the PIDDosome did not cause any obvious phenotypes (O'Reilly et al. 2002; Berube et al. 2005; Manzl et al. 2009). Thus, the idea that caspase-2 or the PIDDosome were needed for DNA damage–induced apoptosis became less palatable. However, loss of caspase-2 accelerates cancer in MYC-transgenic (Ho et al. 2009; Manzl et al. 2012), Her2-transgenic (Parsons et al. 2013), as well as ATM-deficient mice (Puccini et al. 2013), so there was residual interest in studying caspase-2 biology. A role for caspase-2 in cell-cycle control was postulated because $Casp2^{-/-}$ MEFs had difficulties arresting the cell cycle in response to DNA damage (Ho et al. 2008) and aneuploid cells were more frequent in MYC-driven lymphomas (Dorstyn et al. 2012) or in SV40 immortalized MEFs lacking caspase-2 (Peintner et al. 2015). A role for caspase-2 in killing aneuploid cells or cells arrested in mitosis by spindle poisons also was reported. Together, these findings pointed toward a role for caspase-2 in surveilling mitotic fidelity.

Looking into a potential link between cell-cycle progression and caspase-2, we were excited to find that caspase-2 becomes selectively activated in a PIDDosome-dependent manner when cells undergo cytokinesis failure (Fava et al. 2017). Specifically, the presence of supernumerary centrosomes once cells return to the G_1 state drives a PIDDosome-dependent, p53-mediated cell-cycle arrest. Polyploidy in the presence of supernumerary centrosomes is considered to occur frequently and early on in tumor formation. Extra centrosomes, priming cells for multipolar division, have been discussed as a cause of cancer for more than 100 years (Holland and Cleveland 2009; Conduit et al. 2015).

Subsequently, we found that the presence of extra centrosomes, rather than the increased DNA content or an aneuploid karyotype, is the direct cue that activates caspase-2 in the

PIDDosome (Fava et al. 2017). Active caspase-2 then processes and inactivates the ubiquitin ligase MDM2, which allows accumulation of p53 and p21-dependent cell-cycle arrest. Whether this function of caspase-2 contributes to tumor suppression in mouse models of cancer remains to be formally tested. Extra centrosomes owing to PLK4 overexpression has recently been shown to promote spontaneous carcinogenesis (Levine et al. 2017) and a PIDDosome-deficient background might accelerate tumor development.

These findings broaden our understanding of how extra centrosomes activate p53, and they open up new lines of research in the field of scheduled polyploidization during organogenesis or regeneration (e.g., in the naturally polyploid liver). Of note, PIDDosome mutant animals show increased liver ploidy and our unpublished findings suggest a tight control of PIDDosome expression during liver development and regeneration after partial hepatectomy involving E2F family proteins. Thus, E2F family members control the ability of hepatocytes to activate p53 in response to cytokinesis failure and extra centrosomes.

Finally, caspase-8, best known for its apoptosis-inducing function downstream of death receptors and as an inhibitor of necroptotic cell death, was linked to faithful chromosome alignment in mitosis. During normal mitosis, a ripoptosome-like complex (Fig. 1) recruited the mitotic kinase PLK1, where it was inactivated by caspase-8-mediated proteolysis. Cleavage of PLK1 limited phosphorylation of the SAC component BUBR1, fine-tuning chromosome segregation fidelity. It is possible that this role of caspase-8 is responsible for its tumor-suppressive activity, rather than its apoptosis-promoting function or its anti-inflammatory role in restraining necroptosis. Of note, low levels of RIPK1 together with high levels of PLK1 correlate with increased aneuploidy in breast, lung, and colorectal cancer (Liccardi et al. 2019). Somewhat at odds with these findings is the fact that mice lacking Casp8 together with the necroptosis executer Mlkl or its upstream activator, Ripk3, are viable. These animals do not seem to develop spontaneous tumors at a higher rate than normal. However, long-term observation is hampered by the fact that these knockouts develop lymphadenopathy and fatal autoimmune disease (Alvarez-Diaz et al. 2016; Zhang et al. 2016). Tissue-specific deletion of Casp8 and Mlkl outside the hematopoietic compartment may help to better address the role of caspase-8 in the maintenance of genomic stability and as a barrier against cancer. Of note, loss of Casp8 in B cells facilitates their transformation (Hakem et al. 2012) and CASP8 is found mutated in HCC, gastric, breast, as well as colorectal cancer (Kim et al. 2003; Soung et al. 2005a,b; Park et al. 2016).

Taken together, these studies suggest that caspases do more than their acknowledged roles in apoptosis and pathogen-driven inflammation. A reevaluation of the interaction of caspases with their substrates linked to cell-cycle control but originally described as targets for proteolysis under apoptotic conditions may open up exciting new lines of research.

OPEN QUESTIONS

Many of the findings discussed here were described in vitro. On the one hand, there are clear examples where these translate into in vivo phenotypes, for example, the B-cell proliferation defect in caspase-6 knockout mice. On the other hand, caspase-8 conditional knockout mice lack obvious cell-cycle defects. While cell-cycle transition and mitotic timing seem unaffected, loss of caspase-8 may facilitate the emergence of genomically unstable cells (Liccardi et al. 2019). If caspase-8 has cell-cycle functions beyond apoptosis promotion and necroptosis suppression, why do caspase-8 mutant mice lacking RIPK3 or MLKL not develop cancer spontaneously, nor have obvious cell-cycle defects (Alvarez-Diaz et al. 2016; Zhang et al. 2016)? Why does the increased ploidy and reported aneuploidy in Casp2-mutant mice not suffice to drive transformation (Dorstyn et al. 2012; Fava et al. 2017)? To what extent do compensatory mechanisms prevent the emergence of cell-cycle defects in animal models (Zheng et al. 2000)? Executioner caspase-3 and -7 show a measure of functional redundancy both in vitro and in vivo. It may be the case that multiple gene knockouts may be necessary for cell-cycle phenotypes to emerge. Combined loss of caspase-3 and -7 is lethal (Lakhani et al. 2006) and, so far, consequences of combined conditional ablation of both caspases have not been reported. Caspase-6 is mainly expressed in the nervous system, and therefore the analysis of caspase-3, -7 double-mutant animals may yield interesting new insights. Of note, even the loss of individual cyclin-dependent kinases, other than CDK1, can be well compensated in vivo (Barrière et al. 2007; Santamaria et al. 2007). Surprisingly, loss of any of the PIDDosome components is insufficient to drive spontaneous cancer, but can enhance tumor development driven by MYC overexpression (Manzl et al. 2012; Peintner et al. 2015). Consequently, there is much work ahead of us to define the role of caspase-2 within and potentially also outside the PIDDosome in tumorigenesis.

COMPETING INTEREST STATEMENT

The authors declare no conflicts of interest.

ACKNOWLEDGMENTS

We want to thank all members of our team for fruitful discussion. We apologize to all those whose valuable contributions were not cited because of space constrains. Work in our laboratories is supported by the County of South Tyrol, the Austrian Science Fund, FWF Grant Nos. P 26856, P 29499, PIR 3, and the ERC-AdG "POLICE."

REFERENCES

Allan LA, Clarke PR. 2007. Phosphorylation of caspase-9 by CDK1/cyclin B1 protects mitotic cells against apoptosis. *Mol Cell* **26:** 301–310. doi:10.1016/j.molcel.2007.03.019

Allan LA, Skowyra A, Rogers KI, Zeller D, Clarke PR. 2018. Atypical APC/C-dependent degradation of Mcl-1 provides an apoptotic timer during mitotic arrest. *EMBO J* **37:** e96831. doi:10.15252/embj.201796831

Alvarez-Diaz S, Dillon CP, Lalaoui N, Tanzer MC, Rodriguez DA, Lin A, Lebois M, Hakem R, Josefsson EC, O'Reilly LA, et al. 2016. The pseudokinase MLKL and the kinase RIPK3 have distinct roles in autoimmune disease caused by loss of death-receptor-induced apoptosis.

Immunity **45:** 513–526. doi:10.1016/j.immuni.2016.07.016

Andersen JL, Johnson CE, Freel CD, Parrish AB, Day JL, Buchakjian MR, Nutt LK, Thompson JW, Moseley MA, Kornbluth S. 2009. Restraint of apoptosis during mitosis through interdomain phosphorylation of caspase-2. *EMBO J* **28:** 3216–3227. doi:10.1038/emboj.2009.253

Arechiga AF, Bell BD, Leverrier S, Weist BM, Porter M, Wu Z, Kanno Y, Ramos SJ, Ong ST, Siegel R, et al. 2007. A Fas-associated death domain protein/caspase-8-signaling axis promotes S-phase entry and maintains S6 kinase activity in T cells responding to IL-2. *J Immunol* **179:** 5291–5300. doi:10.4049/jimmunol.179.8.5291

Baek KH, Shin HJ, Jeong SJ, Park JW, McKeon F, Lee CW, Kim CM. 2005. Caspases-dependent cleavage of mitotic checkpoint proteins in response to microtubule inhibitor. *Oncol Res* **15:** 161–168. doi:10.3727/096504005776367906

Barrière C, Santamaría D, Cerqueira A, Galán J, Martín A, Ortega S, Malumbres M, Dubus P, Barbacid M. 2007. Mice thrive without Cdk4 and Cdk2. *Mol Oncol* **1:** 72–83. doi:10.1016/j.molonc.2007.03.001

Berube C, Boucher LM, Ma W, Wakeham A, Salmena L, Hakem R, Yeh WC, Mak TW, Benchimol S. 2005. Apoptosis caused by p53-induced protein with death domain (PIDD) depends on the death adapter protein RAIDD. *Proc Natl Acad Sci* **102:** 14314–14320. doi:10.1073/pnas.0506475102

Beyaert R, Kidd VJ, Cornelis S, Van de Craen M, Denecker G, Lahti JM, Gururajan R, Vandenabeele P, Fiers W. 1997. Cleavage of PITSLRE kinases by ICE/CASP-1 and CPP32/CASP-3 during apoptosis induced by tumor necrosis factor. *J Biol Chem* **272:** 11694–11697. doi:10.1074/jbc.272.18.11694

Bratton SB, Salvesen GS. 2010. Regulation of the Apaf-1-caspase-9 apoptosome. *J Cell Sci* **123:** 3209–3214. doi:10.1242/jcs.073643

Brito DA, Rieder CL. 2006. Mitotic checkpoint slippage in humans occurs via cyclin B destruction in the presence of an active checkpoint. *Curr Biol* **16:** 1194–1200. doi:10.1016/j.cub.2006.04.043

Brokatzky D, Dörflinger B, Haimovici A, Weber A, Kirschnek S, Vier J, Metz A, Henschel J, Steinfeldt T, Gentle IE, et al. 2019. A non-death function of the mitochondrial apoptosis apparatus in immunity. *EMBO J* **38:** e100907. doi:10.15252/embj.2018100907

Burgess RC, Misteli T. 2015. Not all DDRs are created equal: non-canonical DNA damage responses. *Cell* **162:** 944–947. doi:10.1016/j.cell.2015.08.006

Bushell M, Poncet D, Marissen WE, Flotow H, Lloyd RE, Clemens MJ, Morley SJ. 2000. Cleavage of polypeptide chain initiation factor eIF4GI during apoptosis in lymphoma cells: characterisation of an internal fragment generated by caspase-3-mediated cleavage. *Cell Death Differ* **7:** 628–636. doi:10.1038/sj.cdd.4400699

Chan KS, Koh CG, Li HY. 2012. Mitosis-targeted anti-cancer therapies: where they stand. *Cell Death Dis* **3:** e411. doi:10.1038/cddis.2012.148

Chou ST, Yen YC, Lee CM, Chen MS. 2010. Pro-apoptotic role of Cdc25A: activation of cyclin B1/Cdc2 by the Cdc25A C-terminal domain. *J Biol Chem* **285:** 17833–17845. doi:10.1074/jbc.M109.078386

Cite this article as *Cold Spring Harb Perspect Biol* doi: 10.1101/cshperspect.a036475

Conduit PT, Wainman A, Raff JW. 2015. Centrosome function and assembly in animal cells. *Nat Rev Mol Cell Biol* **16:** 611–624. doi:10.1038/nrm4062

Connolly PF, Jäger R, Fearnhead HO. 2014. New roles for old enzymes: killer caspases as the engine of cell behavior changes. *Front Physiol* **5:** 149. doi:10.3389/fphys.2014.00149

Crawford ED, Wells JA. 2011. Caspase substrates and cellular remodeling. *Annu Rev Biochem* **80:** 1055–1087. doi:10.1146/annurev-biochem-061809-121639

Cursi S, Rufini A, Stagni V, Condò I, Matafora V, Bachi A, Bonifazi AP, Coppola L, Superti-Furga G, Testi R, et al. 2006. Src kinase phosphorylates Caspase-8 on Tyr380: a novel mechanism of apoptosis suppression. *EMBO J* **25:** 1895–1905. doi:10.1038/sj.emboj.7601085

Degterev A, Huang Z, Boyce M, Li Y, Jagtap P, Mizushima N, Cuny GD, Mitchison TJ, Moskowitz MA, Yuan J. 2005. Chemical inhibitor of nonapoptotic cell death with therapeutic potential for ischemic brain injury. *Nat Chem Biol* **1:** 112–119. doi:10.1038/nchembio711

Díaz-Martínez LA, Karamysheva ZN, Warrington R, Li B, Wei S, Xie XJ, Roth MG, Yu H. 2014. Genome-wide siRNA screen reveals coupling between mitotic apoptosis and adaptation. *EMBO J* **33:** 1960–1976. doi:10.15252/embj.201487826

Dix MM, Simon GM, Cravatt BF. 2008. Global mapping of the topography and magnitude of proteolytic events in apoptosis. *Cell* **134:** 679–691. doi:10.1016/j.cell.2008.06.038

Dorstyn L, Puccini J, Wilson CH, Shalini S, Nicola M, Moore S, Kumar S. 2012. Caspase-2 deficiency promotes aberrant DNA-damage response and genetic instability. *Cell Death Differ* **19:** 1411. doi:10.1038/cdd.2012.76

Duncan JS, Turowec JP, Duncan KE, Vilk G, Wu C, Luscher B, Li SS, Gloor GB, Litchfield DW. 2011. A peptide-based target screen implicates the protein kinase CK2 in the global regulation of caspase signaling. *Sci Signal* **4:** ra30. doi:10.1126/scisignal.2001682

Eichhorn JM, Sakurikar N, Alford SE, Chu R, Chambers TC. 2013. Critical role of anti-apoptotic Bcl-2 protein phosphorylation in mitotic death. *Cell Death Dis* **4:** e834. doi:10.1038/cddis.2013.360

Erhardt P, Tomaselli KJ, Cooper GM. 1997. Identification of the MDM2 oncoprotein as a substrate for CPP32-like apoptotic proteases. *J Biol Chem* **272:** 15049–15052. doi:10.1074/jbc.272.24.15049

Etemad B, Kops GJ. 2016. Attachment issues: kinetochore transformations and spindle checkpoint silencing. *Curr Opin Cell Biol* **39:** 101–108. doi:10.1016/j.ceb.2016.02.016

Falk M, Ussat S, Reiling N, Wesch D, Kabelitz D, Adam-Klages S. 2004. Caspase inhibition blocks human T cell proliferation by suppressing appropriate regulation of IL-2, CD25, and cell cycle-associated proteins. *J Immunol* **173:** 5077–5085. doi:10.4049/jimmunol.173.8.5077

Fattman CL, Delach SM, Dou QP, Johnson DE. 2001. Sequential two-step cleavage of the retinoblastoma protein by caspase-3/-7 during etoposide-induced apoptosis. *Oncogene* **20:** 2918–2926. doi:10.1038/sj.onc.1204414

Fava LL, Villunger A. 2014. Stop competing, start talking! *EMBO J* **33:** 1849–1851. doi:10.15252/embj.201489466

Fava LL, Bock FJ, Geley S, Villunger A. 2012. Caspase-2 at a glance. *J Cell Sci* **125:** 5911–5915. doi:10.1242/jcs.115105

Fava LL, Schuler F, Sladky V, Haschka MD, Soratroi C, Eiterer L, Demetz E, Weiss G, Geley S, Nigg EA, et al. 2017. The PIDDosome activates p53 in response to supernumerary centrosomes. *Genes Dev* **31:** 34–45. doi:10.1101/gad.289728.116

Fernando P, Megeney LA. 2007. Is caspase-dependent apoptosis only cell differentiation taken to the extreme? *FASEB J* **21:** 8–17. doi:10.1096/fj.06-5912hyp

Fernando P, Kelly JF, Balazsi K, Slack RS, Megeney LA. 2002. Caspase 3 activity is required for skeletal muscle differentiation. *Proc Natl Acad Sci* **99:** 11025–11030. doi:10.1073/pnas.162172899

Gascoigne KE, Taylor SS. 2008. Cancer cells display profound intra- and interline variation following prolonged exposure to antimitotic drugs. *Cancer Cell* **14:** 111–122. doi:10.1016/j.ccr.2008.07.002

Gentiletti F, Mancini F, D'Angelo M, Sacchi A, Pontecorvi A, Jochemsen AG, Moretti F. 2002. MDMX stability is regulated by p53-induced caspase cleavage in NIH3T3 mouse fibroblasts. *Oncogene* **21:** 867–877. doi:10.1038/sj.onc.1205137

Gervais JL, Seth P, Zhang H. 1998. Cleavage of CDK inhibitor p21$^{Cip1/Waf1}$ by caspases is an early event during DNA damage–induced apoptosis. *J Biol Chem* **273:** 19207–19212. doi:10.1074/jbc.273.30.19207

Gilot D, Serandour AL, Ilyin GP, Lagadic-Gossmann D, Loyer P, Corlu A, Coutant A, Baffet G, Peter ME, Fardel O, et al. 2005. A role for caspase-8 and c-FLIPL in proliferation and cell-cycle progression of primary hepatocytes. *Carcinogenesis* **26:** 2086–2094. doi:10.1093/carcin/bgi187

Gutteridge RE, Ndiaye MA, Liu X, Ahmad N. 2016. Plk1 inhibitors in cancer therapy: from laboratory to clinics. *Mol Cancer Ther* **15:** 1427–1435. doi:10.1158/1535-7163.MCT-15-0897

Hagiwara H, Sunada Y. 2004. Mechanism of taxane neurotoxicity. *Breast Cancer* **11:** 82–85. doi:10.1007/BF02968008

Hain KO, Colin DJ, Rastogi S, Allan LA, Clarke PR. 2016. Prolonged mitotic arrest induces a caspase-dependent DNA damage response at telomeres that determines cell survival. *Sci Rep* **6:** 26766. doi:10.1038/srep26766

Hakem A, El Ghamrasni S, Maire G, Lemmers B, Karaskova J, Jurisicova A, Sanchez O, Squire J, Hakem R. 2012. Caspase-8 is essential for maintaining chromosomal stability and suppressing B-cell lymphomagenesis. *Blood* **119:** 3495–3502. doi:10.1182/blood-2011-07-367532

Harding SM, Benci JL, Irianto J, Discher DE, Minn AJ, Greenberg RA. 2017. Mitotic progression following DNA damage enables pattern recognition within micronuclei. *Nature* **548:** 466–470. doi:10.1038/nature23470

Haschka MD, Soratroi C, Kirschnek S, Häcker G, Hilbe R, Geley S, Villunger A, Fava LL. 2015. The NOXA-MCL1-BIM axis defines lifespan on extended mitotic arrest. *Nat Commun* **6:** 6891. doi:10.1038/ncomms7891

Haschka M, Karbon G, Fava LL, Villunger A. 2018. Perturbing mitosis for anti-cancer therapy: is cell death the only answer? *EMBO Rep* **19:** e45440. doi:10.15252/embr.201745440

Hashimoto T, Yamauchi L, Hunter T, Kikkawa U, Kamada S. 2008. Possible involvement of caspase-7 in cell cycle progression at mitosis. *Genes Cells* **13:** 609–621. doi:10.1111/j.1365-2443.2008.01192.x

Ho MY, Mackey JR. 2014. Presentation and management of docetaxel-related adverse effects in patients with breast cancer. *Cancer Manag Res* **6:** 253–259.

Ho LH, Read SH, Dorstyn L, Lambrusco L, Kumar S. 2008. Caspase-2 is required for cell death induced by cytoskeletal disruption. *Oncogene* **27:** 3393–3404. doi:10.1038/sj.onc.1211005

Ho LH, Taylor R, Dorstyn L, Cakouros D, Bouillet P, Kumar S. 2009. A tumor suppressor function for caspase-2. *Proc Natl Acad Sci* **106:** 5336–5341. doi:10.1073/pnas.0811928106

Holland AJ, Cleveland DW. 2009. Boveri revisited: chromosomal instability, aneuploidy and tumorigenesis. *Nat Rev Mol Cell Biol* **10:** 478–487. doi:10.1038/nrm2718

Holler N, Zaru R, Micheau O, Thome M, Attinger A, Valitutti S, Bodmer JL, Schneider P, Seed B, Tschopp J. 2000. Fas triggers an alternative, caspase-8-independent cell death pathway using the kinase RIP as effector molecule. *Nat Immunol* **1:** 489–495. doi:10.1038/82732

Hsu SL, Yu CT, Yin SC, Tang MJ, Tien AC, Wu YM, Huang CY. 2006. Caspase 3, periodically expressed and activated at G_2/M transition, is required for nocodazole-induced mitotic checkpoint. *Apoptosis* **11:** 765–771. doi:10.1007/s10495-006-5880-x

Huang HC, Shi J, Orth JD, Mitchison TJ. 2009. Evidence that mitotic exit is a better cancer therapeutic target than spindle assembly. *Cancer Cell* **16:** 347–358. doi:10.1016/j.ccr.2009.08.020

Huang HC, Mitchison TJ, Shi J. 2010. Stochastic competition between mechanistically independent slippage and death pathways determines cell fate during mitotic arrest. *PLoS One* **5:** e15724. doi:10.1371/journal.pone.0015724

Ichim G, Lopez J, Ahmed SU, Muthalagu N, Giampazolias E, Delgado ME, Haller M, Riley JS, Mason SM, Athineos D, et al. 2015. Limited mitochondrial permeabilization causes DNA damage and genomic instability in the absence of cell death. *Mol Cell* **57:** 860–872. doi:10.1016/j.molcel.2015.01.018

Jänicke RU, Walker PA, Lin XY, Porter AG. 1996. Specific cleavage of the retinoblastoma protein by an ICE-like protease in apoptosis. *EMBO J* **15:** 6969–6978. doi:10.1002/j.1460-2075.1996.tb01089.x

Julien O, Wells JA. 2017. Caspases and their substrates. *Cell Death Differ* **24:** 1380–1389. doi:10.1038/cdd.2017.44

Kawadler H, Gantz MA, Riley JL, Yang X. 2008. The paracaspase MALT1 controls caspase-8 activation during lymphocyte proliferation. *Mol Cell* **31:** 415–421. doi:10.1016/j.molcel.2008.06.008

Kelly AE, Funabiki H. 2009. Correcting aberrant kinetochore microtubule attachments: an Aurora B-centric view. *Curr Opin Cell Biol* **21:** 51–58. doi:10.1016/j.ceb.2009.01.004

Kennedy NJ, Kataoka T, Tschopp J, Budd RC. 1999. Caspase activation is required for T cell proliferation. *J Exp Med* **190:** 1891–1896. doi:10.1084/jem.190.12.1891

Kim HS, Lee JW, Soung YH, Park WS, Kim SY, Lee JH, Park JY, Cho YG, Kim CJ, Jeong SW, et al. 2003. Inactivating mutations of caspase-8 gene in colorectal carcinomas. *Gastroenterology* **125:** 708–715. doi:10.1016/S0016-5085(03)01059-X

Kim M, Murphy K, Liu F, Parker SE, Dowling ML, Baff W, Kao GD. 2005. Caspase-mediated specific cleavage of BubR1 is a determinant of mitotic progression. *Mol Cell Biol* **25:** 9232–9248. doi:10.1128/MCB.25.21.9232-9248.2005

Komlodi-Pasztor E, Sackett DL, Fojo AT. 2012. Inhibitors targeting mitosis: tales of how great drugs against a promising target were brought down by a flawed rationale. *Clin Cancer Res* **18:** 51–63. doi:10.1158/1078-0432.CCR-11-0999

Krippner-Heidenreich A, Talanian RV, Sekul R, Kraft R, Thole H, Ottleben H, Lüscher B. 2001. Targeting of the transcription factor Max during apoptosis: phosphorylation-regulated cleavage by caspase-5 at an unusual glutamic acid residue in position P1. *Biochem J* **358:** 705–715. doi:10.1042/bj3580705

Kumar S. 2009. Caspase 2 in apoptosis, the DNA damage response and tumour suppression: enigma no more? *Nat Rev Cancer* **9:** 897–903. doi:10.1038/nrc2745

LaBaer J, Garrett MD, Stevenson LF, Slingerland JM, Sandhu C, Chou HS, Fattaey A, Harlow E. 1997. New functional activities for the p21 family of CDK inhibitors. *Genes Dev* **11:** 847–862. doi:10.1101/gad.11.7.847

Lakhani SA, Masud A, Kuida K, Porter GJ, Booth CJ, Mehal WZ, Inayat I, Flavell RA. 2006. Caspases 3 and 7: key mediators of mitochondrial events of apoptosis. *Science* **311:** 847–851. doi:10.1126/science.1115035

Lara-Gonzalez P, Westhorpe FG, Taylor SS. 2012. The spindle assembly checkpoint. *Curr Biol* **22:** R966–980. doi:10.1016/j.cub.2012.10.006

Larsen BD, Rampalli S, Burns LE, Brunette S, Dilworth FJ, Megeney LA. 2010. Caspase 3/caspase-activated DNase promote cell differentiation by inducing DNA strand breaks. *Proc Natl Acad Sci* **107:** 4230–4235. doi:10.1073/pnas.0913089107

Lee K, Kenny AE, Rieder CL. 2011. Caspase activity is not required for the mitotic checkpoint or mitotic slippage in human cells. *Mol Biol Cell* **22:** 2470–2479. doi:10.1091/mbc.e11-03-0228

Levine MS, Bakker B, Boeckx B, Moyett J, Lu J, Vitre B, Spierings DC, Lansdorp PM, Cleveland DW, Lambrechts D, et al. 2017. Centrosome amplification is sufficient to promote spontaneous tumorigenesis in mammals. *Dev Cell* **40:** 313–322.e5. doi:10.1016/j.devcel.2016.12.022

Levkau B, Koyama H, Raines EW, Clurman BE, Herren B, Orth K, Roberts JM, Ross R. 1998. Cleavage of p21[Cip1/Waf1] and p27[Kip1] mediates apoptosis in endothelial cells through activation of Cdk2: role of a caspase cascade. *Mol Cell* **1:** 553–563. doi:10.1016/S1097-2765(00)80055-6

Li X, Wen W, Liu K, Zhu F, Malakhova M, Peng C, Li T, Kim HG, Ma W, Cho YY, et al. 2011. Phosphorylation of caspase-7 by p21-activated protein kinase (PAK) 2 inhibits chemotherapeutic drug-induced apoptosis of breast cancer cell lines. *J Biol Chem* **286:** 22291–22299. doi:10.1074/jbc.M111.236596

Liccardi G, Ramos Garcia L, Tenev T, Annibaldi A, Legrand AJ, Robertson D, Feltham R, Anderton H, Darding M, Peltzer N, et al. 2019. RIPK1 and caspase-8 ensure chromosome stability independently of their role in cell death and inflammation. *Mol Cell* **73:** 413–428.e7. doi:10.1016/j .molcel.2018.11.010

Lin Y, Ma W, Benchimol S. 2000. Pidd, a new death-domain-containing protein, is induced by p53 and promotes apoptosis. *Nat Genet* **26:** 124–127.

Liu W, Linn S. 2000. Proteolysis of the human DNA polymerase ε catalytic subunit by caspase-3 and calpain specifically during apoptosis. *Nucleic Acids Res* **28:** 4180–4188. doi:10.1093/nar/28.21.4180

Mackenzie KJ, Carroll P, Martin CA, Murina O, Fluteau A, Simpson DJ, Olova N, Sutcliffe H, Rainger JK, Leitch A, et al. 2017. cGAS surveillance of micronuclei links genome instability to innate immunity. *Nature* **548:** 461–465. doi:10.1038/nature23449

Mahrus S, Trinidad JC, Barkan DT, Sali A, Burlingame AL, Wells JA. 2008. Global sequencing of proteolytic cleavage sites in apoptosis by specific labeling of protein N termini. *Cell* **134:** 866–876. doi:10.1016/j.cell.2008.08.012

Manzl C, Krumschnabel G, Bock F, Sohm B, Labi V, Baumgartner F, Logette E, Tschopp J, Villunger A. 2009. Caspase-2 activation in the absence of PIDDosome formation. *J Cell Biol* **185:** 291–303. doi:10.1083/jcb.200811105

Manzl C, Peintner L, Krumschnabel G, Bock F, Labi V, Drach M, Newbold A, Johnstone R, Villunger A. 2012. PIDDosome-independent tumor suppression by Caspase-2. *Cell Death Differ* **19:** 1722–1732. doi:10.1038/cdd.2012.54

Marissen WE, Guo Y, Thomas AA, Matts RL, Lloyd RE. 2000. Identification of caspase 3-mediated cleavage and functional alteration of eukaryotic initiation factor 2α in apoptosis. *J Biol Chem* **275:** 9314–9323. doi:10.1074/jbc .275.13.9314

Matsuura K, Wakasugi M, Yamashita K, Matsunaga T. 2008. Cleavage-mediated activation of Chk1 during apoptosis. *J Biol Chem* **283:** 25485–25491. doi:10.1074/jbc.M803 111200

Matthess Y, Raab M, Sanhaji M, Lavrik IN, Strebhardt K. 2010. Cdk1/cyclin B1 controls Fas-mediated apoptosis by regulating caspase-8 activity. *Mol Cell Biol* **30:** 5726–5740. doi:10.1128/MCB.00731-10

Matthess Y, Raab M, Knecht R, Becker S, Strebhardt K. 2014. Sequential Cdk1 and Plk1 phosphorylation of caspase-8 triggers apoptotic cell death during mitosis. *Mol Oncol* **8:** 596–608. doi:10.1016/j.molonc.2013.12.013

Mazars A, Fernandez-Vidal A, Mondesert O, Lorenzo C, Prévost G, Ducommun B, Payrastre B, Racaud-Sultan C, Manenti S. 2009. A caspase-dependent cleavage of CDC25A generates an active fragment activating cyclin-dependent kinase 2 during apoptosis. *Cell Death Differ* **16:** 208–218. doi:10.1038/cdd.2008.142

Mazumder S, Gong B, Chen Q, Drazba JA, Buchsbaum JC, Almasan A. 2002. Proteolytic cleavage of cyclin E leads to inactivation of associated kinase activity and amplification of apoptosis in hematopoietic cells. *Mol Cell Biol* **22:** 2398–2409. doi:10.1128/MCB.22.7.2398-2409.2002

McArthur K, Whitehead LW, Heddleston JM, Li L, Padman BS, Oorschot V, Geoghegan ND, Chappaz S, Davidson S, San Chin H, et al. 2018. BAK/BAX macropores facilitate mitochondrial herniation and mtDNA efflux during apoptosis. *Science* **359:** eaao6047. doi:10.1126/science .aao6047

McGrogan BT, Gilmartin B, Carney DN, McCann A. 2008. Taxanes, microtubules and chemoresistant breast cancer. *Biochim Biophys Acta* **1785:** 96–132.

Mitchison TJ. 2012. The proliferation rate paradox in anti-mitotic chemotherapy. *Mol Biol Cell* **23:** 1–6. doi:10.1091/ mbc.e10-04-0335

Ning X, Wang Y, Jing M, Sha M, Lv M, Gao P, Zhang R, Huang X, Feng JM, Jiang Z. 2019. Apoptotic caspases suppress type I interferon production via the cleavage of cGAS, MAVS, and IRF3. *Mol Cell* **74:** 19–31.e7. doi:10 .1016/j.molcel.2019.02.013

Nutt LK, Buchakjian MR, Gan E, Darbandi R, Yoon SY, Wu JQ, Miyamoto YJ, Gibbon JA, Andersen JL, Freel CD, et al. 2009. Metabolic control of oocyte apoptosis mediated by 14-3-3ζ-regulated dephosphorylation of caspase-2. *Dev Cell* **16:** 856–866. doi:10.1016/j.devcel.2009.04.005

Oberst A, Dillon CP, Weinlich R, McCormick LL, Fitzgerald P, Pop C, Hakem R, Salvesen GS, Green DR. 2011. Catalytic activity of the caspase-8-FLIP$_L$ complex inhibits RIPK3-dependent necrosis. *Nature* **471:** 363–367. doi:10.1038/nature09852

Okita N, Yoshimura M, Watanabe K, Minato S, Kudo Y, Higami Y, Tanuma S. 2013. CHK1 cleavage in programmed cell death is intricately regulated by both caspase and non-caspase family proteases. *Biochim Biophys Acta* **1830:** 2204–2213. doi:10.1016/j.bbagen.2012.10.009

Olson NE, Graves JD, Shu GL, Ryan EJ, Clark EA. 2003. Caspase activity is required for stimulated B lymphocytes to enter the cell cycle. *J Immunol* **170:** 6065–6072. doi:10 .4049/jimmunol.170.12.6065

O'Reilly LA, Ekert P, Harvey N, Marsden V, Cullen L, Vaux DL, Hacker G, Magnusson C, Pakusch M, Cecconi F, et al. 2002. Caspase-2 is not required for thymocyte or neuronal apoptosis even though cleavage of caspase-2 is dependent on both Apaf-1 and caspase-9. *Cell Death Differ* **9:** 832–841. doi:10.1038/sj.cdd.4401033

Orth JD, Loewer A, Lahav G, Mitchison TJ. 2012. Prolonged mitotic arrest triggers partial activation of apoptosis, resulting in DNA damage and p53 induction. *Mol Biol Cell* **23:** 567–576. doi:10.1091/mbc.e11-09-0781

Park HL, Ziogas A, Chang J, Desai B, Bessonova L, Garner C, Lee E, Neuhausen SL, Wang SS, Ma H, et al. 2016. Novel polymorphisms in caspase-8 are associated with breast cancer risk in the California Teachers Study. *BMC Cancer* **16:** 14. doi:10.1186/s12885-015-2036-9

Parsons MJ, McCormick L, Janke L, Howard A, Bouchier-Hayes L, Green DR. 2013. Genetic deletion of caspase-2 accelerates MMTV/c-neu-driven mammary carcinogenesis in mice. *Cell Death Differ* **20:** 1174–1182. doi:10 .1038/cdd.2013.38

Pászty K, Verma AK, Padányi R, Filoteo AG, Penniston JT, Enyedi A. 2002. Plasma membrane Ca^{2+}ATPase isoform 4b is cleaved and activated by caspase-3 during the early phase of apoptosis. *J Biol Chem* **277:** 6822–6829. doi:10 .1074/jbc.M109548200

Peintner L, Dorstyn L, Kumar S, Aneichyk T, Villunger A, Manzl C. 2015. The tumor-modulatory effects of caspase-2 and Pidd1 do not require the scaffold protein Raidd. *Cell Death Differ* **22:** 1803–1811. doi:10.1038/cdd.2015.31

Pelizon C, d'Adda di Fagagna F, Farrace L, Laskey RA. 2002. Human replication protein Cdc6 is selectively cleaved by caspase 3 during apoptosis. *EMBO Rep* **3**: 780–784. doi:10 .1093/embo-reports/kvf161

Podmirseg SR, Jäkel H, Ranches GD, Kullmann MK, Sohm B, Villunger A, Lindner H, Hengst L. 2016. Caspases uncouple p27^{Kip1} from cell cycle regulated degradation and abolish its ability to stimulate cell migration and invasion. *Oncogene* **35**: 4580–4590. doi:10.1038/onc.2015.524

Powley IR, Hughes MA, Cain K, MacFarlane M. 2016. Caspase-8 tyrosine-380 phosphorylation inhibits CD95 DISC function by preventing procaspase-8 maturation and cycling within the complex. *Oncogene* **35**: 5629–5640. doi:10.1038/onc.2016.99

Puccini J, Shalini S, Voss AK, Gatei M, Wilson CH, Hiwase DK, Lavin MF, Dorstyn L, Kumar S. 2013. Loss of caspase-2 augments lymphomagenesis and enhances genomic instability in Atm-deficient mice. *Proc Natl Acad Sci* **110**: 19920–19925. doi:10.1073/pnas.1311947110

Rieder CL, Maiato H. 2004. Stuck in division or passing through: what happens when cells cannot satisfy the spindle assembly checkpoint. *Dev Cell* **7**: 637–651. doi:10 .1016/j.devcel.2004.09.002

Riley JS, Quarato G, Cloix C, Lopez J, O'Prey J, Pearson M, Chapman J, Sesaki H, Carlin LM, Passos JF, et al. 2018. Mitochondrial inner membrane permeabilisation enables mtDNA release during apoptosis. *EMBO J* **37**: e99238. doi:10.15252/embj.201899238

Rongvaux A, Jackson R, Harman CC, Li T, West AP, de Zoete MR, Wu Y, Yordy B, Lakhani SA, Kuan CY, et al. 2014. Apoptotic caspases prevent the induction of type I interferons by mitochondrial DNA. *Cell* **159**: 1563–1577. doi:10.1016/j.cell.2014.11.037

Rudel T, Bokoch GM. 1997. Membrane and morphological changes in apoptotic cells regulated by caspase-mediated activation of PAK2. *Science* **276**: 1571–1574. doi:10.1126/ science.276.5318.1571

Salmena L, Lemmers B, Hakem A, Matysiak-Zablocki E, Murakami K, Au PY, Berry DM, Tamblyn L, Shehabeldin A, Migon E, et al. 2003. Essential role for caspase 8 in T-cell homeostasis and T-cell-mediated immunity. *Genes Dev* **17**: 883–895. doi:10.1101/gad.1063703

Santaguida S, Amon A. 2015. Short- and long-term effects of chromosome mis-segregation and aneuploidy. *Nat Rev Mol Cell Biol* **16**: 473–485. doi:10.1038/nrm4025

Santamaria D, Barriere C, Cerqueira A, Hunt S, Tardy C, Newton K, Caceres JF, Dubus P, Malumbres M, Barbacid M. 2007. Cdk1 is sufficient to drive the mammalian cell cycle. *Nature* **448**: 811–815. doi:10.1038/nature06046

Sayan BS, Sayan AE, Knight RA, Melino G, Cohen GM. 2006. p53 is cleaved by caspases generating fragments localizing to mitochondria. *J Biol Chem* **281**: 13566– 13573. doi:10.1074/jbc.M512467200

Schories B, Engel K, Dörken B, Gossen M, Bommert K. 2004. Characterization of apoptosis-induced Mcm3 and Cdc6 cleavage reveals a proapoptotic effect for one Mcm3 fragment. *Cell Death Differ* **11**: 940–942. doi:10.1038/sj.cdd .4401411

Schwab BL, Leist M, Knippers R, Nicotera P. 1998. Selective proteolysis of the nuclear replication factor MCM3 in apoptosis. *Exp Cell Res* **238**: 415–421. doi:10.1006/excr .1997.3850

Serrano BP, Szydlo HS, Alfandari D, Hardy JA. 2017. Active site-adjacent phosphorylation at Tyr-397 by c-Abl kinase

inactivates caspase-9. *J Biol Chem* **292**: 21352–21365. doi:10.1074/jbc.M117.811976

Shi J, Mitchison TJ. 2017. Cell death response to anti-mitotic drug treatment in cell culture, mouse tumor model and the clinic. *Endocr Relat Cancer* **24**: T83–T96. doi:10.1530/ ERC-17-0003

Shi J, Orth JD, Mitchison T. 2008. Cell type variation in responses to antimitotic drugs that target microtubules and kinesin-5. *Cancer Res* **68**: 3269–3276. doi:10.1158/ 0008-5472.CAN-07-6699

Slee EA, Harte MT, Kluck RM, Wolf BB, Casiano CA, Newmeyer DD, Wang HG, Reed JC, Nicholson DW, Alnemri ES, et al. 1999. Ordering the cytochrome c-initiated caspase cascade: hierarchical activation of caspases-2, -3, -6, -7, -8, and -10 in a caspase-9-dependent manner. *J Cell Biol* **144**: 281–292. doi:10.1083/jcb.144.2.281

Soto M, Raaijmakers JA, Medema RH. 2019. Consequences of genomic diversification induced by segregation errors. *Trends Genet* **35**: 279–291. doi:10.1016/j.tig.2019.01.003

Soung YH, Lee JW, Kim SY, Jang J, Park YG, Park WS, Nam SW, Lee JY, Yoo NJ, Lee SH. 2005a. CASPASE-8 gene is inactivated by somatic mutations in gastric carcinomas. *Cancer Res* **65**: 815–821.

Soung YH, Lee JW, Kim SY, Sung YJ, Park WS, Nam SW, Kim SH, Lee JY, Yoo NJ, Lee SH. 2005b. Caspase-8 gene is frequently inactivated by the frameshift somatic mutation 1225_1226delTG in hepatocellular carcinomas. *Oncogene* **24**: 141–147. doi:10.1038/sj.onc.1208244

Stack JH, Newport JW. 1997. Developmentally regulated activation of apoptosis early in *Xenopus* gastrulation results in cyclin A degradation during interphase of the cell cycle. *Development* **124**: 3185–3195.

Tang HL, Tang HM, Mak KH, Hu S, Wang SS, Wong KM, Wong CS, Wu HY, Law HT, Liu K, et al. 2012. Cell survival, DNA damage, and oncogenic transformation after a transient and reversible apoptotic response. *Mol Biol Cell* **23**: 2240–2252. doi:10.1091/mbc.e11-11-0926

Tinel A, Tschopp J. 2004. The PIDDosome, a protein complex implicated in activation of caspase-2 in response to genotoxic stress. *Science* **304**: 843–846. doi:10.1126/sci ence.1095432

Topham CH, Taylor SS. 2013. Mitosis and apoptosis: how is the balance set? *Curr Opin Cell Biol* **25**: 780–785. doi:10 .1016/j.ceb.2013.07.003

Tu S, Cerione RA. 2001. Cdc42 is a substrate for caspases and influences Fas-induced apoptosis. *J Biol Chem* **276**: 19656–19663. doi:10.1074/jbc.M009838200

Upton JP, Austgen K, Nishino M, Coakley KM, Hagen A, Han D, Papa FR, Oakes SA. 2008. Caspase-2 cleavage of BID is a critical apoptotic signal downstream of endoplasmic reticulum stress. *Mol Cell Biol* **28**: 3943–3951. doi:10 .1128/MCB.00013-08

Van Opdenbosch N, Lamkanfi M. 2019. Caspases in cell death, inflammation, and disease. *Immunity* **50**: 1352– 1364. doi:10.1016/j.immuni.2019.05.020

van Raam BJ, Ehrnhoefer DE, Hayden MR, Salvesen GS. 2013. Intrinsic cleavage of receptor-interacting protein kinase-1 by caspase-6. *Cell Death Differ* **20**: 86–96. doi:10.1038/cdd.2012.98

Watanabe C, Shu GL, Zheng TS, Flavell RA, Clark EA. 2008. Caspase 6 regulates B cell activation and differentiation into plasma cells. *J Immunol* **181:** 6810–6819. doi:10 .4049/jimmunol.181.10.6810

White MJ, McArthur K, Metcalf D, Lane RM, Cambier JC, Herold MJ, van Delft MF, Bedoui S, Lessene G, Ritchie ME, et al. 2014. Apoptotic caspases suppress mtDNA-induced STING-mediated type I IFN production. *Cell* **159:** 1549–1562. doi:10.1016/j.cell.2014.11.036

Woo M, Hakem R, Furlonger C, Hakem A, Duncan GS, Sasaki T, Bouchard D, Lu L, Wu GE, Paige CJ, et al. 2003. Caspase-3 regulates cell cycle in B cells: a consequence of substrate specificity. *Nat Immunol* **4:** 1016–1022. doi:10.1038/ni976

Yim H, Jin YH, Park BD, Choi HJ, Lee SK. 2003. Caspase-3-mediated cleavage of Cdc6 induces nuclear localization of p49-truncated Cdc6 and apoptosis. *Mol Biol Cell* **14:** 4250–4259. doi:10.1091/mbc.e03-01-0029

Yim H, Hwang IS, Choi JS, Chun KH, Jin YH, Ham YM, Lee KY, Lee SK. 2006. Cleavage of Cdc6 by caspase-3 promotes ATM/ATR kinase-mediated apoptosis of HeLa cells. *J Cell Biol* **174:** 77–88. doi:10.1083/jcb.200509141

Zasadil LM, Andersen KA, Yeum D, Rocque GB, Wilke LG, Tevaarwerk AJ, Raines RT, Burkard ME, Weaver BA.

2014. Cytotoxicity of paclitaxel in breast cancer is due to chromosome missegregation on multipolar spindles. *Sci Transl Med* **6:** 229ra43. doi:10.1126/scitranslmed .3007965

Zhang X, Fan C, Zhang H, Zhao Q, Liu Y, Xu C, Xie Q, Wu X, Yu X, Zhang J, et al. 2016. MLKL and FADD are critical for suppressing progressive lymphoproliferative disease and activating the NLRP3 inflammasome. *Cell Rep* **16:** 3247–3259. doi:10.1016/j.celrep.2016.06.103

Zheng TS, Hunot S, Kuida K, Momoi T, Srinivasan A, Nicholson DW, Lazebnik Y, Flavell RA. 2000. Deficiency in caspase-9 or caspase-3 induces compensatory caspase activation. *Nat Med* **6:** 1241–1247. doi:10.1038/81343

Zhou BB, Li H, Yuan J, Kirschner MW. 1998. Caspase-dependent activation of cyclin-dependent kinases during Fas-induced apoptosis in Jurkat cells. *Proc Natl Acad Sci* **95:** 6785–6790. doi:10.1073/pnas.95.12.6785

Zhou L, Cai X, Han X, Xu N, Chang DC. 2014. CDK1 switches mitotic arrest to apoptosis by phosphorylating Bcl-2/Bax family proteins during treatment with microtubule interfering agents. *Cell Biol Int* **38:** 737–746. doi:10 .1002/cbin.10259

Cell Death in Plant Immunity

Eugenia Pitsili,[1] Ujjal J. Phukan,[1] and Nuria S. Coll

Centre for Research in Agricultural Genomics (CRAG), CSIC-IRTA-UAB-UB, Bellaterra 08193, Barcelona, Spain

Correspondence: nuria.sanchez-coll@cragenomica.es

Pathogen recognition by the plant immune system leads to defense responses that are often accompanied by a form of regulated cell death known as the hypersensitive response (HR). HR shares some features with regulated necrosis observed in animals. Genetically, HR can be uncoupled from local defense responses at the site of infection and its role in immunity may be to activate systemic responses in distal parts of the organism. Recent advances in the field reveal conserved cell death–specific signaling modules that are assembled by immune receptors in response to pathogen-derived effectors. The structural elucidation of the plant resistosome—an inflammasome-like structure that may attach to the plasma membrane on activation—opens the possibility that HR cell death is mediated by the formation of pores at the plasma membrane. Necrotrophic pathogens that feed on dead tissue have evolved strategies to trigger the HR cell death pathway as a survival strategy. Ectopic activation of immunomodulators during autoimmune reactions can also promote HR cell death. In this perspective, we discuss the role and regulation of HR in these different contexts.

To detect potential invaders and respond appropriately, plants have evolved a complex and fine-tuned immune system. Current models have both extracellular and intracellular plant immune receptors initiating signaling cascades in response to invasion (Cook et al. 2015). In turn, potential invaders have developed diverse virulence strategies to evade or subvert plant immunity.

A form of regulated cell death known as the hypersensitive response (HR) is a frequent consequence of pathogen recognition by the plant immune system. The term hypersensitivity stems from the abnormally rapid death of plant cells encountering biotrophic pathogens, which rely on plant living tissue for their survival (Stakman 1915). HR can be manipulated genetically and is under tight control to avoid runaway cell death beyond the site of infection. HR cell death resembles forms of regulated necrosis in mammals, such as necroptosis and pyroptosis, but it also features some apoptosis-like traits (Vanden et al. 2014; Dickman et al. 2017; Galluzzi et al. 2018; Salguero-Linares and Coll 2019). Cell contents leaked during HR cell death may alert other cells to a potential invasion.

HR cell death has been studied mostly in the context of plant defense against biotrophic pathogens or hemibiotrophic pathogens, the latter having an initial biotrophic phase followed by a necrotrophic phase. However, necrotrophic pathogens that feed on dead or dying tissue can hijack HR cell death for their own benefit. Here, we provide a perspective on HR cell death

---

[1]These authors contributed equally to this work.

Cite this article as *Cold Spring Harb Perspect Biol* doi: 10.1101/cshperspect.a036483

signaling based on recent advances in the molecular interactions between plant and pathogens, plus we discuss autoimmunity as a trigger of HR cell death in the context of certain mutations or during hybrid necrosis.

IMMUNE HR CELL DEATH AS A CONSEQUENCE OF PATHOGEN RECOGNITION

The plant immune system is constantly evolving to detect invasive microbes or their effects on the plant. Initially, plasma membrane pattern-recognition receptors (PRRs) were thought to recognize conserved microbe-associated molecular patterns, whereas cytoplasmic NLRs (nucleotide-binding domain leucine-rich repeat (LRR)-containing gene family) sensed pathogenic virulence factors or their perturbations to the cell (Jones and Dangl 2006). However, as our knowledge of plant immunity has advanced, it has become evident that PRRs also respond to virulence effectors. NLRs may also "guard" conserved molecules that act as rheostats in plant immune responses (Cook et al. 2015).

In terms of domain architecture, plant NLRs resemble animal NLRs, with a variable amino-terminal domain, a central nucleotide-binding domain, and a highly polymorphic carboxy-terminal leucine-rich domain (Fig. 1). Plant NLRs are classified according to their amino-terminal domains as Toll/interleukin-1 receptor (TIR) domain NLRs (also known as TNLs) or coiled-coil (CC) domain NLRs (or CNLs) (Cui et al. 2014; Zhang et al. 2017). NLRs recognize effector molecules deployed by pathogens, either directly or indirectly, and then initiate signaling cascades that culminate in the expression of genes mediating host defense (Cui et al. 2014). An emerging model in plant immunity is that NLRs work in functionally specialized pairs or even more complex networks, with sensor NLRs perceiving pathogen effectors and helper NLRs initiating downstream signaling (Bonardi et al. 2012; Wu et al. 2017).

Recognition of adapted biotrophic or hemibiotrophic pathogens by the plant immune system often leads to HR cell death. Thus, HR cell death is frequently described as an immune

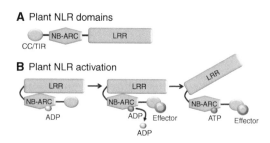

Figure 1. Plant NLRs (nucleotide-binding domain leucine-rich repeat (LRR)-containing gene family). (*A*) Schematic representation of a plant NLR protein. The amino-terminal region usually contains a Toll/interleukin-1 receptor (TIR) homology or a coiled-coil (CC) domain. The central region is composed of a nucleotide-binding APAF-1, R proteins, and CED-4 (NB-ARC) domain. The carboxy-terminal region contains an LRR domain. (*B*) NLR activation. In the inactive, closed state, ADP is bound to the NB-ARC domain. Direct or indirect effector recognition, results in ADP release and ATP binding. This results in a conformational change that renders an open, active NLR.

strategy to block pathogen colonization. However, this is not always the case, because there are numerous examples of HR cell death and inhibition of pathogen growth being genetically uncoupled (Yu et al. 1998; Greenberg et al. 2000; Balagué et al. 2003; Jurkowski et al. 2004; Coll et al. 2010; Sheikh et al. 2014; Menna et al. 2015; Lapin et al. 2019). As shown in Figure 2, HR cell death at the site of infection is crucial to initiate systemic signals that activate immunity in distal parts of the plant and eventually lead to resistance. This phenomenon is known as systemic acquired resistance (SAR) (Fu and Dong 2013; Shine et al. 2019).

Although we are far from an integrated view of HR signaling, research in the last 30 years has substantially increased our understanding of the molecular mechanisms controlling HR. Downstream from NLR activation, HR involves a series of events that include calcium influxes, oxidative bursts originating in different cellular compartments, hormone signaling, mitogen-activated protein kinases, and transcriptional reprogramming (Adachi and Tsuda 2019). Most of these elements are shared between PRR and NLR signaling, and HR cell death has often been

Cite this article as *Cold Spring Harb Perspect Biol* doi: 10.1101/cshperspect.a036483

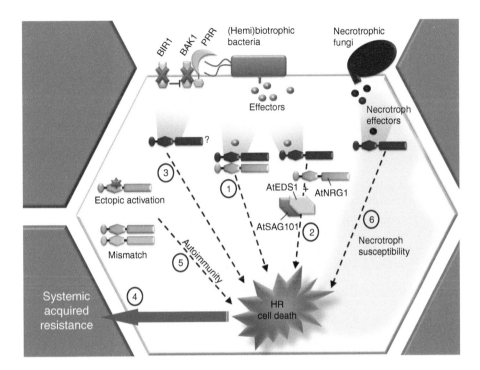

Figure 2. Pathways leading to hypersensitive response (HR) cell death in plant immunity. (*1*) HR can be triggered on recognition of a biotrophic or hemibiotrophic pathogen via direct or indirect effector recognition by NLR (nucleotide-binding domain leucine-rich repeat-containing gene family) immune receptors, often operating in pairs (sensor NLR + helper NLR). (*2*) Cell death–specific modules have been identified, which translate the signal generated by effector perception via Toll/interleukin-1 receptor domain NLR (TIR-NLR) activation, into HR cell death. (*3*) Pattern-recognition receptors (PRRs) signaling at the plasma membrane may be monitored by NLRs, with PRR signaling disturbance leading to HR cell death. (*4*) HR cell death can be genetically uncoupled from local defense responses, but may have a role in activating systemic resistance responses. (*5*) HR can occur as a result of autoimmune reactions, owing to the ectopic activation of NLRs or other defense signaling modulators or an NLR mismatch. (*6*) Necrotrophic fungi can cause disease by hijacking the host HR cell death. A common strategy is activation of NLR receptors by toxins secreted by the fungi into the plant cytoplasm.

regarded as a consequence of surpassing certain signaling thresholds, rather than as a highly regulated phenomenon. However, this view is challenged by recent findings that shed light on HR-specific signaling.

Cell Death Signaling Hubs and the Resistosome

Recent work indicates the importance of signaling hubs downstream from NLR activation, which may partition cell death and immune responses (Wu et al. 2016, 2017; Qi et al. 2018; Castel et al. 2019; Lapin et al. 2019). The lipase-like protein ENHANCED DISEASE SUSCEPTIBILITY1 (AtEDS1) mediates all resistance outputs downstream from activated TNLs (Wiermer et al. 2005). As shown in Figure 2, AtEDS1 interacts with SENESCENCE-ASSOCIATED GENE101 (AtSAG101), and this heterodimer functions together with the helper CNL family member N Requirement Gene 1 (AtNRG1) to form a cell death signaling module in *Arabidopsis thaliana* that can be transferred to unrelated plant species. In parallel, transcriptional reprogramming to enhance the basal defense response is mediated by the interaction of EDS1 with PHYTOALEXIN-DEFICIENT 4 (AtPAD4) and a different helper CNL, ACCELERATED DISEASE RESISTANCE 1 (AtADR1) (Lapin et al. 2019). Helper NLRs have a high degree of redundancy in plant genomes, which may allow functional diversification and expansion of their corresponding sensor NLRs. For example, functionally redundant members of the helper NLR family NRC (NLR required for cell death) may contribute to immunity against different types of pathogens via their interactions with particular sensor NLRs (Wu et al. 2017). Studying interactions and outputs between the components of all these signaling modules is complex because they vary between plant species and according to the pathogen under study. In fact, we still do not know how the signals emanating from these modules execute cell death.

Clues were provided earlier this year by the reconstitution of an NLR supramolecular structure termed the resistosome (Wang et al. 2019a, b). The resistosome has been hypothesized to directly induce HR by forming pores in the plasma membrane, an exciting idea that awaits testing. This immune complex, with stunning structural and mechanistic similarities to mammalian inflammasomes, is composed of the NLR HOPZ-ACTIVATED RESISTANCE 1 (ZAR1) and two receptor-like cytoplasmic kinases (RLCKs) (Fig. 3). In its resting state, ZAR1 is bound to ADP and the RLCK RESISTANCE-RELATED KINASE 1 (RKS1). RKS1 (RLCK XII) is a pseudokinase that interacts with the LRR domain of ZAR1 (Roux et al. 2014). The bacterial pathogen *Xanthomonas campestris* uses a type III secretion system to deliver the bacterial effector AvrAC into the

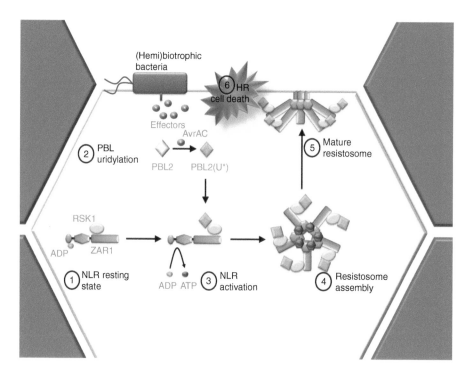

Figure 3. Mechanism of resistosome activation. (*1*) In its resting state, the NLR HOPZ-ACTIVATED RESISTANCE 1 (ZAR1) is bound to ADP and the RLCK RESISTANCE-RELATED KINASE 1 (RKS1). (*2*) *Xanthomonas campestris* secretes the effector AvrAC into the host plant cells, which uridylates the RLCK PBS1-LIKE PROTEIN 2 (PBL2). (*3*) Uridylated PBL2 binds to RKS1, causing conformational changes to the ZAR1–RKS1 dimer that release ADP and prime the complex for activation. (*4*) Subsequent ATP binding results in formation of the resistosome via pentamerization of the ZAR1–RKS1–PBL2 complex. (*5*) Conformational changes expose a funnel-like structure essential for accumulation of the complex in the plasma membrane, bacterial resistance, and (*6*) cell death, which has been hypothesized to be mediated by pore formation at the plasma membrane on insertion of the resistosome.

plant cytoplasm, where it uridylates a decoy RLCK, PBS1-LIKE PROTEIN 2 (PBL2) (Wang et al. 2015a). Unlike RKS1, PBL2 is an active kinase, but its catalytic activity appears dispensable for immune defense. Instead, modified PBL2 (RLCK VII) binds to RKS1 in the ZAR1-RKS1 dimer, causing conformational changes that release ADP and prime the complex for activation. Subsequent ATP binding drives formation of the resistosome via pentamerization of the ZAR1-RKS1-PBL2 complex. Intriguingly, formation of the resistosome exposes a funnel-like structure that is essential both for resistance to bacteria and for accumulation of the complex in the plasma membrane (Fig. 3). This "death-fold switch" may act in an analogous manner to the membrane pores and ion channels formed by mixed lineage kinase domain-like (MLKL) or gasdermins in mammals, or during NLR activation in fungi, potentially suggesting a common evolutionary origin of NLRs from plants and animals (Adachi et al. 2019).

Rather than being the direct cause of cell death, these potential pores could mediate specific ion influxes that activate HR-specific downstream signaling, such as activation of cell death executioner proteases (Dangl and Jones 2019; Feng and Tang 2019). For example, the metacaspase AtMC4 is rapidly activated by calcium that enters the cell on loss of membrane integrity (Huang et al. 2018; Hander et al. 2019). Activation of AtMC4 results in cleavage of the precursor protein PROPEP1, which releases the danger peptide Pep1 to trigger wound-induced defense signaling. This program shares many components with pathogen-induced defense responses. Whether AtMC4 or other proteases are activated by resistosome pores will certainly be worth analyzing in the coming years.

PRR Perturbation as an HR Trigger

Plasma membrane signaling may have a very important role in HR signaling. When PRRs in the plasma membrane sense certain microbial molecular patterns, they team up with coreceptors in specific nanodomains that initiate signaling cascades (Bücherl et al. 2017). For example, knocking out or overexpressing *AtBAK1 (BRAS-*

SINOSTEROID INSENSITIVE1-ASSOCIATED RECEPTOR KINASE1), a coreceptor of several different PRRs, leads to a potent HR cell death response and enhanced resistance to hemibiotrophic pathogens (Kemmerling et al. 2007; Domínguez-Ferreras et al. 2015). The fact that overexpression or elimination of a required element for PRR signaling leads to the same HR phenotype may indicate that perturbation or damage to components of PRR signaling is also monitored (Tang and Zhou 2016). This strategy would allow plant cells to defend against pathogen-mediated inhibition of PRR pathways. Accordingly, inactivation of another PRR regulator, the plasma membrane receptor-like kinase AtBIR1, also results in HR cell death (Liu et al. 2016a).

Proteolytic Pathways Associated with HR

Signaling downstream from NLRs may impact finely tuned proteolytic pathways, including (selective?) autophagy and the concerted action of several proteases (Hofius et al. 2017; Salguero-Linares and Coll 2019). Various proteases in the cytoplasm (metacaspases, phytaspase, or the proteasome subunit PBA1), in the vacuole (vacuolar processing enzyme [VPE]), and those secreted to the extracellular space (cathepsin B, saspase, Rcr3, Pip1) have been shown to be essential for HR cell death (Salguero-Linares and Coll 2019). In fact, they need to be tightly controlled to limit cell death beyond the HR site. Hence, the multiple levels of negative regulation exerted on, for example, the HR cell death protease METACASPASE1 (AtMC1) by the protease inhibitor SERPIN1, the scaffold protein LESION SIMULATING DISEASE 1 (AtLSD1), and the metacaspase AtMC2 (Coll et al. 2010; Lema Asqui et al. 2018). Moreover, AtMC1 has been shown to act additively to autophagy in controlling HR cell death (Coll et al. 2014). Although it is clear that autophagy promotes HR cell death, the mechanism and precise function (trigger or executioner?) remain unknown (Hofius et al. 2009; Coll et al. 2014; Munch et al. 2014). Intriguingly, to date, no canonical proteolytic cascade has ever been characterized in plants (Paulus and Van der Hoorn 2019). The coming years will hopefully provide a deeper insight into this HR-related proteolysis as the study of plant autophagy in plant–pathogen interactions has witnessed a tremendous expansion in the last few years (Avin-Wittenberg et al. 2018) and plant proteostasis is becoming a fully-fledged field of study.

Local versus Peripheral Regulation of HR

It will be important to pay closer attention to the spatiotemporal magnitude of HR in the coming years. This aspect has often been disregarded, with many studies of infected tissue not discriminating between HR versus non-HR cells. There are several examples of differential or antagonistic signaling between the cells undergoing HR and those in the surrounding area. This is true for the metacaspases AtMC1 and AtMC2, which antagonistically regulate HR cell death, and are expressed at the site of HR (AtMC1) or in the cells surrounding the HR zone (AtMC2) (Coll et al. 2010). The transcription factor AtMYB30, which mediates HR cell death and immune responses, has also been shown to be differentially regulated within HR and non-HR zones (Raffaele and Rivas 2013). Finally, signaling pathways downstream from the defense hormones salicylic acid and jasmonic acid are activated in spatially different domains during HR, with salicylic acid in the cell death zone and jasmonic acid in the surrounding area (Betsuyaku et al. 2018). Thus, it will be extremely important to define spatiotemporal markers of HR cell death, so that in the future we can time and characterize the events leading to HR cell death. These markers will help discriminate cells undergoing HR cell death from the surrounding tissue, which needs to activate protective mechanisms to survive, while integrating and transmitting danger/immune signals from dying cells to protect the organism against invasion.

MANIPULATION OF IMMUNE HR CELL DEATH BY NECROTROPHS AS A VIRULENCE STRATEGY

Necrotrophic pathogens have been regarded as generalists, but it is now evident that their inter-

action with the plant host is complex and highly regulated. Necrotrophs secrete toxins that kill plant cells and leave remnants from which the pathogen can feed. These pathogens have evolved very sophisticated strategies to trigger cell death. The most common strategy seems to be hijacking HR cell death pathways by subverting components of the plant immune system.

Secreted toxins, also known as necrotroph effectors (NEs), are recognized by the so-called NE-sensitive genes and trigger HR cell death (Fig. 2). Several NE-sensitive genes possess classical nucleotide-binding and LRR domains, and they often have roles in defense against biotrophic or hemibiotrophic pathogens (Lorang et al. 2007, 2012; Faris et al. 2010). Thus, NE genes appear to be a double-edged sword, being effective at eliciting an HR response to contain biotrophic pathogens, but able to be hijacked by NEs to confer plant susceptibility. A classic example is LOV1 (LONG VEGETATIVE PHASE1), an NLR from *A. thaliana* that confers susceptibility to *Cochliobolus victoriae* (Lorang et al. 2007). This necrotrophic fungus secretes the effector victorin, which activates LOV1 and triggers a resistance-like response that culminates in HR cell death and proliferation of the pathogen (Lorang et al. 2012).

The intricate mechanisms regulating necrotroph–host interactions have also been showcased by the study of *Sclerotinia sclerotiorum*. This necrotrophic fungus triggers HR by secreting oxalic acid into plant cells (Kim et al. 2008). During the initial phases of the infection, oxalic acid reduces levels of reactive oxygen species and creates a reducing environment that favors pathogen proliferation. At the same time, host defenses are dampened and the infection progresses unnoticed. At later stages, and once the infection is well established, oxalic acid triggers an increase in reactive oxygen species that causes cell death (Williams et al. 2011). Oxalic acid has also been shown to inhibit autophagy-mediated cell death, which could provide an additional mechanism to camouflage infection and prevent activation of defense responses (Kabbage et al. 2013).

New NEs and their plant susceptibility targets are emerging from the interaction between

Cite this article as *Cold Spring Harb Perspect Biol* doi: 10.1101/cshperspect.a036483

wheat and the necrotrophic fungus *Parastagonospora nodorum*. For example, the effector ToxA is recognized indirectly by the NLR Tsn1 from wheat, which results in HR cell death and disease (Faris et al. 2010). Another *P. nodorum* effector, Tox1, remains in the extracellular space and is proposed to have a dual role in infection. It binds to chitin in the fungal cell wall to protect it from degradation by host chitinases, while also inducing an HR-like response via its recognition by Snn1, a wall-associated receptor kinase (Liu et al. 2016b; Shi et al. 2016). Adding to the complexity, the susceptibility triggered by an NE can vary depending on the genetic backgrounds of the host and pathogen (Peters Haugrud et al. 2019). The identification of new susceptibility gene candidates in the host holds great potential for the generation of plants that are more resistant to necrotrophic fungi, which are a serious threat to agriculture. Understanding precisely how NE genes interact with their corresponding plant susceptibility genes may allow engineering of new plant protein targets that evade the effectors without compromising plant fitness and yield.

HR CELL DEATH AS A CONSEQUENCE OF AUTOIMMUNITY

HR cell death can also be observed in plants in the absence of pathogens. This autoimmunity leads to ectopic defense activation and spontaneous cell death in the form of macroscopic disease-like lesions (Chakraborty et al. 2018). Plant autoimmunity can be triggered by gain or loss-of-function of plant immune modulators (NLRs and non-NLRs), autophagy, and impaired metabolic processes. In the 1990s, lesion mimic mutants (LMMs), which are plants with spontaneous or mutagenesis-induced mutations showing HR-like cell death in the absence of pathogen, emerged as a promising tool to characterize HR cell death. Characterization of these genes, mostly in *A. thaliana* and rice, has highlighted the importance of several cellular compartments and pathways in HR signaling, including chloroplasts and light energy, sphingolipids and fatty acids, ROS and ion fluxes, autophagy, and plasma membrane signal perception (Brugge-

man et al. 2015). Forward genetic screens targeting LMM revertants have identified additional components of defense signaling pathways, which has led to the idea that LMM phenotypes can be caused by loss of a pathogen effector target that is guarded by an NLR. Subsequent activation of the NLR promotes HR cell death (Rodriguez et al. 2016; Lolle et al. 2017).

The study of autoactive NLR alleles has also been informative. For example, the *snc1-1* (Suppressor of NPR1, Constitutive 1) mutant is a constitutively active variant of the SNC1 TLR that causes autoimmunity and HR cell death (Li et al. 2001). Autoactive SNC1 has been shown to activate immune responses in the nucleus, where it represses small RNAs involved in NLR silencing (Cai et al. 2018), and it associates with a transcriptional corepressor that blocks expression of negative regulators of immunity (Zhu et al. 2010). To ensure appropriate activation of SNC1-dependent immunity, multiple repression mechanisms directed toward this protein have been shown at the transcriptional level as well as posttranscriptionally (Zhu et al. 2010; Cheng et al. 2011; Huang et al. 2014; Johnson et al. 2015, 2017; Dong et al. 2016; Gou et al. 2017; Wang et al. 2017a,b, 2019c; Cai et al. 2018; Zhang et al. 2018; Niu et al. 2019). The *Rp1-D21* gene in maize, which derives from an intergenic recombination event between two NLR genes, *Rp1-D* and *Rp1-dp2*, provides another example of NLR autoactivation resulting in HR cell death (Chintamanani et al. 2010). Intramolecular interactions drive activation of *Rp1-D21*, although HR cell death requires light and temperatures below a certain threshold (Negeri et al. 2013; Wang et al. 2015b). Recently, it was shown that two key enzymes of the lignin biosynthetic pathway form complexes with this hybrid NLR and modulate its activity (Wang and Balint-Kurti 2016).

Autoimmunity leading to ectopic HR cell death can also be a consequence of hybrid necrosis, which is a common type of incompatibility found in the progeny of many crosses within and between species (Fig. 2). In contrast to hybrid vigor, hybrid incompatibility challenges plant fitness and can result from mismatched NLRs (Chen et al. 2016; Vaid and Laitinen

2019). Indeed, genes causing hybrid necrosis are often associated with plant defense responses (Alcazar et al. 2009). Allelic interactions at the ACD6 (ACCELERATED CELL DEATH 6) locus in *A. thaliana* lead to hybrid necrosis and the enhanced expression of defense genes (Świadek et al. 2017). In fact, ACD6 acts as a quantitative resistance gene that balances growth and pathogen resistance in natural populations of *A. thaliana,* and it has been shown that deleterious autoimmune ACD6 alleles are modulated by natural variants of SNC1 (Zhu et al. 2018).

The hybrid necrosis hot spots in the *A. thaliana* genome are often densely populated with NLRs. These immune receptor loci act as hypermodulated complexes that recombine between natural genetic variants and cause imbalanced NLR activity (Chae et al. 2014). The Bateson–Dobzhansky–Muller model explains pairwise heteromeric interactions between distinct unlinked NLR loci that lead to hybrid necrosis and enhanced defense (Phadnis and Malik 2014). On the one hand, the polymorphic nature (high levels of sequence divergence) of these immune loci gives an advantage during the host response to pathogen challenges, while on the other hand it positively correlates to hybrid necrosis impacting plant fitness. Different NLR pairs have been involved in the hybrid necrosis phenomena (Bomblies and Weigel 2007; Tran et al. 2017; Atanasov et al. 2018). Many questions regarding hybrid necrosis remain unanswered. How is the NLR-mediated defense response propagated without pathogen challenge? What is the role of environmental factors and genetic distance in hybrid necrosis induction? How can the deleterious fitness effects be mitigated during interspecific crossing while preserving the resistance trait? Our understanding of the mechanisms regulating HR cell death triggered by autoimmunity is still very limited. A deeper understanding of NLR activation and signal transduction will help us integrate and advance the current knowledge.

CONCLUDING REMARKS

In plant immunity, HR cell death is often used to score resistance to pathogens. However, the mechanisms regulating this complex phenomenon are far from understood. The intricate interplay between sensor and helper immune receptors is starting to emerge, and will help shed light on how cell death is triggered and executed on pathogen perception. Cell death–specific modules are being unveiled as integrators of signals emanating from activation of diverse sensor NLRs. HR cell death is an important part of the immune response to protect distal parts of the plant against future invasions.

The plant resistosome has been described as an inflammasome-like supramolecular structure that assembles on recognition of pathogenic effectors to initiate defense responses. The activated resistosome features a funnel-like structure that is required for insertion of the complex into the plasma membrane and HR cell death. It has been speculated that this structure creates pores in the membrane, which could mediate ion fluxes that activate cell death enzymes. Perturbation of the plasma membrane or its signaling components—including PRRs—may also be monitored by NLRs that can trigger HR cell death.

In addition to immunity against biotrophic and hemibiotrophic pathogens, HR cell death can also mediate susceptibility to necrotrophic pathogens. There are several examples of necrotrophic fungi secreting toxins (also known as NEs) that directly or indirectly activate specific NLRs and cause HR cell death. These NLRs were probably selected in the course of interactions between a plant and biotrophic pathogen, and then hijacked by a necrotrophic fungus for its own benefit. This is an emerging area of research with great potential, because susceptibility genes can serve as targets for genome-editing technologies aimed at increasing resistance against fungi in commercial cultivars.

The analysis of autoimmune phenotypes in plants is also providing a better understanding of the mechanisms regulating HR cell death. Autoactive or mismatched NLR alleles confer constitutive immunity and ectopic HR-like cell death phenotypes, highlighting the importance of a multilayered and finely tuned regulation of immune modulators to avoid deleterious fitness costs for the plant. The booming field of plant

Cite this article as *Cold Spring Harb Perspect Biol* doi: 10.1101/cshperspect.a036483

immunity will surely deepen our insight into the mechanisms regulating pathogen-triggered HR cell death, helping us understand to what extent it is programmed.

ACKNOWLEDGMENTS

We thank Mary-Paz González García, Marc Valls, and Jose Salguero-Linares for comments on the manuscript. We also thank Jane Parker and Hans Thordal-Christensen for inspiring discussions on the topic. We apologize to all authors whose work has been omitted because of space limitations. This work was supported by the Spanish Ministry of Economy and Competitiveness with Grant Nos. RyC 2014-16158 and AGL2016-78002-R and through the Severo Ochoa Programme for Centres of Excellence in R&D (SEV-2015-0533). We also acknowledge financial support from the CERCA Programme/Generalitat de Catalunya.

REFERENCES

Adachi H, Tsuda K. 2019. Convergence of cell-surface and intracellular immune receptor signalling. *New Phytol* **221:** 1676–1678. doi:10.1111/nph.15634

Adachi H, Kamoun S, Maqbool A. 2019. A resistosome-activated "death switch." *Nat Plants* **5:** 457–458. doi:10.1038/s41477-019-0425-9

Alcazar R, Garcia AV, Parker JE, Reymond M. 2009. Incremental steps toward incompatibility revealed by *Arabidopsis* epistatic interactions modulating salicylic acid pathway activation. *Proc Natl Acad Sci* **106:** 334–339. doi:10.1073/pnas.0811734106

Atanasov KE, Liu C, Erban A, Kopka J, Parker JE, Alcázar R. 2018. *NLR* mutations suppressing immune hybrid incompatibility and their effects on disease resistance. *Plant Physiol* **177:** 1152–1169. doi:10.1104/pp.18.00462

Avin-Wittenberg T, Baluška F, Bozhkov P V, Elander PH, Fernie AR, Galili G, Hassan A, Hofius D, Isono E, Le Bars R, et al. 2018. Autophagy-related approaches for improving nutrient use efficiency and crop yield protection. *J Exp Bot* **69:** 1335–1353. doi:10.1093/jxb/ery069

Balagué C, Lin B, Alcon C, Flottes G, Malmström S, Köhler C, Neuhaus G, Pelletier G, Gaymard F, Roby D. 2003. HLM1, an essential signaling component in the hypersensitive response, is a member of the cyclic nucleotide-gated channel ion channel family. *Plant Cell* **15:** 365–379.

Betsuyaku S, Katou S, Takebayashi Y, Sakakibara H, Nomura N, Fukuda H. 2018. Salicylic acid and jasmonic acid pathways are activated in spatially different domains around the infection site during effector-triggered immunity in *Arabidopsis thaliana*. *Plant Cell Physiol* **59:** 439. doi:10.1093/pcp/pcy008

Bomblies K, Weigel D. 2007. Hybrid necrosis: autoimmunity as a potential gene-flow barrier in plant species. *Nat Rev Genet* **8:** 382–393. doi:10.1038/nrg2082

Bonardi V, Cherkis K, Nishimura MT, Dangl JL. 2012. A new eye on NLR proteins: focused on clarity or diffused by complexity? *Curr Opin Immunol* **24:** 41–50. doi:10.1016/j.coi.2011.12.006

Bruggeman Q, Raynaud C, Benhamed M, Delarue M. 2015. To die or not to die? Lessons from lesion mimic mutants. *Front Plant Sci* **6:** 24. doi:10.3389/fpls.2015.00024

Bücherl CA, Jarsch IK, Schudoma C, Segonzac C, Mbengue M, Robatzek S, MacLean D, Ott T, Zipfe C. 2017. Plant immune and growth receptors share common signalling components but localise to distinct plasma membrane nanodomains. *eLife* **6:** e25114. doi:10.7554/eLife.25114

Cai Q, Liang C, Wang S, Hou Y, Gao L, Liu L, He W, Ma W, Mo B, Chen X. 2018. The disease resistance protein SNC1 represses the biogenesis of microRNAs and phased siRNAs. *Nat Commun* **9:** 5080. doi:10.1038/s41467-018-07516-z

Castel B, Ngou PM, Cevik V, Redkar A, Kim DS, Yang Y, Ding P, Jones JDG. 2019. Diverse NLR immune receptors activate defence via the RPW8-NLR NRG1. *New Phytol* **222:** 966–980. doi:10.1111/nph.15659

Chae E, Bomblies K, Kim ST, Karelina D, Zaidem M, Ossowski S, Martín-Pizarro C, Laitinen RAE, Rowan BA, Tenenboim H, et al. 2014. Species-wide genetic incompatibility analysis identifies immune genes as hot spots of deleterious epistasis. *Cell* **159:** 1341–1351. doi:10.1016/j.cell.2014.10.049

Chakraborty J, Ghosh P, Das S. 2018. Autoimmunity in plants. *Planta* **248:** 751–767. doi:10.1007/s00425-018-2956-0

Chen C EZ, Lin HX. 2016. Evolution and molecular control of hybrid incompatibility in plants. *Front Plant Sci* **7:** 1–10.

Cheng YT, Li Y, Huang S, Huang Y, Dong X, Zhang Y, Li X. 2011. Stability of plant immune-receptor resistance proteins is controlled by SKP1-CULLIN1-F-box (SCF)-mediated protein degradation. *Proc Natl Acad Sci* **108:** 14694–14699. doi:10.1073/pnas.1105685108

Chintamanani S, Hulbert SH, Johal GS, Balint-Kurti PJ. 2010. Identification of a maize locus that modulates the hypersensitive defense response, using mutant-assisted gene identification and characterization. *Genetics* **184:** 813–825. doi:10.1534/genetics.109.111880

Coll NS, Vercammen D, Smidler A, Clover C, Van Breusegem F, Dangl JL, Epple P. 2010. *Arabidopsis* type I metacaspases control cell death. *Science* **330:** 1393–1397. doi:10.1126/science.1194980

Coll NS, Smidler A, Puigvert M, Popa C, Valls M, Dangl JL. 2014. The plant metacaspase AtMC1 in pathogen-triggered programmed cell death and aging: functional linkage with autophagy. *Cell Death Differ* **21:** 1399–1408. doi:10.1038/cdd.2014.50

Cook DE, Mesarich CH, Thomma BPHJ. 2015. Understanding plant immunity as a surveillance system to detect invasion. *Annu Rev Phytopathol* **53:** 541–563. doi:10.1146/annurev-phyto-080614-120114

Cui H, Tsuda K, Parker JE. 2014. Effector-triggered immunity: from pathogen perception to robust defense. *Annu*

Rev Plant Biol **66:** 487–511. doi:10.1146/annurev-ar plant-050213-040012

Dangl JL, Jones JDG. 2019. A pentangular plant inflamma-some. *Science* **364:** 31–32. doi:10.1126/science.aax0174

Dickman M, Williams B, Li Y, Figueiredo P, Wolpert T. 2017. Reassessing apoptosis in plants. *Nat Plants* **3:** 773–779. doi:10.1038/s41477-017-0020-x

Domínguez-ferreras A, Kiss-papp M, Jehle AK, Felix G, Chinchilla D. 2015. An overdose of the *Arabidopsis* coreceptor BRASSINOSTEROID INSENSITIVE1-ASSOCIATED RECEPTOR KINASE1 or its ectodomain causes autoimmunity in a SUPPRESSOR OF BIR1-1-dependent manner. *Plant Physiol* **168:** 1106–1121.

Dong OX, Meteignier LV, Plourde MB, Ahmed B, Wang M, Jensen C, Jin H, Moffett P, Li X, Germain H. 2016. *Arabidopsis* TAF15b localizes to RNA processing bodies and contributes to *snc1*-mediated autoimmunity. *Mol Plant Microbe Interact* **29:** 247–257. doi:10.1094/MPMI-11-15-0246-R

Faris JD, Zhang Z, Lu H, Lu S, Reddy L, Cloutier S, Fellers JP, Meinhardt SW, Rasmussen JB, Xu SS, et al. 2010. A unique wheat disease resistance-like gene governs effector-triggered susceptibility to necrotrophic pathogens. *Proc Natl Acad Sci* **107:** 13544–13549. doi:10.1073/pnas.1004090107

Feng B, Tang D. 2019. Mechanism of plant immune activation and signaling: insight from the first solved plant resistosome structure. *J Integr Plant Biol* **61:** 902–907 doi:10.1111/jipb.12814

Fu ZQ, Dong X. 2013. Systemic acquired resistance: turning local infection into global defense. *Annu Rev Plant Biol* **64:** 839–863. doi:10.1146/annurev-arplant-042811-105606

Galluzzi L, Vitale I, Aaronson SA, Abrams JM, Adam D, Agostinis P, Alnemri ES, Altucci L, Amelio I, Andrews DW, et al. 2018. Molecular mechanisms of cell death: recommendations of the nomenclature committee on cell death 2018. *Cell Death Differ* **25:** 486–541. doi:10.1038/s41418-017-0012-4

Gou M, Huang Q, Qian W, Zhang Z, Jia Z, Hua J. 2017. Sumoylation E3 ligase SIZ1 modulates plant immunity partly through the immune receptor gene *SNC1* in *Arabidopsis*. *Mol Plant Microbe Interact* **30:** 334–342. doi:10.1094/MPMI-02-17-0041-R

Greenberg JT, Silverman FP, Liang H. 2000. Uncoupling salicylic acid-dependent cell death and defense-related responses from disease resistance in the *Arabidopsis* mutant *acd5*. *Genetics* **156:** 341–350.

Hander T, Fernández-fernández ÁD, Kumpf RP, Willems P, Schatowitz H, Rombaut D, Staes A, Nolf J, Pottie R, Yao P, et al. 2019. Damage on plants activates Ca^{2+}-dependent metacaspases for release of immunomodulatory peptides. *Science* **363:** eaar7486. doi:10.1126/science.aar7486

Hofius D, Schultz-Larsen T, Joensen J, Tsitsigiannis DI, Petersen NHT, Mattsson O, Jørgensen LB, Jones JDG, Mundy J, Petersen M. 2009. Autophagic components contribute to hypersensitive cell death in *Arabidopsis*. *Cell* **137:** 773–783. doi:10.1016/j.cell.2009.02.036

Hofius D, Li L, Hafrén A, Coll NS. 2017. Autophagy as an emerging arena for plant–pathogen interactions. *Curr Opin Plant Biol* **38:** 117–123. doi:10.1016/j.pbi.2017.04.017

Huang Y, Minaker S, Roth C, Huang S, Hieter P, Lipka V, Wiermer M, Li X. 2014. An E4 ligase facilitates polyubiquitination of plant immune receptor resistance proteins in *Arabidopsis*. *Plant Cell* **26:** 485–496. doi:10.1105/tpc.113.119057

Huang Y, Cui Y, Hou X, Huang T. 2018. The AtMC4 regulates the stem cell homeostasis in *Arabidopsis* by catalyzing the cleavage of AtLa1 protein in response to environmental hazards. *Plant Sci* **266:** 64–75. doi:10.1016/j.plantsci.2017.10.008

Johnson KCM, Xia S, Feng X, Li X. 2015. The chromatin remodeler SPLAYED negatively regulates SNC1-mediated immunity. *Plant Cell Physiol* **56:** 1616–1623. doi:10.1093/pcp/pcv087

Johnson KCM, Zhao J, Wu Z, Roth C, Lipka V, Wiermer M, Li X. 2017. The putative kinase substrate MUSE7 negatively impacts the accumulation of NLR proteins. *Plant J* **89:** 1174–1183. doi:10.1111/tpj.13454

Jones JDG, Dangl JL. 2006. The plant immune system. *Nature* **444:** 323–329. doi:10.1038/nature05286

Jurkowski GI, Smith RK, Yu I, Ham JH, Sharma SB, Klessig DF, Fengler KA, Bent AF. 2004. *Arabidopsis* DND2, a second cyclic nucleotide-gated ion channel gene for which mutation causes the "*defense, no death*" phenotype. *Mol Plant Microbe Interact* **17:** 511–520. doi:10.1094/MPMI.2004.17.5.511

Kabbage M, Williams B, Dickman MB. 2013. Cell death control: the interplay of apoptosis and autophagy in the pathogenicity of *Sclerotinia sclerotiorum*. *PLoS Pathog* **9:** e1003287. doi:10.1371/journal.ppat.1003287

Kemmerling B, Schwedt A, Rodriguez P, Mazzotta S, Frank M, Qamar SA, Mengiste T, Betsuyaku S, Parker JE, Müssig C, et al. 2007. The BRI1-associated kinase 1, BAK1, has a Brassinolide-independent role in plant cell-death control. *Curr Biol* **17:** 1116–1122. doi:10.1016/j.cub.2007.05.046

Kim KS, Min JY, Dickman MB. 2008. Oxalic acid is an elicitor of plant programmed cell death during *Sclerotinia sclerotiorum* disease development. *Mol Plant Microbe Interact* **21:** 605–612. doi:10.1094/MPMI-21-5-0605

Lapin D, Kovacova V, Sun X, Dongus JA, Bhandari DD, von Born P, Bautor J, Guarneri N, Rzemieniewski J, Stuttmann J, et al. 2019. A coevolved EDS1-SAG101-NRG1 module mediates cell death signaling by TIR-domain immune receptors. *Plant Cell*. doi:10.1105/tpc.19.00118

Lema Asqui S, Vercammen D, Serrano I, Valls M, Rivas S, Van Breusegem F, Conlon FL, Dangl JL, Coll NS. 2018. AtSERPIN1 is an inhibitor of the metacaspase AtMC1-mediated cell death and autocatalytic processing *in planta*. *New Phytol* **218:** 1156–1166. doi:10.1111/nph.14446

Li X, Clarke JD, Zhang Y, Dong X. 2001. Activation of an EDS1-mediated *R*-gene pathway in the *snc1* mutant leads to constitutive, NPR1-independent pathogen resistance. *Mol Plant Microbe Interact* **14:** 1131–1139. doi:10.1094/MPMI.2001.14.10.1131

Liu Y, Huang X, Li M, He P, Zhang Y. 2016a. Loss-of-function of *Arabidopsis* receptor-like kinase BIR1 activates cell death and defense responses mediated by BAK1 and SOBIR1. *New Phytol* **212:** 637–645. doi:10.1111/nph.14072

Liu Z, Gao Y, Kim YM, Faris JD, Shelver WL, de Wit PJGM, Xu SS, Friesen TL. 2016b. SnTox1, a *Parastagonospora nodorum* necrotrophic effector, is a dual-function protein that facilitates infection while protecting from wheat-produced chitinases. *New Phytol* **211:** 1052–1064. doi:10.1111/nph.13959

Lolle S, Greeff C, Petersen K, Roux M, Jensen MK, Bressendorff S, Rodriguez E, Sømark K, Mundy J, Petersen M. 2017. Matching NLR immune receptors to autoimmunity in camta3 mutants using antimorphic NLR alleles. *Cell Host Microbe* **21:** 518–529.e4. doi:10.1016/j.chom.2017.03.005

Lorang JM, Sweat TA, Wolpert TJ. 2007. Plant disease susceptibility conferred by a "resistance" gene. *Proc Natl Acad Sci* **104:** 14861–14866. doi:10.1073/pnas.0702572104

Lorang J, Kidarsa T, Bradford CS, Gilbert B, Curtis M, Tzeng SC, Maier CS, Wolpert TJ. 2012. Tricking the guard: exploiting plant defense for disease susceptibility. *Science* **338:** 659–662. doi:10.1126/science.1226743

Menna A, Nguyen D, Guttman DS, Desveaux D. 2015. Elevated temperature differentially influences effector-triggered immunity outputs in *Arabidopsis*. *Front Plant Sci* **6:** 1–7. doi:10.3389/fpls.2015.00995

Munch D, Rodriguez E, Bressendorff S, Park OK, Hofius D, Petersen M. 2014. Autophagy deficiency leads to accumulation of ubiquitinated proteins, ER stress, and cell death in *Arabidopsis*. *Autophagy* **10:** 1579–1587. doi:10.4161/auto.29406

Negeri A, Wang GF, Benavente L, Kibiti CM, Chaikam V, Johal G, Balint-Kurti P. 2013. Characterization of temperature and light effects on the defense response phenotypes associated with the maize *Rp1-D21* autoactive resistance gene. *BMC Plant Biol* **13:** 106. doi:10.1186/1471-2229-13-106

Niu D, Lin XL, Kong X, Qu GP, Cai B, Lee J, Jin JB. 2019. SIZ1-mediated SUMOylation of TPR1 suppresses plant immunity in *Arabidopsis*. *Mol Plant* **12:** 215–228. doi:10.1016/j.molp.2018.12.002

Paulus JK, Van der Hoorn RAL. 2019. Do proteolytic cascades exist in plants? *J Exp Bot* **70:** 1997–2002. doi:10.1093/jxb/erz016

Peters Haugrud AR, Zhang Z, Richards JK, Friesen TL, Faris JD. 2019. Genetics of variable disease expression conferred by inverse gene-for-gene interactions in the wheat–*Parastagonospora nodorum* pathosystem. *Plant Physiol* **180:** 420–434. doi:10.1104/pp.19.00149

Phadnis N, Malik HS. 2014. Speciation via autoimmunity: a dangerous mix. *Cell* **159:** 1247–1249. doi:10.1016/j.cell.2014.11.028

Qi T, Seong K, Thomazella DPT, Kim JR, Pham J, Seo E, Cho MJ, Schultink A, Staskawicz BJ. 2018. NRG1 functions downstream of EDS1 to regulate TIR-NLR-mediated plant immunity in *Nicotiana benthamiana*. *Proc Natl Acad Sci* **115:** E10979–E10987. doi:10.1073/pnas.1814856115

Raffaele S, Rivas S. 2013. Regulate and be regulated: integration of defense and other signals by the AtMYB30 transcription factor. *Front Plant Sci* **4:** 98. doi:10.3389/fpls.2013.00098

Rodriguez E, El Ghoul H, Mundy J, Petersen M. 2016. Making sense of plant autoimmunity and "negative regulators." *FEBS J* **283:** 1385–1391. doi:10.1111/febs.13613

Roux F, Noël L, Rivas S, Roby D. 2014. ZRK atypical kinases: emerging signaling components of plant immunity. *New Phytol* **203:** 713–716. doi:10.1111/nph.12841

Salguero-Linares J, Coll NS. 2019. Plant proteases in the control of the hypersensitive response. *J Exp Bot* **70:** 2087–2095. doi:10.1093/jxb/erz030

Sheikh AH, Raghuram B, Eschen-Lippold L, Scheel D, Lee J, Sinha AK. 2014. Agroinfiltration by cytokinin-producing *Agrobacterium* sp. strain GV3101 primes defense responses in *Nicotiana tabacum*. *Mol Plant Microbe Interact* **27:** 1175–1185. doi:10.1094/MPMI-04-14-0114-R

Shi G, Zhang Z, Friesen TL, Raats D, Fahima T, Brueggeman RS, Lu S, Trick HN, Liu Z, Chao W, et al. 2016. The hijacking of a receptor kinase-driven pathway by a wheat fungal pathogen leads to disease. *Sci Adv* **2:** e1600822. doi:10.1126/sciadv.1600822

Shine MB, Xiao X, Kachroo P, Kachroo A. 2019. Signaling mechanisms underlying systemic acquired resistance to microbial pathogens. *Plant Sci* **279:** 81–86. doi:10.1016/j.plantsci.2018.01.001

Stakman EC. 1915. Relation between *Puccina graminis* and plants highly resistant to its attack. *J Agric Res* **4:** 193–199.

Świadek M, Proost S, Sieh D, Yu J, Todesco M, Jorzig C, Rodriguez Cubillos AE, Plötner B, Nikoloski Z, Chae E, et al. 2017. Novel allelic variants in *ACD6* cause hybrid necrosis in local collection of *Arabidopsis thaliana*. *New Phytol* **213:** 900–915. doi:10.1111/nph.14155

Tang D, Zhou JM. 2016. PEPRs spice up plant immunity. *EMBO J* **35:** 4–5. doi:10.15252/embj.201593434

Tran DTN, Chung EH, Habring-Müller A, Demar M, Schwab R, Dangl JL, Weigel D, Chae E. 2017. Activation of a plant NLR complex through heteromeric association with an autoimmune risk variant of another NLR. *Curr Biol* **27:** 1148–1160. doi:10.1016/j.cub.2017.03.018

Vaid N, Laitinen RAE. 2019. Diverse paths to hybrid incompatibility in *Arabidopsis*. *Plant J* **97:** 199–213. doi:10.1111/tpj.14061

Vanden BT, Linkermann A, Jouan-Lanhouet S, Walczak H, Vandenabeele P. 2014. Regulated necrosis: the expanding network of non-apoptotic cell death pathways. *Nat Rev Mol Cell Biol* **15:** 135–147.

Wang GF, Balint-Kurti PJ. 2016. Maize homologs of CCoAOMT and HCT, two key enzymes in lignin biosynthesis, form complexes with the NLR Rp1 protein to modulate the defense response. *Plant Physiol* **171:** 2166–2177. doi:10.1104/pp.16.00224

Wang G, Roux B, Feng F, Guy E, Li L, Li N, Zhang X, Lautier M, Jardinaud MF, Chabannes M, et al. 2015a. The decoy substrate of a pathogen effector and a pseudokinase specify pathogen-induced modified-self recognition and immunity in plants. *Cell Host Microbe* **18:** 285–295. doi:10.1016/j.chom.2015.08.004

Wang GF, Ji J, EI-Kasmi F, Dangl JL, Johal G, Balint-Kurti PJ. 2015b. Molecular and functional analyses of a maize autoactive NB-LRR protein identify precise structural requirements for activity. *PLoS Pathog* **11:** e1004830. doi:10.1371/journal.ppat.1004830

Wang S, Wang S, Sun Q, Yang L, Zhu Y, Yuan Y, Hua J. 2017a. A role of cytokinin transporter in *Arabidopsis* immunity. *Mol Plant Microbe Interact* **30:** 325–333. doi:10.1094/MPMI-01-17-0011-R

Wang Z, Cui D, Liu J, Zhao J, Liu C, Xin W, Li Y, Liu N, Ren D, Tang D, et al. 2017b. *Arabidopsis* ZED1-related kinases mediate the temperature-sensitive intersection of immune response and growth homeostasis. *New Phytol* **215:** 711–724. doi:10.1111/nph.14585

Wang J, Hu M, Wang J, Qi J, Han Z, Wang G, Qi Y, Wang HW, Zhou JM, Chai J. 2019a. Reconstitution and structure of a plant NLR resistosome conferring immunity. *Science* **364:** eaav5870. doi:10.1126/science.aav5870

Wang J, Wang J, Hu M, Wu S, Qi J, Wang G, Han Z, Qi Y, Gao N, Wang HW, et al. 2019b. Ligand-triggered allosteric ADP release primes a plant NLR complex. *Science* **364:** eaav5868. doi:10.1126/science.aav5868

Wang Z, Cui D, Liu C, Zhao J, Liu J, Liu N, Tang D, Hu Y. 2019c. TCP transcription factors interact with ZED1-related kinases as components of the temperature-regulated immunity. *Plant Cell Environ* **42:** 2045–2056. doi:10.1111/pce.13515

Wiermer M, Feys BJ, Parker JE. 2005. Plant immunity: the EDS1 regulatory node. *Curr Opin Plant Biol* **8:** 383–389. doi:10.1016/j.pbi.2005.05.010

Williams B, Kabbage M, Kim HJ, Britt R, Dickman MB. 2011. Tipping the balance: *Sclerotinia sclerotiorum* secreted oxalic acid suppresses host defenses by manipulating the host redox environment. *PLoS Pathog* **7:** e1002107. doi:10.1371/journal.ppat.1002107

Wu CH, Belhaj K, Bozkurt TO, Kamoun S. 2016. Helper NLR proteins NRC2a/b and NRC3 but not NRC1 are required for Pto-mediated cell death and resistance in *Nicotiana benthamiana*. *New Phytol* **209:** 1344–1352. doi:10.1111/nph.13764

Wu CH, Abd-El-Haliem A, Bozkurt TO, Belhaj K, Terauchi R, Vossen JH, Kamoun S. 2017. NLR network mediates immunity to diverse plant pathogens. *Proc Natl Acad Sci* **114:** 8113–8118. doi:10.1073/pnas.1702041114

Yu I, Parker J, Bent AF. 1998. Gene-for-gene disease resistance without the hypersensitive response in *Arabidopsis dnd1* mutant. *Proc Natl Acad Sci* **95:** 7819–7824. doi:10.1073/pnas.95.13.7819

Zhang X, Dodds PN, Bernoux M. 2017. What do we know about NOD-like receptors in plant immunity? *Annu Rev Phytopathol* **55:** 205–229. doi:10.1146/annurev-phyto-080516-035250

Zhang N, Wang Z, Bao Z, Yang L, Wu D, Shu X, Hua J. 2018. MOS1 functions closely with TCP transcription factors to modulate immunity and cell cycle in *Arabidopsis*. *Plant J* **93:** 66–78. doi:10.1111/tpj.13757

Zhu Z, Xu F, Zhang Y, Cheng YT, Wiermer M, Li X, Zhang Y. 2010. *Arabidopsis* resistance protein SNC1 activates immune responses through association with a transcriptional corepressor. *Proc Natl Acad Sci* **107:** 13960–13965. doi:10.1073/pnas.1002828107

Zhu W, Zaidem M, Van de Weyer AL, Gutaker RM, Chae E, Kim ST, Bemm F, Li L, Todesco M, Schwab R, et al. 2018. Modulation of *ACD6* dependent hyperimmunity by natural alleles of an *Arabidopsis thaliana* NLR resistance gene. *PLoS Genet* **14:** e1007628. doi:10.1371/journal.pgen.1007628

Dysregulation of Cell Death in Human Chronic Inflammation

Yue Li, Christoph Klein, and Daniel Kotlarz

Dr. von Hauner Children's Hospital, Department of Pediatrics, University Hospital,
Ludwig-Maximilians-Universität (LMU) Munich, 80337 Munich, Bavaria, Germany

Correspondence: daniel.kotlarz@med.uni-muenchen.de

Inflammation is a fundamental biological process mediating host defense and wound healing during infections and tissue injury. Perpetuated and excessive inflammation may cause autoinflammation, autoimmunity, degenerative disorders, allergies, and malignancies. Multimodal signaling by tumor necrosis factor receptor 1 (TNFR1) plays a crucial role in determining the transition between inflammation, cell survival, and programmed cell death. Targeting TNF signaling has been proven as an effective therapeutic in several immune-related disorders. Mouse studies have provided critical mechanistic insights into TNFR1 signaling and its potential role in a broad spectrum of diseases. The characterization of patients with monogenic primary immunodeficiencies (PIDs) has highlighted the importance of TNFR1 signaling in human disease. In particular, patients with PIDs have revealed paradoxical connections between immunodeficiency, chronic inflammation, and dysregulated cell death. Importantly, studies on PIDs may help to predict beneficial effects and side-effects of therapeutic targeting of TNFR1 signaling.

Inflammation is a protective mechanism in host defense and wound healing during tissue damage or infection (Medzhitov 2008). The degree of inflammation depends on the infectious or toxic triggers and on host susceptibility. Inflammatory responses are complex processes involving vascular permeability, inflammatory mediators (e.g., chemokines, adhesion molecules, cytokines, enzymes), detecting sensors, and extracellular matrix components, as well as recruitment of circulating inflammatory cells, activation of resident immune cells, and adaptive immunity.

Inflammatory mediators, danger-associated molecular patterns (DAMPs), and hypoxia lead to recruitment and degranulation of platelets and resident mast cells as well as activation of tissue-resident immune cells. The release of chemoattractants orchestrates leucocyte migration to the site of inflammation (Medzhitov 2008). Neutrophils with phagocytotic and microbicidal functions are recruited from the circulation as well. Initially, neutrophils potentiate the proinflammatory environment to eliminate inflammatory agents, but apoptosis and clearance of neutrophils are central processes in the resolution of inflammation (Mantovani et al. 2011). Circulating monocytes enter the site of inflammation and differentiate into tissue macrophages that phagocytose foreign particles, debris, and apoptotic cells. The clearance of apoptotic neutrophils triggers a switch from a pro- to an anti-inflammatory program in macrophages. In the late phase of inflammation, lymphocytes will be recruited and mediate adaptive immunity (Serhan and Savill 2005). Coordinated networks are required to resolve and control inflammatory processes. Excessive and uncontrolled inflammation caused by failure to remove noxious materials and apoptotic inflammatory cells may contribute to autoinflammation, autoimmunity, degenerative diseases, allergy, and malignancies (Silva et al. 2008).

Inflammation and cell death are intertwined biological processes sharing many receptors and effector molecules. The release of proinflammatory factors by dying cells may facilitate recovery or extension of inflammation, but accumulating evidence suggests that perturbed cell-death responses may actively contribute to inflammation (Rock and Kono 2008). Whereas necroptosis and pyroptosis release DAMPs (for example, ATP, DNA, and uric acid) through permeabilized membranes and are primarily considered to enhance inflammation, apoptosis contains cytoplasmic content and is thought to be critical in the termination process (Rock and Kono 2008). While different forms of cell death share morphological and biochemical similarities, the molecular characteristics and host responses can be drastically different depending on the biological context. The fate decision of cell death versus inflammation is tightly controlled by multiple pathways, including proinflammatory tumor necrosis factor receptor 1 (TNFR1) signaling. Mouse studies have unveiled mechanistic insights on the regulation of TNFR1 signaling and how it may contribute to disease (Fig. 1; Silke et al. 2015). The characterization of patients with monogenic primary immunodeficiencies (PIDs) has shown the critical role of TNFR1 signaling in human disease and highlighted paradoxical links between immunodeficiency and dysregulation of cell death in chronic inflammation (Table 1). Here, we review recent insights with a focus on novel inherited errors of human immunity.

MULTIMODAL TNFR1-DEPENDENT SIGNALING DETERMINES INFLAMMATORY AND CELL-DEATH FATES

TNF plays a critical role in regulating host defense, but can also be pathogenic in several inflammatory conditions (Monaco et al. 2015). TNFR1 signaling intertwines inflammation and cell death by engaging IKK/NF-κB and caspase-8/receptor interacting protein kinase 1 (RIPK1)/RIPK3 signaling (Fig. 1; Kalliolias and Ivashkiv 2016). TNF is produced by several immune, epithelial, endothelial, and stromal cell types (Grivennikov et al. 2005). Upon binding of TNF to trimeric TNFR1, a membrane-associated complex I is formed by recruitment of the adaptor protein TNFR1-associated death domain protein (TRADD), TNFR1-associated factor 2 (TRAF2), cellular inhibitor of apoptosis proteins 1 and 2 (cIAP1/2), RIPK1, and linear ubiquitin chain assembly complex (LUBAC) (Micheau and Tschopp 2003; Kirisako et al. 2006). The latter is composed of heme-oxidized IRP2 ubiquitin ligase 1 (HOIL-1), HOIL-1-interacting protein (HOIP), and SHANK-associated RH domain-interacting protein (SHARPIN) (Kirisako et al. 2006; Gerlach et al. 2011; Ikeda et al. 2011; Tokunaga et al. 2011). Modification of RIPK1 and possibly other complex I components with Lys63-linked polyubiquitin assembled by cIAP1/2, and Met1-linked ubiquitin assembled by LUBAC, mediates activation of TGF-β-activated kinase 1 (TAK1) and IκB kinase (IKK) (Micheau and Tschopp 2003; Wang et al. 2008). Activated TAK1 and IKK induce MAPK signaling and ubiquitin–protein system-mediated degradation of IκB leading to NF-κB activation.

Compromised prosurvival signaling emanating from complex I results in the formation of alternative cytosolic TNF-induced complexes mediating apoptosis and necroptosis (Van Antwerp et al. 1996). Proinflammatory NF-κB signaling can be terminated by disassembly of complex I through A20- and cylindromatosis (CYLD)-mediated deubiquitylation of RIPK1 and TRAF2 (Wertz et al. 2004; Wang et al. 2008). Formation of cytosolic complexes containing TRADD, Fas-associated protein with death domain (FADD), RIPK1, and procas-

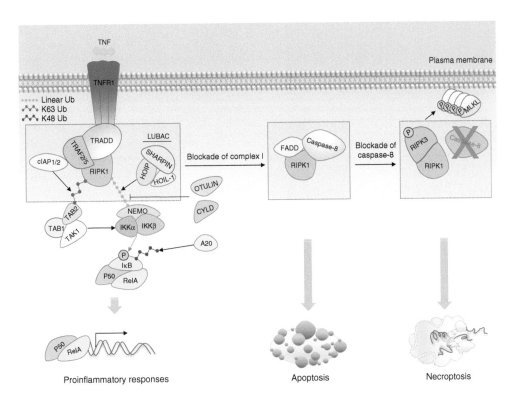

Figure 1. The tumor necrosis factor receptor 1 (TNFR1) signaling pathway as a master regulator of inflammation and cell death. TNFR1 encountering TNF nucleates complex I, which includes TNFR1-associated death domain protein (TRADD), receptor-interacting protein kinase 1 (RIPK1), TNFR1-associated factor 2 (TRAF2), cellular inhibitor of apoptosis proteins 1 and 2 (cIAP1/2), and the linear ubiquitin chain assembly complex (LUBAC) composed of heme-oxidized IRP2 ubiquitin ligase 1 (HOIL-1), HOIL1-interacting protein (HOIP), and SHANK-associated RH domain-interacting protein (SHARPIN). Polyubiquitinated RIPK1 recruits the IκB kinase (IKK) complex (composed of the NF-κB essential modulator [NEMO], IKKα, and IKKβ) and the TAK1 complex, which mediate NF-κB and MAPK signaling. Degradation of phosphorylated IκB mediates translocation of p50 and RelA to the nucleus and transcription of proinflammatory and prosurvival NF-κB target genes. The stability of complex I is regulated by deubiquitinating enzymes such as A20, cylindromatosis (CYLD), and OTU deubiquitinase with linear linkage specificity (OTULIN). Formation of complex II containing Fas-associated protein with death domain (FADD), caspase-8, TRADD, and RIPK1 can trigger apoptosis. If the activity of caspase-8 is compromised, RIPK3 interacts with RIPK1 via its RHIM domain. Autophosphorylated RIPK3 leads to recruitment, phosphorylation, and oligomerization of the pseudokinase mixed lineage kinase domain-like (MLKL). Translocation of activated MLKL to the plasma membrane results in necroptosis. Proteins highlighted by red frames indicate that mutations in the corresponding genes have been reported as monogenic causes for primary immunodeficiencies.

pase-8 (Micheau and Tschopp 2003; Wang et al. 2008) can lead to homodimerization and activation of caspase-8, with subsequent cleavage of caspase-3 and -7 mediating extrinsic apoptosis (Boatright et al. 2003; Micheau and Tschopp 2003). Apoptosis is the best-defined form of programmed cell death with characteristic morphological and biochemical changes such as

nuclear envelope disassembly, cytoplasmic condensation and fragmentation, membrane blebbing, and formation of membrane-bound bodies (Green et al. 2009). Apoptosis plays a pivotal role in controlling immune cell development and homeostasis, by eliminating self-reactive, overactivated, and infected immune cells (Green et al. 2009). Apoptotic cells are ingested

Table 1. Phenotypes of constitutive knockout mouse models and patients with monogenic immune-related disorders affecting tumor necrosis factor receptor 1 (TNFR1) signaling

Genes (mouse/human)	Mouse	Human
Casp8/CASP8	Embryonic lethality; cardiac deformations; neural tube defects; hematopoietic progenitor dysfunctions (Varfolomeev et al. 1998)	ALPS-like disorder (lymphadenopathy, splenomegaly, immunodeficiency with defective activation of T, B, and NK cells (Chun et al. 2002) Late-onset multiorgan lymphocytic infiltrations with granulomas (Niemela et al. 2015) Immunodeficiency (increased susceptibility to viral and bacterial infections, defects in T and B cells) and VEO-IBD (Lehle et al. 2019)
Chuk/CHUK (IKKα)	Perinatal lethality; impaired limb outgrowth; skeletal morphogenesis; epidermal defects (Hu et al. 1999; Takeda et al. 1999)	Abortions; multiple fetal malformations (e.g., craniofacial abnormalities and absent limbs) (Lahtela et al. 2010)
Cyld/CYLD	Autoimmunity; abnormal thymocyte development; impaired lymphocyte activation; B-cell hyperplasia (Reiley et al. 2006; Zhang et al. 2006)	Phenotypic heterogeneity including cylindromatosis, multiple familial trichepithelioma type I, and Brooke–Spiegler syndrome (Bignell et al. 2000; Mathis et al. 2015)
Fadd/FADD	Embryonic lethality as a result of defective vascular development (Yeh et al. 1998)	Immunodeficiency (bacterial and viral susceptibility); functional hyposplenism; febrile episodes; encephalopathy; developmental abnormalities (Bolze et al. 2010)
Ikbkb/IKBKB (IKKβ)	Embryonic lethality; TNFR1-dependent hepatocyte apoptosis and degeneration (Li et al. 1999; Tanaka et al. 1999)	Severe combined immunodeficiency (hypogammaglobulinemia or agammaglobulinemia, peripheral T and B cells are exclusively of naïve phenotype, absence of regulatory T cells and γδ T cells, impaired lymphocyte activation) (Pannicke et al. 2013)
Ikbkg/IKBKG (NEMO)	Males: embryonic lethality; liver degeneration; defective generation and/ or persistence of lymphocytes Females: severe skin lesions with extensive granulocyte infiltration and hyperproliferation; hepatocyte and keratinocyte apoptosis (Schmidt-Supprian et al. 2000)	Loss-of-function mutations: incontinentia pigmenti (Smahi et al. 2000) Hypomorphic mutations: X-linked ectodermal dysplasia with immunodeficiency and diverse clinical manifestations (e.g., life-threatening infections, inflammatory diseases, osteopetrosis, lymphedema) (Zonana et al. 2000; Döffinger et al. 2001)
Mlkl/MLKL	No detectable abnormality in the development of immune cells (Wu et al. 2013)	No human disease identified
Nfkb1/NFKB1	Intestinal inflammation; B-cell dysfunction; defective adaptive immunity in response to infections (Sha et al. 1995)	Haploinsufficiency: common variable immunodeficiency with recurrent respiratory infections; hypogammaglobulinemia; autoimmunity; progressing pulmonary disease (Chen et al. 2013; Fliegauf et al. 2015) Loss of function: lymphadenopathy; splenomegaly; autoimmunity; defects in B-cell differentiation (Tuijnenburg et al. 2018)

Cite this article as *Cold Spring Harb Perspect Biol* doi: 10.1101/cshperspect.a037036

Gene	Mouse phenotype	Human disease
Nfkbia/*NFKBIA* (1κBα)	Early neonatal lethality; severe inflammatory dermatitis; enhanced granulopoiesis (Beg and Baltimore 1996; Klement et al. 1996) Knockin mice (Ser32Ile): immunodeficiency; defective lymphoid organogenesis (Mooster et al. 2015)	Gain-of-function mutations: EDA-ID; T- and B-cell deficiencies with increased susceptibility to infections (Courtois et al. 2003; Boisson et al. 2017)
Otulin/*OTULIN*	Embryonic lethality; compromised craniofacial and neuronal development; impaired angiogenesis (Rivkin et al. 2013)	Fatal autoinflammation; recurrent nodular panniculitis; lipodystrophy; diarrhea; joint swelling; failure to thrive (Damgaard et al. 2016; Zhou et al. 2016b)
Rbck1/*RBCK1* (HOIL)	Embryonic lethality; disrupted vascular architecture and cell death in the yolk sac endothelium (Peltzer et al. 2018)	Immunodeficiency (susceptibility to pyogenic bacterial infections) and autoinflammation (hyperresponsiveness to IL-1β); amylopectinosis (Boisson et al. 2012)
Rela/*RELA*	Embryonic lethality; TNF-mediated cell death of hepatocytes, macrophages, and fibroblasts (Beg and Baltimore 1996)	Chronic mucocutaneous ulceration; increased apoptosis of fibroblasts in response to TNF; impaired NF-κB activation in fibroblasts and PBMCs; impaired stromal cell survival (Badran et al. 2017)
Ripk1/*RIPK1*	Perinatal lethality; massive apoptosis in lymphoid and adipose tissue; multiorgan hyperinflammation (Kelliher et al. 1998)	Life-threatening immunodeficiency (lymphopenia, recurrent infections, defective differentiation of T and B cells); VEO-IBD; arthritis (Cuchet-Lourenço et al. 2018; Li et al. 2019)
Ripk3/*RIPK3*	Viable and fertile (Newton et al. 2004)	No human disease identified
Rnf31/*RNF31* (HOIP)	Embryonic lethality; defective vascularization caused by aberrant endothelial cell death (Peltzer et al. 2014)	Multiorgan autoinflammation (hyperreactive monocytes in response to IL-1β); combined immunodeficiency (recurrent viral and bacterial infections, lymphopenia, antibody deficiency, impaired B-cell activation, and differentiation in response to CD40, impaired T-cell distribution and functions); subclinical amylopectinosis; systemic lymphangiectasia (Boisson et al. 2015)
Sharpin/*SHARPIN*	Liver inflammation; splenomegaly; severe eosinophilic skin inflammation; defective lymphoid organogenesis associated with excessive TNFR1-mediated death (Kumari et al. 2014; Rickard et al. 2014a)	No human disease identified
Tnfaip3/*TNFAIP3* (A20)	Perinatal lethality or lethality shortly after birth; severe multiorgan inflammation (e.g., liver, kidneys, intestines, joints, and bone marrow) (Lee et al. 2000)	Early-onset systemic autoinflammatory syndrome resembling Behçet's disease (Zhou et al. 2016a)

by phagocytes before they can release immunogenic intracellular contents, and this prevents activation of the innate immune system (Green et al. 2016). Impaired apoptosis has been implicated in the pathogenesis of immune-related disease conditions, as exemplified by autoimmune lymphoproliferative syndrome (ALPS) (Fisher et al. 1995; Rieux-Laucat et al. 1995; Drappa et al. 1996).

When the activity of caspase-8 is compromised, necroptosis is initiated by heterodimerization of RIPK1 and RIPK3 via their RIP homotypic interaction motif (RHIM) (Cho et al. 2009; He et al. 2009). Oligomerization and autophosphorylation of RIPK3 result in the recruitment and phosphorylation of the pseudokinase MLKL (Murphy et al. 2013). Subsequent oligomerization and translocation of MLKL to the plasma membrane lead to cell rupture (Petrie et al. 2019). The release of DAMPs from necroptotic cells may be highly immunogenic (Oberst et al. 2011). Altered necroptosis has been implicated in malignancies as well as several pathological inflammatory conditions, including infectious, cardiovascular, neurological, renal, and hepatic diseases (Weinlich et al. 2017). Moreover, several mouse studies have shown that deletion of *Mlkl* can partially ameliorate inflammation (Rickard et al. 2014a,b; Alvarez-Diaz et al. 2016), indicating that necroptosis may contribute to the pathogenesis of inflammatory disorders. In contrast, necroptosis may benefit host defense against viruses such as herpes simplex virus 1 by eliminating infected cells (Huang et al. 2015; Guo et al. 2018).

Multimodal TNFR1 signaling governs the transition between inflammation, survival, and programmed cell death. However, the exact physiological mechanisms triggering the transition from prosurvival to prodeath responses are still unclear. The complexity of TNFR1 signaling will be further modulated by cross talk with other signaling pathways that can engage with inflammatory and cell-death modules. Dysregulation of components involved in TNFR1 signaling can lead to chronic inflammation. Correspondingly, inhibition of TNF is an effective treatment for several autoinflammatory and autoimmune disorders.

INFLAMMATION IN MONOGENIC DISORDERS AFFECTING TNFR1-MEDIATED SIGNALING

Caspase-8 Deficiency

Caspase-8 is an initiator cysteinyl aspartate-specific protease critical for receptor-mediated apoptosis induced by TNF, TRAIL, and Fas ligand (FASL) (Boldin et al. 1996; Muzio et al. 1996; Ashkenazi and Dixit 1998). The zymogen procaspase-8 consists of two amino-terminal death effector domains (DEDs) and a carboxy-terminal protease domain with two catalytic subunits (p10 and p18) (Earnshaw et al. 1999). Procaspase-8 dimerization via the DED promotes proteolytic cleavage that generates active caspase-8 heterotetramers ($p10_2$-$p18_2$) (Earnshaw et al. 1999). Active caspase-8 then cleaves and activates the executioner caspases-3 and -7 to induce apoptosis (Earnshaw et al. 1999).

The essential role of caspase-8 in death receptor-induced apoptosis was shown using cells from *Casp8* knockout (KO) mice (Varfolomeev et al. 1998). These mice exhibited embryonic lethality associated with cardiac deformations, neural tube defects, and hematopoietic progenitor dysfunctions (Varfolomeev et al. 1998; Sakamaki et al. 2002). Conditional KOs of *Casp8* have revealed critical roles for caspase-8 in the response to tissue damage or infection. For example, caspase-8-deficient hepatocytes exhibited impaired proliferation after injury to the liver and this prompted chronic inflammation of the liver (Ben Moshe et al. 2007). Loss of *Casp8* in the epidermis also caused inflammation with hyperactive responses to activators of interferon regulatory factor (IRF)3 (Kovalenko et al. 2009). Furthermore, mice lacking caspase-8 in IECs developed spontaneous ileitis that was associated with TNF-induced necroptotic cell death (Günther et al. 2011).

When the activity of caspase-8 is hampered, RIPK1 initiates RIPK3/MLKL-dependent necroptosis (Cho et al. 2009; He et al. 2009; Zhang et al. 2009). Interestingly, most disease phenotypes associated with caspase-8 deficiency in mice were attributed to aberrant necroptosis because they were rescued by loss of *Ripk3* or *Mlkl*

(Kaiser et al. 2011; Oberst et al. 2011; Alvarez-Diaz et al. 2016). Mouse studies have also implicated caspase-8 in lymphocyte differentiation and function (Salmena et al. 2003; Kang et al. 2004; Beisner et al. 2005). T-cell-specific deletion of *Casp8* resulted in profound depletion of peripheral T cells associated with defective activation and/or survival upon engagement of the T-cell receptor (TCR) (Salmena et al. 2003). These defects impaired CD8[+] T-cell-mediated antiviral immunity. Proliferation of caspase-8-deficient T cells could be restored by inhibition of RIPK1 or genetic ablation of *Ripk3*, implying that caspase-8 suppresses necroptosis during T-cell activation (Bell et al. 2008; Ch'en et al. 2011; Kaiser et al. 2011; Oberst et al. 2011). B-cell-specific deletion of caspase-8 did not impact B-cell development but compromised B-cell activation by Toll-like receptor (TLR) agonists (Lemmers et al. 2007).

The relevance of caspase-8 for human immunity was originally recognized by studies involving two siblings with germline homozygous missense mutations in *CASP8* (Chun et al. 2002). Similar to ALPS patients with loss-of-function mutations in genes encoding Fas, FASL, and caspase-10, the patients with germline mutation in *CASP8* presented with lymphadenopathy and splenomegaly that was associated with defective Fas-mediated apoptosis in T cells (Chun et al. 2002). The homozygous mutation in *CASP8* (Arg248Trp) was located in the p18 protease subunit and it reduced protein stability and enzymatic activity. Unlike typical ALPS, the caspase-8-deficient patients also had defects in the activation of their T-, B-, and natural killer cells causing immunodeficiency (Chun et al. 2002). T-cell dysfunction was associated with impaired TCR-induced nuclear translocation of NF-κB (Su et al. 2005), but given the later studies in mice (Bell et al. 2008; Ch'en et al. 2011; Kaiser et al. 2011; Oberst et al. 2011), the question became whether the defects in NF-κB signaling were a consequence of aberrant necroptosis.

The clinical spectrum of caspase-8 deficiency was further broadened by the description of two patients with the mutation Arg248Trp. These patients presented with late-onset multiorgan lymphocytic infiltrations with granulomas (Niemela et al. 2015). By contrast, Lehle et al. recently described patients with homozygous missense mutations in *CASP8* (Gln237Arg) that affect the cleavage and activation of caspase-8 (Lehle et al. 2019). These patients had life-threatening very early-onset inflammatory bowel disease (VEO-IBD) and immunodeficiency that was accompanied by increased susceptibility to viral and bacterial infections, marked lymphadenopathy, reduced TCR-dependent T-cell proliferation and activation, and impaired B-cell maturation (Lehle et al. 2019). Mouse studies have previously shown that myeloid cells lacking *Casp8* exhibited increased NLRP3-dependent inflammasome activity with enhanced secretion of proinflammatory cytokines IL-1β and IL-18 (Kang et al. 2013). Correspondingly, caspase-8-deficient patient monocytes secreted more proinflammatory IL-1β than control monocytes in response to lipopolysaccharide (LPS) (Lehle et al. 2019). In caspase-8-deficient human BLaER1 monocyte/macrophage models, blockade of either NLPR3-dependent inflammasome activity or MLKL-dependent necroptosis attenuated IL-1β secretion (Gaidt et al. 2016; Lehle et al. 2019). Thus, both pathways are implicated in proinflammatory cytokine responses. These findings are consistent with the notion that necroptosis can activate the NLRP3 inflammasome (Vince and Silke 2016). Targeting necroptosis might present an attractive therapeutic approach in caspase-8-deficient patients with VEO-IBD, but more detailed mechanistic studies are required.

The identification of caspase-8-deficient patients with VEO-IBD underscores the critical function of caspase-8 in maintaining human intestinal epithelial homeostasis (Lehle et al. 2019). Whereas TRAIL triggered cell death in healthy donor-derived intestinal organoids, caspase-8-deficient cells were unresponsive to TRAIL. In contrast to mouse organoids with loss of *Casp8* (Günther et al. 2011), patient-derived caspase-8-deficient intestinal organoids did not exhibit a marked increase in TNF-induced cell death (Lehle et al. 2019). Further studies are needed to determine genotype–phenotype correlations of the mutations in human

CASP8. The physiological triggers of intestinal inflammation in human caspase-8 deficiency need to be further defined to identify targeted therapies.

FADD Deficiency

FADD is an adaptor protein that recruits caspase-8 to death receptors (Wilson et al. 2009). Mice lacking *Fadd* show RIPK3- and MLKL-dependent embryonic lethality similar to mice lacking caspase-8 (Yeh et al. 1998; Alvarez-Diaz et al. 2016). T-cell-specific loss of *Fadd,* similar to caspase-8 deficiency, caused defective T-cell proliferation that was rescued by inhibition of RIPK1 (Osborn et al. 2010). In addition, Osborn et al. observed enlarged lymph nodes and spleen with increased B cells and red blood cells, respectively. Mice lacking *Fadd* in epidermal keratinocytes or intestinal epithelial cells (IECs) developed severe inflammation (Bonnet et al. 2011; Welz et al. 2011), indicating that FADD is essential for homeostasis in the skin and intestine. Skin inflammation was triggered by RIPK3-dependent necroptosis, and was partially dependent on the catalytic activity of the deubiquitinating enzyme CYLD and/or TNFR1 signaling (Bonnet et al. 2011). Loss of *Fadd* in IECs caused spontaneous RIPK3-dependent colitis with epithelial erosions and crypt abscesses (Welz et al. 2011). Disease was prevented by deletion of *Cyld* or *Myd88,* or by elimination of the microbiota. Thus, TLR signaling activated by bacteria was a key driver of colitis (Welz et al. 2011).

In humans, a homozygous loss-of-function mutation in the death domain of *FADD* (Cys105Trp) was reported to impair Fas-induced apoptosis, as in patients with ALPS (Bolze et al. 2010). However, the related patients presented with immunodeficiency, bacterial susceptibility, and developmental abnormalities rather than autoimmunity (Bolze et al. 2010). In contrast to KO mouse models, FADD-deficient patients showed normal T-cell proliferation, but impaired interferon-dependent antiviral immunity, leading to increased susceptibility to viral diseases (e.g., varicella zoster, parainfluenza virus, and Epstein–Barr virus).

RIPK1 Deficiency

RIPK1 is a key molecule controlling both inflammation and cell-death responses via scaffolding-dependent and kinase-specific functions (Ofengeim and Yuan 2013). In particular, RIPK1 mediates multimodal TNFR1 signaling depending on the cell type and biological context. While TNF-induced NF-κB nuclear translocation promotes cell survival and inflammation, modulation of signaling cascades can induce caspase-8-mediated apoptosis or RIPK3-dependent necroptosis, as reviewed in Pasparakis and Vandenabeele (2015). RIPK1-deficient mice exhibited perinatal lethality because of multiorgan hyperinflammation that is driven by aberrant caspase-8-dependent apoptosis and MLKL-dependent necroptosis (Kelliher et al. 1998; Dillon et al. 2014; Kaiser et al. 2014; Rickard et al. 2014b). Conditional KO mice have demonstrated the essential role of RIPK1 in controlling immune and intestinal homeostasis. Mice with loss of *Ripk1* in IECs developed severe inflammation in the gut because of FADD/caspase-8-dependent apoptosis (Dannappel et al. 2014; Takahashi et al. 2014), whereas keratinocyte-specific RIPK1 KOs developed skin inflammation associated with ZBP1/RIPK3/MLKL-dependent necroptosis (Lin et al. 2016). T-cell-specific deletion of *Ripk1* in mice caused severe lymphopenia and defective T-cell proliferation (Dowling et al. 2016). RIPK1 also contributes to the maintenance of peripheral B cells (Zhang et al. 2011). RIPK1-deficient fetal liver-derived mouse macrophages exhibited enhanced inflammasome activity upon LPS priming (Lawlor et al. 2015).

In contrast to RIPK1-deficient mice, knock-in mice expressing catalytically inactive RIPK1 D138N or K45A showed no signs of tissue pathology and are protected from TNF-induced acute shock (Berger et al. 2014; Newton et al. 2014; Polykratis et al. 2014). Thus, the kinase activity of RIPK1 is dispensable for suppressing cell death. Necrostatin-1, a small molecule inhibitor of the kinase activity of RIPK1, has been shown to protect mice from retinal degeneration (Murakami et al. 2014), retinitis pigmentosa (Sato et al. 2013), ischemic brain injury (Deg-

Cite this article as *Cold Spring Harb Perspect Biol* doi: 10.1101/cshperspect.a037036

terev et al. 2005; Northington et al. 2011), neurodegeneration (Zhu et al. 2011), myocardial infarction, cardiac hypoxia (Smith et al. 2007; Oerlemans et al. 2012), and renal ischemia-reperfusion injury (Lau et al. 2013).

Recently, studies on patients with monogenic defects of *RIPK1* have provided critical insights into the role of RIPK1 in human disease. The patients presented with early-onset, life-threatening immunodeficiency and intestinal inflammation (Cuchet-Lourenço et al. 2018; Li et al. 2019; Uchiyama et al. 2019). Some patients showed arthritis (Cuchet-Lourenço et al. 2018), but skin inflammation was not observed, which is in contrast to RIPK1-deficient mice. Human RIPK1 deficiency was associated with impaired T- and B-cell maturation, defective TNF-mediated activation of the NF-κB pathway, and dysregulated cytokine signaling in immune cells. Cuchet-Lourenço et al. (2018) suggested that inflammation in *RIPK1*-deficient patients was caused by altered cytokine secretion and necroptosis of immune cells. In parallel and independent experiments, Li et al. studied six pedigrees and demonstrated that *RIPK1*-deficient macrophages exhibited increased inflammasome activity in response to LPS. Inhibition of MLKL- and NLRP3-dependent pathways by small molecule inhibitors attenuated secretion of proinflammatory IL-1β (Li et al. 2019), but the underlying mechanisms are not completely understood. Blockade of IL-1 has not yet been tested in RIPK1-deficient patients.

Since histological examination of gastrointestinal biopsies revealed only occasional apoptotic morphology, Cuchet-Lourenço et al. proposed that dysfunction of the immune system was critical for disease development. The authors concluded that allogeneic hematopoietic stem cell transplantation (HSCT) may constitute a curative therapy and showed resolution of clinical symptoms in one patient (Cuchet-Lourenço et al. 2018). Li et al. (2019) studied RIPK1-deficient IECs as well as hematopoietic cells. RIPK1-deficient IECs were resistant to killing by TNF, suggesting that RIPK1 also plays a critical intrinsic role in controlling epithelial homeostasis. Differences in the observed phenotypes might be because of the treatment of patients with anti-inflammatory drugs and antibiotics, their genetic background, or environmental factors. HSCT might cure cytokine production defects in immune cells, but not intrinsic epithelial defects, similar to NEMO-deficient patients (Miot et al. 2017). The in vivo triggers perturbing epithelial integrity in mice or humans lacking RIPK1 have not been defined. Moreover, the currently reported RIPK1-deficient patients provided no insights into the role of the kinase domain of RIPK1, because the patient-specific mutations reduced expression of RIPK1 protein. Further studies are needed to define genotype–phenotype correlations, triggers, and molecular consequences of human RIPK1 deficiency.

MONOGENIC DEFECTS OF THE NF-κB SIGNALING PATHWAY

NF-κB is a master transcriptional regulator of cell survival and proliferation, innate and adaptive immunity, and inflammation. Consequently, NF-κB signaling must be tightly regulated for tissue and immune homeostasis (Hayden and Ghosh 2011). Abnormal NF-κB signaling might cause defective immune activation, immunodeficiency, autoimmunity, or lymphoid malignancies (Courtois and Gilmore 2006). Human monogenic defects in NF-κB signaling components have been shown to cause severe immune disorders (Hayden and Ghosh 2008) that may vary from phenotypes in mouse models.

The IKK complex is composed of catalytic subunits (IKK1/IKKα, IKK2/IKKβ) and a regulatory subunit (NF-κB essential modulator [NEMO]) (Chen et al. 1996; DiDonato et al. 1997; Yamaoka et al. 1998). Mice lacking IKKβ were embryonic lethal (Li et al. 1999; Tanaka et al. 1999), whereas impaired degradation of IκBα and delayed NF-κB signaling in IKKβ-deficient patients caused severe combined immunodeficiency (Pannicke et al. 2013). Mice lacking IKKα died at birth because of multiple severe malformations and skin defects (Hu et al. 1999; Takeda et al. 1999). Patients with IKKα deficiency manifested with similar phenotypes, but showed more severe craniofacial abnormalities (Lahtela et al. 2010).

Several mouse and human studies have documented that loss of NEMO, the regulatory subunit of the IKK complex, causes defective NF-κB activation. Loss of X-linked *Nemo/Ikbkg* caused embryonic lethality in male mice, whereas severe skin lesions were observed in heterozygous females (Schmidt-Supprian et al. 2000). Mice with IEC-specific KO of *Nemo* developed spontaneous colitis with enhanced apoptosis of Paneth cells and impaired expression of antimicrobial factors, which was dependent on the kinase activity of RIPK1 (Vlantis et al. 2016). Mutations in human X-linked *NEMO/IKBKG* cause varying phenotypes, in particular anhidrotic ectodermal dysplasia with immunodeficiency (EDA-ID) (Zonana et al. 2000; Döffinger et al. 2001) or incontinentia pigmenti (Smahi et al. 2000). Notably, about 25% of patients develop colitis associated with poor HSCT outcome (Hanson et al. 2008; Kawai et al. 2012).

Similar to *NEMO* deficiency, autosomal-dominant gain-of-function mutations in *NFKBIA* (encoding IκBα) caused sustained inhibition of NF-κB signaling that leads to EDA-ID, T- and B-cell deficiency, and increased susceptibility to infections (Courtois et al. 2003; Boisson et al. 2017). Knockin mice that are heterozygous for the human *NFKBIA* mutation (Ser32Ile) developed EDA-ID and lacked lymph nodes, Peyer's patches, splenic marginal zones, and follicular dendritic cells. They also failed to develop contact hypersensitivity or form germinal centers, which are features characteristic of defective noncanonical NF-κB signaling through NF-κB2/RelB (Mooster et al. 2015).

Haploinsufficiency of *NFKB1* (p105/p50) or *NFKB2* (p100/p52) can cause common variable immunodeficiency with recurrent respiratory infections, hypogammaglobulinemia, and autoimmunity (Chen et al. 2013; Fliegauf et al. 2015). In addition, patients with loss-of-function mutations in *NFKB1* demonstrated noninfective complications, including lymphadenopathy, splenomegaly, and autoimmunity (Tuijnenburg et al. 2018). It remains to be shown whether the phenotype of pyoderma gangrenosum in patients with monoallelic *NFKB1* mutations is caused by dominant-active effects, loss-of-function, or haploinsufficiency. Of note,

all patients showed defective B-cell differentiation. Similarly, *Nfkb1*-deficient mice developed intestinal inflammation that was associated with profound B-cell dysfunction, including defects in proliferation, class-switch recombination, maturation, humoral immunity, cytokine secretion, and susceptibility to infection (Sha et al. 1995; Bendall et al. 1999).

MONOGENIC DISORDERS OF UBIQUITINATION AND DEUBIQUITINATION IN THE TNFR1 SIGNALING CASCADE

The ubiquitin system plays a crucial role in balancing gene activation and cell death (Aksentijevich and Zhou 2017). Perturbed ubiquitination or deubiquitination can result in dysregulation of the immune system (Aksentijevich and Zhou 2017). LUBAC, the E3 ligase composed of HOIL-1, HOIP, and SHARPIN, inhibits its TNFR1-mediated cell death by generating linear polyubiquitin chains on NEMO and other complex I components (Peltzer et al. 2014, 2018; Rickard et al. 2014a). Loss of *Rnf31* (encoding HOIP) caused embryonic lethality in mice (Peltzer et al. 2014) as a result of aberrant cell death (Peltzer et al. 2018), whereas excessive cell death in *Sharpin*-deficient mice caused severe eosinophilic skin inflammation and defective lymphoid organogenesis (Kumari et al. 2014; Rickard et al. 2014a). Mice with keratinocyte-specific depletion of LUBAC components developed severe dermatitis caused by FASL-, TRAIL-, and TNF-induced cell death (Taraborrelli et al. 2018).

No human loss-of-function mutations in *SHARPIN* have been reported yet, but HOIP or HOIL-1 deficiencies cause PID and autoinflammation with overlapping phenotypes such as susceptibility to infections and amylopectinosis (Boisson et al. 2012, 2015). Mutations in *RNF31* or *RBCK1* (encoding HOIL-1) that impaired the stability of LUBAC attenuated NF-κB signaling in fibroblasts or B cells treated with IL-1β or TNF. However, patient-derived monocytes were hyperresponsive to IL-1β, leading to up-regulation of inflammatory cytokines and chemokines. TNF-inhibitory treatment has been shown to ameliorate pathology temporarily, but autoinflammation was controlled by HSCT in one HOIL-1-deficient patient (Boisson et al. 2012). It is unclear why HOIL-1 and HOIP are essential for embryogenesis in mice, but not humans. Heterogeneity in the genetic background of humans may be a factor, or there may be physiological differences between species.

The deubiquitinases A20, OTULIN, and CYLD are negative regulators of NF-κB signaling (Lork et al. 2017). However, emerging data have also suggested unexpected roles of these deubiquitinases in regulating cell death independent of NF-κB signaling (Draber et al. 2015; Heger et al. 2018; Polykratis et al. 2019). Defects in these genes lead to increased proinflammatory cytokine profiles (Lork et al. 2017). A20 can cleave Lys63-linked polyubiquitin chains on target proteins, such as RIPK1 and NEMO, to inhibit NF-κB signaling, but it is the binding of A20 to Met1-linked ubiquitin chains that appears to limit the formation of complexes that trigger proinflammatory cell death. For example, mice lacking A20 in myeloid cells developed arthritis that was driven by necroptosis and activation of the NLRP3 inflammasome. Analyses of A20 knockin mice indicated that the ubiquitin-binding ZnF7 domain in A20 is critical for preventing arthritis, whereas the deubiquitinating activity of A20 is dispensable (Draber et al. 2015; Polykratis et al. 2019).

A20-deficient mice die shortly after birth showing severe multiorgan inflammation (Lee et al. 2000). Tissue-specific deletion of *Tnfaip3* (encoding A20) in lymphocytes, enterocytes, dendritic cells, keratinocytes, mast cells, hepatocytes, and microglial cells has further demonstrated the crucial role of A20 in maintaining immune homeostasis and inhibiting inflammation (Cox et al. 1992; Tavares et al. 2010; Hammer et al. 2011; Wang et al. 2013; Vereecke et al. 2014; Drennan et al. 2016; Maelfait et al. 2016). Genetic variants of human *TNFAIP3* are associated with a broad range of inflammatory and autoimmune diseases such as systemic lupus erythematosus, rheumatoid arthritis, psoriasis, type I diabetes, celiac disease, Crohn's disease, coronary artery disease in type 2 diabetes, and

systemic sclerosis (Ma and Malynn 2012; Zhou et al. 2016a). Mutations causing *TNFAIP3* haploinsufficiency led to early-onset systemic autoinflammatory syndrome, resembling Behcet's disease, because of increased NF-κB-mediated proinflammatory cytokine production (Zhou et al. 2016a). The authors did not specifically study cell-death responses, but patient cells treated with LPS showed enhanced cleavage of caspase-1 and secretion of mature IL-1. These findings are reminiscent of RIPK1 and CASP8 deficiencies, and thus it is tempting to speculate that enhanced inflammasome activation is mediated by aberrant necroptosis.

OTULIN cleaves Met1-linked polyubiquitin chains conjugated by LUBAC (Keusekotten et al. 2013). Recent studies have shown that OTULIN promotes rather than counteracts LUBAC activity. Specifically, OTULIN limits autoubiquitination of LUBAC, which would otherwise lead to RIPK1-dependent cell death (Heger et al. 2018). Consequently, *Otulin* KO mice were embryonic lethal (Rivkin et al. 2013) similar to mice lacking HOIP or HOIL-1 (Peltzer et al. 2014, 2018). Homozygous missense mutations in human *OTULIN* caused cell-type-specific alterations in NF-κB signaling, fatal autoinflammation with recurrent nodular panniculitis, lipodystrophy, diarrhea, joint swelling, and failure to thrive (Damgaard et al. 2016; Zhou et al. 2016b). Patient-derived monocytes and fibroblasts exhibited increased sensitivity to TNF-induced cell death (Damgaard et al. 2019). Moreover, treatment with anti-TNF neutralizing antibodies could ameliorate inflammation, whereas HSCT induced sustained remission in OTULIN-deficient patients (Damgaard et al. 2019).

CYLD has been extensively studied for its role in removing Lys63- or Met1-linked polyubiquitin chains from proteins mediating NF-κB signaling. For example, CYLD deubiquitinates proteins in TNFR1 complex I, which limits NF-κB signaling and promotes the assembly of cell-death signaling complexes (Draber et al. 2015). It has been suggested that CYLD regulates innate and adaptive immune responses via its negative regulation of NF-κB signaling components, but dysfunctional cell death might also contribute to immune-related phenotypes. Mice lacking CYLD showed autoimmunity associated with abnormal thymocyte development, lymphocyte activation, and B-cell hyperplasia (Reiley et al. 2006, 2007; Zhang et al. 2006; Jin et al. 2007). CYLD deficiency in humans can lead to distinct phenotypes with cylindromatosis and skin manifestations such as multiple familial trichoepithelioma, type I (Mathis et al. 2015; Farkas et al. 2016). The phenotypic heterogeneity of human *CYLD* deficiency is likely the result of its diverse roles in controlling other NF-κB-independent pathways such as cell-death responses, cell-cycle progression, and microtubule dynamics (Sun 2010; Zhang et al. 2017a).

TNFR1 SIGNALING AS A THERAPEUTIC TARGET—WHAT DO WE LEARN FROM MONOGENIC DISEASES?

Several mouse and human studies indicate the critical role of TNFR1 signaling in health and disease, as reviewed in Brenner et al. (2015). TNF inhibition has proven effective as treatment for several autoinflammatory and autoimmune conditions (Kalliolias and Ivashkiv 2016). However, many patients with inflammatory disorders are refractory to anti-TNF therapy or develop side-effects (Kalliolias and Ivashkiv 2016). Thus, alternative strategies targeting the TNFR1 pathway are needed to expand the therapeutic armamentarium.

RIPK1/RIPK3/MLKL-dependent necroptosis has been implicated in malignancies and several pathological inflammatory conditions (Weinlich et al. 2017). Small-molecule inhibitors targeting RIPK1 kinase activity present attractive therapeutic potential, because mice expressing catalytically inactive RIPK1 develop normally without inflammatory phenotypes (Berger et al. 2014; Newton et al. 2014; Polykratis et al. 2014). The therapeutic potential of RIPK1 inhibitors has been demonstrated in various mouse disease models (Silke et al. 2015). Based on these studies RIPK1 inhibitor programs have successfully passed clinical phase I trials for the treatment of chronic psoriasis, rheumatoid arthritis, and ulcerative colitis (GSK2982772, DNL747) (Harris et al. 2017; Mullard 2018).

Targeting of RIPK3 is a new idea to treat inflammatory diseases, particularly since mice lacking *Ripk3* are viable (Newton et al. 2004). However, knockin mice expressing catalytically inactive RIPK3 D161N exhibited caspase-8-dependent embryonic lethality (Newton et al. 2014), raising concerns about the toxic effects of targeting RIPK3. Indeed, inhibitors of RIPK3 (GSK'840, GSK'843, and GSK'872) trigger RIPK3- and caspase-8-dependent apoptosis reminiscent of that seen in RIPK3 D161N mice (Kaiser et al. 2013; Mandal et al. 2014). Thus, further refinement of RIPK3-based therapies is needed.

Blockade of MLKL has been considered as a means of selectively inhibiting necroptosis. For example, necrosulfonamide (NSA), which is a compound that modifies Cys86 of human MLKL to block its oligomerization, has been suggested as a potential therapeutic for neurodegenerative diseases (Zhang et al. 2017b), but it has not been tested in clinical trials. The MLKL inhibitor compound 1 caused cell toxicity at high concentrations, and thus has not been used in clinical applications (Hildebrand et al. 2014). Recently, a new inhibitor (TC13172) targeting Cys86 of MLKL was demonstrated to block the translocation of MLKL to cell membranes in cell lines (Yan et al. 2017).

Inhibitors of caspase-8 have been proposed for patients with dysregulated cell death and/or inflammation. The pancaspase inhibitor Emricasan has antiapoptotic and anti-inflammatory effects, and has been explored for the treatment of liver disease (Frenette et al. 2019; Garcia-Tsao et al. 2019), renal disease, and diabetes (Kudelova et al. 2015). However, inhibition of caspase-8 might induce necroptosis in some cell types and thereby promote inflammation.

As a master regulator of immunity, NF-κB has been implicated in various autoimmune diseases (Herrington et al. 2016). Selective targeting of NF-κB activity presents another line of therapeutic modulation, but specificity is a major challenge. Commonly used anti-inflammatory agents, such as antirheumatic drugs, nonsteroidal anti-inflammatory drugs, and glucocorticoids have been shown to partly modulate NF-κB signaling at various levels (Yama-

Cite this article as *Cold Spring Harb Perspect Biol* doi: 10.1101/cshperspect.a037036

moto and Gaynor 2001; Herrington et al. 2016). Specific inhibitors of NF-κB, such as caffeic acid phenethyl ester and carfilzomib, are now available for treatment of myeloma (Kane et al. 2003; Herndon et al. 2013), but remain to be evaluated for autoimmunity.

Mouse models have been exquisite tools for studying the pathomechanisms of diseases and for drug development. However, mice may respond differently from humans to therapies, and show distinct phenotypes from patients with monogenic disorders in orthologous genes. The characterization of PID provides critical molecular insights into key factors mediating TNFR1 signaling. Further studies on PID are required to explore genotype–phenotype correlations and the molecular mechanisms of disease in detail. These studies lay the groundwork for the development of targeted therapies for both rare and common immune and inflammatory diseases. Furthermore, patients with monogenic disorders affecting the TNFR1 pathway help to predict the therapeutic efficacy and side-effects of available therapies targeting TNFR1 signaling.

CONCLUDING REMARKS

TNFR1 signaling is a crucial "command center" controlling immunity, inflammation, and cell death. Dysregulation of these pathways may cause immunodeficiency and/or autoinflammation. Advances in genomic technologies have facilitated the identification of patients with life-threatening PID. The characterization of these patients has provided critical and unexpected insights into the essential role of human TNFR1 signaling in controlling inflammation. The identified candidate genes at the intersection of prosurvival and cell-death pathways have shown that modulation of the TNFR1 pathway can contribute to both severe immunodeficiency and chronic inflammation. Further mechanistic studies in mice and especially advanced human preclinical models will provide critical understanding of imbalanced inflammation and cell death in PID. This knowledge on rare monogenic diseases will help to optimize personalized treatments for children with devastating conditions, but will also prioritize new targets for drug development of common autoimmunity and autoinflammation.

ACKNOWLEDGMENTS

Together with patients suffering from inborn errors of immunity, we have a chance to investigate basic principles of human immunity and autoinflammation. We dedicate this work to all patients with inborn errors of immunity and their families, and thank them for their support of our translational studies. Furthermore, we are grateful to the interdisciplinary medical teams and all global collaboration partners, in particular, the international VEO-IBD consortium. We thank the Care-for-Rare Foundation and The Leona M. and Harry B. Helmsley Charitable Trust for financial support.

REFERENCES

Aksentijevich I, Zhou Q. 2017. NF-κB pathway in autoinflammatory diseases: dysregulation of protein modifications by ubiquitin defines a new category of autoinflammatory diseases. *Front Immunol* **8:** 399. doi:10.3389/fimmu.2017.00399

Alvarez-Diaz S, Dillon CP, Lalaoui N, Tanzer MC, Rodriguez DA, Lin A, Lebois M, Hakem R, Josefsson EC, O'Reilly LA, et al. 2016. The pseudokinase MLKL and the kinase RIPK3 have distinct roles in autoimmune disease caused by loss of death-receptor-induced apoptosis. *Immunity* **45:** 513–526. doi:10.1016/j.immuni.2016.07.016

Ashkenazi A, Dixit VM. 1998. Death receptors: signaling and modulation. *Science* **281:** 1305–1308. doi:10.1126/science.281.5381.1305

Badran YR, Dedeoglu F, Leyva Castillo JM, Bainter W, Ohsumi TK, Bousvaros A, Goldsmith JD, Geha RS, Chou J. 2017. Human *RELA* haploinsufficiency results in autosomal-dominant chronic mucocutaneous ulceration. *J Exp Med* **214:** 1937–1947. doi:10.1084/jem.20160724

Beg AA, Baltimore D. 1996. An essential role for NF-κB in preventing TNF-α-induced cell death. *Science* **274:** 782–784. doi:10.1126/science.274.5288.782

Beisner DR, Ch'en IL, Kolla RV, Hoffmann A, Hedrick SM. 2005. Cutting edge: innate immunity conferred by B cells is regulated by caspase-8. *J Immunol* **175:** 3469–3473. doi:10.4049/jimmunol.175.6.3469

Bell BD, Leverrier S, Weist BM, Newton RH, Arechiga AF, Luhrs KA, Morrissette NS, Walsh CM. 2008. FADD and caspase-8 control the outcome of autophagic signaling in proliferating T cells. *Proc Natl Acad Sci* **105:** 16677–16682. doi:10.1073/pnas.0808597105

Ben Moshe T, Barash H, Kang TB, Kim JC, Kovalenko A, Gross E, Schuchmann M, Abramovitch R, Galun E,

Wallach D. 2007. Role of caspase-8 in hepatocyte response to infection and injury in mice. *Hepatology* 45: 1014–1024. doi:10.1002/hep.21495

Bendall HH, Sikes ML, Ballard DW, Oltz EM. 1999. An intact NF-κB signaling pathway is required for maintenance of mature B cell subsets. *Mol Immunol* 36: 187–195. doi:10.1016/S0161-5890(99)00031-0

Berger SB, Kasparcova V, Hoffman S, Swift B, Dare L, Schaeffer M, Capriotti C, Cook M, Finger J, Hughes-Earle A, et al. 2014. Cutting edge: RIP1 kinase activity is dispensable for normal development but is a key regulator of inflammation in SHARPIN-deficient mice. *J Immunol* 192: 5476–5480. doi:10.4049/jimmunol.1400499

Bignell GR, Warren W, Seal S, Takahashi M, Rapley E, Barfoot R, Green H, Brown C, Biggs PJ, Lakhani SR, et al. 2000. Identification of the familial cylindromatosis tumour-suppressor gene. *Nat Genet* 25: 160–165. doi:10.1038/76006

Boatright KM, Renatus M, Scott FL, Sperandio S, Shin H, Pedersen IM, Ricci JE, Edris WA, Sutherlin DP, Green DR, et al. 2003. A unified model for apical caspase activation. *Mol Cell* 11: 529–541. doi:10.1016/S1097-2765(03)00051-0

Boisson B, Laplantine E, Prando C, Giliani S, Israelsson E, Xu Z, Abhyankar A, Israël L, Trevejo-Nunez G, Bogunovic D, et al. 2012. Immunodeficiency, autoinflammation and amylopectinosis in humans with inherited HOIL-1 and LUBAC deficiency. *Nat Immunol* 13: 1178–1186. doi:10.1038/ni.2457

Boisson B, Laplantine E, Dobbs K, Cobat A, Tarantino N, Hazen M, Lidov HG, Hopkins G, Du L, Belkadi A, et al. 2015. Human HOIP and LUBAC deficiency underlies autoinflammation, immunodeficiency, amylopectinosis, and lymphangiectasia. *J Exp Med* 212: 939–951. doi:10.1084/jem.20141130

Boisson B, Puel A, Picard C, Casanova JL. 2017. Human IκBα gain of function: a severe and syndromic immunodeficiency. *J Clin Immunol* 37: 397–412. doi:10.1007/s10875-017-0400-z

Boldin MP, Goncharov TM, Goltsev YV, Wallach D. 1996. Involvement of MACH, a novel MORT1/FADD-interacting protease, in Fas/APO-1- and TNF receptor-induced cell death. *Cell* 85: 803–815. doi:10.1016/S0092-8674(00)81265-9

Bolze A, Byun M, McDonald D, Morgan NV, Abhyankar A, Premkumar L, Puel A, Bacon CM, Rieux-Laucat F, Pang K, et al. 2010. Whole-exome-sequencing-based discovery of human FADD deficiency. *Am J Hum Genet* 87: 873–881. doi:10.1016/j.ajhg.2010.10.028

Bonnet MC, Preukschat D, Welz PS, van Loo G, Ermolaeva MA, Bloch W, Haase I, Pasparakis M. 2011. The adaptor protein FADD protects epidermal keratinocytes from necroptosis in vivo and prevents skin inflammation. *Immunity* 35: 572–582. doi:10.1016/j.immuni.2011.08.014

Brenner D, Blaser H, Mak TW. 2015. Regulation of tumour necrosis factor signalling: live or let die. *Nat Rev Immunol* 15: 362–374. doi:10.1038/nri3834

Chen ZJ, Parent L, Maniatis T. 1996. Site-specific phosphorylation of IκBα by a novel ubiquitination-dependent protein kinase activity. *Cell* 84: 853–862. doi:10.1016/S0092-8674(00)81064-8

Ch'en IL, Tsau JS, Molkentin JD, Komatsu M, Hedrick SM. 2011. Mechanisms of necroptosis in T cells. *J Exp Med* 208: 633–641. doi:10.1084/jem.20110251

Chen K, Coonrod EM, Kumánovics A, Franks ZF, Durtschi JD, Margraf RL, Wu W, Heikal NM, Augustine NH, Ridge PG, et al. 2013. Germline mutations in NFKB2 implicate the noncanonical NF-κB pathway in the pathogenesis of common variable immunodeficiency. *Am J Hum Genet* 93: 812–824. doi:10.1016/j.ajhg.2013.09.009

Cho YS, Challa S, Moquin D, Genga R, Ray TD, Guildford M, Chan FK. 2009. Phosphorylation-driven assembly of the RIP1-RIP3 complex regulates programmed necrosis and virus-induced inflammation. *Cell* 137: 1112–1123. doi:10.1016/j.cell.2009.05.037

Chun HJ, Zheng L, Ahmad M, Wang J, Speirs CK, Siegel RM, Dale JK, Puck J, Davis J, Hall CG, et al. 2002. Pleiotropic defects in lymphocyte activation caused by caspase-8 mutations lead to human immunodeficiency. *Nature* 419: 395–399. doi:10.1038/nature01063

Courtois G, Gilmore TD. 2006. Mutations in the NF-κB signaling pathway: implications for human disease. *Oncogene* 25: 6831–6843. doi:10.1038/sj.onc.1209939

Courtois G, Smahi A, Reichenbach J, Döffinger R, Cancrini C, Bonnet M, Puel A, Chable-Bessia C, Yamaoka S, Feinberg J, et al. 2003. A hypermorphic IκBα mutation is associated with autosomal dominant anhidrotic ectodermal dysplasia and T cell immunodeficiency. *J Clin Invest* 112: 1108–1115. doi:10.1172/JCI18714

Cox IH, Erickson SJ, Foley WD, Dewire DM. 1992. Ureteric jets: evaluation of normal flow dynamics with color Doppler sonography. *AJR Am J Roentgenol* 158: 1051–1055. doi:10.2214/ajr.158.5.1566665

Cuchet-Lourenço D, Eletto D, Wu C, Plagnol V, Papapietro O, Curtis J, Ceron-Gutierrez L, Bacon CM, Hackett S, Alsaleem B, et al. 2018. Biallelic *RIPK1* mutations in humans cause severe immunodeficiency, arthritis, and intestinal inflammation. *Science* 361: 810–813. doi:10.1126/science.aar2641

Damgaard RB, Walker JA, Marco-Casanova P, Morgan NV, Titheradge HL, Elliott PR, McHale D, Maher ER, McKenzie ANJ, Komander D. 2016. The deubiquitinase OTULIN is an essential negative regulator of inflammation and autoimmunity. *Cell* 166: 1215–1230.e20. doi:10.1016/j.cell.2016.07.019

Damgaard RB, Elliott PR, Swatek KN, Maher ER, Stepensky P, Elpeleg O, Komander D, Berkun Y. 2019. OTULIN deficiency in ORAS causes cell type-specific LUBAC degradation, dysregulated TNF signalling and cell death. *EMBO Mol Med* 11: e9324. doi:10.15252/emmm.201809324

Dannappel M, Vlantis K, Kumari S, Polykratis A, Kim C, Wachsmuth L, Eftychi C, Lin J, Corona T, Hermance N, et al. 2014. RIPK1 maintains epithelial homeostasis by inhibiting apoptosis and necroptosis. *Nature* 513: 90–94. doi:10.1038/nature13608

Degterev A, Huang Z, Boyce M, Li Y, Jagtap P, Mizushima N, Cuny GD, Mitchison TJ, Moskowitz MA, Yuan J. 2005. Chemical inhibitor of nonapoptotic cell death with therapeutic potential for ischemic brain injury. *Nat Chem Biol* 1: 112–119. doi:10.1038/nchembio711

DiDonato JA, Hayakawa M, Rothwarf DM, Zandi E, Karin M. 1997. A cytokine-responsive IκB kinase that activates the transcription factor NF-κB. *Nature* **388**: 548–554. doi:10.1038/41493

Dillon CP, Weinlich R, Rodriguez DA, Cripps JG, Quarato G, Gurung P, Verbist KC, Brewer TL, Llambi F, Gong YN, et al. 2014. RIPK1 blocks early postnatal lethality mediated by caspase-8 and RIPK3. *Cell* **157**: 1189–1202. doi:10.1016/j.cell.2014.04.018

Döffinger R, Smahi A, Bessia C, Geissmann F, Feinberg J, Durandy A, Bodemer C, Kenwrick S, Dupuis-Girod S, Blanche S, et al. 2001. X-linked anhidrotic ectodermal dysplasia with immunodeficiency is caused by impaired NF-κB signaling. *Nat Genet* **27**: 277–285. doi:10.1038/85837

Dowling JP, Cai Y, Bertin J, Gough PJ, Zhang J. 2016. Kinase-independent function of RIP1, critical for mature T-cell survival and proliferation. *Cell Death Dis* **7**: e2379. doi:10.1038/cddis.2016.307

Draber P, Kupka S, Reichert M, Draberova H, Lafont E, de Miguel D, Spilgies L, Surinova S, Taraborrelli L, Hartwig T, et al. 2015. LUBAC-recruited CYLD and A20 regulate gene activation and cell death by exerting opposing effects on linear ubiquitin in signaling complexes. *Cell Rep* **13**: 2258–2272. doi:10.1016/j.celrep.2015.11.009

Drappa J, Vaishnaw AK, Sullivan KE, Chu JL, Elkon KB. 1996. *Fas* gene mutations in the Canale–Smith syndrome, an inherited lymphoproliferative disorder associated with autoimmunity. *N Engl J Med* **335**: 1643–1649. doi:10.1056/NEJM199611283352204

Drennan MB, Govindarajan S, Verheugen E, Coquet JM, Staal J, McGuire C, Taghon T, Leclercq G, Beyaert R, van Loo G, et al. 2016. NKT sublineage specification and survival requires the ubiquitin-modifying enzyme TNFAIP3/A20. *J Exp Med* **213**: 1973–1981. doi:10.1084/jem.20151065

Earnshaw WC, Martins LM, Kaufmann SH. 1999. Mammalian caspases: structure, activation, substrates, and functions during apoptosis. *Annu Rev Biochem* **68**: 383–424. doi:10.1146/annurev.biochem.68.1.383

Farkas K, Deák BK, Sánchez LC, Martínez AM, Corell JJ, Botella AM, Benito GM, López RR, Vanecek T, Kazakov DV, et al. 2016. The CYLD p.R758X worldwide recurrent nonsense mutation detected in patients with multiple familial trichoepithelioma type 1, Brooke–Spiegler syndrome and familial cylindromatosis represents a mutational hotspot in the gene. *BMC Genet* **17**: 36. doi:10.1186/s12863-016-0346-9

Fisher GH, Rosenberg FJ, Straus SE, Dale JK, Middelton LA, Lin AY, Strober W, Lenardo MJ, Puck JM. 1995. Dominant interfering Fas gene mutations impair apoptosis in a human autoimmune lymphoproliferative syndrome. *Cell* **81**: 935–946. doi:10.1016/0092-8674(95)90013-6

Fliegauf M, Bryant VL, Frede N, Slade C, Woon ST, Lehnert K, Winzer S, Bulashevska A, Scerri T, Leung E, et al. 2015. Haploinsufficiency of the NF-κB1 subunit p50 in common variable immunodeficiency. *Am J Hum Genet* **97**: 389–403. doi:10.1016/j.ajhg.2015.07.008

Frenette CT, Morelli G, Shiffman ML, Frederick RT, Rubin RA, Fallon MB, Cheng JT, Cave M, Khaderi SA, Massoud O, et al. 2019. Emricasan improves liver function in patients with cirrhosis and high model for end-stage liver disease scores compared with placebo. *Clin Gastroenterol Hepatol* **17**: 774–783.e4. doi:10.1016/j.cgh.2018.06.012

Gaidt MM, Ebert TS, Chauhan D, Schmidt T, Schmid-Burgk JL, Rapino F, Robertson AA, Cooper MA, Graf T, Hornung V. 2016. Human monocytes engage an alternative inflammasome pathway. *Immunity* **44**: 833–846. doi:10.1016/j.immuni.2016.01.012

Garcia-Tsao G, Fuchs M, Shiffman M, Borg BB, Pyrsopoulos N, Shetty K, Gallegos-Orozco JF, Reddy KR, Feyssa E, Chan JL, et al. 2019. Emricasan (IDN-6556) lowers portal pressure in patients with compensated cirrhosis and severe portal hypertension. *Hepatology* **69**: 717–728. doi:10.1002/hep.30199

Gerlach B, Cordier SM, Schmukle AC, Emmerich CH, Rieser E, Haas TL, Webb AI, Rickard JA, Anderton H, Wong WW, et al. 2011. Linear ubiquitination prevents inflammation and regulates immune signalling. *Nature* **471**: 591–596. doi:10.1038/nature09816

Green DR, Ferguson T, Zitvogel L, Kroemer G. 2009. Immunogenic and tolerogenic cell death. *Nat Rev Immunol* **9**: 353–363. doi:10.1038/nri2545

Green DR, Oguin TH, Martinez J. 2016. The clearance of dying cells: table for two. *Cell Death Differ* **23**: 915–926. doi:10.1038/cdd.2015.172

Grivennikov SI, Tumanov AV, Liepinsh DJ, Kruglov AA, Marakusha BI, Shakhov AN, Murakami T, Drutskaya LN, Förster I, Clausen BE, et al. 2005. Distinct and nonredundant in vivo functions of TNF produced by T cells and macrophages/neutrophils: protective and deleterious effects. *Immunity* **22**: 93–104.

Günther C, Martini E, Wittkopf N, Amann K, Weigmann B, Neumann H, Waldner MJ, Hedrick SM, Tenzer S, Neurath MF, et al. 2011. Caspase-8 regulates TNF-α-induced epithelial necroptosis and terminal ileitis. *Nature* **477**: 335–339. doi:10.1038/nature10400

Guo H, Gilley RP, Fisher A, Lane R, Landsteiner VJ, Ragan KB, Dovey CM, Carette JE, Upton JW, Mocarski ES, et al. 2018. Species-independent contribution of ZBP1/DAI/DLM-1-triggered necroptosis in host defense against HSV1. *Cell Death Dis* **9**: 816. doi:10.1038/s41419-018-0868-3

Hammer GE, Turer EE, Taylor KE, Fang CJ, Advincula R, Oshima S, Barrera J, Huang EJ, Hou B, Malynn BA, et al. 2011. Expression of A20 by dendritic cells preserves immune homeostasis and prevents colitis and spondyloarthritis. *Nat Immunol* **12**: 1184–1193. doi:10.1038/ni.2135

Hanson EP, Monaco-Shawver L, Solt LA, Madge LA, Banerjee PP, May MJ, Orange JS. 2008. Hypomorphic nuclear factor-κB essential modulator mutation database and reconstitution system identifies phenotypic and immunologic diversity. *J Allergy Clin Immunol* **122**: 1169–1177. e16. doi:10.1016/j.jaci.2008.08.018

Harris PA, Berger SB, Jeong JU, Nagilla R, Bandyopadhyay D, Campobasso N, Capriotti CA, Cox JA, Dare L, Dong X, et al. 2017. Discovery of a first-in-class receptor interacting protein 1 (RIP1) kinase specific clinical candidate (GSK2982772) for the treatment of inflammatory diseases. *J Med Chem* **60**: 1247–1261. doi:10.1021/acs.jmedchem.6b01751

Hayden MS, Ghosh S. 2008. Shared principles in NF-κB signaling. *Cell* **132**: 344–362. doi:10.1016/j.cell.2008.01.020

Hayden MS, Ghosh S. 2011. NF-κB in immunobiology. *Cell Res* **21**: 223–244. doi:10.1038/cr.2011.13

He S, Wang L, Miao L, Wang T, Du F, Zhao L, Wang X. 2009. Receptor interacting protein kinase-3 determines cellular necrotic response to TNF-α. *Cell* **137:** 1100–1111. doi:10.1016/j.cell.2009.05.021

Heger K, Wickliffe KE, Ndoja A, Zhang J, Murthy A, Dugger DL, Maltzman A, de Sousa e Melo F, Hung J, Zeng Y, et al. 2018. OTULIN limits cell death and inflammation by deubiquitinating LUBAC. *Nature* **559:** 120–124. doi:10.1038/s41586-018-0256-2

Herndon TM, Deisseroth A, Kaminskas E, Kane RC, Koti KM, Rothmann MD, Habtemariam B, Bullock J, Bray JD, Hawes J, et al. 2013. U.S. Food and Drug Administration approval: carfilzomib for the treatment of multiple myeloma. *Clin Cancer Res* **19:** 4559–4563. doi:10.1158/1078-0432.CCR-13-0755

Herrington FD, Carmody RJ, Goodyear CS. 2016. Modulation of NF-κB signaling as a therapeutic target in autoimmunity. *J Biomol Screen* **21:** 223–242. doi:10.1177/1087057115617456

Hildebrand JM, Tanzer MC, Lucet IS, Young SN, Spall SK, Sharma P, Pierotti C, Garnier JM, Dobson RC, Webb AI, et al. 2014. Activation of the pseudokinase MLKL unleashes the four-helix bundle domain to induce membrane localization and necroptotic cell death. *Proc Natl Acad Sci* **111:** 15072–15077. doi:10.1073/pnas.1408987111

Hu Y, Baud V, Delhase M, Zhang P, Deerinck T, Ellisman M, Johnson R, Karin M. 1999. Abnormal morphogenesis but intact IKK activation in mice lacking the IKKα subunit of IκB kinase. *Science* **284:** 316–320. doi:10.1126/science.284.5412.316

Huang Z, Wu SQ, Liang Y, Zhou X, Chen W, Li L, Wu J, Zhuang Q, Chen C, Li J, et al. 2015. RIP1/RIP3 binding to HSV-1 ICP6 initiates necroptosis to restrict virus propagation in mice. *Cell Host Microbe* **17:** 229–242. doi:10.1016/j.chom.2015.01.002

Ikeda F, Deribe YL, Skånland SS, Stieglitz B, Grabbe C, Franz-Wachtel M, van Wijk SJ, Goswami P, Nagy V, Terzic J, et al. 2011. SHARPIN forms a linear ubiquitin ligase complex regulating NF-κB activity and apoptosis. *Nature* **471:** 637–641. doi:10.1038/nature09814

Jin W, Reiley WR, Lee AJ, Wright A, Wu X, Zhang M, Sun SC. 2007. Deubiquitinating enzyme CYLD regulates the peripheral development and naive phenotype maintenance of B cells. *J Biol Chem* **282:** 15884–15893. doi:10.1074/jbc.M609952200

Kaiser WJ, Upton JW, Long AB, Livingston-Rosanoff D, Daley-Bauer LP, Hakem R, Caspary T, Mocarski ES. 2011. RIP3 mediates the embryonic lethality of caspase-8-deficient mice. *Nature* **471:** 368–372. doi:10.1038/nature09857

Kaiser WJ, Sridharan H, Huang C, Mandal P, Upton JW, Gough PJ, Sehon CA, Marquis RW, Bertin J, Mocarski ES. 2013. Toll-like receptor 3-mediated necrosis via TRIF, RIP3, and MLKL. *J Biol Chem* **288:** 31268–31279. doi:10.1074/jbc.M113.462341

Kaiser WJ, Daley-Bauer LP, Thapa RJ, Mandal P, Berger SB, Huang C, Sundararajan A, Guo H, Roback L, Speck SH, et al. 2014. RIP1 suppresses innate immune necrotic as well as apoptotic cell death during mammalian parturition. *Proc Natl Acad Sci* **111:** 7753–7758. doi:10.1073/pnas.1401857111

Kalliolias GD, Ivashkiv LB. 2016. TNF biology, pathogenic mechanisms and emerging therapeutic strategies. *Nat Rev Rheumatol* **12:** 49–62. doi:10.1038/nrrheum.2015.169

Kane RC, Bross PF, Farrell AT, Pazdur R. 2003. Velcade: U.S. FDA approval for the treatment of multiple myeloma progressing on prior therapy. *Oncologist* **8:** 508–513. doi:10.1634/theoncologist.8-6-508

Kang TB, Ben-Moshe T, Varfolomeev EE, Pewzner-Jung Y, Yogev N, Jurewicz A, Waisman A, Brenner O, Haffner R, Gustafsson E, et al. 2004. Caspase-8 serves both apoptotic and nonapoptotic roles. *J Immunol* **173:** 2976–2984. doi:10.4049/jimmunol.173.5.2976

Kang TB, Yang SH, Toth B, Kovalenko A, Wallach D. 2013. Caspase-8 blocks kinase RIPK3-mediated activation of the NLRP3 inflammasome. *Immunity* **38:** 27–40. doi:10.1016/j.immuni.2012.09.015

Kawai T, Nishikomori R, Heike T. 2012. Diagnosis and treatment in anhidrotic ectodermal dysplasia with immunodeficiency. *Allergol Int* **61:** 207–217. doi:10.2332/allergolint.12-RAI-0446

Kelliher MA, Grimm S, Ishida Y, Kuo F, Stanger BZ, Leder P. 1998. The death domain kinase RIP mediates the TNF-induced NF-κB signal. *Immunity* **8:** 297–303. doi:10.1016/S1074-7613(00)80535-X

Keusekotten K, Elliott PR, Glockner L, Fiil BK, Damgaard RB, Kulathu Y, Wauer T, Hospenthal MK, Gyrd-Hansen M, Krappmann D, et al. 2013. OTULIN antagonizes LUBAC signaling by specifically hydrolyzing Met1-linked polyubiquitin. *Cell* **153:** 1312–1326. doi:10.1016/j.cell.2013.05.014

Kirisako T, Kamei K, Murata S, Kato M, Fukumoto H, Kanie M, Sano S, Tokunaga F, Tanaka K, Iwai K. 2006. A ubiquitin ligase complex assembles linear polyubiquitin chains. *EMBO J* **25:** 4877–4887. doi:10.1038/sj.emboj.7601360

Klement JF, Rice NR, Car BD, Abbondanzo SJ, Powers GD, Bhatt PH, Chen CH, Rosen CA, Stewart CL. 1996. IκBα deficiency results in a sustained NF-κB response and severe widespread dermatitis in mice. *Mol Cell Biol* **16:** 2341–2349. doi:10.1128/MCB.16.5.2341

Kovalenko A, Kim JC, Kang TB, Rajput A, Bogdanov K, Dittrich-Breiholz O, Kracht M, Brenner O, Wallach D. 2009. Caspase-8 deficiency in epidermal keratinocytes triggers an inflammatory skin disease. *J Exp Med* **206:** 2161–2177. doi:10.1084/jem.20090616

Kudelova J, Fleischmannova J, Adamova E, Matalova E. 2015. Pharmacological caspase inhibitors: research towards therapeutic perspectives. *J Physiol Pharmacol* **66:** 473–482.

Kumari S, Redouane Y, Lopez-Mosqueda J, Shiraishi R, Romanowska M, Lutzmayer S, Kuiper J, Martinez C, Dikic I, Pasparakis M, et al. 2014. Sharpin prevents skin inflammation by inhibiting TNFR1-induced keratinocyte apoptosis. *eLife* **3:** e03422. doi:10.7554/eLife.03422

Lahtela J, Nousiainen HO, Stefanovic V, Tallila J, Viskari H, Karikoski R, Gentile M, Saloranta C, Varilo T, Salonen R, et al. 2010. Mutant *CHUK* and severe fetal encasement malformation. *N Engl J Med* **363:** 1631–1637. doi:10.1056/NEJMoa0911698

Lau A, Wang S, Jiang J, Haig A, Pavlosky A, Linkermann A, Zhang ZX, Jevnikar AM. 2013. RIPK3-mediated necroptosis promotes donor kidney inflammatory injury and reduces allograft survival. *Am J Transplant* **13**: 2805–2818. doi:10.1111/ajt.12447

Lawlor KE, Khan N, Mildenhall A, Gerlic M, Croker BA, D'Cruz AA, Hall C, Kaur Spall S, Anderton H, Masters SL, et al. 2015. RIPK3 promotes cell death and NLRP3 inflammasome activation in the absence of MLKL. *Nat Commun* **6**: 6282. doi:10.1038/ncomms7282

Lee EG, Boone DL, Chai S, Libby SL, Chien M, Lodolce JP, Ma A. 2000. Failure to regulate TNF-induced NF-κB and cell death responses in A20-deficient mice. *Science* **289**: 2350–2354. doi:10.1126/science.289.5488.2350

Lehle AS, Farin HF, Marquardt B, Michels BE, Magg T, Li Y, Liu Y, Ghalandary M, Lammens K, Hollizeck S, et al. 2019. Intestinal inflammation and dysregulated immunity in patients with inherited caspase-8 deficiency. *Gastroenterology* **156**: 275–278. doi:10.1053/j.gastro.2018.09.041

Lemmers B, Salmena L, Bidère N, Su H, Matysiak-Zablocki E, Murakami K, Ohashi PS, Jurisicova A, Lenardo M, Hakem R, et al. 2007. Essential role for caspase-8 in Toll-like receptors and NFκB signaling. *J Biol Chem* **282**: 7416–7423. doi:10.1074/jbc.M606721200

Li Q, Van Antwerp D, Mercurio F, Lee KF, Verma IM. 1999. Severe liver degeneration in mice lacking the IκB kinase 2 gene. *Science* **284**: 321–325. doi:10.1126/science.284.5412.321

Li Y, Führer M, Bahrami E, Socha P, Klaudel-Dreszler M, Bouzidi A, Liu Y, Lehle AS, Magg T, Hollizeck S, et al. 2019. Human RIPK1 deficiency causes combined immunodeficiency and inflammatory bowel diseases. *Proc Natl Acad Sci* **116**: 970–975. doi:10.1073/pnas.1813582116

Lin J, Kumari S, Kim C, Van TM, Wachsmuth L, Polykratis A, Pasparakis M. 2016. RIPK1 counteracts ZBP1-mediated necroptosis to inhibit inflammation. *Nature* **540**: 124–128. doi:10.1038/nature20558

Lork M, Verhelst K, Beyaert R. 2017. CYLD, A20 and OTULIN deubiquitinases in NF-κB signaling and cell death: so similar, yet so different. *Cell Death Differ* **24**: 1172–1183. doi:10.1038/cdd.2017.46

Ma A, Malynn BA. 2012. A20: Linking a complex regulator of ubiquitylation to immunity and human disease. *Nat Rev Immunol* **12**: 774–785. doi:10.1038/nri3313

Maelfait J, Roose K, Vereecke L, Mc Guire C, Sze M, Schuijs MJ, Willart M, Ibañez LI, Hammad H, Lambrecht BN, et al. 2016. A20 deficiency in lung epithelial cells protects against influenza A virus infection. *PLoS Pathog* **12**: e1005410. doi:10.1371/journal.ppat.1005410

Mandal P, Berger SB, Pillay S, Moriwaki K, Huang C, Guo H, Lich JD, Finger J, Kasparcova V, Votta B, et al. 2014. RIP3 induces apoptosis independent of pronecrotic kinase activity. *Mol Cell* **56**: 481–495. doi:10.1016/j.molcel.2014.10.021

Mantovani A, Cassatella MA, Costantini C, Jaillon S. 2011. Neutrophils in the activation and regulation of innate and adaptive immunity. *Nat Rev Immunol* **11**: 519–531. doi:10.1038/nri3024

Mathis BJ, Lai Y, Qu C, Janicki JS, Cui T. 2015. CYLD-mediated signaling and diseases. *Curr Drug Targets* **16**: 284–294. doi:10.2174/1389450115666141024152421

Medzhitov R. 2008. Origin and physiological roles of inflammation. *Nature* **454**: 428–435. doi:10.1038/nature07201

Micheau O, Tschopp J. 2003. Induction of TNF receptor I-mediated apoptosis via two sequential signaling complexes. *Cell* **114**: 181–190. doi:10.1016/S0092-8674(03)00521-X

Miot C, Imai K, Imai C, Mancini AJ, Kucuk ZY, Kawai T, Nishikomori R, Ito E, Pellier I, Dupuis Girod S, et al. 2017. Hematopoietic stem cell transplantation in 29 patients hemizygous for hypomorphic *IKBKG*/NEMO mutations. *Blood* **130**: 1456–1467. doi:10.1182/blood-2017-03-771600

Monaco C, Nanchahal J, Taylor P, Feldmann M. 2015. Anti-TNF therapy: past, present and future. *Int Immunol* **27**: 55–62. doi:10.1093/intimm/dxu102

Mooster JL, Le Bras S, Massaad MJ, Jabara H, Yoon J, Galand C, Heesters BA, Burton OT, Mattoo H, Manis J, et al. 2015. Defective lymphoid organogenesis underlies the immune deficiency caused by a heterozygous S32I mutation in IκBα. *J Exp Med* **212**: 185–202. doi:10.1084/jem.20140979

Mullard A. 2018. Microglia-targeted candidates push the Alzheimer drug envelope. *Nat Rev Drug Discov* **17**: 303–305. doi:10.1038/nrd.2018.65

Murakami Y, Matsumoto H, Roh M, Giani A, Kataoka K, Morizane Y, Kayama M, Thanos A, Nakatake S, Notomi S, et al. 2014. Programmed necrosis, not apoptosis, is a key mediator of cell loss and DAMP-mediated inflammation in dsRNA-induced retinal degeneration. *Cell Death Differ* **21**: 270–277. doi:10.1038/cdd.2013.109

Murphy JM, Czabotar PE, Hildebrand JM, Lucet IS, Zhang JG, Alvarez-Diaz S, Lewis R, Lalaoui N, Metcalf D, Webb AI, et al. 2013. The pseudokinase MLKL mediates necroptosis via a molecular switch mechanism. *Immunity* **39**: 443–453. doi:10.1016/j.immuni.2013.06.018

Muzio M, Chinnaiyan AM, Kischkel FC, O'Rourke K, Shevchenko A, Ni J, Scaffidi C, Bretz JD, Zhang M, Gentz R, et al. 1996. FLICE, a novel FADD-homologous ICE/CED-3-like protease, is recruited to the CD95 (Fas/APO-1) death–inducing signaling complex. *Cell* **85**: 817–827. doi:10.1016/S0092-8674(00)81266-0

Newton K, Sun X, Dixit VM. 2004. Kinase RIP3 is dispensable for normal NF-κBs, signaling by the B-cell and T-cell receptors, tumor necrosis factor receptor 1, and Toll-like receptors 2 and 4. *Mol Cell Biol* **24**: 1464–1469. doi:10.1128/MCB.24.4.1464-1469.2004

Newton K, Dugger DL, Wickliffe KE, Kapoor N, de Almagro MC, Vucic D, Komuves L, Ferrando RE, French DM, Webster J, et al. 2014. Activity of protein kinase RIPK3 determines whether cells die by necroptosis or apoptosis. *Science* **343**: 1357–1360. doi:10.1126/science.1249361

Niemela J, Kuehn HS, Kelly C, Zhang M, Davies J, Melendez J, Dreiling J, Kleiner D, Calvo K, Oliveira JB, et al. 2015. Caspase-8 deficiency presenting as late-onset multi-organ lymphocytic infiltration with granulomas in two adult siblings. *J Clin Immunol* **35**: 348–355. doi:10.1007/s10875-015-0150-8

Northington FJ, Chavez-Valdez R, Graham EM, Razdan S, Gauda EB, Martin LJ. 2011. Necrostatin decreases oxidative damage, inflammation, and injury after neonatal HI. *J Cereb Blood Flow Metab* **31**: 178–189. doi:10.1038/jcbfm.2010.72

Oberst A, Dillon CP, Weinlich R, McCormick LL, Fitzgerald P, Pop C, Hakem R, Salvesen GS, Green DR. 2011. Catalytic activity of the caspase-8-FLIP(L) complex inhibits RIPK3-dependent necrosis. *Nature* **471:** 363–367. doi:10 .1038/nature09852

Oerlemans MI, Liu J, Arslan F, den Ouden K, van Middelaar BJ, Doevendans PA, Sluijter JP. 2012. Inhibition of RIP1-dependent necrosis prevents adverse cardiac remodeling after myocardial ischemia-reperfusion in vivo. *Basic Res Cardiol* **107:** 270. doi:10.1007/s00395-012-0270-8

Ofengeim D, Yuan J. 2013. Regulation of RIP1 kinase signalling at the crossroads of inflammation and cell death. *Nat Rev Mol Cell Biol* **14:** 727–736. doi:10.1038/nrm3683

Osborn SL, Diehl G, Han SJ, Xue L, Kurd N, Hsieh K, Cado D, Robey EA, Winoto A. 2010. Fas-associated death domain (FADD) is a negative regulator of T-cell receptor-mediated necroptosis. *Proc Natl Acad Sci* **107:** 13034–13039. doi:10.1073/pnas.1005997107

Pannicke U, Baumann B, Fuchs S, Henneke P, Rensing-Ehl A, Rizzi M, Janda A, Hese K, Schlesier M, Holzmann K, et al. 2013. Deficiency of innate and acquired immunity caused by an *IKBKB* mutation. *N Engl J Med* **369:** 2504–2514. doi:10.1056/NEJMoa1309199

Pasparakis M, Vandenabeele P. 2015. Necroptosis and its role in inflammation. *Nature* **517:** 311–320. doi:10 .1038/nature14191

Peltzer N, Rieser E, Taraborrelli L, Draber P, Darding M, Pernaute B, Shimizu Y, Sarr A, Draberova H, Montinaro A, et al. 2014. HOIP deficiency causes embryonic lethality by aberrant TNFR1-mediated endothelial cell death. *Cell Rep* **9:** 153–165. doi:10.1016/j.celrep.2014.08.066

Peltzer N, Darding M, Montinaro A, Draber P, Draberova H, Kupka S, Rieser E, Fisher A, Hutchinson C, Taraborrelli L, et al. 2018. LUBAC is essential for embryogenesis by preventing cell death and enabling haematopoiesis. *Nature* **557:** 112–117. doi:10.1038/s41586-018-0064-8

Petrie EJ, Czabotar PE, Murphy JM. 2019. The structural basis of necroptotic cell death signaling. *Trends Biochem Sci* **44:** 53–63. doi:10.1016/j.tibs.2018.11.002

Polykratis A, Hermance N, Zelic M, Roderick J, Kim C, Van TM, Lee TH, Chan FKM, Pasparakis M, Kelliher MA. 2014. Cutting edge: RIPK1 kinase inactive mice are viable and protected from TNF-induced necroptosis in vivo. *J Immunol* **193:** 1539–1543. doi:10.4049/jimmunol .1400590

Polykratis A, Martens A, Eren RO, Shirasaki Y, Yamagishi M, Yamaguchi Y, Uemura S, Miura M, Holzmann B, Kollias G, et al. 2019. A20 prevents inflammasome-dependent arthritis by inhibiting macrophage necroptosis through its ZnF7 ubiquitin-binding domain. *Nat Cell Biol* **21:** 731–742. doi:10.1038/s41556-019-0324-3

Reiley WW, Zhang M, Jin W, Losiewicz M, Donohue KB, Norbury CC, Sun SC. 2006. Regulation of T cell development by the deubiquitinating enzyme CYLD. *Nat Immunol* **7:** 411–417. doi:10.1038/ni1315

Reiley WW, Jin W, Lee AJ, Wright A, Wu X, Tewalt EF, Leonard TO, Norbury CC, Fitzpatrick L, Zhang M, et al. 2007. Deubiquitinating enzyme CYLD negatively regulates the ubiquitin-dependent kinase Tak1 and prevents abnormal T cell responses. *J Exp Med* **204:** 1475–1485. doi:10.1084/jem.20062694

Rickard JA, Anderton H, Etemadi N, Nachbur U, Darding M, Peltzer N, Lalaoui N, Lawlor KE, Vanyai H, Hall C, et al. 2014a. TNFR1-dependent cell death drives inflamma-

tion in Sharpin-deficient mice. *eLife* **3.** doi:10.7554/eLife .03464

Rickard JA, O'Donnell JA, Evans JM, Lalaoui N, Poh AR, Rogers T, Vince JE, Lawlor KE, Ninnis RL, Anderton H, et al. 2014b. RIPK1 regulates RIPK3-MLKL-driven systemic inflammation and emergency hematopoiesis. *Cell* **157:** 1175–1188. doi:10.1016/j.cell.2014.04.019

Rieux-Laucat F, Le Deist F, Hivroz C, Roberts IA, Debatin KM, Fischer A, de Villartay JP. 1995. Mutations in Fas associated with human lymphoproliferative syndrome and autoimmunity. *Science* **268:** 1347–1349. doi:10 .1126/science.7539157

Rivkin E, Almeida SM, Ceccarelli DF, Juang YC, MacLean TA, Srikumar T, Huang H, Dunham WH, Fukumura R, Xie G, et al. 2013. The linear ubiquitin-specific deubiquitinase gumby regulates angiogenesis. *Nature* **498:** 318–324. doi:10.1038/nature12296

Rock KL, Kono H. 2008. The inflammatory response to cell death. *Annu Rev Pathol* **3:** 99–126. doi:10.1146/annurev .pathmechdis.3.121806.151456

Sakamaki K, Inoue T, Asano M, Sudo K, Kazama H, Sakagami J, Sakata S, Ozaki M, Nakamura S, Toyokuni S, et al. 2002. Ex vivo whole-embryo culture of caspase-8-deficient embryos normalize their aberrant phenotypes in the developing neural tube and heart. *Cell Death Differ* **9:** 1196–1206. doi:10.1038/sj.cdd.4401090

Salmena L, Lemmers B, Hakem A, Matysiak-Zablocki E, Murakami K, Au PY, Berry DM, Tamblyn L, Shehabeldin A, Migon E, et al. 2003. Essential role for caspase 8 in T-cell homeostasis and T-cell-mediated immunity. *Genes Dev* **17:** 883–895. doi:10.1101/gad.1063703

Sato K, Li S, Gordon WC, He J, Liou GI, Hill JM, Travis GH, Bazan NG, Jin M. 2013. Receptor interacting protein kinase-mediated necrosis contributes to cone and rod photoreceptor degeneration in the retina lacking interphotoreceptor retinoid-binding protein. *J Neurosci* **33:** 17458–17468. doi:10.1523/jneurosci.1380-13.2013

Schmidt-Supprian M, Bloch W, Courtois G, Addicks K, Israël A, Rajewsky K, Pasparakis M. 2000. NEMO/IKK γ-deficient mice model incontinentia pigmenti. *Mol Cell* **5:** 981–992. doi:10.1016/S1097-2765(00)80263-4

Serhan CN, Savill J. 2005. Resolution of inflammation: the beginning programs the end. *Nat Immunol* **6:** 1191–1197. doi:10.1038/ni1276

Sha WC, Liou HC, Tuomanen EI, Baltimore D. 1995. Targeted disruption of the p50 subunit of NF-κB leads to multifocal defects in immune responses. *Cell* **80:** 321–330. doi:10.1016/0092-8674(95)90415-8

Silke J, Rickard JA, Gerlic M. 2015. The diverse role of RIP kinases in necroptosis and inflammation. *Nat Immunol* **16:** 689–697. doi:10.1038/ni.3206

Silva MT, do Vale A, dos Santos NM. 2008. Secondary necrosis in multicellular animals: an outcome of apoptosis with pathogenic implications. *Apoptosis* **13:** 463–482. doi:10.1007/s10495-008-0187-8

Smahi A, Courtois G, Vabres P, Yamaoka S, Heuertz S, Munnich A, Israël A, Heiss NS, Klauck SM, Kioschis P, et al. 2000. Genomic rearrangement in NEMO impairs NF-κB activation and is a cause of incontinentia pigmenti. The International Incontinentia Pigmenti (IP) Consortium. *Nature* **405:** 466–472. doi:10.1038/ 35013114

Smith CC, Davidson SM, Lim SY, Simpkin JC, Hothersall JS, Yellon DM. 2007. Necrostatin: a potentially novel cardio-protective agent? *Cardiovasc Drugs Ther* **21:** 227–233. doi:10.1007/s10557-007-6035-1

Su H, Bidère N, Zheng L, Cubre A, Sakai K, Dale J, Salmena L, Hakem R, Straus S, Lenardo M. 2005. Requirement for caspase-8 in NF-κB activation by antigen receptor. *Science* **307:** 1465–1468. doi:10.1126/science.1104765

Sun SC. 2010. CYLD: a tumor suppressor deubiquitinase regulating NF-κB activation and diverse biological processes. *Cell Death Differ* **17:** 25–34. doi:10.1038/cdd.2009.43

Takahashi N, Vereecke L, Bertrand MJ, Duprez L, Berger SB, Divert T, Gonçalves A, Sze M, Gilbert B, Kourula S, et al. 2014. RIPK1 ensures intestinal homeostasis by protecting the epithelium against apoptosis. *Nature* **513:** 95–99. doi:10.1038/nature13706

Takeda K, Takeuchi O, Tsujimura T, Itami S, Adachi O, Kawai T, Sanjo H, Yoshikawa K, Terada N, Akira S. 1999. Limb and skin abnormalities in mice lacking IKKα. *Science* **284:** 313–316. doi:10.1126/science.284.5412.313

Tanaka M, Fuentes ME, Yamaguchi K, Durnin MH, Dalrymple SA, Hardy KL, Goeddel DV. 1999. Embryonic lethality, liver degeneration, and impaired NF-κB activation in IKK-β-deficient mice. *Immunity* **10:** 421–429. doi:10.1016/S1074-7613(00)80042-4

Taraborrelli L, Peltzer N, Montinaro A, Kupka S, Rieser E, Hartwig T, Sarr A, Darding M, Draber P, Haas TL, et al. 2018. LUBAC prevents lethal dermatitis by inhibiting cell death induced by TNF, TRAIL and CD95L. *Nat Commun* **9:** 3910. doi:10.1038/s41467-018-06155-8

Tavares RM, Turer EE, Liu CL, Advincula R, Scapini P, Rhee L, Barrera J, Lowell CA, Utz PJ, Malynn BA, et al. 2010. The ubiquitin modifying enzyme A20 restricts B cell survival and prevents autoimmunity. *Immunity* **33:** 181–191. doi:10.1016/j.immuni.2010.07.017

Tokunaga F, Nakagawa T, Nakahara M, Saeki Y, Taniguchi M, Sakata S, Tanaka K, Nakano H, Iwai K. 2011. SHARPIN is a component of the NF-κB-activating linear ubiquitin chain assembly complex. *Nature* **471:** 633–636. doi:10.1038/nature09815

Tuijnenburg P, Lango Allen H, Burns SO, Greene D, Jansen MH, Staples E, Stephens J, Carss KJ, Biasci D, Baxendale H, et al. 2018. Loss-of-function nuclear factor κB subunit 1 (NFKB1) variants are the most common monogenic cause of common variable immunodeficiency in Europeans. *J Allergy Clin Immunol* **142:** 1285–1296. doi:10.1016/j.jaci.2018.01.039

Uchiyama Y, Kim CA, Pastorino AC, Ceroni J, Lima PP, de Barros Dorna M, Honjo RS, Bertola D, Hamanaka K, Fujita A, et al. 2019. Primary immunodeficiency with chronic enteropathy and developmental delay in a boy arising from a novel homozygous RIPK1 variant. *J Hum Genet* **64:** 955–960. doi:10.1038/s10038-019-0631-3

Van Antwerp DJ, Martin SJ, Kafri T, Green DR, Verma IM. 1996. Suppression of TNF-α-induced apoptosis by NF-κB. *Science* **274:** 787–789. doi:10.1126/science.274.5288.787

Varfolomeev EE, Schuchmann M, Luria V, Chiannilkulchai N, Beckmann JS, Mett IL, Rebrikov D, Brodianski VM, Kemper OC, Kollet O, et al. 1998. Targeted disruption of the mouse caspase 8 gene ablates cell death induction by the TNF receptors, Fas/Apo1, and DR3 and is lethal prenatally. *Immunity* **9:** 267–276. doi:10.1016/S1074-7613(00)80609-3

Vereecke L, Vieira-Silva S, Billiet T, van Es JH, Mc Guire C, Slowicka K, Sze M, van den Born M, De Hertogh G, Clevers H, et al. 2014. A20 controls intestinal homeostasis through cell-specific activities. *Nat Commun* **5:** 5103. doi:10.1038/ncomms6103

Vince JE, Silke J. 2016. The intersection of cell death and inflammasome activation. *Cell Mol Life Sci* **73:** 2349–2367. doi:10.1007/s00018-016-2205-2

Vlantis K, Wullaert A, Polykratis A, Kondylis V, Dannappel M, Schwarzer R, Welz P, Corona T, Walczak H, Weih F, et al. 2016. NEMO prevents RIP kinase 1-mediated epithelial cell death and chronic intestinal inflammation by NF-κB-dependent and -independent functions. *Immunity* **44:** 553–567. doi:10.1016/j.immuni.2016.02.020

Wang L, Du F, Wang X. 2008. TNF-α induces two distinct caspase-8 activation pathways. *Cell* **133:** 693–703. doi:10.1016/j.cell.2008.03.036

Wang X, Deckert M, Xuan NT, Nishanth G, Just S, Waisman A, Naumann M, Schlüter D. 2013. Astrocytic A20 ameliorates experimental autoimmune encephalomyelitis by inhibiting NF-κB- and STAT1-dependent chemokine production in astrocytes. *Acta Neuropathol* **126:** 711–724. doi:10.1007/s00401-013-1183-9

Weinlich R, Oberst A, Beere HM, Green DR. 2017. Necroptosis in development, inflammation and disease. *Nat Rev Mol Cell Biol* **18:** 127–136. doi:10.1038/nrm.2016.149

Welz PS, Wullaert A, Vlantis K, Kondylis V, Fernández-Majada V, Ermolaeva M, Kirsch P, Sterner-Kock A, van Loo G, Pasparakis M. 2011. FADD prevents RIP3-mediated epithelial cell necrosis and chronic intestinal inflammation. *Nature* **477:** 330–334. doi:10.1038/nature10273

Wertz IE, O'Rourke KM, Zhou H, Eby M, Aravind L, Seshagiri S, Wu P, Wiesmann C, Baker R, Boone DL, et al. 2004. De-ubiquitination and ubiquitin ligase domains of A20 downregulate NF-κB signalling. *Nature* **430:** 694–699. doi:10.1038/nature02794

Wilson NS, Dixit V, Ashkenazi A. 2009. Death receptor signal transducers: nodes of coordination in immune signaling networks. *Nat Immunol* **10:** 348–355. doi:10.1038/ni.1714

Wu J, Huang Z, Ren J, Zhang Z, He P, Li Y, Ma J, Chen W, Zhang Y, Zhou X, et al. 2013. Mlkl knockout mice demonstrate the indispensable role of Mlkl in necroptosis. *Cell Res* **23:** 994–1006. doi:10.1038/cr.2013.91

Yamamoto Y, Gaynor RB. 2001. Therapeutic potential of inhibition of the NF-κB pathway in the treatment of inflammation and cancer. *J Clin Invest* **107:** 135–142. doi:10.1172/JCI11914

Yamaoka S, Courtois G, Bessia C, Whiteside ST, Weil R, Agou F, Kirk HE, Kay RJ, Israël A. 1998. Complementation cloning of NEMO, a component of the IκB kinase complex essential for NF-κB activation. *Cell* **93:** 1231–1240. doi:10.1016/S0092-8674(00)81466-X

Yan B, Liu L, Huang S, Ren Y, Wang H, Yao Z, Li L, Chen S, Wang X, Zhang Z. 2017. Discovery of a new class of highly potent necroptosis inhibitors targeting the mixed lineage kinase domain-like protein. *Chem Commun (Camb)* **53:** 3637–3640. doi:10.1039/C7CC00667E

Yeh WC, de la Pompa JL, McCurrach ME, Shu HB, Elia AJ, Shahinian A, Ng M, Wakeham A, Khoo W, Mitchell K, et al. 1998. FADD: essential for embryo development and signaling from some, but not all, inducers of apoptosis. *Science* **279**: 1954–1958. doi:10.1126/science.279.5358 .1954

Zhang J, Stirling B, Temmerman ST, Ma CA, Fuss IJ, Derry JM, Jain A. 2006. Impaired regulation of NF-κB and increased susceptibility to colitis-associated tumorigenesis in CYLD-deficient mice. *J Clin Invest* **116**: 3042–3049. doi:10.1172/JCI28746

Zhang DW, Shao J, Lin J, Zhang N, Lu BJ, Lin SC, Dong MQ, Han J. 2009. RIP3, an energy metabolism regulator that switches TNF-induced cell death from apoptosis to necrosis. *Science* **325**: 332–336. doi:10.1126/science .1172308

Zhang H, Zhou X, McQuade T, Li J, Chan FK, Zhang J. 2011. Functional complementation between FADD and RIP1 in embryos and lymphocytes. *Nature* **471**: 373–376. doi:10 .1038/nature09878

Zhang Q, Lenardo MJ, Baltimore D. 2017a. 30 years of NF-κB: a blossoming of relevance to human pathobiology. *Cell* **168**: 37–57. doi:10.1016/j.cell.2016.12.012

Zhang S, Tang MB, Luo HY, Shi CH, Xu YM. 2017b. Necroptosis in neurodegenerative diseases: a potential thera-peutic target. *Cell Death Dis* **8**: e2905. doi:10.1038/cddis .2017.286

Zhou Q, Wang H, Schwartz DM, Stoffels M, Park YH, Zhang Y, Yang D, Demirkaya E, Takeuchi M, Tsai WL, et al. 2016a. Loss-of-function mutations in TNFAIP3 leading to A20 haploinsufficiency cause an early-onset autoin-flammatory disease. *Nat Genet* **48**: 67–73. doi:10.1038/ ng.3459

Zhou Q, Yu X, Demirkaya E, Deuitch N, Stone D, Tsai WL, Kuehn HS, Wang H, Yang D, Park YH, et al. 2016b. Biallelic hypomorphic mutations in a linear deubiquiti-nase define otulipenia, an early-onset autoinflammatory disease. *Proc Natl Acad Sci* **113**: 10127–10132. doi:10 .1073/pnas.1612594113

Zhu S, Zhang Y, Bai G, Li H. 2011. Necrostatin-1 ameliorates symptoms in R6/2 transgenic mouse model of Hunting-ton's disease. *Cell Death Dis* **2**: e115. doi:10.1038/cddis .2010.94

Zonana J, Elder ME, Schneider LC, Orlow SJ, Moss C, Golabi M, Shapira SK, Farndon PA, Wara DW, Emmal SA, et al. 2000. A novel X-linked disorder of immune deficiency and hypohidrotic ectodermal dysplasia is allelic to incontinentia pigmenti and due to mutations in IKK-γ (NEMO). *Am J Hum Genet* **67**: 1555–1562. doi:10.1086/ 316914

Neutrophil Extracellular Traps in Host Defense

Sabrina Sofia Burgener and Kate Schroder

Institute for Molecular Bioscience (IMB), and IMB Centre for Inflammation and Disease Research, The University of Queensland, St Lucia 4072, Australia

Correspondence: K.Schroder@imb.uq.edu.au

Neutrophils are produced in the bone marrow and then patrol blood vessels from which they can be rapidly recruited to a site of infection. Neutrophils bind, engulf, and efficiently kill invading microbes via a suite of defense mechanisms. Diverse extracellular and intracellular microbes induce neutrophils to extrude neutrophil extracellular traps (NETs) through the process of NETosis. Here, we review the signaling mechanisms and cell biology underpinning the key NETosis pathways during infection and the antimicrobial functions of NETs in host defense.

In 1880, Paul Ehrlich described cells with granules and a lobulated nucleus, and called these cells polymorphonuclear leukocytes. Polymorphonuclear leukocytes were then further classified into eosinophils, basophils, and neutrophils. Neutrophils are now recognized as essential effector cells of innate immunity and key regulators of both innate and adaptive immune responses. Neutrophils are produced in the bone marrow and then patrol blood vessels, from which they can be rapidly recruited to a site of infection. Neutrophil activation by pathogen products or the local inflammatory milieu prolongs neutrophil life span and arms these cells with antimicrobial effector functions (Dale et al. 2008; Nauseef and Borregaard 2014).

Neutrophils combat microbes using a suite of defense mechanisms. Neutrophils internalize microbes through the process of phagocytosis after which the antimicrobial arsenal of the neutrophil granules is delivered to the phagosome to mediate microbial killing. Neutrophils can also externalize their granule content to neutralize extracellular microbes in a process called degranulation. The key granule proteins involved in microbial killing are α-defensins, lysozymes, and the three neutrophil serine proteases, neutrophil elastase (NE), cathepsin G (CatG), and proteinase-3 (PR-3). Phagocytosis and degranulation also induce the assembly of the nicotinamide adenine dinucleotide phosphate (NADPH) oxidase complex. NAPDH oxidase converts oxygen into superoxide, which can be further catalyzed to H_2O_2, hydroxyl anion, or peroxynitrite anion, to produce an array of reactive oxygen species (ROS) (Faurschou and Borregaard 2003; Borregaard 2014).

Although neutrophils have evolved multiple strategies to inactivate and eliminate invading microbes, pathogens have also developed strategies to escape the neutrophil immune defense machinery described above. Neutrophils can, thus, exert an additional antimicrobial defense program when they encounter microbes that are either too large for phagocytosis, or have evaded phagosomal destruction by escaping into the cytosol. Here, neutrophils extrude a physical barrier to pathogen dissemination, called a neutrophil extracellular trap (NET) (Amulic et al. 2012; Papayannopoulos 2018). This review summarizes the major pathways that lead to the extrusion of NETs during infection and their functions in host defense.

NEUTROPHIL EXTRACELLULAR TRAPS

NETs are webs of neutrophil DNA coated with histones and antimicrobial proteins that entrap microbes. Neutrophils cast their NETs in a multistep process called NETosis (Brinkmann et al. 2004; Fuchs et al. 2007; Papayannopoulos et al. 2010). NETosis is a distinct cellular program from apoptosis or necroptosis, and true NET structures are not usually cast during these latter forms of cell death. NETosis involves several distinct and sequential morphological changes in the neutrophil (Fig. 1; Fuchs et al. 2007; Yipp and Kubes 2013): (1) The characteristic lobulated architecture of the nucleus is lost, (2) the nuclear and granular membranes become permeable, (3) histone inactivation leads to chromatin expansion into the cytosol, (4) chromatin mixes with granule content, (5) the loss of internal membranes leads to the disappearance of cytosolic organelles, and (6) the plasma membrane becomes permeable and the NETs are released into the extracellular space.

Within NETs, the DNA is decorated with a conserved set of nuclear, granular, and cytoplasmic proteins. The key antimicrobial effector proteins of NETs are histones (H2A, H2B, H3, and H4) and granule proteases (such as NE, CatG, and PR-3), as well as myeloperoxidase (MPO) and lactotransferrin. The proteases remain enzymatically active even when the NET is exposed to endogenous protease inhibitors. Other granule proteins, such as azurocidin, lysozyme C, and α-defensins, are also reported within NETs (Dwyer et al. 2014). NETs also contain cytoplasmic and cytoskeleton proteins such as S100 calcium-binding proteins, actin, myosin, and cytokeratin (Urban et al. 2009; O'Donoghue et al. 2013; Dwyer et al. 2014). Proteomics stud-

ies identified 24 proteins within PMA-induced NETs (Urban et al. 2009; O'Donoghue et al. 2013), and 80 proteins within NETs induced by nonmucoid and mucoid strains of *Pseudomonas aeruginosa* (Dwyer et al. 2014). It is, thus, possible that the composition of the NET is influenced by its eliciting stimulus.

THE NEUTROPHIL'S DECISION TO CAST A NET IS INFORMED BY MICROBE SIZE AND LOCATION

When encountering a microbe, the neutrophil needs to determine which antimicrobial defense function to deploy. A microbe size-sensing mechanism allows neutrophils to decide whether or not to launch a NET in response to extracellular pathogens (Branzk et al. 2014). Small microbes that are readily taken up by phagocytosis are poor NET stimulants because the phagosome fuses with neutrophil azurophil granules, and neutrophil granule proteases are, thus, unavailable for initiating NETosis (Branzk et al. 2014). Phagosome formation, hence, serves as a checkpoint to prevent NETosis. Some microbes are too large to be engulfed by phagocytosis, for example, the fungus *Candida albicans* that forms large filamentous hyphae, or aggregates of *Mycobacterium bovis*. If a neutrophil encounters these large microbes, the azurophil granule protein NE is released into the cytoplasm to initiate NET formation. Neutrophils deficient in dectin-1, a receptor that mediates phagocytosis of fungi, are hindered in their ability to determine which antimicrobial defense mechanism to apply, resulting in uncontrolled NET release (Branzk et al. 2014). Microbe size and location also affect other neutrophil antimicrobial functions. Neutrophil interaction with large microbes, such as *C. albicans* hyphae, leads to the extracellular deployment of ROS and strong interleukin (IL)-1β secretion with resultant neutrophil clustering wherein about eight neutrophils engage with one 100-μm hyphal filament. In contrast, small phagocytosed microbes trigger the intracellular deployment of ROS, suppressing cytokine release and neutrophil clustering, and thereby preventing neutro-

Cite this article as *Cold Spring Harb Perspect Biol* doi: 10.1101/cshperspect.a037028

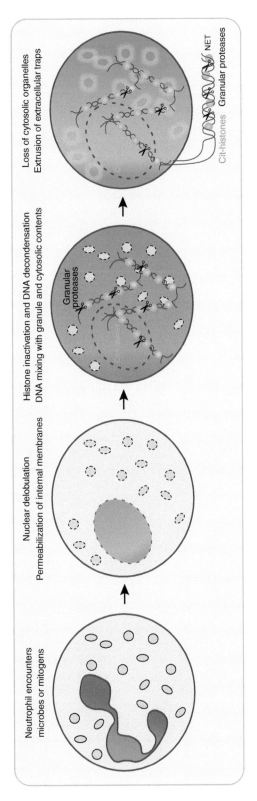

Figure 1. NETosis is a distinct cellular program in neutrophils. Neutrophil NETosis involves sequential morphological changes: (1) The characteristic lobulated architecture of the nucleus (purple) is lost, (2) the membranes of the nucleus and granules (pink) become permeable, (3) histone inactivation by clipping and/or citrullination (Cit-histones) leads to chromatin expansion into the cytosol, (4) chromatin mixes with granule content, (5) the loss of internal membranes leads to the disappearance of cytosolic organelles, and (6) the plasma membrane breaks and neutrophil extracellular traps (NETs) are released into the extracellular space.

phil accumulation and tissue damage during microbial clearance (Warnatsch et al. 2017).

Microbe size may not be the only factor that determines the antimicrobial response of neutrophils. Some microbes can manipulate the neutrophil response by evading phagocytosis, and may thereby induce NET formation (Sumby et al. 2005; Döhrmann et al. 2014; Eby et al. 2014; Storisteanu et al. 2017; Eisenbeis et al. 2018). For example, Gram-negative bacteria that escape the phagosome to translocate to the cytosol induce NETs in a newly described pathway that involves the noncanonical inflammasome machinery (Chen et al. 2018).

In the 15 years since their discovery, NETs have been an area of intense research, which has provided important insights into this novel antimicrobial mechanism, and revealed several distinct pathways for trap release. In this review, we describe three key mechanisms by which neutrophils extrude NETs in host defense through the process of suicidal NETosis, noncanonical NETosis, and vital NETosis. Whereas all of these NETosis pathways ultimately lead to neutrophil death, NET extrusion occurs during cell rupture in suicidal and noncanonical NETosis while NETs are expelled from neutrophils that can continue to perform cell functions (e.g., migration) in vital NETosis.

SUICIDAL NETosis

The first NETosis pathway was described in 2004, in which NET extrusion and cell death (suicidal NETosis) was observed in response to high doses of phorbol-12-myristate-13-acetate (PMA) (Brinkmann et al. 2004). The strong and robust NET responses induced by PMA have led to the adoption of PMA as a key in vitro stimulant for the study of NETosis. PMA is a potent activator of several signaling pathways within neutrophils such as protein kinase C (PKC) (Steinberg 2008). PMA mediates suicidal NETosis by inducing the assembly of NADPH oxidase, which triggers ROS production, MPO activation, and the cytoplasmic release of NE, and leads to chromatin decondensation in the nucleus (Fig. 2). Following this, the nuclear membrane breaks down, resulting in the release

of the decondensed chromatin into the cytosol. Chromatin is then expelled into the extracellular space via GSDMD pores or GSDMD-driven membrane tears, and results in neutrophil death (Brinkmann et al. 2004; Fuchs et al. 2007; Sollberger et al. 2018). Below, we will focus on this well-characterized pathway of NE-, MPO-, and ROS-driven NETosis as an exemplar for suicidal NETosis. However, ionophores (e.g., A23187, nigericin) also trigger suicidal NETosis through a poorly characterized mechanism that is independent of NE, MPO, and ROS (Neeli and Radic 2013; Kenny et al. 2017), highlighting the complexity of NETosis pathways.

NADPH Oxidase

On activation by PMA, neutrophils up-regulate glycolysis (Rodríguez-Espinosa et al. 2015), and this is required for NET expulsion (Rodríguez-Espinosa et al. 2015; Siler et al. 2017). Glucose uptake induces signaling by extracellular signal-regulated kinase (ERK) (Rodríguez-Espinosa et al. 2015), leading to the phosphorylation of one of the three cytoplasmic regulatory subunits of the NADPH oxidase, $p47^{phox}$ (El Benna et al. 1996; Hakkim et al. 2011). On phosphorylation, $p47^{phox}$ is recruited to the granule and plasma membranes, in which it interacts with other subunits to assemble the NADPH oxidase complex and generate superoxide (Dang et al. 1999; Lopes et al. 1999; Kleniewska et al. 2012; Nunes et al. 2013; Dinauer 2016). Inhibitors of the ERK pathway, thus, block NADPH oxidase function and PMA-induced NET release (Hakkim et al. 2011). PMA- and microbe-induced NET release is also suppressed in neutrophils deficient in Rac2, a small GTPase required for NADPH oxidase function (Lim et al. 2011; Gavillet et al. 2018), and in neutrophils from chronic granulomatous disease (CGD) patients that are deficient in NADPH oxidase function (Bianchi et al. 2009; Röhm et al. 2014). Gene therapy restored NADPH oxidase function in a CGD patient, and consequently reinstated the capacity of their neutrophils to produce NETs, which were able to inhibit *Aspergillus nidulans* growth (Bianchi et al. 2009). As invasive aspergillosis is a leading cause of death in CGD patients, this suggests

Cite this article as *Cold Spring Harb Perspect Biol* doi: 10.1101/cshperspect.a037028

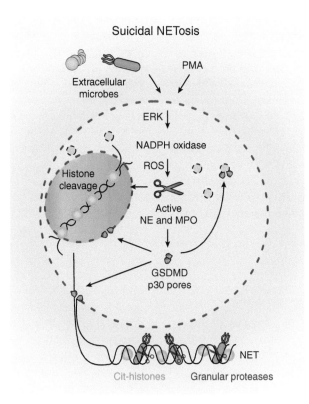

Figure 2. Suicidal NETosis. Phorbol-12-myristate-13-acetate (PMA) or extracellular microbes induce suicidal NETosis by inducing the extracellular signal-regulated kinase (ERK)-dependent assembly of the nicotinamide adenine dinucleotide phosphate (NADPH) oxidase, which generates reactive oxygen species (ROS) and the cytoplasmic release of granule proteins, and leads to chromatin decondensation in the nucleus. Chromatin is then expelled into the extracellular space via gasdermin D (GSDMD) pores or GSDMD-driven membrane tears, resulting in neutrophil death. (NE) neutrophil elastase, (MPO) myeloperoxidase, (NET) neutrophil extracellular trap.

that ROS-dependent NETs may be critical for host defense against *Aspergilli* species.

Granule Proteases

Under resting conditions, the neutrophil's antimicrobial arsenal is stored in granules. ROS generation by NAPDH oxidase triggers the release of these proteins into the cytosol. The protease NE is particularly critical for suicidal NETosis, as murine NE deficiency prevents NET formation during in vivo challenge with *Klebsiella pneumoniae* (Papayannopoulos et al. 2010).

The exact mechanism by which NE is released into the cytosol to initiate NETosis is unclear, but it is proposed to occur without granule rupture (Metzler et al. 2014) and involve co-operation between granule proteins (O'Donoghue et al. 2013). In resting neutrophils, NE clusters together with MPO, azurocidin, CatG, defensin-1, lysozyme, and lactoferrin to form a complex called an "azurosome" in the azurophil granule (Brinkmann et al. 2004; Fuchs et al. 2007; Papayannopoulos et al. 2010; Metzler et al. 2011). On neutrophil NADPH oxidase activation, ROS triggers MPO activation and the dissociation of MPO and NE from the azurosome (Metzler et al. 2011, 2014). This induces the release of NE into the cytosol through an unclear mechanism that requires MPO activity within the cell (Metzler et al. 2014). Human mutations in the *MPO* locus can decrease or ablate MPO function (Nauseef et al. 1996; Dinauer 2014). Whereas neutrophils from patients with a partial MPO deficiency can still form NETs on exposure to *C. albicans*, those from MPO-deficient patients cannot (Metzler et al. 2011).

In the cytoplasm, NE binds to the actin cytoskeleton where it degrades F-actin, leading to actin disassembly and neutrophil immobilization (Metzler et al. 2014). In the nucleus, NE cleaves and inactivates histones such as H4 and H2B, leading to chromatin relaxation and DNA decondensation (Papayannopoulos et al. 2010). MPO functions synergistically with NE to mediate chromatin decondensation during NETosis (Papayannopoulos et al. 2010). The exact requirement for MPO in this process is unclear, but may simply involve NE liberation from granules as described above.

PAD4-Dependent Histone Citrullination

Suicidal NETosis is often accompanied by histone citrullination by peptidylarginine deiminase type 4 (PAD4), which can contribute to chromatin relaxation. PAD4 is activated by a cytosolic spike in calcium as is induced by ionophores (Neeli et al. 2008) but not by PMA (Neeli and Radic 2013; Damgaard et al. 2016). Whereas PAD4-dependent histone citrullination often accompanies suicidal NETosis, it is often dispensable for NET extrusion via this pathway (Claushuis et al. 2018; Guiducci et al. 2018), perhaps because histones are additionally inactivated by NE-dependent cleavage, providing redundancy. PAD4 is, however, crucial for chromatin decondensation during vital NETosis.

Gasdermin D: A NETosis Executioner

Gasdermin D (GSDMD) was recently identified as an executioner of suicidal and noncanonical NETosis (Chen et al. 2018; Sollberger et al. 2018). GSDMD usually resides in the cytosol of resting cells. On cleavage by specific proteases, a pore-forming fragment (GSDMD-p30) is generated. During PMA-induced suicidal NETosis, NE cleaves GSDMD to generate GSDMD-p30 pores in the plasma and granule membranes, leading to rupture of these membranes during NETosis (Sollberger et al. 2018). GSDMD pores permeabilize the nuclear mem-

brane during noncanonical NETosis (Chen et al. 2018) and may similarly contribute to nuclear permeabilization during suicidal NETosis. A GSDMD inhibitor blocked PMA-induced NETosis without affecting the function of NADPH oxidase, NE or MPO, highlighting the requirement for membrane permeabilization in this pathway (Sollberger et al. 2018).

Model for Suicidal NETosis

The above studies together suggest the following model for suicidal NETosis (Fig. 2). On neutrophil activation, ROS triggers the activation of MPO and azurosome dissociation within azurophil granules, thereby facilitating the MPO-driven release of NE into the cytoplasm. This small amount of NE appears to be sufficient to cleave GSDMD to GSDMD-p30. This, in turn, induces the formation of GSDMD pores within the granule membrane, leading to granule rupture, further NE release into the cytoplasm, and further GSDMD cleavage in a feedforward loop. NE then translocates into the nucleus, where it cleaves and inactivates histones, leading to chromatin expansion and mixing with cytoplasmic and granule contents. GSDMD-p30 pores then induce plasma membrane rupture to enable NET release (Sollberger et al. 2018).

NONCANONICAL NETosis

We recently described a new pathway of suicidal NETosis, which we here propose to term noncanonical NETosis, which is deployed when neutrophils detect Gram-negative bacteria in their cytosol. This pathway relies on bacterial sensing by the noncanonical inflammasome (Fig. 3), leading to caspase-4/11- and GSDMD-driven NET extrusion (Chen et al. 2018) through a signaling pathway that resembles cell lysis by pyroptosis but results in the morphological features of neutrophil death by NETosis.

Signaling by the Noncanonical Inflammasome

The prefix "noncanonical" in this NETosis pathway indicates the involvement of the noncanonical inflammasome. Inflammasomes can be

Cite this article as *Cold Spring Harb Perspect Biol* doi: 10.1101/cshperspect.a037028

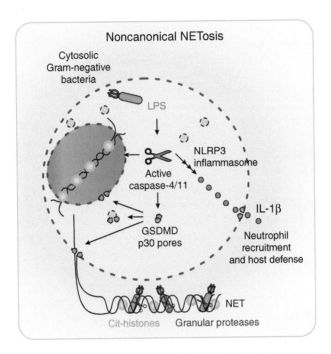

Figure 3. Noncanonical NETosis. The lipopolysaccharide (LPS) of cytosolic Gram-negative bacteria triggers the assembly of the noncanonical inflammasome, leading to caspase-4/11 activation and gasdermin D (GSDMD) cleavage to p30. GSDMD-p30 pores rupture the nuclear and granule membranes and, eventually, the plasma membrane. Caspase-11 gains access to the nucleus, where it cleaves histones to mediate DNA decondensation. Chromatin is then expelled into the extracellular space during plasma membrane rupture, resulting in neutrophil death. During NETosis, GSDMD-driven NLRP3 inflammasome activation triggers the release of interleukin (IL)-1β to induce neutrophil recruitment, activation, and killing of the neutrophil extracellular trap (NET)-entrapped bacterium.

broadly categorized into those that activate caspase-1 ("canonical inflammasomes") and those that activate caspase-11 in mouse or caspase-4/5 in humans ("noncanonical inflammasomes"). The noncanonical inflammasomes are sensors for cytosolic Gram-negative bacteria. On detection of bacterial lipopolysaccharide (LPS), these inflammasomes assemble to activate caspase-4/5 in humans and caspase-11 in mice. Signaling by both canonical and noncanonical inflammasomes leads to caspase-1- or caspase-4/5/11-dependent cleavage of GSDMD to trigger GSDMD-p30 plasma membrane pores, the release of proinflammatory cytokines (e.g., IL-1β and IL-18), and cell lysis by pyroptosis (Kayagaki et al. 2015; Shi et al. 2015; Liu et al. 2016). Pyroptosis has been intensely studied in macrophages but is less characterized in other cell types.

NET Extrusion via the Noncanonical Inflammasome

Compared to macrophages, neutrophils signal via specialized inflammasome pathways that induce distinct outcomes for cell death or viability. We, and others, previously showed that neutrophils do not undergo caspase-1-dependent pyroptosis during canonical inflammasome signaling (Chen et al. 2014, 2018; Karmakar et al. 2015; Monteleone et al. 2018). Neutrophils do, however, die on activation of the noncanonical inflammasome pathway (Chen et al. 2018). Human and murine neutrophils infected with Gram-negative bacteria that access the cytosol (ΔsifA Salmonella enterica, Citrobacter rodentium) died and extruded NETs through a cell death pathway displaying the hallmark features of NETosis, including nuclear delobulation,

DNA extrusion, DNA-MPO colocalization, and histone citrullination (Chen et al. 2018). Neutrophil NET extrusion required the pyroptotic machinery, caspase-4/11, and GSDMD, but proceeded independently of NE or MPO. PAD4-dependent histone citrullination accompanied noncanonical NETosis but was not required for death or NET extrusion. Mechanistically, GSDMD-p30 pores target several neutrophil membranes to mediate permeabilization of the granule and nuclear membranes and, eventually, rupture of the plasma membrane (Chen et al. 2018; Sollberger et al. 2018). GSDMD-p30 nuclear pores appear to enable caspase-11 to access the chromatin, in which caspase-11 performs an analogous function to NE to degrade histones and thereby relax chromatin (Chen et al. 2018).

Noncanonical NETosis suppresses bacterial residence in the neutrophil cytosol and prevents in vivo microbial dissemination. During in vitro infection, the noncanonical NETosis pathway suppressed neutrophil cytosolic infection with $\Delta sifA$ *Salmonella* (Chen et al. 2018), presumably by inducing the death of the infected neutrophil. Murine deficiency in *Casp11* or *Gsdmd* promoted microbial dissemination to a secondary infection site during murine in vivo challenge with $\Delta sifA$ *Salmonella* (Chen et al. 2018), confirming previous reports that the noncanonical inflammasome pathway mediates host defense (Aachoui et al. 2013; Storek and Monack 2015; Thurston et al. 2016). Interestingly, when NETs were dismantled at a primary infection site by applying DNase I, this promoted bacterial dissemination in wild-type (WT) but not $Casp11^{-/-}$ or $Gsdmd^{-/-}$ mice (Chen et al. 2018). The NETs themselves thus mediate host defense, likely via immobilizing bacteria to enable their destruction by newly recruited neutrophils.

VITAL NETosis

In contrast to the above, relatively slow, pathway of suicidal NETosis, vital NETosis involves the rapid release of NETs on neutrophil stimulation with specific bacteria, bacterial products, Toll-like receptor (TLR) 4-activated platelets or complement proteins (Fig. 4). Vital NETosis, also called "leukotoxic hypercitrullination" (Konig and Andrade 2016), mimics features of suicidal NETosis but does not immediately induce neutrophil death. During vital NETosis, neutrophils are still capable of functions such as migration, phagocytosis, and killing of bacteria while releasing their NET (Clark et al. 2007; Pilsczek et al. 2010; Yipp et al. 2012). The key features of vital NETosis are rapid histone citrullination, nuclear blebbing, and the vesicular transport of nuclear blebs to the plasma membrane, in which chromatin is externalized without permeabilizing the plasma membrane.

Vital NETosis is often observed during thrombosis and autoimmune diseases such as rheumatoid arthritis, systemic lupus erythematosus, and autoimmune vasculitis (Clark et al. 2007; Kessenbrock et al. 2009; Fuchs et al. 2010; Pilsczek et al. 2010; Yipp et al. 2012; Martinod et al. 2016). Here, we will focus on neutrophil vital NETosis during interaction with microbes, and summarize the key unique features of vital NETosis that are not observed in suicidal or noncanonical NETosis.

Receptor-Mediated Rapid Vital NETosis

Several surface receptors appear to be required for vital NETosis. TLR2 was required for vital NETosis during murine in vivo skin infection with *Staphylococcus aureus* (Yipp et al. 2012), and TLR2 or TLR4 receptor blockade suppressed NET formation induced by in vitro exposure to *S. aureus* (Wan et al. 2017). Neutrophil exposure to bacterial LPS in whole blood triggers vital NETosis. Here, NET extrusion requires TLR4 expression on platelets and platelet–neutrophil interaction via P-selectin (Pieterse et al. 2016). The complement system is also implicated in vital NETosis. When immobilized by the extracellular matrix component fibronectin, the β-glucan of *C. albicans* is recognized by complement receptor 3, inducing robust NET formation (Byrd et al. 2013). These findings highlight the complexity of host–pathogen interaction during a coordinated immune response.

PAD4 Functions in Vital NETosis

Calcium signaling, PAD4 activation, and histone citrullination appear to be common re-

Cite this article as *Cold Spring Harb Perspect Biol* doi: 10.1101/cshperspect.a037028

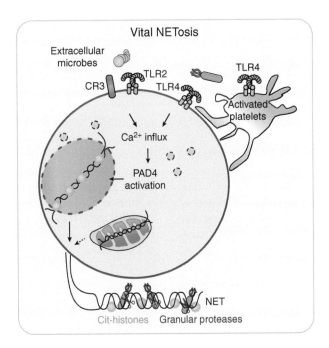

Figure 4. Vital NETosis. Microbial products or activated platelets stimulate vital NETosis through cell-surface receptors (e.g., TLR2, TLR4, complement receptor-3 [CR3]). The resulting spike in cytosolic calcium activates neutrophil PAD4, triggering histone citrullination and DNA decondensation. Chromatin derived from nuclear blebs or the mitochondria is expelled to generate a neutrophil extracellular trap (NET) while the neutrophil is still able to perform cellular functions such as cell migration.

quirements for vital NETosis. Whereas most stimulators of suicidal NETosis require NADPH oxidase function and induce NE-dependent histone clipping, NADPH oxidase function is dispensable for vital NETosis (Pilsczek et al. 2010; Douda et al. 2015), which proceeds without apparent histone clipping (Pieterse et al. 2018). Ionophores, bacterial products, or an increase in extracellular pH trigger the generation of mitochondrial ROS, calcium influx, and calcium-activated PAD4 enzymatic function in neutrophils (Douda et al. 2015; Naffah de Souza et al. 2017). PAD4 can then citrullinate histones to mediate DNA decondensation. PAD4 deficiency in *Shigella flexneri*–infected neutrophils rendered these cells unable to citrullinate histones, undergo vital NETosis, or mediate bacterial killing (Li et al. 2010). In line with an essential function for PAD4 in this pathway, a selective PAD4 inhibitor abrogated vital NETosis induced by ionomycin or *S. aureus* (Lewis et al. 2015).

DNA Source during Vital NETosis

Whereas suicidal NETosis clearly involves the expulsion of nuclear DNA, the source of extruded DNA during vital NETosis is less clear. A number of studies report nuclear blebbing leading to the release of histone-rich nuclear DNA (Pilsczek et al. 2010; Yipp et al. 2012; Gupta et al. 2018). Other studies suggest that NETs contain mitochondrial DNA (Yousefi et al. 2009; Keshari et al. 2012; Itagaki et al. 2015; Wang et al. 2015; Lood et al. 2016). Several studies indicate that NETs containing mitochondrial DNA elicit proinflammatory cytokine production from immune cells (Keshari et al. 2012; Wang et al. 2015; Lood et al. 2016).

MICROBIAL TRIGGERS OF NETosis

A wide variety of microbes are reported to induce NETosis. NET-inducing bacteria include *S. flexneri* (Brinkmann et al. 2004), *Escherichia coli* (Grinberg et al. 2008), *Streptococcus pneumoniae* (Beiter et al. 2006), *Streptococcus pyogenes* (Sumby et al. 2005), and *Mycobacterium tuberculosis* (Ramos-Kichik et al. 2009). Fungi such as *C. albicans* (Urban et al. 2006, 2009) and *Aspergillus fumigatus* (Bruns et al. 2010) are well characterized to induce NETosis. Even parasites are reported to induce NETosis, including *Leishmania* spp. (Guimaraes-Costa et al. 2009), *Plasmodium falciparum* (Baker et al. 2008), and *Toxoplasma gondii* (Abi Abdallah and Denkers 2012; Abi Abdallah et al. 2012). NET extrusion thus appears to be a conserved immune defense strategy in response to many different microbes.

FUNCTIONS FOR NETS IN ANTIMICROBIAL DEFENSE

Most NET studies agree that these structures trap extracellular microbes such as *C. albicans* or *T. gondii* (Urban et al. 2006, 2009). Imaging techniques that use flow chamber systems or intravital microscopy elegantly showed *E. coli* capture by NETs in vitro and in vivo (Clark et al. 2007). For example, confocal intravital microscopy imaged NETs capturing circulating *E. coli* in the hepatic sinusoids (McDonald et al. 2012).

There is growing evidence that NETs can also directly kill some pathogens such as *E. coli* (Grinberg et al. 2008) or *Leishamania* spp. (Guimaraes-Costa et al. 2009). The direct antimicrobial properties of NETs remain a topic of debate, and it is possible that NETs exert variable efficiency for killing different microbial species. NETs contain a selection of proteins with known antimicrobial properties such as histones (Hirsch 1958), and granule-resident antimicrobial proteins such as NE and MPO. NE kills bacteria by cleaving proteins on the bacterial outer membrane and targeting virulence factors of colonic enterobacteria, such as *Shigella* or *Yersinia* (Belaaouaj et al. 1998, 2000; Weinrauch et al. 2002), and is crucial in host defense against Gram-negative bacteria such as *E. coli* or *K. pneumoniae* (Belaaouaj et al. 1998). MPO remains active upon the extruded NET, where it generates ROS, such as hypochlorous acid from H_2O_2 derived from NAPDH oxidase or

the bacterium itself, to mediate bacteria killing (Allen and Stephens 2011; Parker et al. 2012a,b).

GSDMD pores may be a key antimicrobial weapon generated during noncanonical and suicidal NETosis (Chen et al. 2018; Sollberger et al. 2018). Both forms of NETosis trigger the cleavage of GSDMD to its pore-forming p30 fragment, which is poorly soluble, and immediately associates with lipids such as cardiolipin, phosphatidylinositol phosphates, and phosphatidylserine (Ding et al. 2016; Liu et al. 2016; Sborgi et al. 2016). GSDMD-p30 is reported to target bacterial membranes to induce the killing of extracellular *E. coli* and *S. aureus* (Ding et al. 2016; Liu et al. 2016) and cytosolic *L. monocytogenes* (Liu et al. 2016). GSDMD-p30 pores in the host cell plasma membrane are also crucial for its antimicrobial functions. Here, GSDMD-p30 pores facilitate plasma membrane rupture to enable NET release, and also the secretion of IL-1β to recruit neutrophils to a site of infection (Monteleone et al. 2018). At an infection site, IL-1β and other factors activate neutrophils and extend their life span to ensure they are able to mediate host-defensive functions (Miller et al. 2007; Rider et al. 2011; Biondo et al. 2014; Chen et al. 2014). Thus, GSDMD may serve several functions in the host defense, in which GSDMD-driven NETs first immobilize the pathogen, GSDMD pores directly target the bacterial membrane, and GSDMD-dependent IL-1β serves as a call-to-arms to recruit circulating neutrophils to kill the entrapped microbe.

MICROBIAL EVASION OF NETs

The effectiveness of NETs in mediating host defense is highlighted by the observation that many microbes have developed NET-evasion mechanisms. Some microbes suppress NETs by blocking the upstream signaling pathways, whereas others encode DNases to degrade the NET or have developed resistance to antimicrobial proteins upon the NET (Sumby et al. 2005; Döhrmann et al. 2014; Eby et al. 2014; Storisteanu et al. 2017; Eisenbeis et al. 2018). For example, *Bordetella pertussis* encodes an adenylate cyclase toxin (ACT) to produce intracellular cAMP to suppress the neutrophil ERK-dependent oxidative burst and, hence, NETosis (Eby et al. 2014). Group A *Streptococcus* has developed several strategies to evade NET-dependent host defense, including the expression of nucleases to degrade NETs (Sumby et al. 2005; Walker et al. 2007), and suppressing MPO release via the M1T1 serotype *streptococcal* collagen-like protein 1 (Scl-1) (Döhrmann et al. 2014). The *Staphylococcal* protein A (*SpA*) is a virulence factor that is secreted into the extrabacterial space to suppress neutrophil phagocytosis of *S. aureus,* but in doing so, promotes NETosis (Hoppenbrouwers et al. 2018). Other *S. aureus* virulence factors, such as the DNA-binding protein extracellular adherence protein (Eap), suppress NET formation and function (Eisenbeis et al. 2018). Such microbial evasion strategies highlight the requirement for neutrophils to have a multitude of microbial killing mechanisms at their disposal for many and varied host defense strategies.

CONCLUDING REMARKS

Diverse extracellular and intracellular microbes induce neutrophils to trigger NETs by suicidal, noncanonical, and vital NETosis. By entrapping microbes, NETs prevent microbial dissemination to a secondary infection site and, in some cases, NETs exert direct antimicrobial functions. In addition to these antimicrobial functions, emerging literature documents pathological roles for NETs in autoimmune diseases. It remains to be elucidated whether distinct NETosis pathways engender differences in immunomodulatory or antimicrobial properties of the resulting NET. The recent discovery that the suicidal and noncanonical NETosis pathways converge on a single cell death executioner protein, GSDMD, suggests that GSDMD inhibitors may have therapeutic potential in the treatment of NET-associated diseases such as thrombosis, rheumatoid arthritis, systemic lupus erythematosus, and autoimmune vasculitis. Further research to elucidate the molecular and cellular mechanisms by which neutrophils cast their NETs will be critical for understanding this important host defense program and the design of new autoimmune disease therapies.

ACKNOWLEDGMENTS

We thank Dr. Madhavi Maddugoda for editorial suggestions. This work was supported by the National Health and Medical Research Council of Australia (Project Grant 1163924 and Fellowship 1141131 to K.S.). K.S. is a coinventor on patent applications for NLRP3 inhibitors that have been licensed to Inflazome, a company headquartered in Dublin, Ireland. Inflazome is developing drugs that target the NLRP3 inflammasome to address unmet clinical needs in inflammatory disease. K.S. served on the Scientific Advisory Board of Inflazome in 2016–2017. The authors have no further conflicts of interest to declare.

REFERENCES

Aachoui Y, Leaf IA, Hagar JA, Fontana MF, Campos CG, Zak DE, Tan MH, Cotter PA, Vance RE, Aderem A, et al. 2013. Caspase-11 protects against bacteria that escape the vacuole. *Science* **339**: 975–978. doi:10.1126/science.1230751

Abi Abdallah DS, Denkers EY. 2012. Neutrophils cast extracellular traps in response to protozoan parasites. *Front Immunol* **3**: 382.

Abi Abdallah DS, Lin C, Ball CJ, King MR, Duhamel GE, Denkers EY. 2012. *Toxoplasma gondii* triggers release of human and mouse neutrophil extracellular traps. *Infect Immun* **80**: 768–777. doi:10.1128/IAI.05730-11

Allen RC, Stephens JT Jr. 2011. Myeloperoxidase selectively binds and selectively kills microbes. *Infect Immun* **79**: 474–485. doi:10.1128/IAI.00910-09

Amulic B, Cazalet C, Hayes GL, Metzler KD, Zychlinsky A. 2012. Neutrophil function: from mechanisms to disease. *Annu Rev Immunol* **30**: 459–489. doi:10.1146/annurev-immunol-020711-074942

Baker VS, Imade GE, Molta NB, Tawde P, Pam SD, Obadofin MO, Sagay SA, Egah DZ, Iya D, Afolabi BB, et al. 2008. Cytokine-associated neutrophil extracellular traps and antinuclear antibodies in *Plasmodium falciparum* infected children under six years of age. *Malar J* **7**: 41. doi:10.1186/1475-2875-7-41

Beiter K, Wartha F, Albiger B, Normark S, Zychlinsky A, Henriques-Normark B. 2006. An endonuclease allows *Streptococcus pneumoniae* to escape from neutrophil extracellular traps. *Curr Biol* **16**: 401–407. doi:10.1016/j.cub.2006.01.056

Belaaouaj A, McCarthy R, Baumann M, Gao Z, Ley TJ, Abraham SN, Shapiro SD. 1998. Mice lacking neutrophil elastase reveal impaired host defense against gram negative bacterial sepsis. *Nat Med* **4**: 615–618. doi:10.1038/nm0598-615

Belaaouaj A, Kim KS, Shapiro SD. 2000. Degradation of outer membrane protein A in *Escherichia coli* killing by neutrophil elastase. *Science* **289**: 1185–1187. doi:10.1126/science.289.5482.1185

Bianchi M, Hakkim A, Brinkmann V, Siler U, Seger RA, Zychlinsky A, Reichenbach J. 2009. Restoration of NET formation by gene therapy in CGD controls aspergillosis. *Blood* **114:** 2619–2622. doi:10.1182/blood-2009-05-221606

Biondo C, Mancuso G, Midiri A, Signorino G, Domina M, Lanza Cariccio V, Mohammadi N, Venza M, Venza I, Teti G, et al. 2014. The interleukin-1β/CXCL1/2/neutrophil axis mediates host protection against group B streptococcal infection. *Infect Immun* **82:** 4508–4517. doi:10.1128/IAI.02104-14

Borregaard N. 2014. What doesn't kill you makes you stronger: the anti-inflammatory effect of neutrophil respiratory burst. *Immunity* **40:** 1–2. doi:10.1016/j.immuni.2013.12.003

Branzk N, Lubojemska A, Hardison SE, Wang Q, Gutierrez MG, Brown GD, Papayannopoulos V. 2014. Neutrophils sense microbe size and selectively release neutrophil extracellular traps in response to large pathogens. *Nat Immunol* **15:** 1017–1025. doi:10.1038/ni.2987

Brinkmann V, Reichard U, Goosmann C, Fauler B, Uhlemann Y, Weiss DS, Weinrauch Y, Zychlinsky A. 2004. Neutrophil extracellular traps kill bacteria. *Science* **303:** 1532–1535. doi:10.1126/science.1092385

Bruns S, Kniemeyer O, Hasenberg M, Aimanianda V, Nietzsche S, Thywißen A, Jeron A, Latgé JP, Brakhage AA, Gunzer M. 2010. Production of extracellular traps against *Aspergillus fumigatus* in vitro and in infected lung tissue is dependent on invading neutrophils and influenced by hydrophobin RodA. *PLoS Pathog* **6:** e1000873. doi:10.1371/journal.ppat.1000873

Byrd AS, O'Brien XM, Johnson CM, Lavigne LM, Reichner JS. 2013. An extracellular matrix–based mechanism of rapid neutrophil extracellular trap formation in response to *Candida albicans*. *J Immunol* **190:** 4136–4148. doi:10.4049/jimmunol.1202671

Chen KW, Groß CJ, Sotomayor FV, Stacey KJ, Tschopp J, Sweet MJ, Schroder K. 2014. The neutrophil NLRC4 inflammasome selectively promotes IL-1β maturation without pyroptosis during acute *Salmonella* challenge. *Cell Rep* **8:** 570–582. doi:10.1016/j.celrep.2014.06.028

Chen KW, Monteleone M, Boucher D, Sollberger G, Ramnath D, Condon ND, von Pein JB, Broz P, Sweet MJ, Schroder K. 2018. Noncanonical inflammasome signaling elicits gasdermin D-dependent neutrophil extracellular traps. *Sci Immunol* **3:** eaar6676. doi:10.1126/sciimmunol.aar6676

Clark SR, Ma AC, Tavener SA, McDonald B, Goodarzi Z, Kelly MM, Patel KD, Chakrabarti S, McAvoy E, Sinclair GD, et al. 2007. Platelet TLR4 activates neutrophil extracellular traps to ensnare bacteria in septic blood. *Nat Med* **13:** 463–469. doi:10.1038/nm1565

Claushuis TAM, van der Donk LEH, Luitse AL, van Veen HA, van der Wel NN, van Vught LA, Roelofs J, de Boer OJ, Lankelma JM, Boon L, et al. 2018. Role of peptidylarginine deiminase 4 in neutrophil extracellular trap formation and host defense during *Klebsiella pneumoniae*–induced pneumonia-derived sepsis. *J Immunol* **201:** 1241–1252. doi:10.4049/jimmunol.1800314

Dale DC, Boxer L, Liles WC. 2008. The phagocytes: neutrophils and monocytes. *Blood* **112:** 935–945. doi:10.1182/blood-2007-12-077917

Damgaard D, Bjørn ME, Steffensen MA, Pruijn GJ, Nielsen CH. 2016. Reduced glutathione as a physiological co-activator in the activation of peptidylarginine deiminase. *Arthritis Res Ther* **18:** 102. doi:10.1186/s13075-016-1000-7

Dang PM, Babior BM, Smith RM. 1999. NADPH dehydrogenase activity of p67^{PHOX}, a cytosolic subunit of the leukocyte NADPH oxidase. *Biochemistry* **38:** 5746–5753. doi:10.1021/bi982750f

Dinauer MC. 2014. Disorders of neutrophil function: an overview. *Methods Mol Biol* **1124:** 501–515. doi:10.1007/978-1-62703-845-4_30

Dinauer MC. 2016. Primary immune deficiencies with defects in neutrophil function. *Hematology Am Soc Hematol Educ Program* **2016:** 43–50. doi:10.1182/asheducation-2016.1.43

Ding J, Wang K, Liu W, She Y, Sun Q, Shi J, Sun H, Wang DC, Shao F. 2016. Pore-forming activity and structural autoinhibition of the gasdermin family. *Nature* **535:** 111–116. doi:10.1038/nature18590

Döhrmann S, Anik S, Olson J, Anderson EL, Etesami N, No H, Snipper J, Nizet V, Okumura CY. 2014. Role for *streptococcal* collagen-like protein 1 in M1T1 group A *Streptococcus* resistance to neutrophil extracellular traps. *Infect Immun* **82:** 4011–4020. doi:10.1128/IAI.01921-14

Douda DN, Khan MA, Grasemann H, Palaniyar N. 2015. SK3 channel and mitochondrial ROS mediate NADPH oxidase-independent NETosis induced by calcium influx. *Proc Natl Acad Sci* **112:** 2817–2822. doi:10.1073/pnas.1414055112

Dwyer M, Shan Q, D'Ortona S, Maurer R, Mitchell R, Olesen H, Thiel S, Huebner J, Gadjeva M. 2014. Cystic fibrosis sputum DNA has NETosis characteristics and neutrophil extracellular trap release is regulated by macrophage migration-inhibitory factor. *J Innate Immun* **6:** 765–779. doi:10.1159/000363242

Eby JC, Gray MC, Hewlett EL. 2014. Cyclic AMP-mediated suppression of neutrophil extracellular trap formation and apoptosis by the *Bordetella pertussis* adenylate cyclase toxin. *Infect Immun* **82:** 5256–5269. doi:10.1128/IAI.02487-14

Eisenbeis J, Saffarzadeh M, Peisker H, Jung P, Thewes N, Preissner KT, Herrmann M, Molle V, Geisbrecht BV, Jacobs K, et al. 2018. The *Staphylococcus aureus* extracellular adherence protein Eap is a DNA binding protein capable of blocking neutrophil extracellular trap formation. *Front Cell Infect Microbiol* **8:** 235. doi:10.3389/fcimb.2018.00235

El Benna J, Han J, Park JW, Schmid E, Ulevitch RJ, Babior BM. 1996. Activation of p38 in stimulated human neutrophils: phosphorylation of the oxidase component p47^{phox} by p38 and ERK but not by JNK. *Arch Biochem Biophys* **334:** 395–400. doi:10.1006/abbi.1996.0470

Faurschou M, Borregaard N. 2003. Neutrophil granules and secretory vesicles in inflammation. *Microbes Infect* **5:** 1317–1327. doi:10.1016/j.micinf.2003.09.008

Fuchs TA, Abed U, Goosmann C, Hurwitz R, Schulze I, Wahn V, Weinrauch Y, Brinkmann V, Zychlinsky A. 2007. Novel cell death program leads to neutrophil extracellular traps. *J Cell Biol* **176:** 231–241. doi:10.1083/jcb.200606027

Fuchs TA, Brill A, Duerschmied D, Schatzberg D, Monestier M, Myers DD Jr, Wrobleski SK, Wakefield TW, Hartwig

Cite this article as *Cold Spring Harb Perspect Biol* doi: 10.1101/cshperspect.a037028

JH, Wagner DD. 2010. Extracellular DNA traps promote thrombosis. *Proc Natl Acad Sci* **107:** 15880–15885. doi:10.1073/pnas.1005743107

Gavillet M, Martinod K, Renella R, Wagner DD, Williams DA. 2018. A key role for Rac and Pak signaling in neutrophil extracellular traps (NETs) formation defines a new potential therapeutic target. *Am J Hematol* **93:** 269–276. doi:10.1002/ajh.24970

Grinberg N, Elazar S, Rosenshine I, Shpigel NY. 2008. β-Hydroxybutyrate abrogates formation of bovine neutrophil extracellular traps and bactericidal activity against mammary pathogenic *Escherichia coli*. *Infect Immun* **76:** 2802–2807. doi:10.1128/IAI.00051-08

Guiducci E, Lemberg C, Küng N, Schraner E, Theocharides APA, LeibundGut-Landmann S. 2018. *Candida albicans*–induced NETosis is independent of peptidylarginine deiminase 4. *Front Immunol* **9:** 1573. doi:10.3389/fimmu.2018.01573

Guimaraes-Costa AB, Nascimento MT, Froment GS, Soares RP, Morgado FN, Conceicao-Silva F, Saraiva EM. 2009. *Leishmania amazonensis* promastigotes induce and are killed by neutrophil extracellular traps. *Proc Natl Acad Sci* **106:** 6748–6753. doi:10.1073/pnas.0900226106

Gupta S, Chan DW, Zaal KJ, Kaplan MJ. 2018. A high-throughput real-time imaging technique to quantify NETosis and distinguish mechanisms of cell death in human neutrophils. *J Immunol* **200:** 869–879. doi:10.4049/jimmunol.1700905

Hakkim A, Fuchs TA, Martinez NE, Hess S, Prinz H, Zychlinsky A, Waldmann H. 2011. Activation of the Raf-MEK-ERK pathway is required for neutrophil extracellular trap formation. *Nat Chem Biol* **7:** 75–77. doi:10.1038/nchembio.496

Hirsch JG. 1958. Bactericidal action of histone. *J Exp Med* **108:** 925–944. doi:10.1084/jem.108.6.925

Hoppenbrouwers T, Sultan AR, Abraham TE, Lemmens-den Toom NA, Hansenová Maňásková S, van Cappellen WA, Houtsmuller AB, van Wamel WJB, de Maat MPM, van Neck JW. 2018. *Staphylococcal* protein A is a key factor in neutrophil extracellular traps formation. *Front Immunol* **9:** 165. doi:10.3389/fimmu.2018.00165

Itagaki K, Kaczmarek E, Lee YT, Tang IT, Isal B, Adibnia Y, Sandler N, Grimm MJ, Segal BH, Otterbein LE, et al. 2015. Mitochondrial DNA released by trauma induces neutrophil extracellular traps. *PLoS ONE* **10:** e0120549. doi:10.1371/journal.pone.0120549

Karmakar M, Katsnelson M, Malak HA, Greene NG, Howell SJ, Hise AG, Camilli A, Kadioglu A, Dubyak GR, Pearlman E. 2015. Neutrophil IL-1β processing induced by pneumolysin is mediated by the NLRP3/ASC inflammasome and caspase-1 activation and is dependent on K^{+} efflux. *J Immunol* **194:** 1763–1775. doi:10.4049/jimmunol.1401624

Kayagaki N, Stowe IB, Lee BL, O'Rourke K, Anderson K, Warming S, Cuellar T, Haley B, Roose-Girma M, Phung QT, et al. 2015. Caspase-11 cleaves gasdermin D for non-canonical inflammasome signalling. *Nature* **526:** 666–671. doi:10.1038/nature15541

Kenny EF, Herzig A, Krüger R, Muth A, Mondal S, Thompson PR, Brinkmann V, Bernuth HV, Zychlinsky A. 2017. Diverse stimuli engage different neutrophil extracellular trap pathways. *eLife* **6:** 24437. doi:10.7554/eLife.24437

Keshari RS, Jyoti A, Kumar S, Dubey M, Verma A, Srinag BS, Krishnamurthy H, Barthwal MK, Dikshit M. 2012. Neutrophil extracellular traps contain mitochondrial as well as nuclear DNA and exhibit inflammatory potential. *Cytometry A* **81A:** 238–247. doi:10.1002/cyto.a.21178

Kessenbrock K, Krumbholz M, Schönermarck U, Back W, Gross WL, Werb Z, Gröne HJ, Brinkmann V, Jenne DE. 2009. Netting neutrophils in autoimmune small-vessel vasculitis. *Nat Med* **15:** 623–625. doi:10.1038/nm.1959

Kleniewska P, Piechota A, Skibska B, Goraca A. 2012. The NADPH oxidase family and its inhibitors. *Arch Immunol Ther Exp (Warsz)* **60:** 277–294. doi:10.1007/s00005-012-0176-z

Konig MF, Andrade F. 2016. A critical reappraisal of neutrophil extracellular traps and NETosis mimics based on differential requirements for protein citrullination. *Front Immunol* **7:** 461.

Lewis HD, Liddle J, Coote JE, Atkinson SJ, Barker MD, Bax BD, Bicker KL, Bingham RP, Campbell M, Chen YH, et al. 2015. Inhibition of PAD4 activity is sufficient to disrupt mouse and human NET formation. *Nat Chem Biol* **11:** 189–191. doi:10.1038/nchembio.1735

Li P, Li M, Lindberg MR, Kennett MJ, Xiong N, Wang Y. 2010. PAD4 is essential for antibacterial innate immunity mediated by neutrophil extracellular traps. *J Exp Med* **207:** 1853–1862. doi:10.1084/jem.20100239

Lim MB, Kuiper JW, Katchky A, Goldberg H, Glogauer M. 2011. Rac2 is required for the formation of neutrophil extracellular traps. *J Leukoc Biol* **90:** 771–776. doi:10.1189/jlb.1010549

Liu X, Zhang Z, Ruan J, Pan Y, Magupalli VG, Wu H, Lieberman J. 2016. Inflammasome-activated gasdermin D causes pyroptosis by forming membrane pores. *Nature* **535:** 153–158. doi:10.1038/nature18629

Lood C, Blanco LP, Purmalek MM, Carmona-Rivera C, De Ravin SS, Smith CK, Malech HL, Ledbetter JA, Elkon KB, Kaplan MJ. 2016. Neutrophil extracellular traps enriched in oxidized mitochondrial DNA are interferogenic and contribute to lupus-like disease. *Nat Med* **22:** 146–153. doi:10.1038/nm.4027

Lopes LR, Hoyal CR, Knaus UG, Babior BM. 1999. Activation of the leukocyte NADPH oxidase by protein kinase C in a partially recombinant cell-free system. *J Biol Chem* **274:** 15533–15537. doi:10.1074/jbc.274.22.15533

Martinod K, Witsch T, Farley K, Gallant M, Remold-O'Donnell E, Wagner DD. 2016. Neutrophil elastase-deficient mice form neutrophil extracellular traps in an experimental model of deep vein thrombosis. *J Thromb Haemost* **14:** 551–558. doi:10.1111/jth.13239

McDonald B, Urrutia R, Yipp BG, Jenne CN, Kubes P. 2012. Intravascular neutrophil extracellular traps capture bacteria from the bloodstream during sepsis. *Cell Host Microbe* **12:** 324–333. doi:10.1016/j.chom.2012.06.011

Metzler KD, Fuchs TA, Nauseef WM, Reumaux D, Roesler J, Schulze I, Wahn V, Papayannopoulos V, Zychlinsky A. 2011. Myeloperoxidase is required for neutrophil extracellular trap formation: implications for innate immunity. *Blood* **117:** 953–959. doi:10.1182/blood-2010-06-290171

Metzler KD, Goosmann C, Lubojemska A, Zychlinsky A, Papayannopoulos V. 2014. A myeloperoxidase-containing complex regulates neutrophil elastase release and ac-

tin dynamics during NETosis. *Cell Rep* **8**: 883–896. doi:10
.1016/j.celrep.2014.06.044

Miller LS, Pietras EM, Uricchio LH, Hirano K, Rao S, Lin H,
O'Connell RM, Iwakura Y, Cheung AL, Cheng G, et al.
2007. Inflammasome-mediated production of IL-1β is
required for neutrophil recruitment against *Staphylococcus aureus* in vivo. *J Immunol* **179**: 6933–6942. doi:10
.4049/jimmunol.179.10.6933

Monteleone M, Stanley AC, Chen KW, Brown DL, Bezbra-
dica JS, von Pein JB, Holley CL, Boucher D, Shakespear
MR, Kapetanovic R, et al. 2018. Interleukin-1β maturation triggers its relocation to the plasma membrane for
gasdermin-D-dependent and -independent secretion.
Cell Rep **24**: 1425–1433. doi:10.1016/j.celrep.2018.07.027

Naffah de Souza C, Breda LCD, Khan MA, de Almeida SR,
Câmara NOS, Sweezey N, Palaniyar N. 2018. Alkaline pH
promotes NADPH oxidase-independent neutrophil extracellular trap formation: a matter of mitochondrial reactive oxygen species generation and citrullination and
cleavage of histone. *Front Immunol* **8**: 1849. doi:10
.3389/fimmu.2017.01849

Nauseef WM, Borregaard N. 2014. Neutrophils at work. *Nat
Immunol* **15**: 602–611. doi:10.1038/ni.2921

Nauseef WM, Cogley M, McCormick S. 1996. Effect of the
R569W missense mutation on the biosynthesis of myeloperoxidase. *J Biol Chem* **271**: 9546–9549. doi:10.1074/jbc
.271.16.9546

Neeli I, Radic M. 2013. Opposition between PKC isoforms
regulates histone deimination and neutrophil extracellular chromatin release. *Front Immunol* **4**: 38. doi:10.3389/
fimmu.2013.00038

Neeli I, Khan SN, Radic M. 2008. Histone deimination as a
response to inflammatory stimuli in neutrophils. *J Immunol* **180**: 1895–1902. doi:10.4049/jimmunol.180.3.1895

Nunes P, Demaurex N, Dinauer MC. 2013. Regulation of the
NADPH oxidase and associated ion fluxes during phagocytosis. *Traffic* **14**: 1118–1131.

O'Donoghue AJ, Jin Y, Knudsen GM, Perera NC, Jenne DE,
Murphy JE, Craik CS, Hermiston TW. 2013. Global substrate profiling of proteases in human neutrophil extracellular traps reveals consensus motif predominantly contributed by elastase. *PLoS ONE* **8**: e75141. doi:10.1371/
journal.pone.0075141

Papayannopoulos V. 2018. Neutrophil extracellular traps in
immunity and disease. *Nat Rev Immunol* **18**: 134–147.
doi:10.1038/nri.2017.105

Papayannopoulos V, Metzler KD, Hakkim A, Zychlinsky A.
2010. Neutrophil elastase and myeloperoxidase regulate
the formation of neutrophil extracellular traps. *J Cell Biol*
191: 677–691. doi:10.1083/jcb.201006052

Parker H, Albrett AM, Kettle AJ, Winterbourn CC. 2012a.
Myeloperoxidase associated with neutrophil extracellular
traps is active and mediates bacterial killing in the presence of hydrogen peroxide. *J Leukoc Biol* **91**: 369–376.
doi:10.1189/jlb.0711387

Parker H, Dragunow M, Hampton MB, Kettle AJ, Winterbourn CC. 2012b. Requirements for NADPH oxidase and
myeloperoxidase in neutrophil extracellular trap formation differ depending on the stimulus. *J Leukoc Biol* **92**:
841–849. doi:10.1189/jlb.1211601

Pieterse E, Rother N, Yanginlar C, Hilbrands LB, van der
Vlag J. 2016. Neutrophils discriminate between lipopoly-

saccharides of different bacterial sources and selectively
release neutrophil extracellular traps. *Front Immunol* **7**:
484. doi:10.3389/fimmu.2016.00484

Pieterse E, Rother N, Yanginlar C, Gerretsen J, Boeltz S,
Munoz LE, Herrmann M, Pickkers P, Hilbrands LB, van
der Vlag J. 2018. Cleaved N-terminal histone tails distinguish between NADPH oxidase (NOX)-dependent and
NOX-independent pathways of neutrophil extracellular
trap formation. *Ann Rheum Dis* **77**: 1790–1798. doi:10
.1136/annrheumdis-2018-213223

Pilsczek FH, Salina D, Poon KK, Fahey C, Yipp BG, Sibley
CD, Robbins SM, Green FH, Surette MG, Sugai M, et al.
2010. A novel mechanism of rapid nuclear neutrophil
extracellular trap formation in response to *Staphylococcus
aureus*. *J Immunol* **185**: 7413–7425. doi:10.4049/jimmu
nol.1000675

Ramos-Kichik V, Mondragón-Flores R, Mondragón-Caste-
lán M, Gonzalez-Pozos S, Muñiz-Hernandez S, Rojas-
Espinosa O, Chacón-Salinas R, Estrada-Parra S, Estrada-
García I. 2009. Neutrophil extracellular traps are induced
by *Mycobacterium tuberculosis*. *Tuberculosis (Edinb)* **89**:
29–37. doi:10.1016/j.tube.2008.09.009

Rider P, Carmi Y, Guttman O, Braiman A, Cohen I, Voronov
E, White MR, Dinarello CA, Apte RN. 2011. IL-1α and
IL-1β recruit different myeloid cells and promote different stages of sterile inflammation. *J Immunol* **187**: 4835–
4843. doi:10.4049/jimmunol.1102048

Rodríguez-Espinosa O, Rojas-Espinosa O, Moreno-Alta-
mirano MM, López-Villegas EO, Sánchez-García FJ.
2015. Metabolic requirements for neutrophil extracellular
traps formation. *Immunology* **145**: 213–224. doi:10.1111/
imm.12437

Röhm M, Grimm MJ, D'Auria AC, Almyroudis NG, Segal
BH, Urban CF. 2014. NADPH oxidase promotes neutrophil extracellular trap formation in pulmonary aspergillosis. *Infect Immun* **82**: 1766–1777. doi:10.1128/IAI
.00096-14

Sborgi L, Rühl S, Mulvihill E, Pipercevic J, Heilig R, Stahlberg
H, Farady CJ, Müller DJ, Broz P, Hiller S. 2016. GSDMD
membrane pore formation constitutes the mechanism of
pyroptotic cell death. *EMBO J* **35**: 1766–1778. doi:10
.15252/embj.201694696

Shi J, Zhao Y, Wang K, Shi X, Wang Y, Huang H, Zhuang Y,
Cai T, Wang F, Shao F. 2015. Cleavage of GSDMD by
inflammatory caspases determines pyroptotic cell death.
Nature **526**: 660–665. doi:10.1038/nature15514

Siler U, Romao S, Tejera E, Pastukhov O, Kuzmenko E,
Valencia RG, Meda Spaccamela V, Belohradsky BH,
Speer O, Schmugge M, et al. 2017. Severe glucose-6-phos-
phate dehydrogenase deficiency leads to susceptibility to
infection and absent NETosis. *J Allergy Clin Immunol*
139: 212–219.e3. doi:10.1016/j.jaci.2016.04.041

Sollberger G, Choidas A, Burn GL, Habenberger P, Di Lucrezia R, Kordes S, Menninger S, Eickhoff J, Nussbaumer
P, Klebl B, et al. 2018. Gasdermin D plays a vital role in the
generation of neutrophil extracellular traps. *Sci Immunol*
3: eaar6689. doi:10.1126/sciimmunol.aar6689

Steinberg SF. 2008. Structural basis of protein kinase C isoform function. *Physiol Rev* **88**: 1341–1378. doi:10.1152/
physrev.00034.2007

Storek KM, Monack DM. 2015. Bacterial recognition pathways that lead to inflammasome activation. *Immunol Rev* **265:** 112–129. doi:10.1111/imr.12289

Storisteanu DM, Pocock JM, Cowburn AS, Juss JK, Nadesalingam A, Nizet V, Chilvers ER. 2017. Evasion of neutrophil extracellular traps by respiratory pathogens. *Am J Respir Cell Mol Biol* **56:** 423–431. doi:10.1165/rcmb.2016-0193PS

Sumby P, Barbian KD, Gardner DJ, Whitney AR, Welty DM, Long RD, Bailey JR, Parnell MJ, Hoe NP, Adams GG, et al. 2005. Extracellular deoxyribonuclease made by group A *Streptococcus* assists pathogenesis by enhancing evasion of the innate immune response. *Proc Natl Acad Sci* **102:** 1679–1684. doi:10.1073/pnas.0406641102

Thurston TL, Matthews SA, Jennings E, Alix E, Shao F, Shenoy AR, Birrell MA, Holden DW. 2016. Growth inhibition of cytosolic *Salmonella* by caspase-1 and caspase-11 precedes host cell death. *Nat Commun* **7:** 13292. doi:10.1038/ncomms13292

Urban CF, Reichard U, Brinkmann V, Zychlinsky A. 2006. Neutrophil extracellular traps capture and kill *Candida albicans* yeast and hyphal forms. *Cell Microbiol* **8:** 668–676. doi:10.1111/j.1462-5822.2005.00659.x

Urban CF, Ermert D, Schmid M, Abu-Abed U, Goosmann C, Nacken W, Brinkmann V, Jungblut PR, Zychlinsky A. 2009. Neutrophil extracellular traps contain calprotectin, a cytosolic protein complex involved in host defense against *Candida albicans*. *PLoS Pathog* **5:** e1000639. doi:10.1371/journal.ppat.1000639

Walker MJ, Hollands A, Sanderson-Smith ML, Cole JN, Kirk JK, Henningham A, McArthur JD, Dinkla K, Aziz RK, Kansal RG, et al. 2007. DNase Sda1 provides selection pressure for a switch to invasive group A streptococcal infection. *Nat Med* **13:** 981–985. doi:10.1038/nm1612

Wan T, Zhao Y, Fan F, Hu R, Jin X. 2017. Dexamethasone inhibits *S. aureus*-induced neutrophil extracellular pathogen-killing mechanism, possibly through Toll-like receptor regulation. *Front Immunol* **8:** 60.

Wang H, Li T, Chen S, Gu Y, Ye S. 2015. Neutrophil extracellular trap mitochondrial DNA and its autoantibody in systemic lupus erythematosus and a proof-of-concept trial of metformin. *Arthritis Rheumatol* **67:** 3190–3200. doi:10.1002/art.39296

Warnatsch A, Tsourouktsoglou TD, Branzk N, Wang Q, Reincke S, Herbst S, Gutierrez M, Papayannopoulos V. 2017. Reactive oxygen species localization programs inflammation to clear microbes of different size. *Immunity* **46:** 421–432. doi:10.1016/j.immuni.2017.02.013

Weinrauch Y, Drujan D, Shapiro SD, Weiss J, Zychlinsky A. 2002. Neutrophil elastase targets virulence factors of enterobacteria. *Nature* **417:** 91–94. doi:10.1038/417091a

Yipp BG, Kubes P. 2013. NETosis: how vital is it? *Blood* **122:** 2784–2794. doi:10.1182/blood-2013-04-457671

Yipp BG, Petri B, Salina D, Jenne CN, Scott BN, Zbytnuik LD, Pittman K, Asaduzzaman M, Wu K, Meijndert HC, et al. 2012. Infection-induced NETosis is a dynamic process involving neutrophil multitasking in vivo. *Nat Med* **18:** 1386–1393. doi:10.1038/nm.2847

Yousefi S, Mihalache C, Kozlowski E, Schmid I, Simon HU. 2009. Viable neutrophils release mitochondrial DNA to form neutrophil extracellular traps. *Cell Death Differ* **16:** 1438–1444. doi:10.1038/cdd.2009.96

Index